Brief Calculus and Its Applications

NINTH EDITION

Brief Calculus and Its Applications

Larry J. Goldstein
Goldstein Educational Technologies

David C. Lay
University of Maryland

David I. Schneider
University of Maryland

PRENTICE HALL
Upper Saddle River, New Jersey 07458

Library of Congress Cataloging-in-Publication Data

Goldstein, Larry Joel.
      Brief calculus and its applications / Larry J. Goldstein, David C. Lay,
      David I. Schneider.—9th ed.
      p.   cm.
      Includes index.
      ISBN 0-13-087303-9
      1. Calculus.   I. Lay, David C.   II. Schneider, David I.
      III. Title.
QA303.G6248   2001
515--dc21

                                   00-026323
                                        CIP

Acquisitions Editor: *Kathleen Boothby Sestak*
Production Editor: *Lynn Savino Wendel*
Assistant Vice President of Production and Manufacturing: *David W. Riccardi*
Executive Managing Editor: *Kathleen Schiaparelli*
Senior Managing Editor: *Linda Mihatov Behrens*
Manufacturing Buyer: *Alan Fischer*
Manufacturing Manager: *Trudy Pisciotti*
Marketing Manager: *Patrice Lumumba Jones*
Marketing Assistant: *Vince Jansen*
Director of Marketing: *John Tweeddale*
Development Editors: *David Chelton and Susan Gerstein*
Senior Project Manager: *Gina M. Huck, Imaginative Solutions*
Associate Editor, Mathematics/Statistics Media: *Audra J. Walsh*
Editorial Assistant: *Joanne Wendelken*
Art Director: *Maureen Eide*
Assistant to Art Director: *John Christiana*
Interior Designer: *Jill Little*
Cover Designer: *Daniel Conte*
Art Editor: *Grace Hazeldine*
Art Manager: *Gus Vibal*
Director of Creative Services: *Paul Belfanti*
Cover Photo: *Super Stock Inc.*
Art Studio: *Academy Artworks*

Printed in the United States of America
10 9 8 7 6 5 4 3 2 1

ISBN 0-13-087303-9

Prentice-Hall International (UK) Limited, *London*
Prentice-Hall of Australia Pty. Limited, *Sydney*
Prentice-Hall Canada, Inc., *Toronto*
Prentice-Hall Hispanoamericana, S.A., *Mexico*
Prentice-Hall of India Private Limited, *New Delhi*
Prentice-Hall of Japan, Inc., *Tokyo*
Pearson Education Asia Pte. Ltd.
Editora Prentice-Hall do Brasil, Ltda., *Rio de Janeiro*

Contents

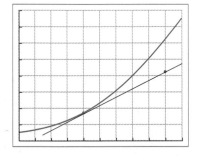
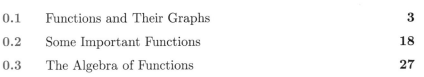
*Sections preceded by a * are optional in the sense that they are not prerequisites for later material.

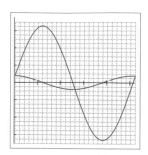

8 The Trigonometric Functions 452

Preface

We have been very pleased with the enthusiastic response to the first seven editions of *Brief Calculus and Its Applications* by teachers and students alike. The present work incorporates many of the suggestions they have put forward.

Although there are many changes, we have preserved the approach and the flavor. Our goals remain the same: to begin the calculus as soon as possible; to present calculus in an intuitive yet intellectually satisfying way; and to illustrate the many applications of calculus to the biological, social, and management sciences.

The distinctive order of topics has proven over the years to be successful—easier for students to learn, and more interesting because students see significant applications early. For instance, the derivative is explained geometrically before the analytic material on limits is presented. This approach gives the students an understanding of the derivative at least as strong as that obtained from the traditional approach. To reach the applications in Chapter 2 quickly, we present only the differentiation rules and the curve sketching needed for those applications. Advanced topics come later when they are needed. Other aspects of this student-oriented approach follow below.

Applications

We provide realistic applications that illustrate the uses of calculus in other disciplines. See the Index of Applications on the inside cover. Wherever possible, we have attempted to use applications to motivate the mathematics.

Examples

The text includes many more worked examples than is customary. Furthermore, we have included computational details to enhance readability by students whose basic skills are weak.

Exercises

The exercises comprise about one-quarter of the text—the most important part of the text in our opinion. The exercises at the ends of the sections are usually arranged in the order in which the text proceeds, so that the homework assignments may easily be made after only part of a section is discussed. Interesting applications and more challenging problems tend to be located near the ends of the exercise sets. Supplementary exercises at the end of each chapter expand the other exercise sets and include problems that require skills from earlier chapters.

Practice Problems

The practice problems have proven to be a popular and useful feature. Practice Problems are carefully selected questions located at the end of each section, just before the exercise set. Complete solutions are given following the exercise set. The practice problems often focus on points that are potentially confusing or are likely to be overlooked. We recommend that the reader seriously attempt the practice problems and study their solutions before moving on to the exercises. In effect, the practice problems constitute a built-in workbook.

Minimal Prerequisites

In Chapter 0, we review those concepts that the reader needs to study calculus. Some important topics, such as the laws of exponents, are reviewed again when they are used in a later chapter. Section 0.6 prepares students for applied problems that appear throughout the text. A reader familiar with the content of Chapter 0 should begin with Chapter 1 and use Chapter 0 as a reference, whenever needed.

New in this Edition

Among the many changes in this edition, the following are the most significant:

1. *Delta Notation* We introduce delta notation in Chapter 0 and use it in our discussioin of the derivative. As in previous editions, we have tried to minimize the use of complicated notation, preferring instead verbal descriptions. However, in the case of the delta notation, we feel that the clarity achieved is worth the extra notation.

2. *Derivative as a Rate of Change* We preview the derivative as a rate of change at the beginning of Chapter 1, anticipating the more detailed discussion in Section 1.8. Since students have difficulty interpreting the derivative as a rate of change, we felt it prudent to allow them to practice repeatedly with the concept.

3. *Analysis of Data* We added a broad theme that might best be described as "calculus for functions defined by data." Throughout the book, we include discussions about real-life applications whose underlying functions are defined by tables of data.

4. *More on Regression (optional)* We added the optional Section 7.6 on multiple and nonlinear regression analysis. The goal in this section is to provide a taste of what a business student will encounter in a course in regression analysis. Our emphasis is on using technology, especially spreadsheets, to do the computations for various flavors of regression (multiple-linear, quadratic, exponential, etc.).

5. *Additional Technology (optional)* The new technology appendix to Chapter 0 includes the graphing calculator material previously found within the chapter, a discussion of calculus and spreadsheets, and a new exercise set testing student technology skills.

6. *Real-Life Data* We have collected spreadsheets containing real-life statistical data and made them available to students and faculty on the Web site `www.prenhall.com/goldstein`.

7. *Projects* Each chapter now includes a project, designed to provide more open-ended problem solving, critical thinking, verbal expression, and integration of mathematical techniques, both manual and technological.

8. *Other Changes* We made improvements throughout the text based on suggestions from students, teachers, reviewers, and editors. Our thanks to all who assisted us with their valuable suggestions.

This edition contains more material than can be covered in most one-semester courses. Optional sections are starred in the table of contents. In addition, the level of theoretical material may be adjusted to the needs of the students. For instance, only the first two pages of Section 1.4 are required in order to introduce the limit notation.

A *Study Guide* for students containing detailed explanations and solutions for every sixth exercise is available. The *Study Guide* also includes helpful hints and strategies for studying that will help students improve their performance in the course. In addition, the *Study Guide* contains a copy of *Visual Calculus*, the popular, easy-to-use software for IBM compatible computers. *Visual Calculus* contains over 20 routines that provide additional insights into the topics discussed in the text. Also, instructors find the software valuable for constructing graphs for exams.

An *Instructor's Solutions Manual* contains worked solutions to every exercise.

TestGen EQ provides nearly 1000 suggested test questions, keyed to chapter and section. *TestGen EQ* is a text-specific testing program networkable for administering tests and capturing grades online. Edit and add your own questions, or use the new "Function Plotter" to create a nearly unlimited number of tests and drill worksheets.

Designed to complement and expand upon the text, the *text Web site* offers a variety of interactive teaching and learning tools. Since many of the text projects use real-life data, we made the data easier to use by making it available in Excel spreadsheets on the Web site. The Web site also includes links to related Web sites, quizzes, Syllabus Builder, and more. For more information, visit `www.prenhall.com/goldstein` or contact your local Prentice Hall representative.

Acknowledgments

The following is a list of reviewers from this and previous editions. We apologize for any omissions. While writing this book, we have received assistance from many persons. And our heartfelt thanks goes out to them all. Especially, we would like to thank the following reviewers, who took the time and energy to share their ideas, preferences, and often their enthusiasm with us.

Russell Lee, Allan Hancock College; Donald Hight, Kansas State College of Pittsburg; Ronald Rose, American River College; W.R. Wilson, Central Piedmont Community College; Bruce Swenson, Foothill College; Samuel Jasper, Ohio University; Carl David Minda, University of Cincinnati; H. Keith Stumpff, Central Missouri State University; Claude Schochet, Wayne State University; James E. Honeycutt, North Carolina University; Charles Himmelberg, University of Kansas; James A. Huckaba, University of Missouri; Joyce Longman, Villanova University; T. Y. Lam, University of California, Berkeley; W. T. Kyner, University of New Mexico; Shirley A. Goldman, University of California, Davis; Dennis White, University of Minnesota; Dennis Bertholf, Oklahoma State University; Wallace A. Wood, Bryant College; James L. Heitsch, University of Illinois, Chicago Circle; John H. Mathews, California State University, Fullerton; Arthur J. Schwartz, University of Michigan; Gordon Lukesh, University of Texas, Austin; William McCord, University of Missouri; W. E. Conway, University of Arizona;

David W. Penico, Virginia Commonwealth University; Howard Frisinger, Colorado State University; Robert Brown, University of California, Los Angeles; Robert Brown, University of Kansas; Carla Wofsky, University of New Mexico; Heath K. Riggs, University of Vermont; James Kaplan, Boston University; Larry Gerstein, University of California, Santa Barbara; Donald E. Myers, University of Arizona, Tempe; Frankl Warner, University of Pennsylvania; Edward Spanier, University of California, Berkeley; David Harbater, University of Pennsylvania; Bruce Edwards, University of Florida; Ann McGaw, University of Texas, Austin; Michael J. Berman, James Madison University; Fred Brauer, University of Wisconsin; Jack R. Barone, Baruch College, CUNY; James W. Brewer, Florida Atlantic University; Alan Candiotti, Drew University; E. John Hornsby, Jr., University of New Orleans; Dennis Brewer, University of Arkansas; Melvin D. Lax, California State University, Long Beach; Lawrence J. Lardy, Syracuse University; Arlene Sherburne, Montgomery College, Rockville; Gabriel Lugo, University of North Carolina, Wilmington; W. R. Hintzman, San Diego State University; Georgia B. Pyrros, University of Delaware; Joan M. Thomas, University of Oregon; Charles Clever, South Dakota State University; James Sochacki, James Madison University; Judy B. Kidd, James Madison University; Jack E. Graves, Syracuse University; Karabi Datta, Northern Illinois University; James V. Balch, Middle Tennessee State University; H. Suey Quan, Golden West College; Albert G. Fadell, SUNY Buffalo; Murray Schechter, Lehigh University; Betty Fein, Oregon State University; Biswa Datta, Northern Illinois University; Dennis DeTurck, University of Pennsylvania; Brenda Diesslin, Iowa State University; Shujuan Ji, Columbia University; Robert A. Miller, City University New York; Geraldine Taiani, Pace University; Janice Epstein, Texas A&M University; Harvey Greenwald, California State Polytechnic University; Robert Seeley, University of Massachusetts, Boston.

Thanks to Laurel Technical Services for its diligent accuracy checking. David Chelton and Susan Gerstein, our developmental editors, provided us with many fine suggestions sure to be appreciated by both students and teachers.

The authors would like to thank the many people at Pearson Education who have contributed to the success of our books over the years. We appreciate the tremendous efforts of the production, art, manufacturing, and marketing departments. Special thanks go to Lynn Savino Wendel, who managed production of this book. The expert skills of our typesetter, Dennis Kletzing, have once again eased the burden of preparing this new edition.

The authors would like to thank our editors, Kathy Boothby Sestak, who helped us plan this edition, and Gina Huck, who filled in while Kathy was on leave.

Larry J. Goldstein
David C. Lay
David I. Schneider

Introduction

O ften it is possible to give a succinct and revealing description of a situation by drawing a graph. For example, Fig. 1 describes the amount of money in a bank account drawing 5% interest, compounded daily. The graph shows that as time passes, the amount of money in the account grows. In Fig. 2 we have drawn a graph that depicts the weekly sales of a breakfast cereal at various times after advertising has ceased. The graph shows that the longer the time since the last advertisement, the fewer the sales. Figure 3 shows the size of a bacteria culture at various times. The culture grows larger as time passes. But there is a maximum size that the culture cannot exceed. This maximum size reflects the restrictions imposed by food supply, space, and similar factors. The graph in Fig. 4 describes the decay of the radioactive isotope iodine 131. As time passes, less and less of the original radioactive iodine remains.

Each of the graphs in Figs. 1 to 4 describes a change that is taking place. The amount of money in the bank is changing as are the sales of cereal, the size of the bacteria culture, and the amount of the iodine. Calculus provides mathematical tools to study each of these changes in a quantitative way.

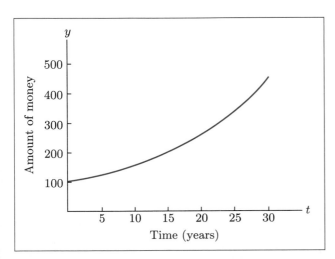

Figure 1. Growth of money in a savings account.

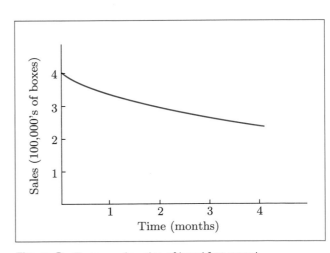

Figure 2. Decrease in sales of breakfast cereal.

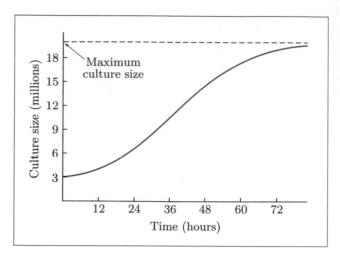

Figure 3. Growth of a bacteria culture.

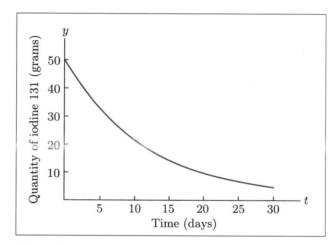

Figure 4. Decay of radioactive iodine.

CHAPTER

Functions

0

E ach of the graphs in Figs. 1 to 4 of the Introduction depicts a relationship between two quantities. For example, Fig. 4 illustrates the relationship between the quantity of iodine (measured in grams) and time (measured in days). The basic quantitative tool for describing such relationships is a *function*. In this preliminary chapter, we develop the concept of a function and review important algebraic operations on functions used later in the text.

0.1 Functions and Their Graphs

Real Numbers

Most applications of mathematics use real numbers. For purposes of such applications (and the discussions in this text), it suffices to think of a real number as a decimal. A *rational* number is one that may be written as a finite or infinite repeating decimal, such as

$$-\frac{5}{2} = -2.5, \qquad 1, \qquad \frac{13}{3} = 4.333\ldots \qquad \text{(rational numbers)}.$$

An *irrational* number has an infinite decimal representation whose digits form no repeating pattern, such as

$$-\sqrt{2} = -1.414214\ldots, \qquad \pi = 3.14159\ldots \qquad \text{(irrational numbers)}.$$

Figure 1. The real number line.

The real numbers are described geometrically by a *number line*, as in Fig. 1. Each number corresponds to one point on the line, and each point determines one real number.

3

We use four types of inequalities to compare real numbers.

$$x < y \qquad x \text{ is less than } y$$
$$x \leq y \qquad x \text{ is less than or equal to } y$$
$$x > y \qquad x \text{ is greater than } y$$
$$x \geq y \qquad x \text{ is greater than or equal to } y$$

The double inequality $a < b < c$ is shorthand for the pair of inequalities $a < b$ and $b < c$. Similar meanings are assigned to other double inequalities, such as $a \leq b < c$. Three numbers in a double inequality, such as $1 < 3 < 4$ or $4 > 3 > 1$, should have the same relative positions on the number line as in the inequality (when read left to right or right to left). Thus $3 < 4 > 1$ is never written because the numbers are "out of order."

Geometrically, the inequality $x \leq b$ means that either x equals b or x lies to the left of b on the number line. The set of real numbers x that satisfy the double inequality $a \leq x \leq b$ corresponds to the line segment between a and b, including the endpoints. This set is sometimes denoted by $[a, b]$ and is called the *closed interval* from a to b. If a and b are removed from the set, the set is written as (a, b) and is called the *open interval* from a to b. The notation for various line segments is listed in Table 1.

The symbols ∞ ("infinity") and $-\infty$ ("minus infinity") do not represent actual real numbers. Rather, they indicate that the corresponding line segment extends infinitely far to the right or left. An inequality that describes such an infinite interval may be written in two ways. For instance, $a \leq x$ is equivalent to $x \geq a$.

▶ **Example 1**   Describe each of the following intervals both graphically and in terms of inequalities.

(a) $(-1, 2)$ (b) $[-2, \pi]$ (c) $(2, \infty)$ (d) $(-\infty, \sqrt{2}]$

Solution   The line segments corresponding to the intervals are shown in Fig. 2(a)–(d). Note that an interval endpoint that is included (e.g., both endpoints of $[a, b]$) is drawn

Table 1	Intervals on the Number Line	
Inequality	**Geometric Description**	**Interval Notation**
$a \leq x \leq b$	●―――● a ... b	$[a, b]$
$a < x < b$	○―――○ a ... b	(a, b)
$a \leq x < b$	●―――○ a ... b	$[a, b)$
$a < x \leq b$	○―――● a ... b	$(a, b]$
$a \leq x$	●――― a	$[a, \infty)$
$a < x$	○――― a	(a, ∞)
$x \leq b$	―――● b	$(-\infty, b]$
$x < b$	―――○ b	$(-\infty, b)$

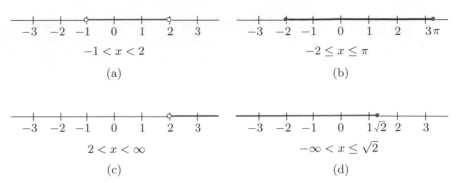

Figure 2. *Line segments.*

as a solid circle, whereas an endpoint not included (e.g., the endpoint a in $(a, b]$) is drawn as an unfilled circle. ◆

▶ Example 2   The variable x describes the profit that a company is anticipated to earn in the current fiscal year. The business plan calls for a profit of at least 5 million dollars. Describe this aspect of the business plan in the language of intervals.

Solution   The phrase "at least" means "greater than or equal to." The business plan requires that $x \geq 5$ (where the units are millions of dollars). This is equivalent to saying that x lies in the infinite interval $[5, \infty)$. ◆

Functions   A *function* of a variable x is a *rule* f that assigns to each value of x a unique number $f(x)$, called *the value of the function at x*. [We read "$f(x)$" as "f of x."] The variable x is called the *independent variable*. The set of values that the independent variable is allowed to assume is called the *domain* of the function. The domain of a function may be explicitly specified as part of the definition of a function or it may be understood from context. (See the following discussion.) The *range* of a function is the set of values that the function assumes.

The functions we shall meet in this book will usually be defined by algebraic formulas. For example, the domain of the function

$$f(x) = 3x - 1$$

consists of all real numbers x. This function is the rule that takes a number, multiplies it by 3, and then subtracts 1. If we specify a value of x, say $x = 2$, then we find the value of the function at 2 by substituting 2 for x in the formula:

$$f(2) = 3(2) - 1 = 5.$$

▶ Example 3   Let f be the function with domain all real numbers x and defined by the formula

$$f(x) = 3x^3 - 4x^2 - 3x + 7.$$

Find $f(2)$ and $f(-2)$.

Solution   To find $f(2)$ we substitute 2 for every occurrence of x in the formula for $f(x)$:

$$\begin{aligned}
f(2) &= 3(2)^3 - 4(2)^2 - 3(2) + 7 \\
&= 3(8) - 4(4) - 3(2) + 7 \\
&= 24 - 16 - 6 + 7 \\
&= 9.
\end{aligned}$$

To find $f(-2)$ we substitute (-2) for each occurrence of x in the formula for $f(x)$. The parentheses ensure that the -2 is substituted correctly. For instance, x^2 must be replaced by $(-2)^2$, not -2^2.

$$\begin{aligned} f(-2) &= 3(-2)^3 - 4(-2)^2 - 3(-2) + 7 \\ &= 3(-8) - 4(4) - 3(-2) + 7 \\ &= -24 - 16 + 6 + 7 \\ &= -27 \end{aligned}$$ ◆

▶ **Example 4**   If x represents the temperature of an object in degrees Celsius, then the temperature in degrees Fahrenheit is a function of x, given by $f(x) = \frac{9}{5}x + 32$.

(a) Water freezes at $0°C$ (C = Celsius) and boils at $100°C$. What are the corresponding temperatures in degrees Fahrenheit?

(b) Aluminum melts at $660°C$. What is its melting point in degrees Fahrenheit?

Solution   (a) $f(0) = \frac{9}{5}(0) + 32 = 32$. Water freezes at $32°F$.

$$f(100) = \frac{9}{5}(100) + 32 = 180 + 32 = 212.$$

Water boils at $212°F$.

(b) $f(660) = \frac{9}{5}(660) + 32 = 1188 + 32 = 1220$. Aluminum melts at $1220°F$. ◆

▶ **Example 5**   (*A Voting Model*) Let x be the proportion of the total popular vote that a Democratic candidate for president receives in a U.S. national election (so x is a number between 0 and 1). Political scientists have observed that a good estimate of the proportion of seats in the House of Representatives going to Democratic candidates is given by the function

$$f(x) = \frac{x^3}{x^3 + (1-x)^3}, \qquad 0 \le x \le 1,$$

whose domain is the interval $[0, 1]$. This formula is called the *cube law*. Compute $f(.6)$ and interpret the result.

Solution   We must substitute .6 for every occurrence of x in $f(x)$:

$$f(.6) = \frac{(.6)^3}{(.6)^3 + (1 - .6)^3} = \frac{(.6)^3}{(.6)^3 + (.4)^3}$$

$$= \frac{.216}{.216 + .064} = \frac{.216}{.280} \approx .77.$$

This calculation shows that the cube law function predicts that if .6 (or 60%) of the total popular vote is for the Democratic candidate for president, then approximately .77 (or 77%) of the seats in the House of Representatives will be won by Democratic candidates; that is, about 335 of the 435 seats will be won by Democrats. Note that for $x = .5$, we have $f(0.5) = .5$. Is this what you would expect? ◆

In the preceding examples, the functions had domains consisting of all real numbers or an interval. For some functions, the domain may consist of several intervals, with a different formula defining the function on each interval. Here is an illustration of this phenomenon.

▶ Example 6　A leading brokerage firm charges a 6% commission on gold purchases in amounts from \$50 to \$300. For purchases exceeding \$300, the firm charges 2% of the amount purchased plus \$12.00. Let x denote the amount of gold purchased (in dollars) and let $f(x)$ be the commission charge as a function of x.

(a) Describe $f(x)$.

(b) Find $f(100)$ and $f(500)$.

Solution　(a) The formula for $f(x)$ depends on whether $50 \leq x \leq 300$ or $300 < x$. When $50 \leq x \leq 300$, the charge is $.06x$ dollars. When $300 < x$, the charge is $.02x + 12$. The domain consists of the values x in one of the two intervals $[50, 300]$ and $(300, \infty)$. In each of these intervals, the function is defined by a separate formula:

$$f(x) = \begin{cases} .06x & \text{for } 50 \leq x \leq 300 \\ .02x + 12 & \text{for } 300 < x. \end{cases}$$

Note that an alternate description of the domain is the interval $[50, \infty)$. That is, the value of x may be any real number greater than or equal to 50.

(b) Since $x = 100$ satisfies $50 \leq x \leq 300$, we use the first formula for $f(x)$: $f(100) = .06(100) = 6$. Since $x = 500$ satisfies $300 < x$, we use the second formula for $f(x)$: $f(500) = .02(500) + 12 = 22$.　　　◆

In calculus, it is often necessary to substitute an algebraic expression for x and simplify the result, as illustrated in the following example.

▶ Example 7　If $f(x) = (4 - x)/(x^2 + 3)$, what is $f(a)$? $f(a + 1)$?

Solution　Here a represents some number. To find $f(a)$, we substitute a for x wherever x appears in the formula defining $f(x)$:

$$f(a) = \frac{4 - a}{a^2 + 3}.$$

To evaluate $f(a + 1)$, substitute $a + 1$ for each occurrence of x in the formula for $f(x)$:

$$f(a + 1) = \frac{4 - (a + 1)}{(a + 1)^2 + 3}.$$

The expression for $f(a + 1)$ may be simplified, using the fact that $(a + 1)^2 = (a + 1)(a + 1) = a^2 + 2a + 1$:

$$f(a + 1) = \frac{4 - (a + 1)}{(a + 1)^2 + 3} = \frac{4 - a - 1}{a^2 + 2a + 1 + 3} = \frac{3 - a}{a^2 + 2a + 4}.$$　　　◆

More About the Domain of a Function　When defining a function, it is necessary to specify the domain of the function, which is the set of acceptable values of the variable. In the preceding examples, we explicitly specified the domains of the functions considered. However, throughout the remainder of the text, we will usually mention functions without specifying domains. In such circumstances, we will understand the intended domain to consist of all numbers for which the defining formula(s) make sense. For example, consider the function

$$f(x) = x^2 - x + 1.$$

The expression on the right may be evaluated for any value of x. So in the absence of any explicit restrictions on x, the domain is understood to consist of all numbers. As a second example, consider the function

$$f(x) = \frac{1}{x}.$$

Here x may be any number except zero. (Division by zero is not permissible.) So the domain intended is the set of nonzero numbers. Similarly, when we write

$$f(x) = \sqrt{x},$$

we understand the domain of $f(x)$ to be the set of all nonnegative numbers, since the square root of a number x is defined if and only if $x \geq 0$.

Graphs of Functions   Often it is helpful to describe a function f geometrically, using a rectangular xy-coordinate system. Given any x in the domain of f, we can plot the point $(x, f(x))$. This is the point in the xy-plane whose y-coordinate is the value of the function at x. The set of *all* such points $(x, f(x))$ usually forms a curve in the xy-plane and is called the *graph of the function $f(x)$*.

It is possible to approximate the graph of $f(x)$ by plotting the points $(x, f(x))$ for a representative set of values of x and joining them by a smooth curve. (See Fig. 3.) The more closely spaced the values of x, the closer the approximation.

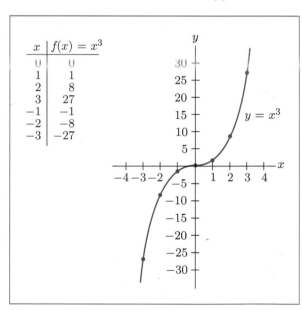

Figure 4.   Graph of f(x) = x³.

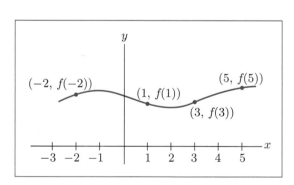

Figure 3.

▶ **Example 8**   Sketch the graph of the function $f(x) = x^3$.

Solution   The domain consists of all numbers x. We choose some representative values of x and tabulate the corresponding values of $f(x)$. We then plot the points $(x, f(x))$ and draw a smooth curve through the points. (See Fig. 4.)   ◆

▶ **Example 9**   Sketch the graph of the function $f(x) = 1/x$.

Solution   The domain of the function consists of all numbers except zero. The table in Fig. 5 lists some representative values of x and the corresponding values of $f(x)$.

A function often has interesting behavior for x near a number not in the domain. So when we chose representative values of x from the domain, we included some values close to zero. The points $(x, f(x))$ are plotted and the graph sketched in Fig. 5. ◆

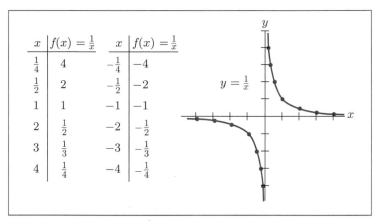

x	$f(x) = \frac{1}{x}$	x	$f(x) = \frac{1}{x}$
$\frac{1}{4}$	4	$-\frac{1}{4}$	-4
$\frac{1}{2}$	2	$-\frac{1}{2}$	-2
1	1	-1	-1
2	$\frac{1}{2}$	-2	$-\frac{1}{2}$
3	$\frac{1}{3}$	-3	$-\frac{1}{3}$
4	$\frac{1}{4}$	-4	$-\frac{1}{4}$

Figure 5. Graph of $f(x) = \dfrac{1}{x}$.

Now that graphing calculators and computer graphing programs are widely available, one seldom needs to sketch graphs by hand-plotting large numbers of points on graph paper. However, to use such a calculator or program effectively, one must know in advance which part of a curve to display. Critical features of a graph may be missed or misinterpreted if, for instance, the scale on the x- or y-axis is inappropriate.

An important use of calculus is to identify key features of a function that should appear in its graph. In many cases, only a few points need be plotted and the general shape of the graph is easy to sketch by hand. For more complicated functions, a graphing program is helpful. Even then, calculus provides a way of checking that the graph on the computer screen has the correct shape. Algebraic calculations are usually part of the analysis. The appropriate algebraic skills are reviewed in this chapter.

Note also that analytic solutions to problems typically provide more precise information than graphing calculators and can provide insight into the behavior of the functions involved in the solution.

The connection between a function and its graph is explored in this section and in Section 0.6.

▶ **Example 10**   Suppose that f is the function whose graph is given in Fig. 6. Notice that the point $(x, y) = (3, 2)$ is on the graph of f.

(a)   What is the value of the function when $x = 3$?

(b)   Find $f(-2)$.

(c)   What is the domain of f?

Solution   (a)   Since $(3, 2)$ is on the graph of f, the y-coordinate 2 must be the value of f at the x-coordinate 3. That is, $f(3) = 2$.

(b)   To find $f(-2)$ we look at the y-coordinate of the point on the graph where $x = -2$. From Fig. 6 we see that $(-2, 1)$ is on the graph of f. Thus $f(-2) = 1$.

(c)   The points on the graph of $f(x)$ all have x-coordinates between -3 and 5 inclusive; and for each value of x between -3 and 5 there is a point $(x, f(x))$

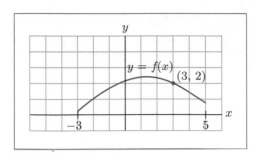

Figure 6.

on the graph. So the domain consists of those x for which $-3 \le x \le 5$. ◆

To every x in the domain, a function assigns one and only one value of y, namely, the function value $f(x)$. This implies, among other things, that not every curve is the graph of a function. To see this, refer first to the curve in Fig. 6, which *is* the graph of a function. It has the following important property: For each x between -3 and 5 inclusive there is a *unique y* such that (x, y) is on the curve. The variable y is called the *dependent variable*, since its value depends on the value of the independent variable x. Refer to the curve in Fig. 7. It cannot be the graph of a function because a function f must assign to each x in its domain a *unique* value $f(x)$. However, for the curve of Fig. 7 there corresponds to $x = 3$ (for example) more than one y-value, namely, $y = 1$ and $y = 4$.

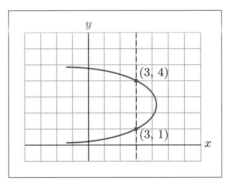

Figure 7. A curve that is **not** the graph of a function.

The essential difference between the curves in Figs. 6 and 7 leads us to the following test.

The Vertical Line Test A curve in the xy-plane is the graph of a function if and only if each vertical line cuts or touches the curve at no more than one point.

▶ Example 11 Which of the curves in Fig. 8 are graphs of functions?

Solution The curve in (a) is the graph of a function. It appears that vertical lines to the left of the y-axis do not touch the curve at all. This simply means that the function represented in (a) is defined only for $x \ge 0$. The curve in (b) is *not* the graph of a function because some vertical lines cut the curve in three places. The curve in (c) is the graph of a function whose domain is all nonzero x. [There is no point on the curve in (c) whose x-coordinate is 0.] ◆

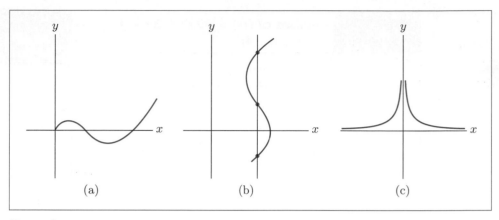

(a) (b) (c)

Figure 8.

There is another notation for functions that we will find useful. Suppose that $f(x)$ is a function. When $f(x)$ is graphed on an xy-coordinate system, the values of $f(x)$ give the y-coordinates of points of the graph. For this reason, the function is often abbreviated by the letter y, and we find it convenient to speak of "the function $y = f(x)$." For example, the function $y = 2x^2 + 1$ refers to the function $f(x)$ for which $f(x) = 2x^2 + 1$. The graph of a function $f(x)$ is often called *the graph of the equation* $y = f(x)$.

Three Views of a Function

There are three methods for describing a function. The first consists of giving a formula for the function along with any limitation on the values of the independent variable. Here is a function specified in this way:

$$f(x) = -2x^2 - 3x + 1 \quad (-3 \le x \le 4).$$

A function specified in terms of a formula is said to be defined *analytically*.

A second method for describing a function is by drawing its graph. Such a function is said to be defined *graphically*. For example, the graph of the preceding function is shown in Fig. 9. Note that by reading the graph, we can, in principle, learn everything that can be derived from the analytic definition.

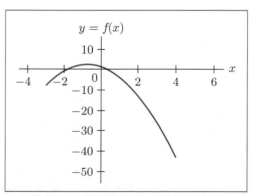

Figure 9. Graph of f(x) = −2x² − 3x + 1
(−3 ≤ x ≤ 4).

Table 2	Values of f(x) = $-2x^2 - 3x + 1$ ($-3 \le x \le 4$).		
x	$f(x)$	x	$f(x)$
-3	-8	1	-4
-2.5	-4	1.5	-8
-2	-1	2	-13
-1.5	1	2.5	-19
-1	2	3	-26
-0.5	2	3.5	-34
0	1	4	-43
0.5	-1		

Table 3	Amount of U.S. domestic mail, 1990–1997	
	Year	Number of pieces (millions)
	1990	166,301
	1992	166,443
	1993	171,220
	1994	178,039
	1995	180,734
	1996	183,440
	1997	190,888

The third method for describing a function is by giving a table of function values, as shown in Table 2. A function described in this way is said to be defined *numerically*.

Newspapers and magazines are filled with examples of data tables and their corresponding functions. For example, Table 3 tabulates the number of pieces of U.S. mail for the years 1990–1997.

Note that all functions up to this point were defined for a continuous range of x-values (say $x < 3$ or $-1 < x < 1$). A table of data, by its very nature, defines a function only for a set of particular values of x, namely, for x equal to a number in the first column. The graph of such a function is a collection of points, called a *scatter plot*. Figure 10 shows the scatter plot for the U.S. mail data in Table 3.

The techniques of calculus apply only to functions defined for a continuous ranges of x. On the other hand, the most common functions arising in daily life are numerically defined. In order to apply calculus to study such functions, it is common practice to extend the numerical definition to include a continuous range of x. There are several ways this can be done. First, we can draw a line

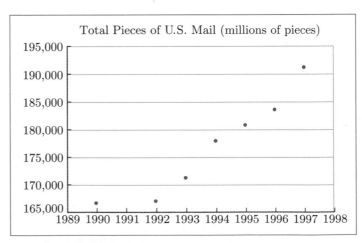

Figure 10. Scatter Plot Graph of Data in Table 3.

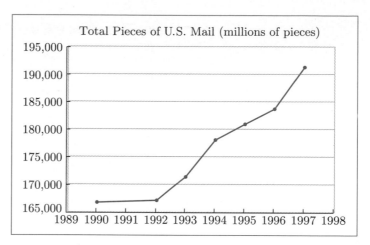

Figure 11. *A piecewise-linear model formed from data points.*

from each point to the next, as in Fig. 11. The resulting graph passes through each of the data points and is called *piecewise linear*, since it is a sequence of straight lines.

To study a graph using calculus, it helps to eliminate any corners. There are many ways to do this. For example, Fig. 12 shows a smooth curve (a curve without the corners of Fig. 11) through the data points of Fig. 10. Although the curve goes through all the data points, it is rather complicated. Figure 13 shows a curve that misses some of the data points, but has a much simpler analytic form (a cubic polynomial) which makes it easy to manipulate using the methods of calculus. We will look further at such approximating curves in Chapter 7.

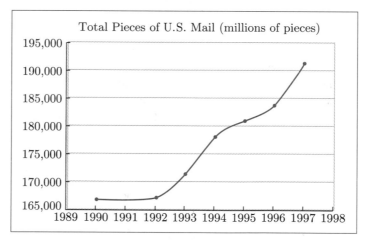

Figure 12. *A sixth-degree model approximating the set of data points of Table 3.*

In analyzing a function, it is useful to study it from all three viewpoints: analytical, graphical, and numerical. This basic principle is called the *rule of three*.

Graphs of Equations   The equations arising in connection with functions are all of the form

$$y = [\text{an expression in } x].$$

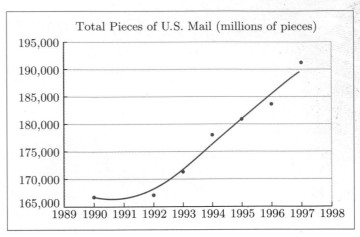

Figure 13. A cubic curve approximating the data points in Fig. 10.

However, not all equations connecting the variables x and y are of this sort. For example, consider these equations:

$$2x + 3y = 5$$
$$x = 3$$
$$x^2 + y^2 = 1.$$

It is possible to graph an equation by plotting points just as for functions. The only difference is that the resulting graph may not satisfy the vertical line test. For example, the graphs of the three preceding equations are shown in Fig. 14. Only the first graph is the graph of a function.

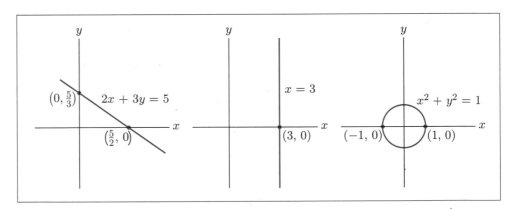

Figure 14.

Graphing calculators cannot display the graphs of all equations. They can only display graphs of equations you can solve for y in terms of x. For instance, in order to graph the first of the preceding equations, it is necessary to first solve for y in terms of x:

$$2x + 3y = 5$$
$$3y = 5 - 2x$$
$$y = \frac{5}{3} - \frac{2}{3}x.$$

The second equation cannot be solved for y in terms of x and so cannot be graphed using a graphing calculator. For the third equation, we may solve for y in terms of x to obtain

$$x^2 + y^2 = 1$$
$$y^2 = 1 - x^2$$
$$y = \pm\sqrt{1 - x^2}.$$

The $\pm$ indicates that the last equation is really two equations, one corresponding to each choice of sign. On a graphing calculator, we must enter each of these equations:

$$y = \sqrt{1 - x^2}, \qquad y = -\sqrt{1 - x^2}.$$

The calculator will draw the graphs of both equations, which together comprise the graph of the original equation.

Letters other than f can be used to denote functions and letters other than x and y can be used to denote variables. This is especially common in applied problems where letters are chosen to suggest the quantities they depict. For instance, the revenue of a company as a function of time might be written $R(t)$.

Incorporating Technology Throughout this text, we include sections describing the use of technology in calculus. The basics of graphing-calculator and spreadsheet technology are given in **Appendix A: Graphing Functions Using Technology** at the end of this chapter. In order to follow the **Incorporating Technology** sections, you should read Appendix A at this point. Henceforth, the **Incorporating Technology** sections will make use of the terminology and the notation of Appendix A. Note, however, that you may skip the **Incorporating Technology** sections without any loss of continuity, in which case, the material in Appendix A will not be needed.

Practice Problems 0.1

1. Is the point $(3, 12)$ on the graph of the function $g(x) = x^2 + 5x - 10$?
2. Sketch the graph of the function $h(t) = t^2 - 2$.

▶ Exercises 0.1*

Draw the following intervals on the number line.

1. $[-1, 4]$
2. $(4, 3\pi)$
3. $[-2, \sqrt{2})$
4. $[1, \frac{3}{2}]$
5. $(-\infty, 3)$
6. $(4, \infty)$

Use intervals to describe the real numbers satisfying the inequalities in Exercises 7–12.

7. $2 \le x < 3$
8. $-1 < x < \frac{3}{2}$
9. $x < 0, x \ge -1$
10. $x \ge -1, x < 8$
11. $x < 3$
12. $x \ge \sqrt{2}$
13. If $f(x) = x^2 - 3x$, find $f(0)$, $f(5)$, $f(3)$, and $f(-7)$.

14. If $f(x) = 9 - 6x + x^2$, find $f(0)$, $f(2)$, $f(3)$, and $f(-13)$.
15. If $f(x) = x^3 + x^2 - x - 1$, find $f(1)$, $f(-1)$, $f(\frac{1}{2})$, and $f(a)$.
16. If $g(t) = t^3 - 3t^2 + t$, find $g(2)$, $g(-\frac{1}{2})$, $g(\frac{2}{3})$, and $g(a)$.
17. If $h(s) = s/(1 + s)$, find $h(\frac{1}{2})$, $h(-\frac{3}{2})$, and $h(a + 1)$.
18. If $f(x) = x^2/(x^2 - 1)$, find $f(\frac{1}{2})$, $f(-\frac{1}{2})$, and $f(a + 1)$.
19. If $f(x) = x^2 - 2x$, find $f(a + 1)$ and $f(a + 2)$.
20. If $f(x) = x^2 + 4x + 3$, find $f(a - 1)$ and $f(a - 2)$.

---

*A complete solution for every third odd-numbered exercise is given in the *Study Guide* for this text.

21. An office supply firm finds that the number of fax machines sold in year x is given approximately by the function $f(x) = 50 + 4x + \frac{1}{2}x^2$, where $x = 0$ corresponds to 1990.

(a) What does $f(0)$ represent?

(b) Find the number of fax machines sold in 1992.

22. When a solution of acetylcholine is introduced into the heart muscle of a frog, it diminishes the force with which the muscle contracts. The data from experiments of the biologist A. J. Clark are closely approximated by a function of the form

$$R(x) = \frac{100x}{b+x}, \qquad x \geq 0,$$

where x is the concentration of acetylcholine (in appropriate units), b is a positive constant that depends on the particular frog, and $R(x)$ is the response of the muscle to the acetylcholine, expressed as a percentage of the maximum possible effect of the drug.

(a) Suppose that $b = 20$. Find the response of the muscle when $x = 60$.

(b) Determine the value of b if $R(50) = 60$ — that is, if a concentration of $x = 50$ units produces a 60% response.

In Exercises 23–26, describe the domain of the function.

23. $f(x) = \dfrac{8x}{(x-1)(x-2)}$

24. $f(t) = \dfrac{1}{\sqrt{t}}$

25. $g(x) = \dfrac{1}{\sqrt{3-x}}$

26. $g(x) = \dfrac{4}{x(x+2)}$

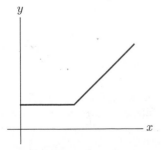

In Exercises 27–32, decide which curves are graphs of functions.

27.

28.

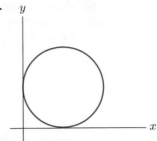

29.

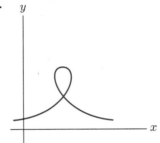

30.

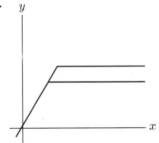

31.

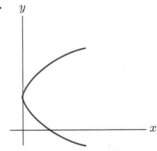

32.

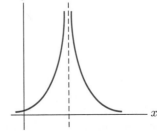

Exercises 33–42 relate to the function whose graph is sketched in Fig. 15.

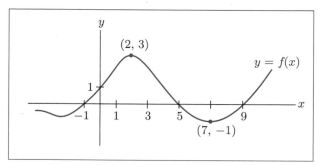

Figure 15.

33. Find $f(0)$.

34. Find $f(7)$.

35. Find $f(2)$.

36. Find $f(-1)$.

37. Is $f(4)$ positive or negative?

38. Is $f(6)$ positive or negative?

39. Is $f(-\frac{1}{2})$ positive or negative?

40. Is $f(1)$ greater than $f(6)$?

41. For what values of x is $f(x) = 0$?

42. For what values of x is $f(x) \geq 0$?

Exercises 43–46 relate to Fig. 16. When a drug is injected into a person's muscle tissue, the concentration y of the drug in the blood is a function of the time elapsed since the injection. The graph of a typical time-concentration function f is given in Fig. 16, where $t = 0$ corresponds to the time of the injection.

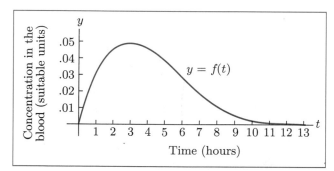

Figure 16. Drug time-concentration curve.

43. What is the concentration of the drug when $t = 1$?

44. What is the value of the time-concentration function f when $t = 6$?

45. Find $f(5)$.

46. At what time does $f(t)$ attain its largest value?

47. Is the point $(3, 12)$ on the graph of the function $f(x) = (x - \frac{1}{2})(x + 2)$?

48. Is the point $(-2, 12)$ on the graph of the function $f(x) = x(5 + x)(4 - x)$?

49. Is the point $(\frac{1}{2}, \frac{2}{5})$ on the graph of the function $g(x) = (3x - 1)/(x^2 + 1)$?

50. Is the point $(\frac{2}{3}, \frac{5}{3})$ on the graph of the function $g(x) = (x^2 + 4)/(x + 2)$?

51. Find the y-coordinate of the point $(a+1, \blacksquare)$ if this point lies on the graph of the function $f(x) = x^3$.

52. Find the y-coordinate of the point $(2+h, \blacksquare)$ if this point lies on the graph of the function $f(x) = (5/x) - x$.

In Exercises 53–56, compute $f(1)$, $f(2)$, and $f(3)$.

53. $f(x) = \begin{cases} \sqrt{x} & \text{for } 0 \leq x < 2 \\ 1 + x & \text{for } 2 \leq x \leq 5 \end{cases}$

54. $f(x) = \begin{cases} 1/x & \text{for } 1 \leq x \leq 2 \\ x^2 & \text{for } 2 < x \end{cases}$

55. $f(x) = \begin{cases} \pi x^2 & \text{for } x < 2 \\ 1 + x & \text{for } 2 \leq x \leq 2.5 \\ 4x & \text{for } 2.5 < x \end{cases}$

56. $f(x) = \begin{cases} 3/(4 - x) & \text{for } x < 2 \\ 2x & \text{for } 2 \leq x < 3 \\ \sqrt{x^2 - 5} & \text{for } 3 \leq x \end{cases}$

57. Suppose that the brokerage firm in Example 6 decides to keep the commission charges unchanged for purchases up to and including $600 but to charge only 1.5% plus $15 for gold purchases exceeding $600. Express the brokerage commission as a function of the amount x of gold purchased.

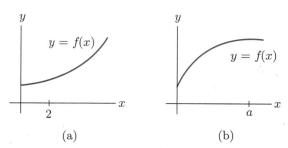

Figure 17.

58. Figure 17(a) shows the number 2 on the x-axis and the graph of a function. Let h represent a positive number and label a possible location for the number $2+h$. Plot the point on the graph whose first coordinate is $2 + h$ and label the point with its coordinates.

59. Figure 17(b) shows the number a on the x-axis and the graph of a function. Let h represent a negative number and label a possible location for the number $a + h$. Plot the point on the graph whose first coordinate is $a + h$ and label the point with its coordinates.

1. If $(3, 12)$ is on the graph of $g(x) = x^2 + 5x - 10$, then we must have $g(3) = 12$. This is not the case, however, because

$$g(3) = 3^2 + 5(3) - 10$$
$$= 9 + 15 - 10 = 14.$$

Thus $(3, 12)$ is *not* on the graph of $g(x)$.

2. Choose some representative values for t, say $t = 0, \pm 1, \pm 2, \pm 3$. For each value of t, calculate $h(t)$ and plot the point $(t, h(t))$. See Fig. 18.

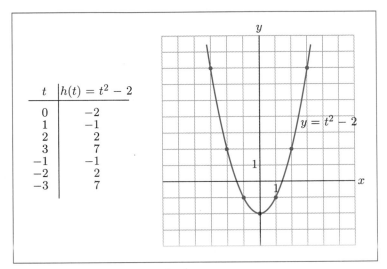

t	$h(t) = t^2 - 2$
0	-2
1	-1
2	2
3	7
-1	-1
-2	2
-3	7

Figure 18. Graph of $h(t) = t^2 - 2$.

0.2 Some Important Functions*

In this section we introduce some of the functions that will play a prominent role in our discussion of calculus.

Linear Functions   As we shall see in the next chapter, a knowledge of the algebraic and geometric properties of straight lines is essential for the study of calculus. Every straight line is the graph of a linear equation of the form

$$cx + dy = e,$$

where c, d, and e are given constants, with c and d not both zero. If $d \neq 0$, then we may solve the equation for y to obtain an equation of the form

$$y = mx + b, \tag{1}$$

for appropriate numbers m and b. If $d = 0$, then we may solve the equation for x to obtain an equation of the form

$$x = a, \tag{2}$$

---

*If you wish to use technology in your course, you should include the introductory material in Appendix A at this point. Many subsequent sections contain (optional) technology discussions and exercises that assume knowledge of Appendix A, without any comment.

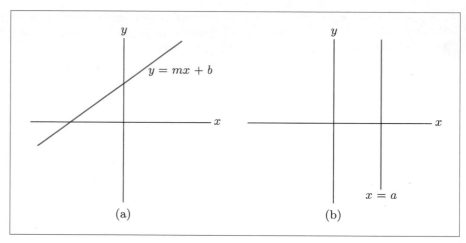

Figure 1.

for an appropriate number a. So every straight line is the graph of an equation of type (1) or (2). The graph of an equation of the form (1) is a nonvertical line [Fig. 1(a)], whereas the graph of (2) is a vertical line [Fig. 1(b)].

The straight line of Fig. 1(a) is the graph of the function $f(x) = mx + b$. Such a function, which is defined for all x, is called a *linear function*. Note that the straight line of Fig. 1(b) is not the graph of a function, since the vertical line test is violated.

An important special case of a linear function occurs if the value of m is zero, that is, $f(x) = b$ for some number b. In this case, $f(x)$ is called a *constant function*, since it assigns the same number b to every value of x. Its graph is the horizontal line whose equation is $y = b$. (See Fig. 2.)

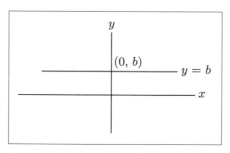

Figure 2. Graph of the constant function f(x) = b.

Linear functions often arise in real-life situations, as the first two examples show.

125000 + X 1000

▶ Example 1   When the U.S. Environmental Protection Agency found a certain company dumping sulfuric acid into the Mississippi River, it fined the company $125,000, plus $1000 per day until the company complied with federal water pollution regulations. Express the total fine as a function of the number x of days the company continued to violate the federal regulations.

Solution   The variable fine for x days of pollution, at $1000 per day, is $1000x$ dollars. The total fine is therefore given by the function

$$f(x) = 125{,}000 + 1000x.$$

◆

Since the graph of a linear function is a line, we may sketch it by locating any two points on the graph and drawing the line through them. For example, to sketch the graph of the function $f(x) = -\frac{1}{2}x + 3$, we may select two convenient values of x, say 0 and 4, and compute $f(0) = -\frac{1}{2}(0) + 3 = 3$ and $f(4) = -\frac{1}{2}(4) + 3 = 1$. The line through the points $(0, 3)$ and $(4, 1)$ is the graph of the function. (See Fig. 3.)

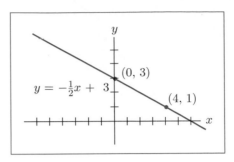

Figure 3.

▶ **Example 2**   A simple cost function for a business consists of two parts—the *fixed costs*, such as rent, insurance, and business loans, which must be paid no matter how many items of a product are produced, and the *variable costs*, which depend on the number of items produced.

Suppose a computer software company produces and sells a new spreadsheet program at a cost of \$25 per copy, and the company has fixed costs of \$10,000 per month. Express the total monthly cost as a function of the number of copies sold, x, and compute the cost when $x = 500$.

Solution   The monthly variable cost is $25x$ dollars. Thus

$$[\text{total cost}] = [\text{fixed costs}] + [\text{variable costs}]$$
$$C(x) = 10{,}000 + 25x.$$

When sales are at 500 copies per month, the cost is

$$C(500) = 10{,}000 + 25(500) = 22{,}500 \text{ (dollars)}.$$

See Fig. 4.   ◆

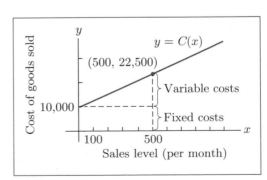

Figure 4. A linear cost function.

The point at which the graph of a linear function intersects the y-axis is called the *y-intercept* of the graph. The point at which the graph intersects the x-axis is called the *x-intercept*. The next example shows how to determine the intercepts of a linear function.

▶ **Example 3**   Determine the intercepts of the graph of the linear function $f(x) = 2x + 5$.

Solution   Since the y-intercept is on the y-axis, its x-coordinate is 0. The point on the line with x-coordinate zero has y-coordinate

$$f(0) = 2(0) + 5 = 5.$$

So the y-intercept is $(0, 5)$. Since the x-intercept is on the x-axis, its y-coordinate is 0. Since $f(x)$ gives the y-coordinate, we must have

$$2x + 5 = 0$$
$$2x = -5$$
$$x = -\tfrac{5}{2}.$$

So $\left(-\tfrac{5}{2}, 0\right)$ is the x-intercept. (See Fig. 5.)   ◆

The function in the next example is described by two expressions. Functions described by more than one expression are said to be *piecewise-defined*.

▶ **Example 4**   Sketch the graph of the following function.

$$f(x) = \begin{cases} \tfrac{5}{2}x - \tfrac{1}{2} & \text{for } -1 \le x \le 1 \\ \tfrac{1}{2}x - 2 & \text{for } x > 1. \end{cases}$$

Solution   This function is defined for $x \ge -1$. But it is defined by means of two distinct linear functions. We graph the two linear functions $\tfrac{5}{2}x - \tfrac{1}{2}$ and $\tfrac{1}{2}x - 2$. Then the graph of $f(x)$ consists of that part of the graph of $\tfrac{5}{2}x - \tfrac{1}{2}$ for which $-1 \le x \le 1$ plus that part of the graph of $\tfrac{1}{2}x - 2$ for which $x > 1$. (See Fig. 6.)   ◆

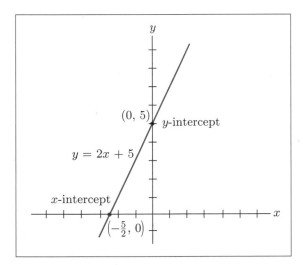

Figure 5.   Graph of f(x) = 2x + 5.

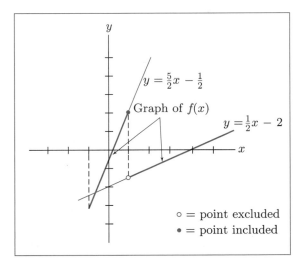

Figure 6.   Graph of a function specified by two expressions.

Quadratic Functions   Economists utilize average cost curves that relate the average unit cost of manufacturing a commodity to the number of units to be produced. (See Fig. 7.) Ecologists use curves that relate the net primary production of nutrients in a plant to the surface area of the foliage. (See Fig. 8.) Each of the curves is bowl-shaped, opening either up or down. The simplest functions whose graphs resemble these curves are the quadratic functions.

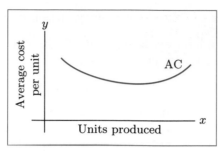

Figure 7. Average cost curve.

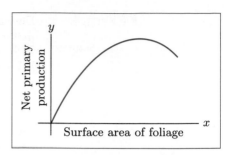

Figure 8. Production of nutrients.

A *quadratic function* is a function of the form

$$f(x) = ax^2 + bx + c,$$

where a, b, and c are constants and $a \neq 0$. The domain of such a function consists of all numbers. The graph of a quadratic function is called a *parabola*. Two typical parabolas are drawn in Figs. 9 and 10. We shall develop techniques for sketching graphs of quadratic functions after we have some calculus at our disposal.

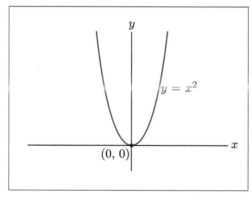

Figure 9. Graph of f(x) = x².

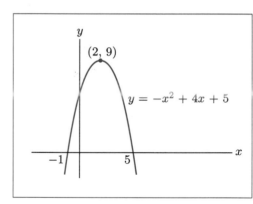

Figure 10. Graph of f(x) = −x² + 4x + 5.

Polynomial and Rational Functions   A *polynomial function* $f(x)$ is one of the form

$$f(x) = a_n x^n + a_{n-1} x^{n-1} + \cdots + a_0,$$

where n is a nonnegative integer and $a_0, a_1, \ldots, a_n$ are given numbers. Some examples of polynomial functions are

$$f(x) = 5x^3 - 3x^2 - 2x + 4$$
$$g(x) = x^4 - x + 1.$$

Of course, linear and quadratic functions are special cases of polynomial functions. The domain of a polynomial function consists of all numbers.

A function expressed as the quotient of two polynomials is called a *rational function*. Some examples are

$$h(x) = \frac{x^2 + 1}{x}$$

$$k(x) = \frac{x + 3}{x^2 - 4}.$$

The domain of a rational function excludes all values of x for which the denominator is zero. For example, the domain of $h(x)$ excludes $x = 0$, whereas the domain of $k(x)$ excludes $x = 2$ and $x = -2$. As we shall see, both polynomial and rational functions arise in applications of calculus.

Rational functions are used in environmental studies as *cost-benefit* models. The cost of removing a pollutant from the atmosphere is estimated as a function of the percentage of the pollutant removed. The higher the percentage removed, the greater the "benefit" to the people who breathe that air. The issues here are complex, of course, and the definition of "cost" is debatable. The cost to remove a small percentage of pollutant may be fairly low. But the removal cost of the final 5% of the pollutant, for example, may be terribly expensive.

▶ **Example 5**   Suppose a cost-benefit function is given by

$$f(x) = \frac{50x}{105 - x}, \qquad 0 \le x \le 100,$$

where x is the percentage of some pollutant to be removed and $f(x)$ is the associated cost (in millions of dollars). See Fig. 11. Find the costs to remove 70%, 95%, and 100% of the pollutant.

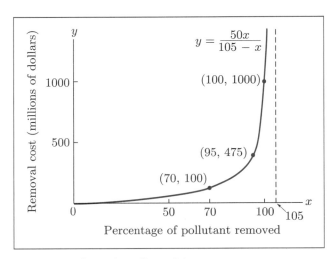

Figure 11. A cost-benefit model.

Solution   The cost to remove 70% is

$$f(70) = \frac{50(70)}{105 - 70} = 100 \text{ (million dollars)}.$$

Similar calculations show that

$$f(95) = 475 \quad \text{and} \quad f(100) = 1000.$$

Observe that the cost to remove the last 5% of the pollutant is $f(100) - f(95) = 1000 - 475 = 525$ million dollars. This is more than five times the cost to remove the first 70% of the pollutant!   ◆

Power Functions   Functions of the form $f(x) = x^r$ are called *power functions*. The meaning of x^r is obvious when r is a positive integer. However, the power function $f(x) = x^r$ may be defined for any number r. We delay until Section 0.5 a discussion of power functions. In that section, we will review the meaning of x^r in case r is a rational number.

The Absolute Value Function   The absolute value of a number x is denoted by $|x|$ and is defined by

$$|x| = \begin{cases} x & \text{if } x \text{ is positive or zero,} \\ -x & \text{if } x \text{ is negative.} \end{cases}$$

For example, $|5| = 5$, $|0| = 0$, and $|-3| = -(-3) = 3$.

The function defined for all numbers x by

$$f(x) = |x|$$

is called the *absolute value function.* Its graph coincides with the graph of the equation $y = x$ for $x \geq 0$ and with the graph of the equation $y = -x$ for $x < 0$. (See Fig. 12.)

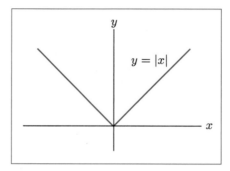

Figure 12.   Graph of the absolute value function.

Incorporating Technology   Inequalities such as $(X \leq 1)$ are either true or false depending on the value of X. Graphing calculators evaluate the inequality as 1 when it is true and 0 when it is false. See Fig. 13. (*Note*: On many calculators, inequality symbols are usually entered from the TEST menu.) Inequalities can be used to obtain expressions for piecewise-defined functions. For instance, the function in Example 4 can be defined by

$$\mathbf{Y_1} = (-1 \leq X) * (X \leq 1) * ((5/2)X - (1/2)) + (X > 1) * ((1/2)X - 2).$$

Piecewise-defined functions are usually best graphed in **Dot** (as opposed to **Connected**) mode. The preceding function is graphed in **Dot** mode in Fig. 14.

If a rational function is undefined at $x = a$, an extraneous nearly vertical line might appear when the function is graphed. Figure 15(a) shows the cost–benefit function from Example 5. This extraneous line can be suppressed by choosing the window setting so that a either appears in the middle or at one end of the screen. See Figs. 15(b) and 15(c).

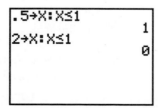

Figure 13.

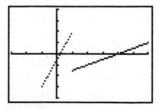

Figure 14.

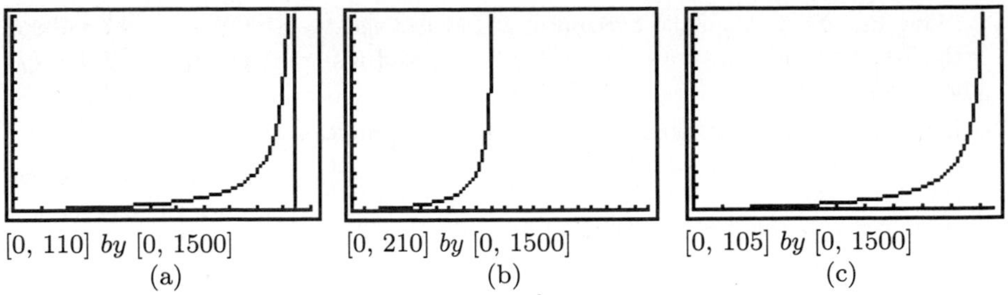

[0, 110] *by* [0, 1500]   [0, 210] *by* [0, 1500]   [0, 105] *by* [0, 1500]
(a)   (b)   (c)

Figure 15.

The absolute value function is displayed as **abs(X)**. The function **abs** is found in the MATH/NUM menu of most TI calculators. (The TI-82 has an ABS key.)

<table>
<tr><td>

Practice Problems 0.2

</td><td>

1. A photocopy service has a fixed cost of $2000 per month (for rent, depreciation of equipment, etc.) and variable costs of $.04 for each page it reproduces for customers. Express its total cost as a (linear) function of the number of pages copied per month.

2. Determine the intercepts of the graph of $f(x) = -\frac{3}{8}x + 6$.

</td></tr>
</table>

▶ Exercises 0.2

Graph the following functions.

1. $f(x) = 2x - 1$

2. $f(t) = 3$

3. $f(x) = 3x + 1$

4. $f(x) = -\frac{1}{2}x - 4$

5. $f(x) = -2x + 3$

6. $f(u) = \frac{1}{4}$

Determine the intercepts of the graphs of the following functions.

7. $f(x) = 9x + 3$

8. $f(x) = -\frac{1}{2}x - 1$

9. $f(v) = 5$

10. $f(x) = 14$

11. $f(x) = -\frac{1}{4}x + 3$

12. $f(w) = 6w - 4$

13. In biochemistry, such as in the study of enzyme kinetics, one encounters a linear function of the form $f(x) = (K/V)x + 1/V$, where K and V are constants.

(a) If $f(x) = .2x + 50$, find K and V so that $f(x)$ may be written in the form $f(x) = (K/V)x + 1/V$.

(b) Find the x-intercept and y-intercept of the line $y = (K/V)x + 1/V$ (in terms of K and V).

14. The constants K and V in Exercise 13 are often determined from experimental data. Suppose that a line is drawn through data points and has x-intercept $(-500, 0)$ and y-intercept $(0, 60)$. Determine K and V so that the line is the graph of the function $f(x) = (K/V)x + 1/V$. [*Hint*: Use Exercise 13(b).]

15. In some cities you can rent a car for $18 per day and $.20 per mile.

(a) Find the cost of renting the car for one day and driving 200 miles.

(b) Suppose that the car is to be rented for one day, and express the total rental expense as a function of the number x of miles driven. (Assume that for each fraction of a mile driven, the same fraction of $.20 is charged.)

16. A gas company will pay a property owner $5000 for the right to drill on the land for natural gas and $.10 for each thousand cubic feet of gas extracted from the land. Express the amount of money the landowner will receive as a function of the amount of gas extracted from the land.

17. In 1998, a patient paid $300 per day for a semiprivate hospital room and $1500 for an appendectomy operation. Express the total amount paid for an appendectomy as a function of the number of days of hospital confinement.

18. When a baseball thrown at 85 miles per hour is hit by a bat swung at x miles per hour, the ball travels $6x - 40$ feet.* (This formula assumes that $50 \leq x \leq 90$ and that the bat is 35 inches long, weighs 32 ounces, and strikes a waist-high pitch so that the plane of the

*Robert K. Adair, *The Physics of Baseball* (New York: Harper & Row, 1990).

swing lies at 35° from the horizontal.) How fast must the bat be swung in order for the ball to travel 350 feet?

19. Let $f(x)$ be the cost-benefit function from Example 5. If 70% of the pollutant has been removed, what is the added cost to remove another 5%? How does this compare with the cost to remove the final 5% of the pollutant? (See Example 5.)

20. Suppose that the cost (in millions of dollars) to remove x percent of a certain pollutant is given by the cost-benefit function

$$f(x) = \frac{20x}{102 - x} \qquad \text{for } 0 \le x \le 100.$$

(a) Find the cost to remove 85% of the pollutant.

(b) Find the cost to remove the final 5% of the pollutant.

Each of the quadratic functions in Exercises 21–26 has the form $y = ax^2 + bx + c$. Identify a, b, and c.

21. $y = 3x^2 - 4x$

22. $y = \dfrac{x^2 - 6x + 2}{3}$

23. $y = 3x - 2x^2 + 1$

24. $y = 3 - 2x + 4x^2$

25. $y = 1 - x^2$

26. $y = \frac{1}{2}x^2 + \sqrt{3}\,x - \pi$

Sketch the graphs of the following functions.

27. $f(x) = \begin{cases} 3x & \text{for } 0 \le x \le 1 \\ \frac{9}{2} - \frac{3}{2}x & \text{for } x > 1 \end{cases}$

28. $f(x) = \begin{cases} 1 + x & \text{for } x \le 3 \\ 4 & \text{for } x > 3 \end{cases}$

29. $f(x) = \begin{cases} 3 & \text{for } x < 2 \\ 2x + 1 & \text{for } x \ge 2 \end{cases}$

30. $f(x) = \begin{cases} \frac{1}{2}x & \text{for } 0 \le x < 4 \\ 2x - 3 & \text{for } 4 \le x \le 5 \end{cases}$

31. $f(x) = \begin{cases} 4 - x & \text{for } 0 \le x < 2 \\ 2x - 2 & \text{for } 2 \le x < 3 \\ x+1 & \text{for } x \ge 3 \end{cases}$

32. $f(x) = \begin{cases} 4x & \text{for } 0 \le x < 1 \\ 8 - 4x & \text{for } 1 \le x < 2 \\ 2x - 4 & \text{for } x \ge 2 \end{cases}$

Evaluate each of the functions in Exercises 33–38 at the given value of x.

33. $f(x) = x^{100}$, $x = -1$

34. $f(x) = x^5$, $x = \frac{1}{2}$

35. $f(x) = |x|$, $x = 10^{-2}$

36. $f(x) = |x|$, $x = \pi$

37. $f(x) = |x|$, $x = -2.5$

38. $f(x) = |x|$, $x = -\frac{2}{3}$

Technology Exercises

In Exercises 39–42, give the graphing calculator expression for the function and display its graph.

39. $f(x) = \begin{cases} 3x - 1 & \text{for } x \le 2 \\ x^2 + 1 & \text{for } x > 2 \end{cases}$

40. $f(x) = \begin{cases} 2x & \text{for } x < 3 \\ \sqrt{x} & \text{for } x \ge 3 \end{cases}$

41. $f(x) = \begin{cases} x^3 + 2x & \text{for } -2 \le x < 2 \\ \sqrt{x - 2} & \text{for } x \ge 2 \end{cases}$

42. $f(x) = \begin{cases} x^4 - 5x & \text{for } x \le 3 \\ 50 - 4x & \text{for } 3 < x \le 8 \end{cases}$

Graph the functions in Exercises 43–46. Select a window setting for which extraneous nearly vertical lines are avoided.

43. $f(x) = \dfrac{x}{x - 2}$ for $x \ge 0$

44. $f(x) = \dfrac{.2x^2}{5 - x}$ for $x \ge 0$

45. $f(x) = \dfrac{2x}{50 - x}$ for $0 \le x < 50$

46. $f(x) = \dfrac{20x}{75 - x}$ for $0 \le x < 75$

47. Graph the function $f(x) = |x|$ for $-10 \le x \le 10$.

48. Graph the function $f(x) = |x - 3|$ for $-10 \le x \le 10$.

| **Solutions to Practice Problems 0.2** | 1. If x represents the number of pages copied per month, then the variable cost is $.04x$ dollars. Now [total cost] = [fixed cost] + [variable cost]. If we define $$f(x) = 2000 + .04x,$$ then $f(x)$ gives the total cost per month. |

2. To find the y-intercept, evaluate $f(x)$ at $x = 0$.

$$f(0) = -\tfrac{3}{8}(0) + 6 = 0 + 6 = 6$$

To find the x-intercept, set $f(x) = 0$ and solve for x.

$$-\tfrac{3}{8}x + 6 = 0$$
$$\tfrac{3}{8}x = 6$$
$$x = \tfrac{8}{3} \cdot 6 = 16$$

Therefore, the y-intercept is $(0, 6)$ and the x-intercept is $(16, 0)$.

0.3 The Algebra of Functions

Many functions we shall encounter later in the text can be viewed as combinations of other functions. For example, let $P(x)$ represent the profit a company makes on the sale of x units of some commodity. If $R(x)$ denotes the revenue received from the sale of x units, and if $C(x)$ is the cost of producing x units, then

$$P(x) = R(x) - C(x)$$

$$[\text{profit}] = [\text{revenue}] - [\text{cost}].$$

Writing the profit function in this way makes it possible to predict the behavior of $P(x)$ from properties of $R(x)$ and $C(x)$. For instance, we can determine when the profit $P(x)$ is positive by observing whether $R(x)$ is greater than $C(x)$. (See Fig. 1.)

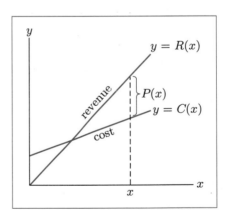

Figure 1. Profit equals revenue minus cost.

The first four examples review the algebraic techniques needed to combine functions by addition, subtraction, multiplication, and division.

▶ Example 1   Let $f(x) = 3x + 4$ and $g(x) = 2x - 6$. Find $f(x) + g(x)$, $f(x) - g(x)$, $\dfrac{f(x)}{g(x)}$, and $f(x)g(x)$.

Solution   For $f(x) + g(x)$ and $f(x) - g(x)$, we add or subtract corresponding terms:

$$f(x) + g(x) = (3x + 4) + (2x - 6) = 3x + 4 + 2x - 6 = 5x - 2$$
$$f(x) - g(x) = (3x + 4) - (2x - 6) = 3x + 4 - 2x + 6 = x + 10.$$

To compute $\dfrac{f(x)}{g(x)}$ and $f(x)g(x)$, we first substitute the formulas for $f(x)$ and $g(x)$:

$$\frac{f(x)}{g(x)} = \frac{3x+4}{2x-6}$$

$$f(x)g(x) = (3x+4)(2x-6).$$

The expression for $\dfrac{f(x)}{g(x)}$ is already in simplest form. To simplify the expression for $f(x)g(x)$, we carry out the multiplication indicated in $(3x+4)(2x-6)$. We must be careful to multiply each term of $3x+4$ by each term of $2x-6$. A common order for multiplying such expressions is (1) the First terms, (2) the Outer terms, (3) the Inner terms, and (4) the Last terms. (This procedure may be remembered by the word FOIL.)

$$f(x)g(x) = (3x+4)(2x-6) = 6x^2 - 18x + 8x - 24$$
$$= 6x^2 - 10x - 24 \qquad \blacklozenge$$

▶ **Example 2**   Let $g(x) = \dfrac{2}{x}$ and $h(x) = \dfrac{3}{x-1}$. Express $g(x) + h(x)$ as a rational function.

Solution   First we have

$$g(x) + h(x) = \frac{2}{x} + \frac{3}{x-1}, \qquad x \neq 0,\ 1.$$

The restriction $x \neq 0,\ 1$ comes from the fact that $g(x)$ is defined only for $x \neq 0$ and $h(x)$ is defined only for $x \neq 1$. (A rational function is not defined for values of the variable for which the denominator is 0.) In order for us to add two fractions, their denominators must be the same. A common denominator for $\dfrac{2}{x}$ and $\dfrac{3}{(x-1)}$ is $x(x-1)$. If we multiply $\dfrac{2}{x}$ by $\dfrac{x-1}{x-1}$, we obtain an equivalent expression whose denominator is $x(x-1)$. Similarly, if we multiply $\dfrac{3}{x-1}$ by $\dfrac{x}{x}$, we obtain an equivalent expression whose denominator is $x(x-1)$. Thus

$$\frac{2}{x} + \frac{3}{x-1} = \frac{2}{x} \cdot \frac{x-1}{x-1} + \frac{3}{x-1} \cdot \frac{x}{x}$$

$$= \frac{2(x-1)}{x(x-1)} + \frac{3x}{x(x-1)}$$

$$= \frac{2(x-1) + 3x}{x(x-1)}$$

$$= \frac{5x-2}{x(x-1)}.$$

So

$$g(x) + h(x) = \frac{5x-2}{x(x-1)}. \qquad \blacklozenge$$

▶ Example 3   Find $f(t)g(t)$, where

$$f(t) = \frac{t}{t-1} \quad \text{and} \quad g(t) = \frac{t+2}{t+1}.$$

Solution   To multiply rational functions, multiply numerator by numerator and denominator by denominator:

$$f(t)g(t) = \frac{t}{t-1} \cdot \frac{t+2}{t+1} = \frac{t(t+2)}{(t-1)(t+1)}.$$

An alternative way of expressing $f(t)g(t)$ is obtained by carrying out the indicated multiplications:

$$f(t)g(t) = \frac{t^2 + 2t}{t^2 + t - t - 1} = \frac{t^2 + 2t}{t^2 - 1}.$$

The choice of which expression to use for $f(t)g(t)$ depends on the particular application.   ◆

▶ Example 4   Find $\dfrac{f(x)}{g(x)}$, where $f(x) = \dfrac{x}{x-3}$ and $g(x) = \dfrac{x+1}{x-5}.$

Solution   The function $f(x)$ is defined only for $x \neq 3$, and $g(x)$ is defined only for $x \neq 5$. The quotient $f(x)/g(x)$ is therefore not defined for $x = 3, 5$. Moreover, the quotient is not defined for values of x for which $g(x)$ is equal to 0, that is, $x = -1$. Thus, the quotient is defined for $x \neq 3, 5, -1$. To divide $f(x)$ by $g(x)$, we multiply $f(x)$ by the reciprocal of $g(x)$:

$$\frac{f(x)}{g(x)} = \frac{x}{x-3} \cdot \frac{x-5}{x+1} = \frac{x(x-5)}{(x-3)(x+1)} = \frac{x^2 - 5x}{x^2 - 2x - 3}, \quad x \neq 3, -1, 5. \quad ◆$$

Composition of Functions   Another important way of combining two functions $f(x)$ and $g(x)$ is to substitute the function $g(x)$ for every occurrence of the variable x in $f(x)$. The resulting function is called the *composition* (or *composite*) of $f(x)$ and $g(x)$ and is denoted by $f(g(x))$.

▶ Example 5   Let $f(x) = x^2 + 3x + 1$ and $g(x) = x - 5$. What is $f(g(x))$?

Solution   We substitute $g(x)$ in place of each x in $f(x)$:

$$\begin{aligned}
f(g(x)) &= [g(x)]^2 + 3g(x) + 1 \\
&= (x-5)^2 + 3(x-5) + 1 \\
&= (x^2 - 10x + 25) + (3x - 15) + 1 \\
&= x^2 - 7x + 11.
\end{aligned}$$ ◆

Later in the text we shall need to study expressions of the form $f(x+h)$, where $f(x)$ is a given function and h represents some number. The meaning of $f(x+h)$ is that $x+h$ is to be substituted for each occurrence of x in the formula for $f(x)$. In fact, $f(x+h)$ is just a special case of $f(g(x))$, where $g(x) = x+h$.

▶ **Example 6**   If $f(x) = x^3$, find $f(x + h) - f(x)$.

Solution

$$f(x + h) = (x + h)^3 = x^3 + 3x^2h + 3xh^2 + h^3$$
$$f(x + h) - f(x) = (x^3 + 3x^2h + 3xh^2 + h^3) - x^3$$
$$= 3x^2h + 3xh^2 + h^3.$$ ◆

▶ **Example 7**   In a certain lake, the bass feed primarily on minnows, and the minnows feed on plankton. Suppose that the size of the bass population is a function $f(n)$ of the number n of minnows in the lake, and the number of minnows is a function $g(x)$ of the amount x of plankton in the lake. Express the size of the bass population as a function of the amount of plankton, if $f(n) = 50 + \sqrt{n/150}$ and $g(x) = 4x + 3$.

Solution   We have $n = g(x)$. Substituting $g(x)$ for n in $f(n)$, we find the size of the bass population is given by

$$f(g(x)) = 50 + \sqrt{\frac{g(x)}{150}} = 50 + \sqrt{\frac{4x + 3}{150}}.$$ ◆

Incorporating Technology   After expressions have been assigned to $\mathbf{Y_1}$ and $\mathbf{Y_2}$, $\mathbf{Y_3}$ can be set to $\mathbf{Y_1} + \mathbf{Y_2}$, $\mathbf{Y_1} - \mathbf{Y_2}$, $\mathbf{Y_1} * \mathbf{Y_2}$, or $\mathbf{Y_1}/\mathbf{Y_2}$. The composite of the two functions is specified by $\mathbf{Y_1}(\mathbf{Y_2})$ on many, but not all, calculators.

Measuring Change in a Variable   As we have mentioned earlier, calculus is about measuring change. As a preliminary to our discussions of the next chapter, let's show how to measure the change in a variable. Consider a variable x whose value is changing. Suppose that it starts from the value $x = a$ and moves to the value $x = b$. We say that $x = a$ is the *initial value* and that $x = b$ is the *terminal value* of x. The change that x undergoes is just the terminal value minus the initial value and is denoted Δx (read: delta x). That is, we have

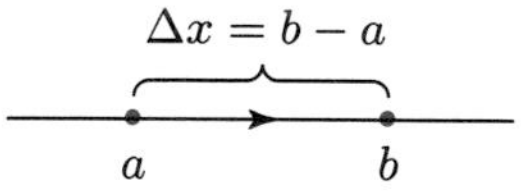

$$\Delta x = [\text{terminal value}] - [\text{initial value}] = b - a.$$

The letter Δ is a Greek letter that is commonly used in mathematics and science to mean "change." The quantity Δx measures the change in x as x varies from its initial value to its terminal value. (See Fig. 2.)

For example, if x starts out at the value 2.0 and ends up at the value 2.05, then

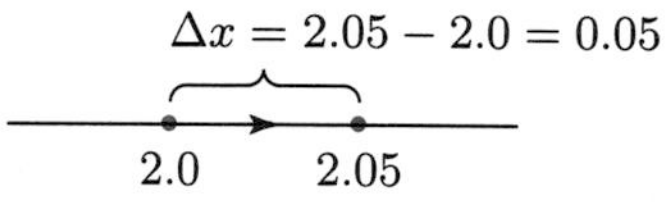

$$\Delta x = 2.05 - 2.0 = 0.05.$$

(See Fig. 3.)

Note that the value of Δx may be negative, as is the case if x starts at 2.0 and ends up at 1.95. In this case

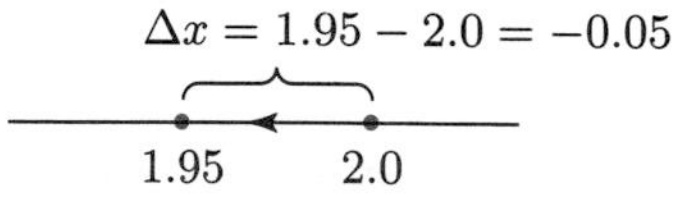

$$\Delta x = 1.95 - 2.0 = -0.05.$$

(See Fig. 4.)

You should view Δx as a directed distance, measuring the movement of x from its initial value to its terminal value; when this movement is in the positive direction, Δx is positive, and when it is in the negative direction, Δx is negative.

Suppose that y is a variable that depends on x, say $y = f(x)$, where f is a function of x. If x changes from an initial point to a terminal point, then the change in y is Δy, where

$$\Delta y = f(\text{terminal point}) - f(\text{initial point}).$$

$\Delta x = b - a$

a b

Figure 2.

$\Delta x = 2.05 - 2.0 = 0.05$

2.0   2.05

Figure 3.

$\Delta x = 1.95 - 2.0 = -0.05$

1.95   2.0

Figure 4.

▶ Example 8   Suppose that $y = x^2$ and that x changes from 2 to $2 + h$.

(a) What is the value of Δx?

(b) What is the value of Δy?

Solution   (a) We have

$$\Delta x = (2 + h) - 2 = h.$$

(b) Substitute $2 + h$ for the terminal point and 2 for the initial point:

$$\Delta y = f(2 + h) - f(2) = (2 + h)^2 - 2^2$$
$$= (4 + 4h + h^2) - 4$$
$$= 4h + h^2.$$ ◆

We may generalize the preceding example as follows. Suppose that x changes from the initial value a to the terminal value $a + h$. Moreover, suppose that $y = f(x)$. Then

$$\Delta x = h$$
$$\Delta y = f(a + h) - f(a).$$

The graphical interpretation of Δx and Δy is shown in Fig. 5.

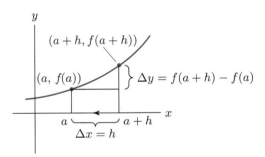

Figure 5.

Practice Problems 0.3

1. Let $f(x) = x^5$, $g(x) = x^3 - 4x^2 + x - 8$.

(a) Find $f(g(x))$.   (b) Find $g(f(x))$.

2. Let $f(x) = x^2$. Calculate $\dfrac{f(1 + h) - f(1)}{h}$ and simplify.

3. Express the quotient in Problem 2 using delta notation.

▶ **Exercises 0.3**

Let $f(x) = x^2 + 1$, $g(x) = 9x$, and $h(x) = 5 - 2x^2$. Calculate the following functions.

1. $f(x) + g(x)$

2. $f(x) - h(x)$

3. $f(x)g(x)$

4. $g(x)h(x)$

5. $\dfrac{f(t)}{g(t)}$

6. $\dfrac{g(t)}{h(t)}$

In Exercises 7–12, express $f(x) + g(x)$ as a rational function. Carry out all multiplications.

7. $f(x) = \dfrac{2}{x - 3}$, $g(x) = \dfrac{1}{x + 2}$

8. $f(x) = \dfrac{3}{x - 6}$, $g(x) = \dfrac{-2}{x - 2}$

9. $f(x) = \dfrac{x}{x - 8}$, $g(x) = \dfrac{-x}{x - 4}$

10. $f(x) = \dfrac{-x}{x+3}$, $g(x) = \dfrac{x}{x+5}$

11. $f(x) = \dfrac{x+5}{x-10}$, $g(x) = \dfrac{x}{x+10}$

12. $f(x) = \dfrac{x+6}{x-6}$, $g(x) = \dfrac{x-6}{x+6}$

Let $f(x) = \dfrac{x}{x-2}$, $g(x) = \dfrac{5-x}{5+x}$, and $h(x) = \dfrac{x+1}{3x-1}$.
Express the following as rational functions.

13. $f(x) - g(x)$

14. $f(t) - h(t)$

15. $f(x)g(x)$

16. $g(x)h(x)$

17. $\dfrac{f(x)}{g(x)}$

18. $\dfrac{h(s)}{f(s)}$

19. $f(x+1)g(x+1)$

20. $f(x+2) + g(x+2)$

21. $\dfrac{g(x+5)}{f(x+5)}$

22. $f\left(\dfrac{1}{t}\right)$

23. $g\left(\dfrac{1}{u}\right)$

24. $h\left(\dfrac{1}{x^2}\right)$

Let $f(x) = x^6$, $g(x) = \dfrac{x}{1-x}$, and $h(x) = x^3 - 5x^2 + 1$.
Calculate the following functions.

25. $f(g(x))$

26. $h(f(t))$

27. $h(g(x))$

28. $g(f(x))$

29. $g(h(t))$

30. $f(h(x))$

31. If $f(x) = x^2$, find $f(x+h) - f(x)$ and simplify.

32. If $f(x) = 1/x$, find $f(x+h) - f(x)$ and simplify.

33. If $g(t) = 4t - t^2$, find $\dfrac{g(t+h) - g(t)}{h}$ and simplify.

34. If $g(t) = t^3 + 5$, find $\dfrac{g(t+h) - g(t)}{h}$ and simplify.

35. After t hours of operation, an assembly line has assembled $A(t) = 20t - \frac{1}{2}t^2$ power lawn mowers, $0 \le t \le 10$. Suppose that the factory's cost of manufacturing x units is $C(x)$ dollars, where $C(x) = 3000 + 80x$.

   (a) Express the factory's cost as a (composite) function of the number of hours of operation of the assembly line.

   (b) What is the cost of the first 2 hours of operation?

36. During the first $\frac{1}{2}$ hour, the employees of a machine shop prepare the work area for the day's work. After that, they turn out 10 precision machine parts per hour, so that the output after t hours is $f(t)$ machine parts, where $f(t) = 10\left(t - \frac{1}{2}\right) = 10t - 5$, $\frac{1}{2} \le t \le 8$. The total cost of producing x machine parts is $C(x)$ dollars, where $C(x) = .1x^2 + 25x + 200$.

   (a) Express the total cost as a (composite) function of t.

   (b) What is the cost of the first 4 hours of operation?

Suppose that $y = -3x + 5$.

37. Calculate Δx and Δy if x has initial value 5 and terminal value 5.1.

38. Calculate Δx and Δy if x has initial value 1 and terminal value .8.

39. Calculate Δy and the terminal value of y provided that the initial value of x is 3 and $\Delta x = .01$.

40. Calculate Δy and the terminal value of y provided that the initial value of x is 3 and $\Delta x = -.01$.

Suppose that $y = \frac{1}{3}x + 2$.

41. Suppose that x has initial value 1. Determine the quotient $\dfrac{\Delta y}{\Delta x}$ for three terminal values for x.

42. Suppose that x has initial value 1 and that $\Delta x = -.1$, $-.01$, $-.001$. Determine the corresponding quotients $\dfrac{\Delta y}{\Delta x}$.

43. Suppose that y is measured in miles and x in hours. What are the units of the quotient $\dfrac{\Delta y}{\Delta x}$?

44. Suppose that y is measured in square meters and x in seconds. What are the units of the quotient $\dfrac{\Delta y}{\Delta x}$?

Consider the following tables of data. Is y a linear function of x? If so, determine the linear function.

45.

x	y
1	5
2	3
3	1
4	-1
5	-3
12	-17

46.

x	y
-2	-1
-1	-1
0	1
1	5
2	11
3	19

Table 1	Conversion Table for Men's Hat Sizes							
Britain	$6\frac{1}{2}$	$6\frac{5}{8}$	$6\frac{3}{4}$	$6\frac{7}{8}$	7	$7\frac{1}{8}$	$7\frac{1}{4}$	$7\frac{3}{8}$
France	53	54	55	56	57	58	59	60
United States	$6\frac{5}{8}$	$6\frac{3}{4}$	$6\frac{7}{8}$	7	$7\frac{1}{8}$	$7\frac{1}{4}$	$7\frac{3}{8}$	$7\frac{1}{2}$

47. Table 1 shows a conversion table for men's hat sizes for three countries. The function $g(x) = 8x + 1$ converts from British sizes to French sizes, and the function $f(x) = \frac{1}{8}x$ converts from French sizes to U.S. sizes. Determine the function $h(x) = f(g(x))$ and give its interpretation.

Technology Exercises

48. Let $f(x) = x^2$. Graph the functions $f(x+1)$, $f(x-1)$, $f(x+2)$, and $f(x-2)$. Make a guess about the relationship between the graph of a general function $f(x)$ and the graph of $f(g(x))$ where $g(x) = x + a$ for some constant a. Test your guess on the functions $f(x) = x^3$ and $f(x) = \sqrt{x}$.

49. Let $f(x) = x^2$. Graph the functions $f(x)+1$, $f(x)-1$, $f(x)+2$, and $f(x)-2$. Make a guess about the relationship between the graph of a general function $f(x)$ and the graph of $f(x) + c$ for some constant c. Test your guess on the functions $f(x) = x^3$ and $f(x) = \sqrt{x}$.

50. Based on the results of Exercises 48 and 49, sketch the graph of $f(x) = (x-1)^2 + 2$ without using a graphing calculator. Check your result with a graphing calculator.

51. Based on the results of Exercises 48 and 49, sketch the graph of $f(x) = (x+2)^2 - 1$ without using a graphing calculator. Check your result with a graphing calculator.

52. Let $f(x) = x^2 + 3x + 1$ and let $g(x) = x^2 - 3x - 1$. Graph the two functions $f(g(x))$ and $g(f(x))$ together in the window $[-4, 4]$ *by* $[-10, 10]$ and determine if they are the same function.

53. Let $f(x) = \dfrac{x}{x-1}$ and graph the function $f(f(x))$ in the window $[-15, 15]$ *by* $[-10, 10]$. Trace to examine the coordinates of several points on the graph and then determine the formula for $f(f(x))$.

Solutions to Practice Problems 0.3

1. (a) $f(g(x)) = [g(x)]^5 = (x^3 - 4x^2 + x - 8)^5$

(b) $g(f(x)) = [f(x)]^3 - 4[f(x)]^2 + f(x) - 8$
$$= (x^5)^3 - 4(x^5)^2 + x^5 - 8$$
$$= x^{15} - 4x^{10} + x^5 - 8$$

2. $\dfrac{f(1+h) - f(1)}{h} = \dfrac{(1+h)^2 - 1^2}{h}$

$$= \frac{1 + 2h + h^2 - 1}{h}$$

$$= \frac{2h + h^2}{h} = 2 + h$$

3. Suppose that x changes from 1 to $1 + h$. Then $\Delta x = (1+h) - 1 = h$, which is the denominator of the given expression in Problem 2. Moreover, the numerator $f(1+h) - f(1)$ is just the change in f as x changes from 1 to $1 + h$. That is, the numerator equals Δf. The expression in Problem 2 is therefore equal to

$$\frac{\Delta f}{\Delta x}.$$

0.4   Zeros of Functions—The Quadratic Formula and Factoring

A *zero* of a function $f(x)$ is a value of x for which $f(x) = 0$. For instance, the function $f(x)$ whose graph is shown in Fig. 1 has $x = -3$, $x = 3$, and $x = 7$ as zeros. Throughout this book we shall need to determine zeros of functions or, what amounts to the same thing, to solve the equation $f(x) = 0$.

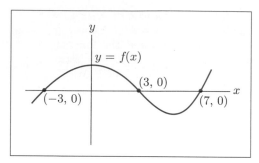

Figure 1. Zeros of a function.

In Section 0.2 we found zeros of linear functions. In this section our emphasis is on zeros of quadratic functions.

The Quadratic Formula   Consider the quadratic function $f(x) = ax^2 + bx + c$, $a \neq 0$. The zeros of this function are precisely the solutions of the quadratic equation

$$ax^2 + bx + c = 0.$$

One way of solving such an equation is via the *quadratic formula*.

The solutions of the equation $ax^2 + bx + c = 0$ are

$$x = \frac{-b \pm \sqrt{b^2 - 4ac}}{2a}.$$

The $\pm$ sign tells us to form two expressions, one with $+$ and one with $-$. The quadratic formula implies that a quadratic equation has at most two roots. It will have none if the expression $b^2 - 4ac$ is negative and one if $b^2 - 4ac$ equals 0. The quadratic formula is derived at the end of the section.

▶ **Example 1**   Solve the quadratic equation $3x^2 - 6x + 2 = 0$.

Solution   Here $a = 3$, $b = -6$, and $c = 2$. Substituting these values into the quadratic formula, we find that

$$\sqrt{b^2 - 4ac} = \sqrt{(-6)^2 - 4(3)(2)} = \sqrt{36 - 24} = \sqrt{12} = \sqrt{4 \cdot 3} = 2\sqrt{3}$$

and

$$x = \frac{-b \pm \sqrt{b^2 - 4ac}}{2a}$$

$$= \frac{-(-6) \pm 2\sqrt{3}}{2(3)}$$

$$= \frac{6 \pm 2\sqrt{3}}{6}$$

$$= 1 \pm \frac{\sqrt{3}}{3}.$$

The solutions of the equation are $1 + \sqrt{3}/3$ and $1 - \sqrt{3}/3$. (See Fig. 2.)   ◆

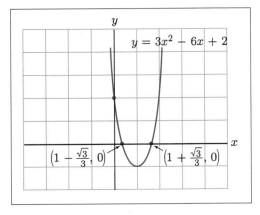

Figure 2.

▶ Example 2   Find the zeros of the following quadratic functions.

(a) $f(x) = 4x^2 - 4x + 1$ 　　　　　　(b) $f(x) = \frac{1}{2}x^2 - 3x + 5$

Solution   (a) We must solve $4x^2 - 4x + 1 = 0$. Here $a = 4$, $b = -4$, and $c = 1$, so that

$$\sqrt{b^2 - 4ac} = \sqrt{(-4)^2 - 4(4)(1)} = \sqrt{0} = 0.$$

Thus there is only one zero, namely,

$$x = \frac{-(-4) \pm 0}{2(4)} = \frac{4}{8} = \frac{1}{2}.$$

The graph of $f(x)$ is sketched in Fig. 3.

(b) We must solve $\frac{1}{2}x^2 - 3x + 5 = 0$. Here $a = \frac{1}{2}$, $b = -3$, and $c = 5$, so that

$$\sqrt{b^2 - 4ac} = \sqrt{(-3)^2 - 4(\tfrac{1}{2})(5)} = \sqrt{9 - 10} = \sqrt{-1}.$$

The square root of a negative number is undefined, so we conclude that $f(x)$ has no zeros. The reason for this is clear from Fig. 4. The graph of $f(x)$ lies entirely above the x-axis and has no x-intercepts.   ◆

The common problem of finding where two curves intersect amounts to finding the zero of a function.

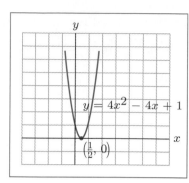

Figure 3.

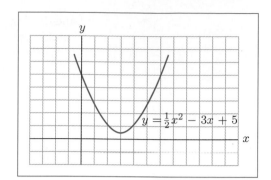

Figure 4.

▶ Example 3   Find the points of intersection of the graphs of the functions $y = x^2 + 1$ and $y = 4x$. (See Fig. 5.)

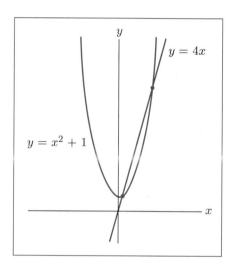

Figure 5. Points of intersection of two graphs.

Solution   If a point (x, y) is on both graphs, then its coordinates must satisfy both equations. That is, x and y must satisfy $y = x^2 + 1$ and $y = 4x$. Equating the two expressions for y, we have

$$x^2 + 1 = 4x.$$

To use the quadratic formula, we rewrite the equation in the form

$$x^2 - 4x + 1 = 0.$$

From the quadratic formula,

$$x = \frac{4 \pm \sqrt{16 - 4}}{2} = \frac{4 \pm \sqrt{12}}{2} = \frac{4 \pm 2\sqrt{3}}{2} = 2 \pm \sqrt{3}.$$

Thus the x-coordinates of the points of intersection are $2+\sqrt{3}$ and $2-\sqrt{3}$. To find the y-coordinates, we substitute these values of x into either equation, $y = x^2 + 1$ or $y = 4x$. The second equation is simpler. We obtain $y = 4(2 + \sqrt{3}) = 8 + 4\sqrt{3}$ and $y = 4(2 - \sqrt{3}) = 8 - 4\sqrt{3}$. Thus the points of intersection are $(2+\sqrt{3}, 8+4\sqrt{3})$ and $(2 - \sqrt{3}, 8 - 4\sqrt{3})$.   ◆

▶ Example 4   A cable television company estimates that with x thousand subscribers, its monthly revenue and cost (in thousands of dollars) are

$$R(x) = 32x - .21x^2,$$

$$C(x) = 195 + 12x.$$

Determine the company's *break-even points*; that is, find the number of subscribers at which the revenue equals the cost. See Fig. 6.

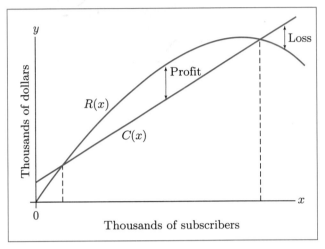

Figure 6. Break-even points.

Solution   Let $P(x)$ be the profit function.

$$
\begin{aligned}
P(x) &= R(x) - C(x) \\
&= (32x - .21x^2) - (195 + 12x) \\
&= -.21x^2 + 20x - 195
\end{aligned}
$$

The break-even points occur where the profit is zero. Thus we must solve

$$-.21x^2 + 20x - 195 = 0.$$

From the quadratic formula,

$$x = \frac{-20 \pm \sqrt{20^2 - 4(-.21)(-195)}}{2(-.21)} = \frac{-20 \pm \sqrt{236.2}}{-.42}$$

$$\approx 47.62 \pm 36.59 = 11.03 \text{ and } 84.21.$$

The break-even points occur where the company has 11,030 or 84,210 subscribers. Between those two levels, the company will be profitable.   ◆

Factoring   If $f(x)$ is a polynomial, we can often write $f(x)$ as a product of linear factors (i.e., factors of the form $ax + b$). If this can be done, then the zeros of $f(x)$ can be determined by setting each of the linear factors equal to zero and solving for x. (The reason is that the product of numbers can be zero only when one of the numbers is zero.)

▶ Example 5   Factor the following quadratic functions.

(a) $x^2 + 7x + 12$ (b) $x^2 - 13x + 12$

(c) $x^2 - 4x - 12$ (d) $x^2 + 4x - 12$

Solution   Note first that for any numbers c and d,

$$(x + c)(x + d) = x^2 + (c + d)x + cd.$$

In the quadratic on the right, the constant term is the product cd, whereas the coefficient of x is the sum $c + d$.

(a) Think of all integers c and d such that $cd = 12$. Then choose the pair that satisfies $c + d = 7$; that is, take $c = 3$, $d = 4$. Thus

$$x^2 + 7x + 12 = (x + 3)(x + 4).$$

(b) We want $cd = 12$. Since 12 is positive, c and d must be both positive or both negative. We must also have $c + d = -13$. These facts lead us to

$$x^2 - 13x + 12 = (x - 12)(x - 1).$$

(c) We want $cd = -12$. Since -12 is negative, c and d must have opposite signs. Also, they must sum to give -4. We find that

$$x^2 - 4x - 12 = (x - 6)(x + 2).$$

(d) This is almost the same as part (c).

$$x^2 + 4x - 12 = (x + 6)(x - 2). \qquad \blacklozenge$$

▶ Example 6   Factor the following polynomials.

(a) $x^2 - 6x + 9$ (b) $x^2 - 25$

(c) $3x^2 - 21x + 30$ (d) $20 + 8x - x^2$

Solution   (a) We look for $cd = 9$ and $c + d = -6$. The solution is $c = d = -3$, and

$$x^2 - 6x + 9 = (x - 3)(x - 3) = (x - 3)^2.$$

In general,

$$x^2 - 2cx + c^2 = (x - c)(x - c) = (x - c)^2.$$

(b) We use the identity

$$x^2 - c^2 = (x + c)(x - c).$$

Hence

$$x^2 - 25 = (x + 5)(x - 5).$$

(c) We first factor out a common factor of 3 and then use the method of Example 5.

$$3x^2 - 21x + 30 = 3(x^2 - 7x + 10)$$
$$= 3(x - 5)(x - 2).$$

(d) We first factor out a -1 in order to make the coefficient of x^2 equal to $+1$.

$$20 + 8x - x^2 = (-1)(x^2 - 8x - 20)$$
$$= (-1)(x - 10)(x + 2). \qquad \blacklozenge$$

▶ **Example 7**   Factor the following polynomials.

    (a) $x^2 - 8x$ (b) $x^3 + 3x^2 - 18x$ (c) $x^3 - 10x$

Solution   In each case we first factor out a common factor of x.

(a) $x^2 - 8x = x(x - 8)$.

(b) $x^3 + 3x^2 - 18x = x(x^2 + 3x - 18) = x(x + 6)(x - 3)$.

(c) $x^3 - 10x = x(x^2 - 10)$. To factor $x^2 - 10$, we use the identity $x^2 - c^2 = (x + c)(x - c)$, where $c^2 = 10$ and $c = \sqrt{10}$. Thus

$$x^3 - 10x = x(x^2 - 10) = x(x + \sqrt{10})(x - \sqrt{10}). \qquad \blacklozenge$$

▶ **Example 8**   Solve the following equations.

    (a) $x^2 - 2x - 15 = 0$ (b) $x^2 - 20 = x$ (c) $\dfrac{x^2 + 10x + 25}{x + 1} = 0$

Solution   (a)   The equation $x^2 - 2x - 15 = 0$ may be written in the form

$$(x - 5)(x + 3) = 0.$$

The product of two numbers is zero if one or the other of the numbers (or both) is zero. Hence

$$x - 5 = 0 \quad \text{or} \quad x + 3 = 0.$$

That is,

$$x = 5 \quad \text{or} \quad x = -3.$$

(b)   First we must rewrite the equation $x^2 - 20 = x$ in the form $ax^2 + bx + c = 0$; that is,

$$x^2 - x - 20 = 0$$
$$(x - 5)(x + 4) = 0.$$

We conclude that

$$x - 5 = 0 \quad \text{or} \quad x + 4 = 0;$$

that is,

$$x = 5 \quad \text{or} \quad x = -4.$$

(c)   A rational function will be zero only if the numerator is zero. Thus

$$x^2 + 10x + 25 = 0$$
$$(x + 5)^2 = 0$$
$$x + 5 = 0.$$

That is,

$$x = -5.$$

Since the denominator is not 0 at $x = -5$, we conclude that $x = -5$ is the solution. $\blacklozenge$

Derivation of the Quadratic Formula

$$ax^2 + bx + c = 0$$

$$ax^2 + bx = -c$$

$$4a^2x^2 + 4abx = -4ac \qquad \text{(both sides multiplied by } 4a\text{)}$$

$$4a^2x^2 + 4abx + b^2 = b^2 - 4ac \qquad (b^2 \text{ added to both sides)}$$

Now note that $4a^2x^2 + 4abx + b^2 = (2ax + b)^2$. To check this, simply multiply out the right-hand side. Therefore,

$$(2ax + b)^2 = b^2 - 4ac$$

$$2ax + b = \pm\sqrt{b^2 - 4ac}$$

$$2ax = -b \pm \sqrt{b^2 - 4ac}$$

$$x = \frac{-b \pm \sqrt{b^2 - 4ac}}{2a}.$$

Incorporating Technology Most graphing calculators have a **zero** (or ROOT) routine that computes zeros of functions, and have an **intersect** (or ISECT) routine that computes the intersection points of graphs. See Appendices A–D for the details for TI calculators. Figure 7 shows a zero of the function from Example 1 and Fig. 8 shows an intersection point from Example 3.

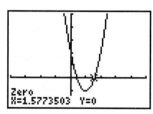

Figure 7.

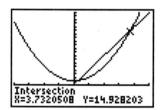

Figure 8.

One of the problems in graphing a function is to find a domain (xMin to xMax) that contains all the zeros. For a polynomial function, there is an easy solution to this problem. Write the polynomial in the form $c(x^n + a_{n-1}x^{n-1} + \cdots + a_0)$ and let M be the number that is one more than the largest magnitude of the coefficients $a_{n-1}, \ldots, a_0$. Then the interval $[-M, M]$ will contain all the zeros of the polynomial. For instance, all the zeros of the polynomial $x^3 - 10x^2 + 9x + 8$ lie in the interval $[-11, 11]$. After you view the polynomial on the domain $[-M, M]$, you can usually find a shorter domain that also contains all the zeros.

**Practice Problems
0.4**

1. Solve the equation $x - 14/x = 5$.

2. Use the quadratic formula to solve $7x^2 - 35x + 35 = 0$.

▶ Exercises 0.4

Use the quadratic formula to find the zeros of the functions in Exercises 1–6.

1. $f(x) = 2x^2 - 7x + 6$

2. $f(x) = 3x^2 + 2x - 1$

3. $f(t) = 4t^2 - 12t + 9$

4. $f(x) = \frac{1}{4}x^2 + x + 1$

5. $f(x) = -2x^2 + 3x - 4$

6. $f(a) = 11a^2 - 7a + 1$

Use the quadratic formula to solve the equations in Exercises 7–12.

7. $5x^2 - 4x - 1 = 0$

8. $x^2 - 4x + 5 = 0$

9. $15x^2 - 135x + 300 = 0$

10. $z^2 - \sqrt{2}\,z - \frac{5}{4} = 0$

11. $\frac{3}{2}x^2 - 6x + 5 = 0$

12. $9x^2 - 12x + 4 = 0$

Factor the polynomials in Exercises 13–24.

13. $x^2 + 8x + 15$

14. $x^2 - 10x + 16$

15. $x^2 - 16$

16. $x^2 - 1$

17. $3x^2 + 12x + 12$

18. $2x^2 - 12x + 18$

19. $30 - 4x - 2x^2$

20. $15 + 12x - 3x^2$

21. $3x - x^2$

22. $4x^2 - 1$

23. $6x - 2x^3$

24. $16x + 6x^2 - x^3$

Find the points of intersection of the pairs of curves in Exercises 25–32.

25. $y = 2x^2 - 5x - 6$, $y = 3x + 4$

26. $y = x^2 - 10x + 9$, $y = x - 9$

27. $y = x^2 - 4x + 4$, $y = 12 + 2x - x^2$

28. $y = 3x^2 + 9$, $y = 2x^2 - 5x + 3$

29. $y = x^3 - 3x^2 + x$, $y = x^2 - 3x$

30. $y = \frac{1}{2}x^3 - 2x^2$, $y = 2x$

31. $y = \frac{1}{2}x^3 + x^2 + 5$, $y = 3x^2 - \frac{1}{2}x + 5$

32. $y = 30x^3 - 3x^2$, $y = 16x^3 + 25x^2$

Solve the equations in Exercises 33–38.

33. $\frac{21}{x} - x = 4$

34. $x + \frac{2}{x - 6} = 3$

35. $x + \frac{14}{x + 4} = 5$

36. $1 = \frac{5}{x} + \frac{6}{x^2}$

37. $\frac{x^2 + 14x + 49}{x^2 + 1} = 0$

38. $\frac{x^2 - 8x + 16}{1 + \sqrt{x}} = 0$

39. Suppose the cable television company's cost function in Example 4 changes to $C(x) = 275 + 12x$. Determine the new break-even points.

40. When a car is moving at x miles per hour and the driver decides to slam on the brakes, the car will travel $x + \frac{1}{20}x^2$ feet.* If a car travels 175 feet after the driver decides to stop, how fast was the car moving?

Technology Exercises

In Exercises 41–44 find the zeros of the function. (Use the specified viewing window.)

41. $f(x) = x^2 - x - 2$; $[-4, 5]$ by $[-4, 10]$

42. $f(x) = x^3 - 3x + 2$; $[-3, 3]$ by $[-10, 10]$

43. $f(x) = \sqrt{x + 2} - x + 2$; $[-2, 7]$ by $[-2, 4]$

44. $f(x) = \frac{x}{x + 2} - x^2 + 1$; $[-1.5, 2]$ by $[-2, 3]$

In Exercises 45–48 find the points of intersection of the graphs of the functions. (Use the specified viewing window.)

45. $f(x) = 2x - 1$; $g(x) = x^2 - 2$; $[-4, 4]$ by $[-6, 10]$

46. $f(x) = -x - 2$; $g(x) = -4x^2 + x + 1$; $[-2, 2]$ by $[-5, 2]$

47. $f(x) = 3x^4 - 14x^3 + 24x - 3$; $g(x) = 2x - 30$; $[-3, 5]$ by $[-80, 30]$

48. $f(x) = \frac{1}{x}$; $g(x) = \sqrt{x^2 - 1}$; $[0, 4]$ by $[-1, 3]$

In Exercises 49–52 find a good window setting for the graph of the function. The graph should show all the zeros of the polynomial.

49. $f(x) = x^3 - 22x^2 + 17x + 19$

50. $f(x) = x^4 - 200x^3 - 100x^2$

51. $f(x) = 3x^3 + 52x^2 - 12x - 12$

52. $f(x) = 2x^5 - 24x^4 - 24x + 2$

Solutions to Practice Problems 0.4	**1.** Multiply both sides of the equation by x. Then $$x^2 - 14 = 5x.$$

---

*The general formula is $f(x) = ax + bx^2$, where the constant a depends on the driver's reaction time and the constant b depends on the weight of the car and the type of tires. This mathematical model was analyzed in D. Burghes, I. Huntley, and J. McDonald, *Applying Mathematics: A Course in Mathematical Modelling* (New York: Halstead Press, 1982), pp. 57–60.

Now, take the term $5x$ to the left side of the equation and solve by factoring.

$$x^2 - 5x - 14 = 0$$
$$(x - 7)(x + 2) = 0$$
$$x = 7 \quad \text{or} \quad x = -2$$

2. In this case, each coefficient is a multiple of 7. To simplify the arithmetic, we divide both sides of the equation by 7 before using the quadratic formula.

$$x^2 - 5x + 5 = 0$$

$$\sqrt{b^2 - 4ac} = \sqrt{(-5)^2 - 4(1)(5)} = \sqrt{5}$$

$$x = \frac{-b \pm \sqrt{b^2 - 4ac}}{2a} = \frac{5 \pm \sqrt{5}}{2 \cdot 1} = \frac{5}{2} \pm \frac{1}{2}\sqrt{5}$$

0.5   Exponents and Power Functions

In this section we review the operations with exponents that occur frequently throughout the text. We begin with the definition of b^r for various types of numbers b and r.

For any nonzero number b and any positive integer n, we have by definition that

$$b^n = \underbrace{b \cdot b \cdot \cdots \cdot b}_{n \text{ times}},$$

$$b^{-n} = \frac{1}{b^n},$$

and

$$b^0 = 1.$$

For example, $2^4 = 2 \cdot 2 \cdot 2 \cdot 2 = 16$, $2^{-4} = \dfrac{1}{2^4} = \dfrac{1}{16}$, and $2^0 = 1$.

Next, we consider numbers of the form $b^{1/n}$, where n is a positive integer. For instance,

$2^{1/2}$ is the positive number whose square is 2: $2^{1/2} = \sqrt{2}$;

$2^{1/3}$ is the positive number whose cube is 2: $2^{1/3} = \sqrt[3]{2}$;

$2^{1/4}$ is the positive number whose fourth power is 2: $2^{1/4} = \sqrt[4]{2}$;

and so on. In general, when b is zero or positive, $b^{1/n}$ is zero or the positive number whose nth power is b.

If n is even, there is no number whose nth power is b if b is negative. Therefore, $b^{1/n}$ is not defined if n is even and $b < 0$. When n is odd, we may permit b to be negative as well as positive. For example, $(-8)^{1/3}$ is the number whose cube is -8; that is,

$$(-8)^{1/3} = -2.$$

Thus, when b is negative and n is odd, we again define $b^{1/n}$ to be the number whose nth power is b.

Finally, let us consider numbers of the form $b^{m/n}$ and $b^{-m/n}$, where m and n are positive integers. We may assume that the fraction m/n is in lowest terms (so that m and n have no common factor). Then we define

$$b^{m/n} = (b^{1/n})^m$$

whenever $b^{1/n}$ is defined, and

$$b^{-m/n} = \frac{1}{b^{m/n}}$$

whenever $b^{m/n}$ is defined and is not zero. For example,

$$8^{5/3} = (8^{1/3})^5 = (2)^5 = 32,$$

$$8^{-5/3} = \frac{1}{8^{5/3}} = \frac{1}{32},$$

$$(-8)^{5/3} = \left[(-8)^{1/3}\right]^5 = [-2]^5 = -32.$$

Exponents may be manipulated algebraically according to the following rules:

Laws of Exponents

1. $b^r b^s = b^{r+s}$

2. $b^{-r} = \dfrac{1}{b^r}$

3. $\dfrac{b^r}{b^s} = b^r \cdot b^{-s} = b^{r-s}$

4. $(b^r)^s = b^{rs}$

5. $(ab)^r = a^r b^r$

6. $\left(\dfrac{a}{b}\right)^r = \dfrac{a^r}{b^r}$

▶ **Example 1** Use the laws of exponents to calculate the following quantities.

(a) $2^{1/2}\,50^{1/2}$ (b) $(2^{1/2}\,2^{1/3})^6$ (c) $\dfrac{5^{3/2}}{\sqrt{5}}$

Solution (a) $2^{1/2}\,50^{1/2} = (2 \cdot 50)^{1/2}$ (Law 5)

$$= \sqrt{100}$$
$$= 10$$

(b) $\left(2^{1/2}\,2^{1/3}\right)^6 = \left(2^{(1/2)+(1/3)}\right)^6$ (Law 1)

$$= \left(2^{5/6}\right)^6$$
$$= 2^{(5/6)6}$$ (Law 4)
$$= 2^5$$
$$= 32$$

(c) $\dfrac{5^{3/2}}{\sqrt{5}} = \dfrac{5^{3/2}}{5^{1/2}}$

$$= 5^{(3/2)-(1/2)}$$ (Law 3)
$$= 5^1$$
$$= 5$$

◆

Now that we know the definition of x^r for r rational, we can examine some calculations and applications involving power functions.

▶ **Example 2**   Simplify the following expressions.

(a) $\dfrac{1}{x^{-4}}$ (b) $\dfrac{x^2}{x^5}$ (c) $\sqrt{x}\left(x^{3/2} + 3\sqrt{x}\right)$

Solution   (a) $\dfrac{1}{x^{-4}} = x^{-(-4)}$ (Law 2 with $r = -4$)

$\qquad = x^4$

(b) $\dfrac{x^2}{x^5} = x^{2-5}$ (Law 3)

$\qquad = x^{-3}$

It is also correct to write this answer as $\dfrac{1}{x^3}$.

(c) $\sqrt{x}\left(x^{3/2} + 3\sqrt{x}\right) = x^{1/2}\left(x^{3/2} + 3x^{1/2}\right)$

$\qquad = x^{1/2}x^{3/2} + 3x^{1/2}x^{1/2}$

$\qquad = x^{(1/2)+(3/2)} + 3x^{(1/2)+(1/2)}$ (Law 1)

$\qquad = x^2 + 3x$ ◆

A *power function* is a function of the form

$$f(x) = x^r,$$

for some number r.

▶ **Example 3**   Let $f(x)$ and $g(x)$ be the power functions

$$f(x) = x^{-1} \quad \text{and} \quad g(x) = x^{1/2}.$$

Determine the following functions.

(a) $\dfrac{f(x)}{g(x)}$ (b) $f(x)g(x)$ (c) $\dfrac{g(x)}{f(x)}$

Solution   (a) $\dfrac{f(x)}{g(x)} = \dfrac{x^{-1}}{x^{1/2}} = x^{-1-(1/2)} = x^{-3/2} = \dfrac{1}{x^{3/2}}$

(b) $f(x)g(x) = x^{-1}x^{1/2} = x^{-1+(1/2)} = x^{-1/2} = \dfrac{1}{x^{1/2}} = \dfrac{1}{\sqrt{x}}$

(c) $\dfrac{g(x)}{f(x)} = \dfrac{x^{1/2}}{x^{-1}} = x^{(1/2)-(-1)} = x^{3/2}$ ◆

Compound Interest   The subject of compound interest provides a significant application of exponents. Let's introduce this topic at this point with a view toward using it as a source of applied problems throughout the book.

When money is deposited in a savings account, interest is paid at stated intervals. If this interest is added to the account and thereafter earns interest itself, then the interest is called *compound interest*. The original amount deposited is called the *principal amount*. The principal amount plus the compound interest is called the *compound amount*. The interval between interest payments is referred to as the *interest period*. In formulas for compound interest, the interest rate is expressed as a decimal rather than a percent. Thus 6% is written as .06.

If \$1000 is deposited at 6% annual interest, compounded annually, the compound amount at the end of the first year will be

$$A_1 = \underset{\text{principal}}{1000} + \underset{\text{interest}}{1000(.06)} = 1000(1 + .06).$$

At the end of the second year the compound amount will be

$$A_2 = \underset{\substack{\text{compound} \\ \text{amount}}}{A_1} + \underset{\text{interest}}{A_1(.06)} = A_1(1 + .06)$$

$$= [1000(1 + .06)](1 + .06) = 1000(1 + .06)^2.$$

At the end of 3 years,

$$A_3 = A_2 + A_2(.06) = A_2(1 + .06)$$
$$= \left[1000(1 + .06)^2\right](1 + .06) = 1000(1 + .06)^3.$$

After n years the compound amount will be

$$A = 1000(1 + .06)^n.$$

In this example the interest period was 1 year. The important point to note, however, is that at the end of each interest period the amount on deposit grew by a factor of $(1 + .06)$. In general, if the interest rate is i instead of .06, the compound amount will grow by a factor of $(1 + i)$ at the end of each interest period.

Suppose that a principal amount P is invested at a compound interest rate i per interest period, for a total of n interest periods. Then the compound amount A at the end of the nth period will be

$$A = P(1 + i)^n. \tag{1}$$

▶ Example 4   Suppose that \$5000 is invested at 8% per year, with interest compounded annually. What is the compound amount after 3 years?

Solution   Substituting $P = 5000$, $i = .08$, and $n = 3$ into formula (1), we have

$$A = 5000(1 + .08)^3 = 5000(1.08)^3$$
$$= 5000(1.259712) = 6298.56 \text{ dollars.} \qquad \blacklozenge$$

It is common practice to state the interest rate as a percentage per year ("per annum"), even though each interest period is often shorter than 1 year. If the annual rate is r and if interest is paid and compounded m times per year, then the interest rate i for each period is given by

$$[\text{rate per period}] = i = \frac{r}{m} = \frac{[\text{annual interest rate}]}{[\text{periods per year}]}.$$

Many banks pay interest quarterly. If the stated annual rate is 5%, then $i = .05/4 = .0125$.

If interest is compounded for t years, with m interest periods each year, there will be a total of mt interest periods. If in formula (1) we replace n by mt and replace i by r/m, we obtain the following formula for the compound amount:

$$A = P \left(1 + \frac{r}{m}\right)^{mt},$$

where P = principal amount,
$\quad r$ = interest rate per annum,
$\quad m$ = number of interest periods per year,
$\quad t$ = number of years.

(2)

▶ Example 5   Suppose that $1000 is deposited in a savings account that pays 6% per annum, compounded quarterly. If no additional deposits or withdrawals are made, how much will be in the account at the end of 1 year?

Solution   We use (2) with $P = 1000$, $r = .06$, $m = 4$, and $t = 1$.

$$A = 1000 \left(1 + \frac{.06}{4}\right)^4 = 1000(1.015)^4$$

$$\approx 1000(1.06136355) \approx 1061.36 \text{ dollars.} \qquad \blacklozenge$$

Note that the $1000 in Example 5 earned a total of $61.36 in (compound) interest. This is 6.136% of $1000. Savings institutions sometimes advertise this rate as the *effective* annual interest rate. That is, the savings institutions mean that *if* they paid interest only once a year, they would have to pay a rate of 6.136% in order to produce the same earnings as their 6% rate compounded quarterly. The stated rate of 6% is often called the *nominal rate*.

The effective annual rate can be increased by compounding the interest more often. Some savings institutions compound interest monthly or even daily.

▶ Example 6   Suppose that the interest in Example 5 were compounded monthly. How much would be in the account at the end of 1 year? What about the case when 6% annual interest is compounded daily?

Solution   For monthly compounding, $m = 12$. From (2) we have

$$A = 1000 \left(1 + \frac{.06}{12}\right)^{12} = 1000(1.005)^{12} \approx 1061.68 \text{ dollars.}$$

The effective rate in this case is 6.168%.

A "bank year" usually consists of 360 days (in order to simplify calculations). So, for daily compounding, we take $m = 360$. Then

$$A = 1000 \left(1 + \frac{.06}{360}\right)^{360} \approx 1000(1.00016667)^{360}$$

$$\approx 1000(1.06183133) \approx 1061.83 \text{ dollars.}$$

With daily compounding, the effective rate is 6.183%. $\blacklozenge$

▶ Example 7   Suppose that a corporation issues a bond costing $200 and paying interest compounded monthly. The interest is accumulated until the bond reaches maturity. (A security of this sort is called a *zero coupon bond*.) Suppose that after 5 years, the bond is worth $500. What is the annual interest rate?

Solution   Let r denote the annual interest rate. The value A of the bond after 5 years $= 60$ months is given by the compound interest formula:

$$A = 200 \left(1 + \frac{r}{12}\right)^{60}.$$

We must find r that satisfies

$$500 = 200 \left(1 + \frac{r}{12}\right)^{60}$$

$$2.5 = \left(1 + \frac{r}{12}\right)^{60}.$$

Raise both sides to the power $\frac{1}{60}$ and apply the laws of exponents to obtain

$$(2.5)^{1/60} = \left[\left(1 + \frac{r}{12}\right)^{60}\right]^{\frac{1}{60}} = \left(1 + \frac{r}{12}\right)^{60 \cdot \frac{1}{60}} = 1 + \frac{r}{12}$$

$$r = 12 \cdot \left((2.5)^{1/60} - 1\right).$$

Using a calculator, we see that $r \approx .18466$. That is, the annual interest rate is 18.466%. (A bond paying a rate of interest this high is generally called a *junk bond*.) ◆

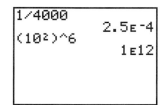

Figure 1.

Incorporating Technology   On graphing calculators, very small numbers and very large numbers are automatically displayed in exponential form. In Fig. 1, **2.5E-4** stands for $2.5 \cdot 10^{-4}$ or .00025, and **1E12** stands for $1 \cdot 10^{12}$ or 1,000,000,000,000. (*Note*: Multiplying a number by 10^{-4} moves the decimal point four places to the left and multiplying by 10^{12} moves the decimal point 12 places to the right.) You may enter numbers in scientific notation using the **EE** (= enter exponent) key found on most calculators.

Practice Problems 0.5

1. Compute the following.
   (a) -5^2 (b) $16^{.75}$

2. Simplify the following.
   (a) $(4x^3)^2$ (b) $\dfrac{\sqrt[3]{x}}{x^3}$ (c) $\dfrac{2 \cdot (x+5)^6}{x^2 + 10x + 25}$

▶ Exercises 0.5

In Exercises 1–28, compute the numbers.

1. 3^3
2. $(-2)^3$
3. 1^{100}
4. 0^{25}
5. $(.1)^4$
6. $(100)^4$
7. -4^2
8. $(.01)^3$
9. $(16)^{1/2}$
10. $(27)^{1/3}$
11. $(.000001)^{1/3}$
12. $\left(\dfrac{1}{125}\right)^{1/3}$
13. 6^{-1}
14. $\left(\dfrac{1}{2}\right)^{-1}$
15. $(.01)^{-1}$
16. $(-5)^{-1}$
17. $8^{4/3}$
18. $16^{3/4}$
19. $(25)^{3/2}$
20. $(27)^{2/3}$
21. $(1.8)^0$
22. $9^{1.5}$
23. $16^{.5}$
24. $(81)^{.75}$
25. $4^{-1/2}$
26. $\left(\dfrac{1}{8}\right)^{-2/3}$
27. $(.01)^{-1.5}$
28. $1^{-1.2}$

In Exercises 29–40, use the laws of exponents to compute the numbers.

29. $5^{1/3} \cdot 200^{1/3}$

30. $(3^{1/3} \cdot 3^{1/6})^6$

31. $6^{1/3} \cdot 6^{2/3}$

32. $(9^{4/5})^{5/8}$

33. $\dfrac{10^4}{5^4}$

34. $\dfrac{3^{5/2}}{3^{1/2}}$

35. $(2^{1/3} \cdot 3^{2/3})^3$

36. $20^{.5} \cdot 5^{.5}$

37. $\left(\dfrac{8}{27}\right)^{2/3}$

38. $(125 \cdot 27)^{1/3}$

39. $\dfrac{7^{4/3}}{7^{1/3}}$

40. $(6^{1/2})^0$

In Exercises 41–70, use the laws of exponents to simplify the algebraic expressions. Your answer should not involve parentheses or negative exponents.

41. $(xy)^6$

42. $(x^{1/3})^6$

43. $\dfrac{x^4 \cdot y^5}{xy^2}$

44. $\dfrac{1}{x^{-3}}$

45. $x^{-1/2}$

46. $(x^3 \cdot y^6)^{1/3}$

47. $\left(\dfrac{x^4}{y^2}\right)^3$

48. $\left(\dfrac{x}{y}\right)^{-2}$

49. $(x^3 y^5)^4$

50. $\sqrt{1+x}\,(1+x)^{3/2}$

51. $x^5 \cdot \left(\dfrac{y^2}{x}\right)^3$

52. $x^{-3} \cdot x^7$

53. $(2x)^4$

54. $\dfrac{-3x}{15x^4}$

55. $\dfrac{-x^3 y}{-xy}$

56. $\dfrac{x^3}{y^{-2}}$

57. $\dfrac{x^{-4}}{x^3}$

58. $(-3x)^3$

59. $\sqrt[3]{x} \cdot \sqrt[3]{x^2}$

60. $(9x)^{-1/2}$

61. $\left(\dfrac{3x^2}{2y}\right)^3$

62. $\dfrac{x^2}{x^5 y}$

63. $\dfrac{2x}{\sqrt{x}}$

64. $\dfrac{1}{yx^{-5}}$

65. $(16x^8)^{-3/4}$

66. $(-8y^9)^{2/3}$

67. $\sqrt{x}\left(\dfrac{1}{4x}\right)^{5/2}$

68. $\dfrac{(25xy)^{3/2}}{x^2 y}$

69. $\dfrac{(-27x^5)^{2/3}}{\sqrt[3]{x}}$

70. $(-32y^{-5})^{3/5}$

The expressions in Exercises 71–74 may be factored as shown. Find the missing factors.

71. $\sqrt{x} - \dfrac{1}{\sqrt{x}} = \dfrac{1}{\sqrt{x}}(\quad)$

72. $2x^{2/3} - x^{-1/3} = x^{-1/3}(\quad)$

73. $x^{-1/4} + 6x^{1/4} = x^{-1/4}(\quad)$

74. $\sqrt{\dfrac{x}{y}} - \sqrt{\dfrac{y}{x}} = \sqrt{xy}\,(\quad)$

75. Explain why $\sqrt{a} \cdot \sqrt{b} = \sqrt{ab}$.

76. Explain why $\sqrt{a}/\sqrt{b} = \sqrt{a/b}$.

In Exercises 77–84, evaluate $f(4)$.

77. $f(x) = x^2$

78. $f(x) = x^3$

79. $f(x) = x^{-1}$

80. $f(x) = x^{1/2}$

81. $f(x) = x^{3/2}$

82. $f(x) = x^{-1/2}$

83. $f(x) = x^{-5/2}$

84. $f(x) = x^0$

Calculate the compound amount from the given data in Exercises 85–92.

85. principal = \$500, compounded annually, 6 years, annual rate = 6%

86. principal = \$700, compounded annually, 8 years, annual rate = 8%

87. principal = \$50,000, compounded quarterly, 10 years, annual rate = 9.5%

88. principal = \$20,000, compounded quarterly, 3 years, annual rate = 12%

89. principal = \$100, compounded monthly, 10 years, annual rate = 5%

90. principal = \$500, compounded monthly, 1 year, annual rate = 4.5%

91. principal = \$1500, compounded daily, 1 year, annual rate = 6%

92. principal = \$1500, compounded daily, 3 years, annual rate = 6%

93. Assume that a couple invests \$1000 upon the birth of their daughter. Assume that the investment earns 6.8% compounded annually. What will the investment be worth on the daughter's 18th birthday?

94. Assume that a couple invests \$4000 each year for four years in an investment that earns 8% compounded annually. What will the value of the investment be 8 years after the first amount is invested?

95. Assume that a \$500 investment earns interest compounded quarterly. Express the value of the investment after one year as a polynomial in the annual rate of interest r.

96. Assume that a \$1000 investment earns interest compounded semiannually. Express the value of the investment after two years as a polynomial in the annual rate of interest r.

97. When a car's brakes are slammed on at a speed of x miles per hour, the stopping distance is $\frac{1}{20}x^2$ feet. Show that when the speed is doubled, the stopping distance increases fourfold.

Technology Exercises

In Exercises 98–101 convert the numbers from graphing calculator form to standard form (that is, without E).

98. `5E-5`

99. `8.103E-4`

100. `1.35E13`

101. `8.23E-6`

<table>
<tr><td>

**Solutions to
Practice Problems
0.5**

</td></tr>
</table>

1. (a) $-5^2 = -25$. [Note that -5^2 is the same as $-(5^2)$. This number is different from $(-5)^2$, which equals 25. Whenever there are no parentheses, apply the exponent first and then apply the other operations.]

(b) Since $.75 = \frac{3}{4}$, $16^{.75} = 16^{3/4} = (\sqrt[4]{16})^3 = 2^3 = 8$.

2. (a) Apply Law 5 with $a = 4$ and $b = x^3$. Then use Law 4.

$$(4x^3)^2 = 4^2 \cdot (x^3)^2 = 16 \cdot x^6$$

[A common error is to forget to square the 4. If that had been our intent, we would have asked for $4\left(x^3\right)^2$.]

(b) $\dfrac{\sqrt[3]{x}}{x^3} = \dfrac{x^{1/3}}{x^3} = x^{(1/3)-3} = x^{-8/3}$. [The answer can also be given as $1/x^{8/3}$.] When simplifying expressions involving radicals, it is usually a good idea to first convert the radicals to exponents.

(c) $\dfrac{2(x+5)^6}{x^2 + 10x + 25} = \dfrac{2 \cdot (x+5)^6}{(x+5)^2} = 2(x+5)^{6-2} = 2(x+5)^4$. [Here the third law of exponents was applied to $(x+5)$. The laws of exponents apply to any algebraic expression.]

0.6 Functions and Graphs in Applications

The key step in solving many applied problems in this text is to construct appropriate functions or equations. Once this is done, the remaining mathematical steps are usually straightforward. This section focuses on representative applied problems and reviews skills needed to set up and analyze functions, equations, and their graphs.

Geometric Problems   Many examples and exercises in the text involve dimensions, areas, or volumes of objects similar to those in Fig. 1. When a problem involves a plane figure, such as a rectangle or circle, one must distinguish between the *perimeter* and the *area* of the figure. The perimeter of a figure, or "distance around" the figure, is a *length* or *sum of lengths*. Typical units, if specified, are inches, feet, centimeters, meters, and so on. Area involves the *product of two lengths*, and the units are *square* inches, *square* feet, *square* centimeters, and so on.

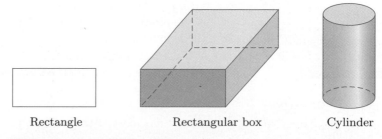

Rectangle   Rectangular box   Cylinder

Figure 1. Geometric figures.

▶ Example 1   Suppose the longer side of the rectangle in Fig. 1 has twice the length of the shorter side, and let x denote the length of the shorter side.

(a) Express the perimeter of the rectangle as a function of x.

(b) Express the area of the rectangle as a function of x.

(c) Suppose the rectangle represents a kitchen countertop to be constructed of a durable material costing $25 per square foot. Write a function $C(x)$ that expresses the cost of the material as a function of x, where lengths are in feet.

Solution   (a) The rectangle is shown in Fig. 2. The length of the longer side is $2x$. If the perimeter is denoted by P, then P is the sum of the lengths of the four sides of the rectangle, namely, $x + 2x + x + 2x$. That is, $P = 6x$.

(b) The area A of the rectangle is the product of the lengths of two adjacent sides. That is, $A = x \cdot 2x = 2x^2$.

(c) Here the area is measured in square feet. The basic principle for this part is

x

$2x$

Figure 2.

$$\begin{bmatrix} \text{cost of} \\ \text{materials} \end{bmatrix} = \begin{bmatrix} \text{cost per} \\ \text{square foot} \end{bmatrix} \cdot \begin{bmatrix} \text{number of} \\ \text{square feet} \end{bmatrix}$$

$$C(x) \quad = \quad 25 \quad \cdot \quad 2x^2$$

$$= 50x^2 \text{ (dollars).} \qquad \blacklozenge$$

When a problem involves a three-dimensional object, such as a box or cylinder, one must distinguish between the *surface area* of the object and the *volume* of the object. Surface area is an area, of course, so it is measured in *square* units. Typically, the surface area is a *sum of areas* (each area is a product of two lengths). The volume of an object is often a *product of three lengths* and is measured in *cubic* units.

▶ Example 2   A rectangular box has a square copper base, wooden sides, and a wooden top. The copper costs $21 per square foot and the wood costs $2 per square foot.

(a) Write an expression giving the surface area (that is, the sum of the areas of the bottom, the top, and the four sides of the box) in terms of the dimensions of the box. Also, write an expression giving the volume of the box.

(b) Write an expression giving the total cost of the materials used to make the box in terms of the dimensions.

Solution   (a) The first step is to assign letters to the dimensions of the box. Denote the length of one (and therefore every) side of the square base by x, and denote the height of the box by h. See Fig. 3.

The top and bottom each has area x^2, and each of the four sides has area xh. Therefore, the surface area is $2x^2 + 4xh$. The volume of the box is the product of the length, width, and height. Because the base is square, the volume is $x^2 h$.

(b) When the various surfaces of the box have different costs per square foot, the cost of each is computed separately:

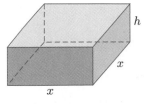

h

x

x

Figure 3. Closed box.

$$[\text{cost of bottom}] = [\text{cost per sq. ft.}] \cdot [\text{area of bottom}] = 21x^2;$$

$$[\text{cost of top}] = [\text{cost per sq. ft.}] \cdot [\text{area of top}] = 2x^2;$$

$$[\text{cost of one side}] = [\text{cost per sq. ft.}] \cdot [\text{area of one side}] = 2xh.$$

The total cost is

$$C = [\text{cost of bottom}] + [\text{cost of top}] + 4 \cdot [\text{cost of one side}]$$
$$= 21x^2 + 2x^2 + 4 \cdot 2xh = 23x^2 + 8xh.$$ ◆

Business Problems Many applications in the text involve cost, revenue, and profit functions.

▶ **Example 3** Suppose a toy manufacturer has fixed costs of $3000 (such as rent, insurance, and business loans) that must be paid no matter how many toys are produced. In addition, there are variable costs of $2 per toy. At a production level of x toys, the variable costs are $2 \cdot x$ (dollars) and the total cost is

$$C(x) = 3000 + 2x \quad \text{(dollars)}.$$

(a) Find the cost of producing 2000 toys.

(b) What additional cost is incurred if the production level is raised from 2000 toys to 2200 toys?

(c) To answer the question "How many toys may be produced at a cost of $5000?" should you compute $C(5000)$ or should you solve the equation $C(x) = 5000$?

Solution (a) $C(2000) = 3000 + 2(2000) = 7000$ (dollars).

(b) The total cost when $x = 2200$ is $C(2200) = 3000 + 2(2200) = 7400$ (dollars). So the *increase* in cost when production is raised from 2000 to 2200 toys is

$$C(2200) - C(2000) = 7400 - 7000 = 400 \quad \text{(dollars)}.$$

(c) This is an important type of question. The phrase "how many toys" implies that the quantity x is unknown. Therefore, the answer is found by solving $C(x) = 5000$ for x:

$$3000 + 2x = 5000$$
$$2x = 2000$$
$$x = 1000 \quad \text{(toys)}.$$

Another way to analyze this problem is to look at the types of units involved. The input x of the cost function is the *quantity* of toys, and the output of the cost function is the *cost*, measured in dollars. Since the question involves 5000 *dollars*, it is the *output* that is specified. The input x is unknown. ◆

▶ **Example 4** Suppose the toys in Example 3 sell for $10 apiece. When x toys are sold, the revenue (amount of money received) $R(x)$ is $10x$ dollars. Given the same cost function, $C(x) = 3000 + 2x$, the profit (or loss) $P(x)$ generated by the x toys will be

$$P(x) = R(x) - C(x)$$
$$= 10x - (3000 + 2x) = 8x - 3000.$$

(a) To determine the revenue generated by 8000 toys, should you compute $R(8000)$ or should you solve the equation $R(x) = 8000$?

(b) If the revenue from the production and sale of some toys is $7000, what is the corresponding profit?

Solution (a) The revenue is unknown, but the input to the revenue function is known. So compute $R(8000)$ to find the revenue.

(b) The profit is unknown, so we want to compute the value of $P(x)$. Unfortunetly, we don't know the value of x. However, the fact that the revenue is $7000 enables us to solve for x. Thus the solution has two steps:

(i) Solve $R(x) = 7000$ to find x.

$$10x = 7000$$
$$x = 700 \quad \text{(toys)}.$$

(ii) Compute $P(x)$ when $x = 700$.

$$P(x) = 8(700) - 3000$$
$$= 2600 \quad \text{(dollars)}. \qquad \blacklozenge$$

Functions and Graphs When a function arises in an applied problem, the graph of the function provides useful information. Every statement or task involving a function corresponds to a feature or task involving its graph. This "graphical" point of view will broaden your understanding of functions and strengthen your ability to work with them.

Modern graphing calculators and calculus computer software provide excellent tools for thinking geometrically about functions. Most popular graphing calculators and programs provide a *cursor* or *cross hairs* that may be moved to any point on the screen, with the x- and y-coordinates of the cursor displayed somewhere on the screen. The next example shows how geometric calculations with the graph of a function correspond to the more familiar numerical computations. This example is worth reading even if a computer (or calculator) is unavailable.

▶ Example 5 To plan for future growth, a company analyzes production costs for one of its products and estimates that the cost (in dollars) of operating at a production level of x units per hour is given by the function

$$C(x) = 150 + 59x - 1.8x^2 + .02x^3.$$

Suppose the graph of this function is available, either displayed on the screen of a graphing utility or perhaps printed on graph paper in a company report. See Fig. 4.

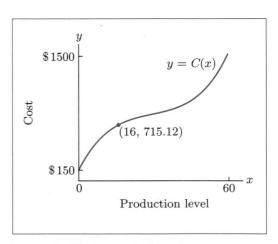

Figure 4. Graph of a cost function.

(a) The point $(16, 715.12)$ is on the graph. What does that say about the cost function $C(x)$?

(b) The equation $C(x) = 900$ may be solved graphically by finding a certain point on the graph and reading its x- and y-coordinates. Describe how to locate the point. How do the coordinates of the point provide the solution to the equation $C(x) = 900$?

(c) The task "Find $C(45)$" may be completed graphically by finding a point on the graph. Describe how to locate the point. How do the coordinates of the point provide the value of $C(45)$?

Solution (a) The fact that $(16, 715.12)$ is on the graph of $C(x)$ means that $C(16) = 715.12$. That is, if the production level is 16 units per hour, then the cost is $715.12.

(b) To solve $C(x) = 900$ graphically, locate 900 on the y-axis and move to the right until you reach the point $(?, 900)$ on the graph of $C(x)$. See Fig. 5. The x-coordinate of the point is the solution of $C(x) = 900$. Estimate x graphically. (With a graphing utility, find the coordinates of the point of intersection of the graph with the horizontal line $y = 900$; on graph paper, use a ruler to find x on the x-axis. To two decimal places, $x = 39.04$.)

(c) To find $C(45)$ graphically, locate 45 on the x-axis and move up until you reach the point $(45, ?)$ on the graph of $C(x)$. See Fig. 6. The y-coordinate of the point is the value of $C(45)$. [In fact, $C(45) = 982.50$.]   ◆

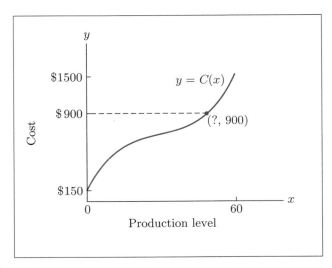

Figure 5.

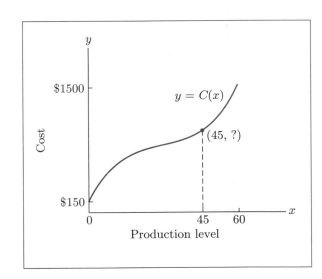

Figure 6.

The two final examples illustrate how to extract information about a function by examining its graph.

▶ **Example 6**   Figure 7 is the graph of the function $R(x)$, the revenue obtained from selling x bicycles.

(a) What is the revenue from the sale of 1000 bicycles?

(b) How many bicycles must be sold to achieve a revenue of $102,000?

(c) What is the revenue from the sale of 1100 bicycles?

(d) What additional revenue is derived from the sale of 100 more bicycles if the current sales level is 1000 bicycles?

Solution (a) Since $(1000, 150,000)$ is on the graph of $R(x)$, the revenue from the sale of 1000 bicycles is $150,000.

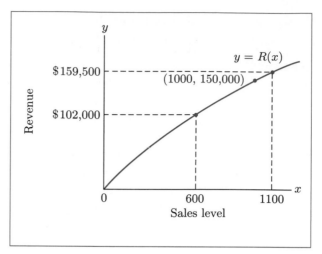

Figure 7. Graph of a revenue function.

(b) The horizontal line at $y = 102{,}000$ intersects the graph at the point with x-coordinate 600. Therefore, the revenue from the sale of 600 bicycles is $102,000.

(c) The vertical line at $x = 1100$ intersects the graph at the point with y-coordinate 159,500. Therefore, $R(1100) = 159{,}500$ and the revenue is $159,500.

(d) When the value of x increases from 1000 to 1100, the revenue increases from 150,000 to 159,500. Therefore, the additional revenue is $9500.   ◆

▶ Example 7   A ball is thrown straight up into the air from the top of a 64-foot tower. The function $h(t)$, the height of the ball (in feet) after t seconds, has the graph shown in Fig. 8. (*Note*: This graph is not a picture of the physical path of the ball; the ball is thrown vertically into the air.)

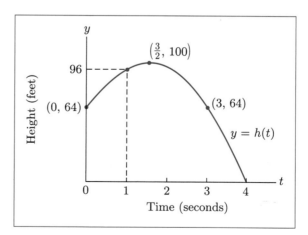

Figure 8. Graph of a height function.

(a) What is the height of the ball after 1 second?
(b) After how many seconds does the ball reach its greatest height, and what is this height?
(c) After how many seconds does the ball hit the ground?

(d) When is the height 64 feet?

Solution   (a) Since the point $(1, 96)$ is on the graph of $h(t)$, $h(1) = 96$. Therefore, the height of the ball after 1 second is 96 feet.

(b) The highest point on the graph of the function has coordinates $\left(\frac{3}{2}, 100\right)$. Therefore, after $\frac{3}{2}$ seconds the ball achieves its greatest height, 100 feet.

(c) The ball hits the ground when the height is 0. This occurs after 4 seconds.

(d) The height of 64 feet occurs twice, at times $t = 0$ and $t = 3$ seconds.   ◆

Table 1 summarizes most of the concepts in Examples 3–7. Although stated here for a profit function, the concepts will arise later for many other types of functions as well. Each statement about the profit is translated into a statement about $f(x)$ and a statement about the graph of $f(x)$. The graph in Fig. 9 illustrates each statement.

Table 1   **Translating an Applied Problem**

Assume that $f(x)$ is the profit in dollars at production level x.

Applied Problem	Function	Graph
When production is at 2 units, the profit is $7.	$f(2) = 7$.	The point $(2, 7)$ is on the graph.
Determine the number of units that generate a profit of $12.	Solve $f(x) = 12$ for x.	Find the x-coordinate(s) of the point(s) on the graph whose y-coordinate is 12.
Determine the profit when the production level is 4 units	Evaluate $f(4)$.	Find the y-coordinate of the point on the graph whose x-coordinate is 4.
Find the production level that maximizes the profit.	Find x such that $f(x)$ is as large as possible.	Find the x-coordinate of the highest point, M, on the graph.
Determine the maximum profit.	Find the maximum value of $f(x)$.	Find the y-coordinate of the highest point on the graph.
Determine the change in profit when the production level is changed from 6 to 7 units.	Find $f(7) - f(6)$.	Determine the difference in heights of the points with x-coordinates 7 and 6.
The profit decreases when the production level is changed from 6 to 7 units.	The function value decreases when x changes from 6 to 7.	The point on the graph with x-coordinate 6 is higher than the point with x-coordinate 7.

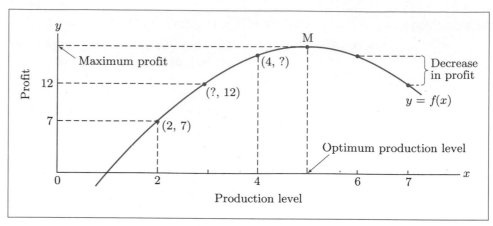

Figure 9. Graph of a profit function.

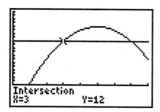

Figure 10.

Incorporating Technology The function in Fig. 9 has the equation $y = -x^2 + 10x - 9$. Most of the tasks from Table 1 can be carried out with a graphing calculator. In Fig. 10, Task 2 is accomplished by finding the intersection of the graph of the profit function and the graph of $y = 12$. In Fig. 11(a), Task 3 is accomplished by TRACE (possibly with some help from **value**), and in Fig. 11(b) Task 3 is accomplished by evaluating the function on the home screen. In Fig. 12, Tasks 4 and 5 are carried out with the **maximum** (or FMAX) routine. The computation for Task 6 can be carried out on the home screen of a TI-82, 83, or 86 as shown in Fig. 13.

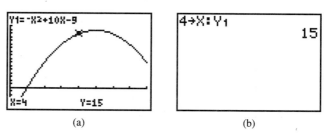

(a)   (b)

Figure 11.

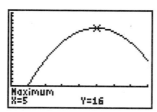

Figure 12.

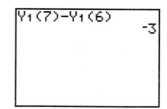

Figure 13.

|---|---|
| **Practice Problems 0.6** | Consider the cylinder shown in Fig. 14. |

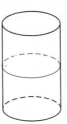

Figure 14.

1. Assign letters to the dimensions of the cylinder.
2. The girth of the cylinder is the circumference of the colored circle in the figure. Express the girth in terms of the dimensions of the cylinder.
3. What is the area of the bottom (or top) of the cylinder?
4. What is the surface area of the side of the cylinder? (*Hint:* Imagine cutting the side of the cylinder and unrolling the cylinder to form a rectangle.)

▶ Exercises 0.6

In Exercises 1–6, assign letters to the dimensions of the geometric object.

1.

Rectangle with height = 3·width

2.

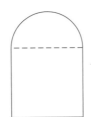

Norman window: Rectangle topped with a semicircle

3.

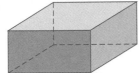

Rectangular box with square base

4.

Rectangular box with height = ½ · length

5.

Cylinder

6.

Cylinder with height = diameter

Exercises 7–14 refer to the letters assigned to the figures in Exercises 1–6.

7. Consider the rectangle in Exercise 1. Write an expression for the perimeter. Suppose the area is 25 square feet, and write this fact as an equation.
8. Consider the rectangle in Exercise 1. Write an expression for the area. Write an equation expressing the fact that the perimeter is 30 centimeters.
9. Consider a circle of radius r. Write an expression for the area. Write an equation expressing the fact that the circumference is 15 centimeters.
10. Consider the Norman window of Exercise 2. Write an expression for the perimeter. Write an equation expressing the fact that the area is 2.5 square meters.
11. Consider the rectangular box in Exercise 3, and suppose that it has no top. Write an expression for the volume. Write an equation expressing the fact that the surface area is 65 square inches.
12. Consider the closed rectangular box in Exercise 4. Write an expression for the surface area. Write an equation expressing the fact that the volume is 10 cubic feet.
13. Consider the cylinder of Exercise 5. Write an equation expressing the fact that the volume is 100 cubic inches.

Suppose the material to construct the left end costs $5 per square inch, the material to construct the right end costs $6 per square inch, and the material to construct the side costs $7 per square inch. Write an expression for the total cost of material for the cylinder.

14. Consider the cylinder of Exercise 6. Write an equation expressing the fact that the surface area is 30π square inches. Write an expression for the volume.

15. Consider a rectangular corral with a partition down the middle, as shown in Fig. 15. Assign letters to the outside dimensions of the corral. Write an equation expressing the fact that 5000 feet of fencing are needed to construct the corral (including the partition). Write an expression for the total area of the corral.

Figure 15.

16. Consider a rectangular corral with two partitions, as in Fig. 16. Assign letters to the outside dimensions of the corral. Write an equation expressing the fact that the corral has a total area of 2500 square feet. Write an expression for the amount of fencing needed to construct the corral (including both partitions).

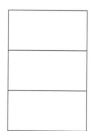

Figure 16.

17. Consider the corral of Exercise 16. Suppose the fencing for the boundary of the corral costs $10 per foot and the fencing for the inner partitions costs $8 per foot. Write an expression for the total cost of the fencing.

18. Consider the rectangular box of Exercise 3. Assume the box has no top, the material needed to construct the base costs $5 per square foot, and the material needed to construct the sides costs $4 per square foot. Write an equation expressing the fact that the total cost of materials is $150. (Use the dimensions assigned in Exercise 3.)

19. Suppose the rectangle in Exercise 1 has a perimeter of 40 cm. Find the area of the rectangle.

20. Suppose the cylinder in Exercise 6 has a volume of 54π cubic inches. Find the surface area of the cylinder.

21. A specialty shop prints custom slogans and designs on T-shirts. The shop's total cost at a daily sales level of x T-shirts is $C(x) = 73 + 4x$ dollars.

(a) At what sales level will the cost be $225?

(b) If the sales level is at 40 T-shirts, how much will the cost rise if the sales level changes to 50 T-shirts?

22. A college student earns income by typing term papers on a computer, which she leases (along with a printer). The student charges $4 per page for her work, and she estimates that her monthly cost when typing x pages is $C(x) = .10x + 75$ dollars.

(a) What is the student's profit if she types 100 pages in one month?

(b) Determine the change in profit when the typing business rises from 100 to 101 pages per month.

23. A frozen yogurt stand makes a profit of $P(x) = .40x - 80$ dollars when selling x scoops of yogurt per day.

(a) Find the break-even sales level, that is, the level at which $P(x) = 0$.

(b) What sales level generates a daily profit of $30?

(c) How many more scoops of yogurt will have to be sold to raise the daily profit from $30 to $40?

24. A cellular telephone company estimates that if it has x thousand subscribers, then its monthly profit is $P(x)$ thousand dollars, where $P(x) = 12x - 200$.

(a) How many subscribers are needed for a monthly profit of 160 thousand dollars?

(b) How many new subscribers would be needed to raise the monthly profit from 160 to 166 thousand dollars?

25. An average sale at a small florist shop is $21, so the shop's weekly revenue function is $R(x) = 21x$, where x is the number of sales in 1 week. The corresponding weekly cost is $C(x) = 9x + 800$ dollars.

(a) What is the florist shop's weekly profit function?

(b) How much profit is made when sales are at 120 per week?

(c) If the profit is $1000 for a week, what is the revenue for the week?

26. A catering company estimates that if it has x customers in a typical week, then its expenses will be approximately $C(x) = 550x + 6500$ dollars, and its revenue will be approximately $R(x) = 1200x$ dollars.

(a) How much profit will the company earn in a week when it has 12 customers?

(b) How much profit is the company making each week if the weekly costs are running at a level of $14,750?

Exercises 27–32 refer to the function $f(r)$, which gives the cost (in cents) of constructing a 100-cubic-inch cylinder of radius r inches. The graph of $f(r)$ is shown in Fig. 17.

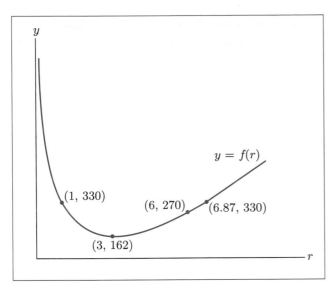

Figure 17. Cost of a cylinder.

27. What is the cost of constructing a cylinder of radius 6 inches?

28. For what value(s) of r is the cost 330 cents?

29. Interpret the fact that the point $(3, 162)$ is on the graph of the function.

30. Interpret the fact that the point $(3, 162)$ is the lowest point on the graph of the function. What does this say in terms of cost versus radius?

31. What is the additional cost of increasing the radius from 3 inches to 6 inches?

32. How much is saved by increasing the radius from 1 inch to 3 inches?

Exercises 33–36 refer to the cost and revenue functions in Fig. 18. The cost of producing x units of goods is $C(x)$ dollars and the revenue from selling x units of goods is $R(x)$ dollars.

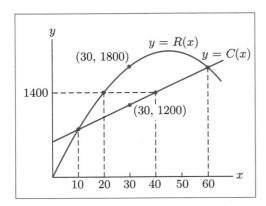

Figure 18. Cost and revenue functions.

33. What are the revenue and cost from the production and sale of 30 units of goods?

34. At what level of production is the revenue $1400?

35. At what level of production is the cost $1400?

36. What is the profit from the manufacture and sale of 30 units of goods?

Exercises 37–40 refer to the cost function in Fig. 19.

37. The point $(1000, 4000)$ is on the graph of the function. Restate this fact in terms of the function $C(x)$.

38. Translate the task "solve $C(x) = 3500$ for x" into a task involving the graph of the function.

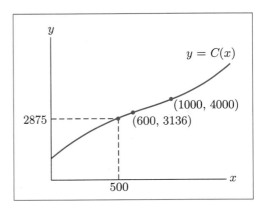

Figure 19. A cost function.

39. Translate the task "find $C(400)$" into a task involving the graph.

40. Suppose 500 units of goods are produced. What is the cost of producing 100 more units of goods?

Exercises 41–44 refer to the profit function in Fig. 20.

41. The point $(2500, 52,500)$ is the highest point on the graph of the function. What does this say in terms of profit versus quantity?

42. The point $(1500, 42,500)$ is on the graph of the function. Restate this fact in terms of the function $P(x)$.

43. Translate the task "solve $P(x) = 30,000$" into a task involving the graph of the function.

44. Translate the task "find $P(2000)$" into a task involving the graph.

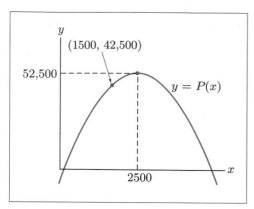

Figure 20. A profit function.

A ball is thrown straight up into the air. The function $h(t)$ gives the height of the ball (in feet) after t seconds. In Exercises 45–50, translate the task into both a statement involving the function and a statement involving the graph of the function.

45. Find the height of the ball after 3 seconds.

46. Find the time at which the ball attains its greatest height.

47. Find the greatest height attained by the ball.

48. Determine when the ball will hit the ground.

49. Determine when the height of the ball is 100 feet.

50. Find the height of the ball when it is first released.

Technology Exercises

51. A ball thrown straight up into the air has height $-16x^2 + 80x$ feet after x seconds.

(a) Graph the function in the window

$$[0, 6] \; by \; [-30, 120].$$

(b) What is the height of the ball after 3 seconds?

(c) At what times will the height be 64 feet?

(d) At what time will the ball hit the ground?

(e) When will the ball reach its greatest height? What is that height?

52. The daily cost (in dollars) of producing x units of a certain product is given by the function $C(x) = 225 + 36.5x - .9x^2 + .01x^3$.

(a) Graph $C(x)$ in the window $[0, 70]$ by $[-400, 2000]$.

(b) What is the cost of producing 50 units of goods?

(c) Consider the situation as in Part (b). What is the additional cost of producing one more unit of goods?

(d) At what production level will the daily cost be $510?

53. A store estimates that the total revenue (in dollars) from the sale of x bicycles per year is given by the function $R(x) = 250x - .2x^2$.

(a) Graph $R(x)$ in the window

$$[200, 500] \; by \; [42000, 75000].$$

(b) What sales level produces a revenue of $63,000?

(c) What revenue is received from the sale of 400 bicycles?

(d) Consider the situation of Part (c). If the sales level were to decrease by 50 bicycles, by how much would revenue fall?

(e) The store believes that if it spends $5000 in advertising, it can raise the total sales from 400 to 450 bicycles next year. Should it spend the $5000? Explain your conclusion.

Solutions to Practice Problems 0.6

1. Let r be the radius of circular base and let h be the height of the cylinder.

2. Girth $= 2\pi r$ (the circumference of the circle).

3. Area of the bottom $= \pi r^2$.

4. The cylinder is a rolled-up rectangle of height h and base $2\pi r$ (the circumference of the circle). The area is $2\pi r h$. See Fig. 21.

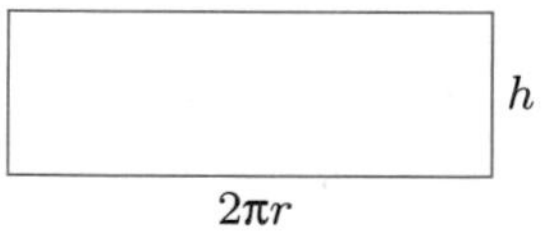

h

$2\pi r$

Figure 21. Unrolled side of a cylinder.

Review of Fundamental Concepts of Chapter 0

1. Explain the relationships and differences among real numbers, rational numbers, and irrational numbers.
2. What are the four types of inequalities and what do they each mean?
3. What is the difference between an open interval and a closed interval from a to b?
4. What is a function?
5. What is meant by "the value of a function at x"?
6. What is meant by the domain and range of a function?
7. What is the graph of a function and how is it related to vertical lines?
8. What is a linear function? Constant function? Give examples.
9. What are the x- and y-intercepts of a function and how are they found?
10. What is a quadratic function? What shape does its graph have?
11. Define and give an example of each of the following types of functions.
    (a) quadratic function (b) polynomial function
    (c) rational function (d) power function
12. What is meant by the absolute value of a number?
13. What five operations on functions are discussed in this chapter? Give an example of each.
14. What is a zero of a function?
15. Give two methods for finding the zeros of a quadratic function.
16. State the six laws of exponents.
17. In the formula $A = P(1+i)^n$, what do A, P, i, and n represent?
18. Explain how to solve $f(x) = b$ geometrically from the graph of $y = f(x)$.
19. Explain how to find $f(a)$ geometrically from the graph of $y = f(x)$.

▶ Chapter 0 Supplementary Exercises

1. Let $f(x) = x^3 + \frac{1}{x}$. Evaluate $f(1)$, $f(3)$, $f(-1)$, $f(-\frac{1}{2})$, and $f(\sqrt{2})$.
2. Let $f(x) = 2x + 3x^2$. Evaluate $f(0)$, $f(-\frac{1}{4})$, and $f(1/\sqrt{2})$.
3. Let $f(x) = x^2 - 2$. Evaluate $f(a-2)$.
4. Let $f(x) = [1/(x+1)] - x^2$. Evaluate $f(a+1)$.

Determine the domains of the following functions.

5. $f(x) = \dfrac{1}{x(x+3)}$
6. $f(x) = \sqrt{x-1}$
7. $f(x) = \sqrt{x^2 + 1}$
8. $f(x) = \dfrac{1}{\sqrt{3x}}$
9. Is the point $(\frac{1}{2}, -\frac{3}{5})$ on the graph of the function $h(x) = (x^2 - 1)/(x^2 + 1)$?
10. Is the point $(1, -2)$ on the graph of the function $k(x) = x^2 + (2/x)$?

Factor the polynomials in Exercises 11–14.

11. $5x^3 + 15x^2 - 20x$
12. $3x^2 - 3x - 60$
13. $18 + 3x - x^2$
14. $x^5 - x^4 - 2x^3$
15. Find the zeros of the quadratic function $y = 5x^2 - 3x - 2$.
16. Find the zeros of the quadratic function $y = -2x^2 - x + 2$.
17. Find the points of intersection of the curves $y = 5x^2 - 3x - 2$ and $y = 2x - 1$.
18. Find the points of intersection of the curves $y = -x^2 + x + 1$ and $y = x - 5$.

Let $f(x) = x^2 - 2x$, $g(x) = 3x - 1$, and $h(x) = \sqrt{x}$. Find the following functions.

19. $f(x) + g(x)$
20. $f(x) - g(x)$
21. $f(x)h(x)$
22. $f(x)g(x)$
23. $f(x)/h(x)$
24. $g(x)h(x)$

Let $f(x) = x/(x^2 - 1)$, $g(x) = (1 - x)/(1 + x)$, and $h(x) = 2/(3x + 1)$. Express the following as rational functions.

25. $f(x) - g(x)$
26. $f(x) - g(x+1)$
27. $g(x) - h(x)$
28. $f(x) + h(x)$
29. $g(x) - h(x-3)$
30. $f(x) + g(x)$

Let $f(x) = x^2 - 2x + 4$, $g(x) = 1/x^2$, and $h(x) = 1/(\sqrt{x}-1)$. Determine the following functions.

31. $f(g(x))$
32. $g(f(x))$
33. $g(h(x))$
34. $h(g(x))$
35. $f(h(x))$
36. $h(f(x))$
37. Simplify $(81)^{3/4}$, $8^{5/3}$, and $(.25)^{-1}$.
38. Simplify $(100)^{3/2}$ and $(.001)^{1/3}$.

39. The population of a city is estimated to be $750 + 25t + .1t^2$ thousand people t years from the present. Ecologists estimate that the average level of carbon monoxide in the air above the city will be $1 + .4x$ ppm (parts per million) when the population is x thousand people. Express the carbon monoxide level as a function of the time t.

40. The revenue $R(x)$ (in thousands of dollars) a company receives from the sale of x thousand units is given by $R(x) - 5x - x^2$. The sales level x is in turn a function $f(d)$ of the number d of dollars spent on advertising, where

$$f(d) = 6\left(1 - \frac{200}{d + 200}\right).$$

Express the revenue as a function of the amount spent on advertising.

In Exercises 41–44, use the laws of exponents to simplify the algebraic expressions.

41. $(\sqrt{x+1})^4$

42. $\dfrac{xy^3}{x^{-5}y^6}$

43. $\dfrac{x^{3/2}}{\sqrt{x}}$

44. $\sqrt[3]{x}\,(8x^{2/3})$

APPENDIX Graphing Functions Using Technology

As you might have already inferred from the preceding examples, sketching the graphs of functions by plotting points can be a tedious procedure. However, it is one that can be carried out using technology—graphing calculators and mathematical software.* Any of these tools will produce accurate graphs of functions. Most of the technology discussed in this text will concern graphing calculators, the most widely available tool for students. Although graphing calculators are not required for study from this text, we will show how they can be used to simplify computations and enhance understanding of the fundamental topics of calculus. Helpful information about calculators and technology will appear at the ends of most sections in subsections entitled "Incorporating Technology."

Actual keystrokes for calculators are not discussed in the main body of the text, but Appendices A through D provide a detailed reference for the keystrokes needed to carry out most calculus tasks on the TI-82, TI-83, TI-85, and TI-86 calculators. A TI-83 generated most of the calculator screens shown in the text.

To graph a function using a graphing calculator, you must perform the following steps:

1. Enter the expression for the function.

2. Enter the specifications for the viewing window.

3. Display the graph.

Let's examine each step in turn.

Functions are normally defined in the function editor (sometimes referred to as the **Y=** editor) and often are given names such as $Y_1, Y_2, Y_3, \ldots$. Figure 1 shows the calculator expressions for the functions

$$Y_1 = x^2 - 2x - 1, \quad Y_2 = x^{1/3}, \quad Y_3 = \frac{x - 1}{x + 1}, \quad \text{and} \quad Y_4 = -\sqrt{x - \frac{1}{x}}.$$

Powers are denoted by a caret (^). With some calculators, multiplication must be explicitly indicated by *. Also, you must use parentheses wherever necessary

---

*Powerful software for calculus includes Derive, Maple, Mathematica, MathCAD, and Visual Calculus. A customized copy of Visual Calculus is supplied with the *Study Guide* for this book.

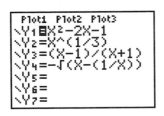

Figure 1.

to clarify the mathematical meaning of an expression. The two minus signs in Y_4 are entered with different keys—the first signifies negation and the second signifies subtraction. In Fig. 1, the highlighted equals sign for Y_1 indicates that this function has been selected to be graphed.

The parameters for the viewing window are set from a screen known as the WINDOW or RANGE screen. Figure 2(a) shows the WINDOW screen for the TI-83 graphing calculator and Fig. 2(b) shows the meanings of the parameters. We write **[a,b]** *by* **[c,d]** to denote the viewing window with setting **Xmin = a**, **Xmax = b**, **Ymin = c**, and **Ymax = d**. One of the most important tasks in using a graphing calculator is to determine the viewing window that shows the features of interest. To determine these values often requires calculus, as we shall learn in later chapters. For now, we will either use the default settings or determine appropriate settings by experimentation.

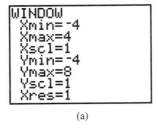

(a)

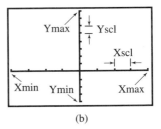

(b)

Figure 2.

After the viewing window has been specified, the graph can be displayed with the press of a single key. Figure 3(a) shows the graph of Y_1 using the setting of Fig. 2(a). If desired, the settings for the viewing window can be altered to see more of the graph or to zoom in on a portion of the graph.

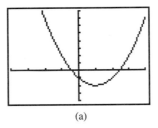

(a)

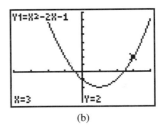

(b)

Figure 3.

Determining Coordinates of Points on a Graph   After drawing a graph, you may determine the coordinates of a point on the graph using the **TRACE** function of the calculator: With the graph displayed, press the **TRACE** button. A cursor is shown on the first graph. By pressing the right and left arrow keys, the cursor moves along the graph. As the cursor moves, the coordinates of the point on which it rests are shown at the bottom of the graph window. See Fig. 4. The up and down arrows are used to move the cursor between graphs, in case several graphs are displayed. Note that if part of the curve is outside the window boundaries, then the cursor will be invisible when it moves over that part. Just keep pressing the right or left arrow keys until the cursor reaches a point within the window boundaries.

Plotting Tabular Data   In many applications, you will want to plot data points as well as the graphs. Statistical plots (also called *scatter* plots) can be used for this task. The x- and y-coordinates of the data points are stored in

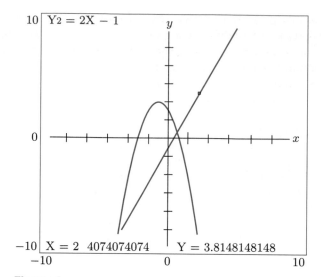

Figure 4.

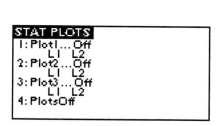

Figure 5.

lists. For example, Fig. 5 shows x- and y-coordinates, respectively, in lists L1 and L2.

1. Press **STAT PLOT =** to display the dialogue as shown in Fig. 6.

2. Turn on the plot you wish to display. For example, to display the points corresponding to **L1** and **L2** in Plot 1, change the **Off** setting for **Plot1** to **On**. Note that the icon for **Plot1** (this is the small picture or the second row of the plot description) shows a scatter plot and the X- and Y-coordinates are taken from lists **L1** and **L2**, respectively.

3. Press **GRAPH** to see Plot1 displayed, as in Fig. 7.

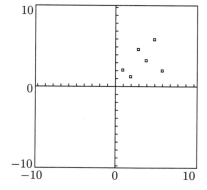

Figure 6.

Figure 7.

Note that if you wish to display both a function graph and certain points, you must display the function graphs and the points as two separate entities, say the function graph defined as **Y1** and the points as **Plot1**. Pressing **GRAPH** displays all function graphs and all statistical plots currently turned on.

You should also note that a set of points from a data table may not define a function, since an x-value may appear more than once in the table, violating the vertical line test for a function. However, if all the x-values in a table are different, then the table does define a function.

Using Spreadsheets in Calculus A *spreadsheet* is a computer program that allows you to manipulate a rectangular grid of data cells. Into each cell, you may enter a number, text, or a formula that computes a cell value in terms of the values in other cells. Spreadsheets, along with word processors, have become one of the fundamental computer applications, and their basic operations are discussed in most computer literacy courses. In our spreadsheet discussions, we will assume an elementary knowledge of spreadsheet operations and will concentrate on the applications of spreadsheets to calculus. In our examples, we will give commands and screen-shots of the most popular spreadsheet program, Microsoft Excel, version 8.0 for Windows.

A spreadsheet allows you to tabulate values of a function. For example, Fig. 8 shows a spreadsheet. The first column contains values of the variable **x**, ranging from **x** = 0 to **x** = 2 in increments of .1. You may generate these entries automatically as follows: Move the cursor to cell **A2** and enter 0.0. Second, highlight enough cells below and including **A2** to contain the desired entries. Then select the command **Edit|Fill...|Series...**. In the dialog box displayed, choose **Columns** and set the **Step Value** to 0.1 and the **Stop Value** to 2.0. Press ENTER. The desired entries are entered down the **A**-column, as shown in Fig. 8.

Figure 8.   Figure 9.

In the **B**-column are the values of the function $y = x^2$. To create these entries, first position the cursor on cell **B2**. The value for this cell is the square of the value in cell **A2**. So enter into this cell:

=A2^2

When you press ENTER, this formula will be evaluated and the cell shows the number 0. Note, however, that when the cursor is on cell **A2**, the data entry field at the top of the grid shows the formula you entered. To enter the remainder of the **B**-column, copy the formula in **B2** to all the other cells besides x-value entries. To do the copying, first put the cursor on **B2** and press **Ctrl-C**. Next, highlight the cells in the **B**-column to which you wish to copy the formula. Then press **Ctrl-V**. The desired formulas are copied and the program evaluates each formula, giving the numerical results shown in Fig. 9.

You may graph tabulated function values. For example, to graph the x-y values shown in Fig. 9, begin by highlighting the data (including the title cells in row 1). Then select the command **Insert|Chart**. As the type of chart, select **X-Y (Scatter Plot)**. You will see a set of pictures showing the various types of plots. To plot just the points, choose the icon in the top row. You will

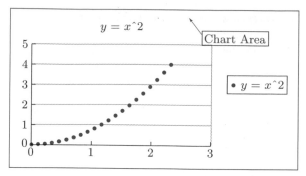

Figure 10.

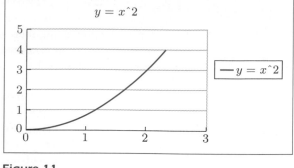

Figure 11.

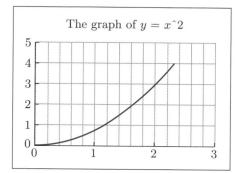

Figure 12.

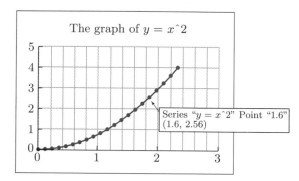

Figure 13.

see a sample graph to which you should respond by pressing Finish. The graph will be displayed on your worksheet as in Fig. 10. To plot a curve without the points, select the second icon on the second row. The graph will be as shown in Fig. 11.

By right clicking on a graph, you are given a window for setting graph options. For example, you may control the grid lines and the tick marks as well as the colors and thickness of lines, and so forth. For example, Fig. 12 shows how Fig. 11 can be modified using these options. By left clicking on the curve, you have access to the original data as shown in Fig. 13.

Many of the data sets in this book are available on Excel spreadsheets at our Web site. By using them, you will be saved the often tedious job of entering the data into your spreadsheet or graphing calculator.

▶ Exercises–Appendix

Use your graphing calculator to display the graphs of the following functions. Use the indicated domain to set the horizontal dimensions of the window. Choose the vertical dimensions so that the graph doesn't "run off the window."

1. $f(x) = -2x + 1$ for $-10 \le x \le 10$

2. $f(x) = \frac{1}{3}x + 1$ for $0 \le x \le 2$

3. $f(x) = 100x - 5000$ for $0 \le x \le 100$

4. $f(x) = -0.05x + 0.03$ for $0 \le x \le 1$

Graph the function $y = -12.1x^2 + 5.1x - 3.7$. Use the TRACE function to determine the following:

5. $f(2.75)$

6. $f(-3)$

7. The solutions of the equation $f(x) = 9.1$

8. The zeros of $f(x)$

Make a table of values of the function $f(x)$, where

9. $f(x) = -3x + 5$, $(x = 0, 1, 2, \dots)$

10. $f(x) = 3x^2 - 2x + 4$ $(x = 0, 1, 2, \dots)$

11. $f(x) = \dfrac{1}{2x}$ $(x = 1, 1.5, 2.0, 2.5, \dots)$

12. $f(x) = \dfrac{x}{x+1}$ $(x = 0, 1, 2, \dots)$

Display a scatter plot containing the following points. Choose the display window so that all points are visible.

13. $(0, 3)$, $(2, -1)$, $(4, 2)$

14. $(0, 0)$, $(1, 1)$, $(2, 3)$, $(3, 5)$

15. $(0, 0.9)$, $(1, 0.99)$, $(2, 1.01)$, $(3, .99)$

16. $(k, 2k - 1)$ $(k = 1, 4, 5, 8)$

Display a scatter plot of the points $(x, f(x))$, $(x = 1, 2, 3, 4)$, where $f(x)$ is as defined in

17. Exercise 9

18. Exercise 10

19. Exercise 11

20. Exercise 12

21. What is wrong with entering the function $f(x) = \dfrac{1}{x+1}$ into a graphing utility as **Y₁ = 1/X + 1**?

22. What is wrong with entering the function $f(x) = x^{3/4}$ into a graphing utility as **Y₁ = X^3/4**?

In Exercises 23–26 graph the function with the specified viewing window setting.

23. $f(x) = x^3 - 33x^2 + 120x + 1500$; $[-8, 30]$ by $[-2000, 2000]$.

24. $f(x) = -x^2 + 2x + 2$; $[-2, 4]$ by $[-8, 5]$

25. $f(x) = \sqrt{x + 1}$; $[0, 10]$ by $[-1, 4]$

26. $f(x) = \dfrac{1}{x^2 + 1}$; $[-4, 4]$ by $[-.5, 1.5]$

If your graphing utility can display tables of function values, in Exercises 66 and 67 create a table similar to the one shown. Then use the table to determine the value of the function when **X** $= 7$. (Appendices A, B, and D explain how to create tables with the TI-82, TI-83, and TI-86 calculators.)

27.

X	Y₁	
0	-1	
1	-2	
2	-1	
3	2	
4	7	
5	14	
6	▓	

Y₁=23

$Y_1 = X^2 - 2X - 1$

28.

X	Y₁	
0	1	
1	1.4142	
2	2.2361	
3	3.1623	
4	4.1231	
5	5.099	
6	▓	

Y₁=6.0827625303

$Y_1 = \sqrt{1 + x^2}$

Chapter Project

The following table records the cumulative percentage distribution of weights of males, ages 18–24. There are two sets of survey data, one taken in the years 1976–80 and a second taken in the years 1988–94.

Table 1	Cumulative Percent Distribution of Population, by Weight for 18- to 24-year-olds	
Source: U.S. National Center for Health Statistics, Vital and Health Statistics, series 11, No. 238.		
Percent Under	**Survey A—1976–1980**	**Survey B—1988–1994**
90 pounds	0.0	0.0
100 pounds	0.2	0.0
110 pounds	0.6	0.5
120 pounds	2.3	1.8
130 pounds	7.7	6.7
140 pounds	20.4	15.8
150 pounds	35.9	29.0
160 pounds	52.4	42.1
170 pounds	66.7	54.9
180 pounds	78.7	66.7
190 pounds	86.2	76.2
200 pounds	90.3	82.1
210 pounds	93.3	86.8
220 pounds	95.2	90.4
230 pounds	97.0	92.9
240 pounds	98.8	95.4
250 pounds	99.2	96.2
260 pounds	100.0	97.2
270 pounds	100.0	97.9
280 pounds	100.0	98.7
290 pounds	100.0	99.0
300 pounds	100.0	99.3
310 pounds	100.0	99.3
320 pounds	100.0	99.5
330 pounds	100.0	99.5
340 pounds	100.0	99.5
350 pounds	100.0	99.6
360 pounds	100.0	99.6
370 pounds	100.0	99.6
380 pounds	100.0	99.6

Chapter Project   (Continued)

(a) Let $w = $ a weight, in pounds. Let $f(w)$ and $g(w)$ denote the piecewise-linear functions defined by the second and third columns of data, respectively. Express, in words, the definitions of the functions $f(w)$ and $g(w)$. Give the range of each function.

(b) Using the first column as x-values, and the second and third columns as y-values, plot the data in the table.

(c) Draw the graphs of $f(w)$ and $g(w)$, By comparing the graphs, what conclusions can you make?

(d) Tabulate the values of Δx and Δy for consecutive rows of the above table. There will be one column of Δx values and two columns with Δy values, one corresponding to each of column 2 and column 3. Pick out one entry from each Δy column and describe in words the meaning of the entry.

CHAPTER

The Derivative

1

The derivative is a mathematical tool used to measure rate of change. To illustrate the sort of change we have in mind, consider the following example. Suppose that a colony of yeast cells grows in a culture dish. Its environment (the dish) imposes limits on the space and nutrients available. Based on experiments,* the culture can have at most 10,000 yeast cells. At time 0 hours, the number of yeast cells is 385. The experimental data in Table 1 list the number of yeast cells present at 5-hour intervals. Let $N(t)$ denote the number of cells in the colony at time t. Then $N(t)$ is a function of t, whose values at particular values of t are given by Table 1.

As t changes, the value of $N(t)$ changes. Let's analyze this change. Refer to Table 1. The third column records changes in the value of t. The first entry records the time elapsed from $t = 0$ until $t = 5$—that is $5 - 0 = 5$ hours. The second entry records the time elapsed from $t = 5$ until $t = 10$—that is $10 - 5 = 5$. And so forth, down the column. To each change in t (from 0 to 5, from 5 to 10, and so on), there is a corresponding change in $N(t)$. For example, if t changes from 0 to 5, then the value of $N(t)$ changes from 385 to 619, a change of 234. The change in $N(t)$ corresponding to each change in t are recorded in the fourth column of Table 1.

Let's assume that the changes in $N(t)$ are uniform during each time interval. That is, the increase of 234 cells occurs uniformly over the interval from time 0 until time 5. Then the rate of increase in the number of cells can be obtained as the quotient:

$$[\text{Rate of change in number of yeast cells}] = \frac{[\text{Change in number of yeast cells}]}{[\text{Change in time}]}$$

$$= \frac{234}{5}$$

$$= 46.8 \text{ cells per hour}$$

Column 5 of Table 1 records the rate of change in the number of yeast cells per hour, for each time interval described by the table. We see that each rate

---

*See C. F. Gause, "Experimental Studies on the Struggle of Existence," *J. Exp. Biology*, 9(1932), 389–402.

Table 1

t	$N(t)$	Change in t	Change in $N(t)$	$\dfrac{\text{Change in } N(t)}{\text{Change in } t}$
0	385			
5	619	5	234	46.8
10	981	5	362	72.4
15	1520	5	539	107.9
20	2281	5	761	152.2
25	3276	5	995	199.0
30	4455	5	1179	235.8
35	5698	5	1243	248.6
40	6859	5	1161	232.2
45	7826	5	967	193.4
50	8558	5	732	146.4
55	9073	5	515	103.0
60	9416	5	343	68.6
65	9638	5	222	44.4
70	9777	5	139	27.8

of change is positive, indicating that the yeast colony is continually increasing in size. However, the rates of change initially grow, then shrink. This indicates that the yeast colony grows at increasingly faster rates, taking advantage of the existing space and nutrients. As the colony becomes more crowded and food becomes more limited (per cell), the rate of increase slows.

We can describe the yeast colony by graphing $N(t)$, as shown in Fig. 1. Note that the colony size increases continually, approaching the maximum colony size of 10,000. Moreover, reading the graph from left to right, we see that the graph initially gets increasingly steep, then decreasingly steep. That is, there appears to be a connection between the rate at which $N(t)$ is changing and the steepness

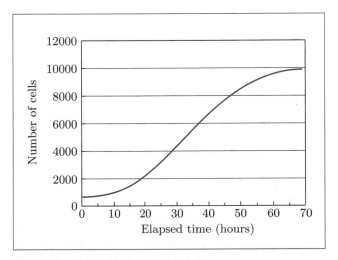

Figure 1. Yeast cells in a culture.

of the graph. This illustrates one of the fundamental ideas of calculus, which may be put roughly as follows: *The rate of change of a function corresponds to the steepness of its graph.*

This chapter is devoted to the *derivative*, which provides a numerical measure of the steepness of a curve at a particular point. By studying the derivative, we will be able to deal numerically with rates of change in applied problems.

1.1   The Slope of a Straight Line

As we shall see later, the study of straight lines is crucial for the study of the steepness of curves. So this section is devoted to a discussion of the geometric and algebraic properties of straight lines.

Let's first concentrate on nonvertical lines.

> **Equations of Nonvertical Lines**   A nonvertical line L has an equation of the form
>
> $$y = mx + b \qquad (1)$$
>
> The number m is called the *slope* of L and the point $(0, b)$ is called the *y-intercept*. The equation (1) is called the *slope-intercept equation* of L.

If we set $x = 0$, we see that $y = b$, so that $(0, b)$ is on the line L. Thus the y-intercept tells us where the line L crosses the y-axis. The slope measures the steepness of the line. In Fig. 1 we give three examples of lines with slope $m = 2$. In Fig. 2 we give three examples of lines with slope $m = -2$.

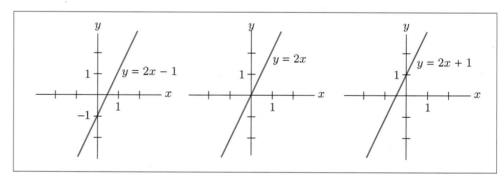

Figure 1. Three lines of slope 2.

To conceptualize the meaning of slope, think of walking along a line from left to right. On lines of positive slope we will be walking uphill; the greater the slope, the steeper the ascent. On lines of negative slope we will be walking downhill; the more negative the slope, the steeper the descent. Walking on lines of zero slope corresponds to walking on level ground. In Fig. 3 we have graphed lines with $m = 3, 1, \frac{1}{3}, 0, -\frac{1}{3}, -1, -3$, all having $b = 0$. The reader can readily verify our conceptualization of slope for these lines.

The slope and y-intercept of a straight line often have physical interpretations, as the following three examples illustrate.

▶ Example 1   A manufacturer finds that the total cost of producing x units of a commodity is $2x + 1000$ dollars. What is the economic significance of the y-intercept and the slope of the line $y = 2x + 1000$? (See Fig. 4.)

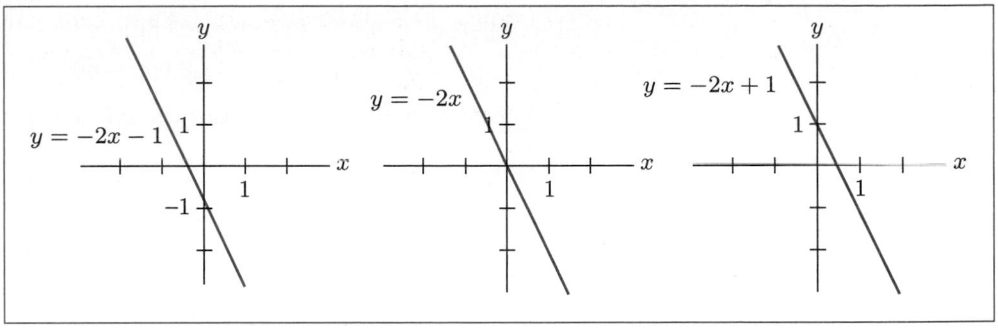

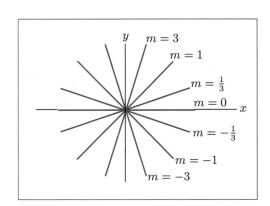

Figure 2. Three lines of slope -2.

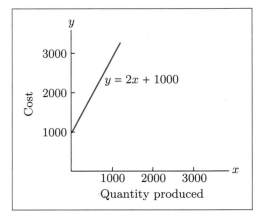

Figure 3.

Figure 4. A cost function.

Solution　The y-intercept is $(0, 1000)$. In other words, when $x = 0$ (no units produced), the cost is still $y = 1000$ dollars. The number 1000 represents the *fixed costs* of the manufacturer—those overhead costs, such as rent and insurance, that must be paid no matter how many items are produced.

　　The slope of the line is 2. This number represents the cost of producing each additional unit. To see this, we can calculate some typical costs.

Quantity Produced	Total Cost
$x = 1500$	$y = 2(1500) + 1000 = 4000$
$x = 1501$	$y = 2(1501) + 1000 = 4002$
$x = 1502$	$y = 2(1502) + 1000 = 4004$

　　Each time x is increased by 1, the value of y increases by 2. The number 2 is called the marginal cost.　　　　　　　　　　　　　　　　　　　　◆

▶ **Example 2**　An apartment complex has a storage tank to hold its heating oil. The tank was filled on January 1, but no more deliveries of oil will be made until some time in March. Let t denote the number of days after January 1 and let y denote the number of gallons of fuel oil in the tank. Current records of the apartment complex show that y and t are related approximately by the equation

$$y = 30{,}000 - 400t. \tag{2}$$

What interpretation can be given to the y-intercept and slope of this line? (See Fig. 5.)

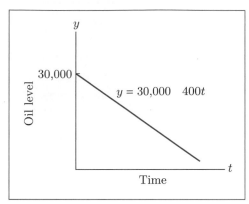

Figure 5. Amount of heating oil in a tank.

Solution   The y-intercept is $(0, 30{,}000)$. This value of y corresponds to $t = 0$, so there were 30,000 gallons of oil in the tank on January 1. Let us examine how fast the oil is removed from the tank.

Days After January 1	Gallons of Oil in the Tank
$t = 0$	$y = 30{,}000 - 400(0) = 30{,}000$
$t = 1$	$y = 30{,}000 - 400(1) = 29{,}600$
$t = 2$	$y = 30{,}000 - 400(2) = 29{,}200$
$t = 3$	$y = 30{,}000 - 400(3) = 28{,}800$
$\vdots$	$\vdots$

The oil level in the tank drops by 400 gallons each day; that is, the oil is being used at the rate of 400 gallons per day. The slope of the line (2) is -400. Thus the slope gives the rate at which the level of oil in the tank is changing. The negative sign on the -400 indicates that the oil level is decreasing rather than increasing.   ◆

▶ Example 3   For tax purposes, businesses are allowed to regard equipment as decreasing in value (or depreciating) each year. The amount of depreciation may be taken as an income tax deduction.

Suppose that the value y of a piece of equipment x years after its purchase is given by

$$y = 500{,}000 - 50{,}000x.$$

Interpret the y-intercept and the slope of the graph.

Solution   The y-intercept is $(0, 500{,}000)$ and corresponds to the value of y when $x = 0$. That is, the y-intercept gives the original value, \$500,000, of the equipment. The slope indicates the rate at which the equipment is changing in value. Thus the value of the equipment is decreasing at the rate of 50,000 dollars per year.   ◆

Properties of the Slope of a Line   Let us now examine several useful properties of the slope of a straight line.

Slope Property 1   Suppose that we start at a point on a line of slope m and move one unit to the right. Then we must move m units in the y-direction in order to return to the line.

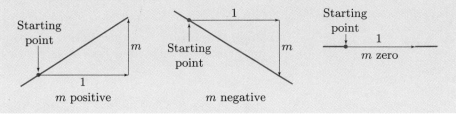

Slope Property 2   We can compute the slope of a line by knowing two points on the line. If (x_1, y_1) and (x_2, y_2) are on the line, then the slope of the line is $\dfrac{y_2 - y_1}{x_2 - x_1}$.

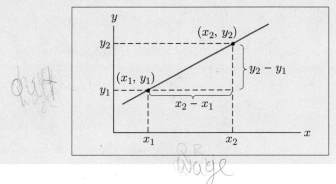

We can state Slope Property 2 in a form that will prepare us for the introduction of the derivative in the next section. Suppose that we move from point (x_1, y_1) [the initial point] to point (x_2, y_2) [the terminal point]. Using the delta notation introduced in Section 0.3, the change in x is

$$\Delta x = [\text{terminal value of } x] - [\text{initial value of } x] = x_2 - x_1.$$

Similarly, the change in y is given by

$$\Delta y = [\text{terminal value of } y] - [\text{initial value of } y] = y_2 - y_1.$$

In the formula for slope given in Slope Property 2, the numerator is Δy and the denominator Δx. So, Slope Property 2 may be expressed in delta notation as follows:

The slope m of a nonvertical line is given by the formula

$$\frac{\Delta y}{\Delta x} = m. \tag{3}$$

Slope Property 3 The equation of a line can be obtained if we know the slope and one point on the line. If the slope is m and if (x_1, y_1) is on the line, then the equation of the line is

$$y - y_1 = m(x - x_1). \tag{4}$$

This equation is called the *point-slope form* of the equation of the line.

Slope Property 4 Distinct lines of the same slope are parallel. Conversely, if two lines are parallel, they have the same slope.

Slope Property 5 When two lines are perpendicular, the product of their slopes is -1.

We omit the proofs of Slope Properties 1–5.

Calculations Involving Slope of a Line

▶ Example 4 Find the slope and the y-intercept of the line whose equation is $2x + 3y = 6$.

Solution We put the equation into slope-intercept form by solving for y in terms of x.

$$3y = -2x + 6$$
$$y = -\frac{2}{3}x + 2$$

The slope is $-\frac{2}{3}$ and the y-intercept is $(0, 2)$. ◆

▶ Example 5 Sketch the graph of the line
(a) passing through $(2, -1)$ with slope 3,
(b) passing through $(2, 3)$ with slope $-\frac{1}{2}$.

Solution We use Slope Property 1. (See Fig. 6.) In each case, we begin at the given point, move one unit to the right, and then move m units in the y-direction (upward for positive m, downward for negative m). The new point reached will also be on the line. Draw the straight line through these two points. ◆

▶ Example 6 Find the slope of the line passing through the points $(6, -2)$ and $(9, 4)$.

Solution We apply Slope Property 2 with $(x_1, y_1) = (6, -2)$ and $(x_2, y_2) = (9, 4)$. Then

$$\frac{y_2 - y_1}{x_2 - x_1} = \frac{4 - (-2)}{9 - 6} = \frac{6}{3} = 2.$$

Thus the slope is 2. [We would have reached the same answer if we had let $(x_1, y_1) = (9, 4)$ and $(x_2, y_2) = (6, -2)$.] The slope is just the difference of the y-coordinates divided by the difference of the x-coordinates, with each difference formed in the same order. ◆

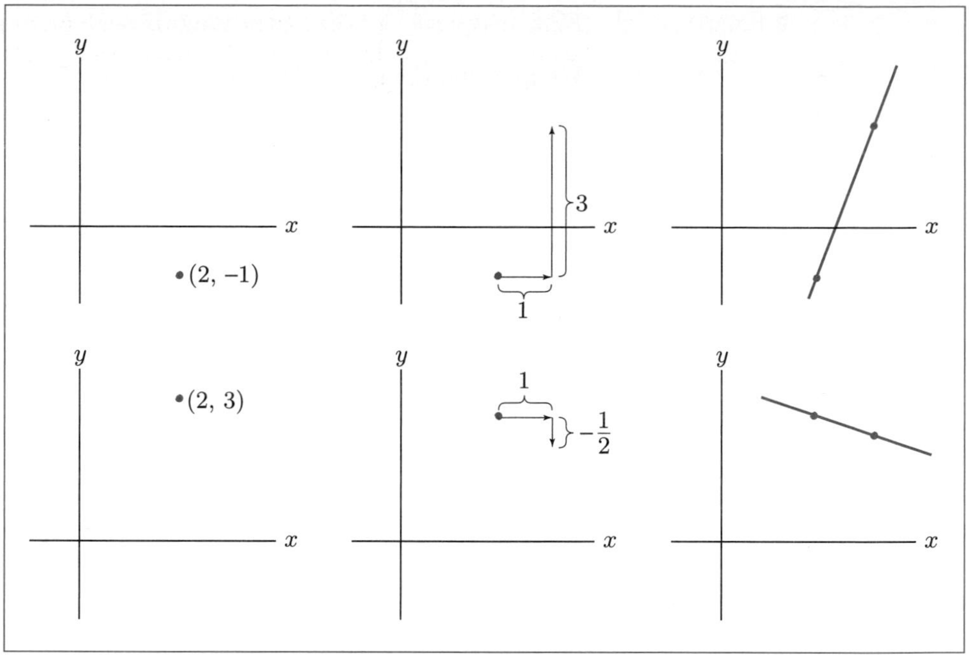

Figure 6.

▶ **Example 7**   Find an equation of the line passing through $(-1, 2)$ with slope 3.

Solution   We let $(x_1, y_1) = (-1, 2)$ and $m = 3$, and we use Slope Property 3. The equation of the line is

$$y - 2 = 3[x - (-1)]$$

or

$$y - 2 = 3(x + 1).$$

If desired, this equation can be put into the form $y = mx + b$:

$$y - 2 = 3(x + 1) = 3x + 3$$
$$y = 3x + 5.$$ ◆

▶ **Example 8**   Find an equation of the line passing through the points $(1, -2)$ and $(2, -3)$.

Solution   By Slope Property 2, the slope of the line is

$$\frac{-3 - (-2)}{2 - 1} = \frac{-3 + 2}{1} = -1.$$

Since $(1, -2)$ is on the line, we can use Property 3 to get the equation of the line:

$$y - (-2) = (-1)(x - 1) \tag{5}$$
$$y + 2 = -x + 1$$
$$y = -x - 1.$$ ◆

▶ Example 9   Find an equation of the line passing through $(5, 3)$ parallel to the line $2x + 5y = 7$.

Solution   We first find the slope of the line $2x + 5y = 7$.

$$2x + 5y = 7$$
$$5y = 7 - 2x$$
$$y = -\frac{2}{5}x + \frac{7}{5}.$$

The slope of this line is $-\frac{2}{5}$. By Slope Property 4, any line parallel to this line will also have slope $-\frac{2}{5}$. Using the given point $(5, 3)$ and Slope Property 3, we get the desired equation:

$$y - 3 = -\frac{2}{5}(x - 5).$$

This equation can also be written as

$$y = -\frac{2}{5}x + 5. \qquad \blacklozenge$$

Finding a Linear Function from Data   Suppose that we are given a table of data. How can we know if the data comes from a linear function? If so, how can we find the linear function? The key to answering both questions is to use equation (3), which says that the quotient

$$\frac{\Delta y}{\Delta x}$$

is always the slope m, independent of the initial and terminal points. As a test for linearity, we can compute the preceding quotient for each pair of consecutive data points. In order for the data to come from a linear function, the quotients $\frac{\Delta y}{\Delta x}$ must have the same value.* The common value is the slope. Moreover, we can determine the linear function, as in the following example.

▶ Example 10   Consider the following table of data.

x	y
-2	-17
0	-7
1	-2
3	8
5	18
6.5	25.5

(a) Does the data come from a linear function?

(b) If the answer to part (a) is yes, determine the linear function.

---

*A linear function must have constant values for $\frac{\Delta y}{\Delta x}$. If the values of the quotient are constant, we cannot necessarily conclude that y is a linear function of x. But in the interests of selecting the simplest model consistent with the data, it is usually valid to conclude that the data comes from a linear function.

Solution   (a) Tabulate the values of Δx, Δy, and $\dfrac{\Delta y}{\Delta x}$ as shown in the following table:

x	y	Δx	Δy	$\dfrac{\Delta y}{\Delta x}$
-2	-17			
0	-7	2	10	5
1	-2	1	5	5
3	8	2	10	5
5	18	2	10	5
6.5	25.5	1.5	7.5	5

Note that the values of $\dfrac{\Delta y}{\Delta x}$ all have the same value 5. We can conclude that the data is linear with slope 5.

(b) The graph must pass through the point $(0, -7)$ and has slope 5. By the point-slope formula, we see that the equation of the line is

$$y - (-7) = 5(x - 0)$$
$$y + 7 = 5x.$$

Solving for y in terms of x gives the desired function $y = 5x - 7$. See Fig. 7.

◆

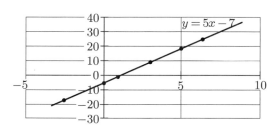

Figure 7.

Incorporating Technology   Here's how to graph nonvertical lines on a graphing calculator.

1. Start from the equation of the line. Write it as a function $y = \ldots$ by solving for y in terms of x.

2. Enter the function found in step 1 into the list of functions in the calculator. (Usually obtained by pressing **Y=**).

3. Press **GRAPH** to display the graph of the function.

Once you have the graph displayed, you may use the calculator to perform various analyses and computations, as illustrated in the following example.

▶ **Example 11** Consider the equation $5y - 15x + 13 = -7$.

(a) Use a graphing calculator to display its graph.

(b) Determine graphically the value of x for which y has the value 3.

Solution
(a) We first express y as a function of x by solving the equation for y in terms of x.

$$5y - 15x + 13 = -7$$
$$5y = 15x - 20$$
$$y = 3x - 4$$

Now we enter this function into the calculator as **Y1**, as shown in Fig. 8. Pressing **GRAPH** yields the graph of the equation, as shown in Fig. 9.

Figure 8.

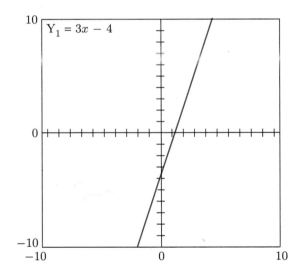

Figure 9.   **Figure 10.**

(b) Use **TRACE** to move the cursor along the graph until the y-coordinate is as close to 3 as you can make it, as shown in Fig. 10. The desired x-value is 2.3148148148. Note that the accuracy of this answer depends on the resolution of your calculator's screen, which determines the numerical "width" of a single step along the graph.   ◆

**Practice Problems
1.1**

Find the slopes of the following lines.

1. The line whose equation is $x = 3y - 7$

2. The line going through the points $(2, 5)$ and $(2, 8)$

▶ Exercises 1.1*

Find the slopes of the following lines.

1. $y = 2 - 5x$

2. $y = -5x$

3. $y - 2$

4. $y - \frac{1}{3}(x + 2)$

5. $y = \dfrac{2x - 1}{7}$

6. $y = \frac{1}{4}$

7. $2x + 3y = 6$

8. $x - y = 2$

Find the equations of the following lines.

9. Slope is 3; y-intercept is $(0, -1)$.

10. Slope is $\frac{1}{2}$; y-intercept is $(0, 0)$.

11. Slope is 1; $(1, 2)$ on line.

12. Slope is $-\frac{1}{3}$; $(6, -2)$ on line.

13. Slope is -7; $(5, 0)$ on line.

14. Slope is $\frac{1}{2}$; $(2, -3)$ on line.

15. Slope is 0; $(7, 4)$ on line.

16. Slope is $-\frac{2}{3}$; $(0, 5)$ on line.

17. $(2, 1)$ and $(4, 2)$ on line.

18. $(5, -3)$ and $(-1, 3)$ on line.

19. $(0, 0)$ and $(1, -2)$ on line.

20. $(2, -1)$ and $(3, -1)$ on line.

21. Parallel to $y = -2x + 1$; $(\frac{1}{2}, 5)$ on line.

22. Parallel to $3x + y = 7$; $(-1, -1)$ on line.

23. Parallel to $3x - 6y = 1$; $(1, 0)$ on line.

24. Parallel to $5x + 2y = -4$; $(0, 17)$ on line.

25. Each of the lines (A), (B), (C), and (D) in Fig. 11 is the graph of one of the equations (a), (b), (c), and (d). Match each equation with its graph.

    (a) $x + y = 1$ (b) $x - y = 1$

    (c) $x + y = -1$ (d) $x - y = -1$

26. Table 1 gives some points on the line $y = mx + b$. Find m and b.

Table 1	Points on a Line				
x	4.8	4.9	5	5.1	5.2
y	3.6	4.8	6	7.2	8.4

In Exercises 27–30, refer to a line of slope m. Suppose you begin at a point on the line and move h units in the x-direction. How many units must you move in the y-direction to return to the line?

27. $m = \frac{1}{2}$, $h = 4$ **28.** $m = 2$, $h = \frac{1}{4}$

29. $m = -3$, $h = .25$ **30.** $m = .2$, $h = 5$

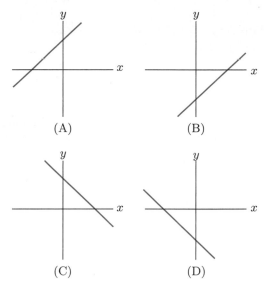

(A)   (B)

(C)   (D)

Figure 11.

In each of Exercises 31–34, we specify a line by giving the slope and one point on the line. We give the first coordinate of some points on the line. Without deriving the equation of the line, find the second coordinate of each of the points.

31. Slope is 2, $(1, 3)$ on line; $(2, \ )$; $(3, \ )$; $(0, \ )$.

32. Slope is -3, $(2, 2)$ on line; $(3, \ )$; $(4, \ )$; $(1, \ )$.

33. Slope is $-\frac{1}{4}$, $(-1, -1)$ on line; $(0, \ )$; $(1, \ )$; $(-2, \ )$.

34. Slope is $\frac{1}{3}$, $(-5, 2)$ on line; $(-4, \ )$; $(-3, \ )$; $(-2, \ )$.

For each pair of lines in the following figures, determine the one with the greater slope.

35.

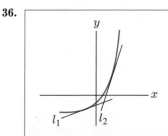

36.

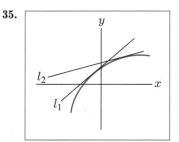

---

*A complete solution for every third odd-numbered exercise is given in the *Study Guide* for this text.

Find the equation and sketch the graph of the following lines.

37. With slope -2 and y-intercept $(0, -1)$.

38. With slope $\frac{1}{3}$ and y-intercept $(0, 1)$.

39. Through $(2, 0)$ with slope $\frac{4}{5}$.

40. Through $(-1, 3)$ with slope 0.

In Exercises 41–46, find an equation of a line with the given property. Each exercise has more than one correct answer.

41. y-intercept is $(0, 5)$.

42. x-intercept is $(9, 0)$.

43. Parallel to the line $4x + 5y = 6$.

44. Horizontal.

45. Slope is -2.

46. Vertical.

In the next section, we shall show that the tangent line to the parabola $y = x^2$ passing through the point with coordinates (x, y) has slope $2x$. (See Fig. 12.) Thus the slopes of the tangent lines at the points $(1, 1)$, $(0, 0)$, and $\left(-\frac{1}{2}, \frac{1}{4}\right)$ are, respectively, 2, 0, and -1. Find an equation of the tangent line through each of the following points.

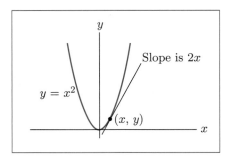

Figure 12.

47. $(1, 1)$ **48.** $(0, 0)$ **49.** $\left(-\frac{1}{2}, \frac{1}{4}\right)$

50. A salesperson's weekly pay depends on the volume of sales. If she sells x units of goods, then her pay is $y = 5x + 60$ dollars. Give an interpretation of the slope and the y-intercept of this straight line.

51. The demand equation for a monopolist is $y = -.02x + 7$, where x is the number of units produced and y is the price. That is, in order to sell x units of goods, the price must be $y = -.02x + 7$ dollars. Interpret the slope and y-intercept of this line.

52. Temperatures of $32°$F and $212°$F correspond to temperatures of $0°$C and $100°$C. Suppose the linear equation $y = mx + b$ converts Fahrenheit temperatures to Celsius temperatures. Find m and b. What is the Celsius equivalent of $98.6°$F?

53. (a) Draw the graph of any function $f(x)$ that passes through the point $(3, 2)$.

    (b) Choose a point to the right of $x = 3$ on the x-axis and label it $3 + h$.

    (c) Draw the straight line through the points $(3, f(3))$ and $(3 + h, f(3 + h))$.

    (d) What is the slope of this straight line (in terms of h)?

54. Prove Slope Property 4 of straight lines. (*Hint:* If $y = mx + b$ and $y = m'x + b'$ are two lines, then they have a point in common if and only if the equation $mx + b = m'x + b'$ has a solution x.)

55. Prove Property 5 of straight lines. (*Hint:* Without loss of generality, assume that both lines pass through the origin. Use Slope Property 1 and the Pythagorean theorem. See Fig. 13.)

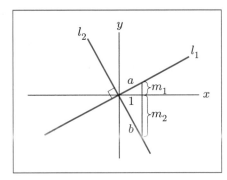

Figure 13.

Technology Exercises

56. Let y denote the average amount claimed for itemized deductions on a tax return reporting x dollars of income. According to Internal Revenue Service data, y is a linear function of x. Moreover, in a recent year, income tax returns reporting \$20,000 of income averaged \$729 in itemized deductions, while returns reporting \$50,000 averaged \$1380.

    (a) Determine y as a function of x.

    (b) Graph this function in the window $[0, 75000]$ by $[0, 2000]$.

    (c) Give an interpretation of the slope in applied terms.

    (d) Determine graphically the average amount of itemized deductions on a return reporting \$75,000.

    (e) Determine graphically the income level at which the average itemized deductions are \$1600.

    (f) Suppose that the income level increases by \$15,000. By how much do the average itemized deductions increase?

57. Let y denote the percentage of the world population that is urban x years after 1980. According to recently published data, y has been a linear function of x since 1980. The percent of the world population that is urban was 39.5 in 1980 and 45.2 in 1995.

(a) Determine y as a function of x.

(b) Graph this function in the window $[0, 40]$ *by* $[0, 100]$.

(c) Interpret the slope in applied terms.

(d) Determine graphically the percent of the world population that was urban in 1990.

(e) Determine graphically the year in which 50% of the world population will be urban.

(f) By what amount does the percent of the world population that is urban increase every 5 years?

Solutions to Practice Problems 1.1

1. We solve for y in terms of x.

$$y = \frac{1}{3}x + \frac{7}{3}.$$

The slope of the line is the coefficient of x, that is, $\frac{1}{3}$.

2. The line passing through these two points is a vertical line; therefore, its slope is undefined.

1.2   The Slope of a Curve at a Point

In order to extend the concept of slope from straight lines to more general curves, we must first discuss the notion of the tangent line to a curve at a point.

We have a clear idea of what is meant by the tangent line to a circle at a point P. It is the straight line that touches the circle at just the one point P. Let us focus on the region near P, designated by the dashed rectangle shown in Fig. 1. The enlarged portion of the circle looks almost straight, and the straight line that it resembles is the tangent line. Further enlargements would make the circle near P look even straighter and have an even closer resemblance to the tangent line. In this sense, the tangent line to the circle at the point P is the straight line through P that best approximates the circle near P. In particular, the tangent line at P reflects the steepness of the circle at P. Thus it seems reasonable to define the *slope* of the circle at P to be the slope of the tangent line at P.

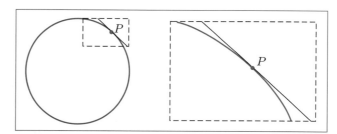

Figure 1. Enlarged portion of a circle.

Similar reasoning leads us to a suitable definition of slope for an arbitrary curve at a point P. Consider the three curves drawn in Fig. 2. We have drawn an enlarged version of the dashed box around each point P. Notice that the portion of each curve lying in the boxed region looks almost straight. If we further magnify the curve near P, it would appear even straighter. Indeed, if we apply higher and higher magnification, the portion of the curve near P would approach a certain straight line more and more exactly. (See Fig. 3.) This straight line is called the *tangent line to the curve at P*. This line best approximates the curve

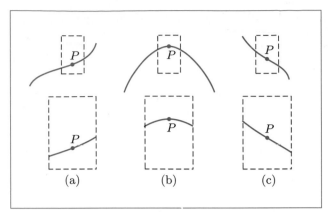

Figure 2. Enlarged portions of curves.

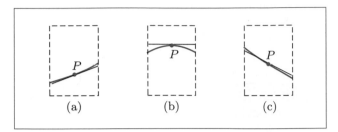

Figure 3. Tangent lines to curves.

near P. We define the *slope of a curve at a point P* to be the slope of the tangent line to the curve at P.

The portion of the curve near P can be, at least within an approximation, replaced by the tangent line at P. Therefore, the slope of the curve at P—that is, the slope of the tangent line at P—measures the rate of increase or decrease of the curve as it passes through P.

▶ Example 1   In Example 2 of Section 1.1 the apartment complex used approximately 400 gallons of oil per day. Suppose that we keep a continuous record of the oil level in the storage tank. The graph for a typical 2-day period appears in Fig. 4. What is the physical significance of the slope of the graph at the point P?

Solution   The curve near P is closely approximated by its tangent line. So think of the curve as replaced by its tangent line near P. Then the slope at P is just the rate of decrease of the oil level at 7 A.M. on March 5.   ◆

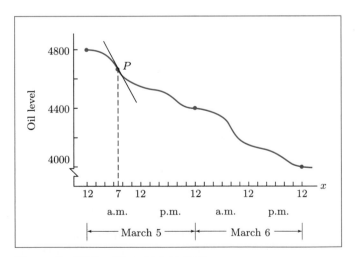

Figure 4. Oil level in a storage tank.

Notice that during the entire day of March 5, the graph in Fig. 4 seems to be the steepest at 7 A.M. That is, the oil level is falling the fastest at that time. This corresponds to the fact that most people awake around 7 A.M., turn up their thermostats, take showers, and so on. Example 1 provides a typical illustration of the manner in which slopes can be interpreted as rates of change. We shall return to this idea in Section 1.8.

In calculus, we can usually compute slopes by using formulas. For instance, in Section 1.3 we shall show that the tangent line to the graph of $y = x^2$ at the point $(1, 1)$ has slope 2; the tangent line at $(3, 9)$ has slope 6; and the tangent line at $\left(-\frac{5}{2}, \frac{25}{4}\right)$ has slope -5. The various tangent lines are shown in Fig. 5. Notice that the slope at each point is two times the x-coordinate of the point. This is a general fact for this graph. In Section 1.3 we shall derive this simple formula (see Fig. 6.):

$$[\text{slope of the graph of } y = x^2 \text{ at the point } (x, y)] = 2x.$$

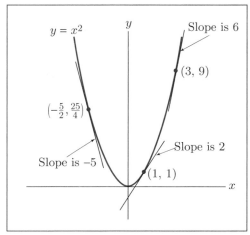

Figure 5. Graph of $y = x^2$.

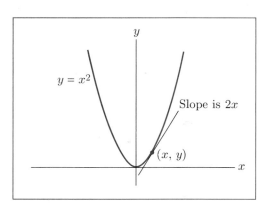

Figure 6. Slope of tangent line to $y = x^2$.

▶ Example 2   (a) What is the slope of the graph of $y = x^2$ at the point $\left(\frac{3}{4}, \frac{9}{16}\right)$?

(b) Write the equation of the tangent line to the graph of $y = x^2$ at the point $\left(\frac{3}{4}, \frac{9}{16}\right)$.

Solution   (a) The x-coordinate of $\left(\frac{3}{4}, \frac{9}{16}\right)$ is $\frac{3}{4}$, so the slope of $y = x^2$ at this point is $2\left(\frac{3}{4}\right) = \frac{3}{2}$.

(b) We shall write the equation of the tangent line in the point-slope form. The point is $\left(\frac{3}{4}, \frac{9}{16}\right)$, and the slope is $\frac{3}{2}$ by part (a). Hence the equation is

$$y - \frac{9}{16} = \frac{3}{2}\left(x - \frac{3}{4}\right).$$ ◆

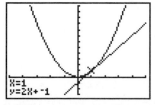

Figure 7. Result of TI-83 **tangent** command.

Incorporating Technology   Many graphing calculators have a command that calculates the slope of a curve at a designated point and automatically draws the tangent line through the point. Some calculators display the slope at the bottom of the screen as $dy/dx = slope$, and others give the x-coordinate of the point and the equation of the tangent line. Figure 7 shows the tangent line of $y = x^2$ at the point $(1, 1)$.

Practice Problems 1.2

1. Refer to Fig. 8.

   (a) What is the slope of the curve at $(3, 4)$?

   (b) What is the equation of the tangent line at the point where $x = 3$?

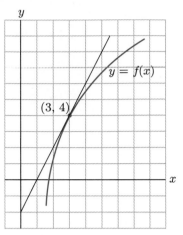

Figure 8.

2. What is the equation of the tangent line to the graph of $y = \frac{1}{2}x + 1$ at the point $(4, 3)$?

▶ Exercises 1.2

Trace the curves in Exercises 1–6 onto another piece of paper and sketch the tangent line in each case at the designated point P.

1.

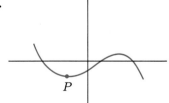

2.

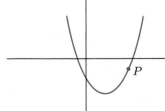

3.

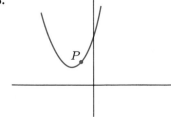

4.

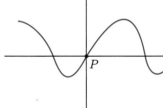

5.

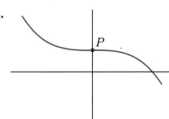

6.

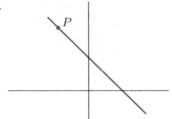

Estimate the slope of each of the following curves at the designated point P.

7.

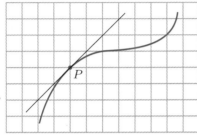

8.

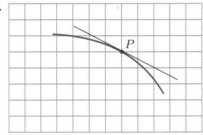

9.

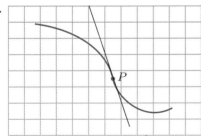

10.

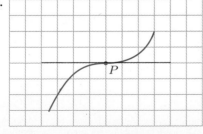

11.

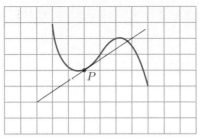

12.

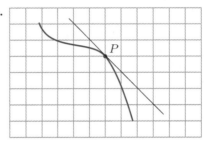

Exercises 13–18 refer to the points in Fig. 9. Assign one of the following descriptors to each point: large positive slope, small positive slope, zero slope, small negative slope, large negative slope.

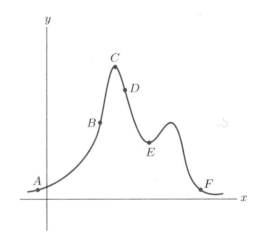

Figure 9.

13. A **14.** B **15.** C

16. D **17.** E **18.** F

In Exercises 19–21, find the slope of the tangent line to the graph of $y = x^2$ at the point indicated and then write the corresponding equation of the tangent line.

19. $(-2, 4)$ **20.** $(-.4, .16)$ **21.** $\left(\frac{4}{3}, \frac{16}{9}\right)$

22. Find the slope of the tangent line to the graph of $y = x^2$ at the point where $x = -\frac{1}{2}$.

23. Write the equation of the tangent line to the graph of $y = x^2$ at the point where $x = 1.5$.

24. Write the equation of the tangent line to the graph of $y = x^2$ at the point where $x = .6$.

25. Find the point on the graph of $y = x^2$ where the curve has slope $\frac{5}{3}$.

26. Find the point on the graph of $y = x^2$ where the curve has slope -4.

27. Find the point on the graph of $y = x^2$ where the tangent line is parallel to the line $x + 2y = 4$.

28. Find the point on the graph of $y = x^2$ where the tangent line is parallel to the line $3x - y = 2$.

In the next section we shall see that the tangent line to the graph of $y = x^3$ at the point (x, y) has slope $3x^2$. See Fig. 10. Using this result, find the slope of the curve at the points in Exercises 29–31.

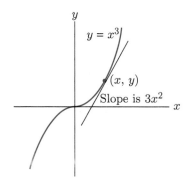

Figure 10. Slope of tangent line to $y = x^3$.

29. $(2, 8)$ **30.** $\left(\frac{3}{2}, \frac{27}{8}\right)$ **31.** $\left(-\frac{1}{2}, -\frac{1}{8}\right)$

32. Find the slope of the curve $y = x^3$ at the point where $x = \frac{1}{4}$.

33. Write the equation of the line tangent to the graph of $y = x^3$ at the point where $x = -1$.

34. Write the equation of the line tangent to the graph of $y = x^3$ at the point where $x = \frac{1}{2}$.

35. Let l be the line through the points P and Q in Fig. 11.
   (a) Suppose $P = (2, 4)$ and $Q = (5, 13)$. Find the slope of the line l and the length of the line segment d.
   (b) As the point Q moves toward P, does the slope of the line l increase or decrease?

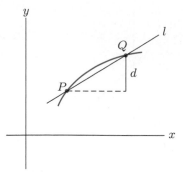

Figure 11. The slope of a secant line.

36. In Fig. 12, h represents a positive number, and $3 + h$ is the number h units to the right of 3. Draw line segments on the graph having the following lengths.
   (a) $f(3)$ (b) $f(3 + h)$
   (c) $f(3 + h) - f(3)$ (d) h
   (e) Draw a line of slope $\dfrac{f(3 + h) - f(3)}{h}$.

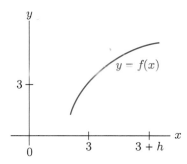

Figure 12. Geometric representation of values.

Technology Exercises

In Exercises 37–40, draw the graph of the function with the given window setting, and then use the tangent-drawing feature of your calculator to find the slope of the function at the given value of x and to draw the tangent line. Write the equation of the tangent line in point-slope form.

37. $f(x) = \sqrt{x + 2}$; $[0, 8]$ *by* $[0, 4]$; $x = 4$

38. $f(x) = \dfrac{1}{x^2 - 2x + 2}$; $[-1, 3]$ *by* $[-.5, 1.5]$; $x = 1$

39. $f(x) = \dfrac{64x + 40}{x + 2}$; $[0, 7]$ *by* $[0, 70]$; $x = 3.5$

40. $f(x) = x^3 - 9x^2 + 28x - 28$; $[-.5, 6.5]$ *by* $[-2, 7]$; $x = 3$

1. (a) The slope of the curve at the point $(3, 4)$ is, by definition, the slope of the tangent line at $(3, 4)$. Note that the point $(4, 6)$ is also on the line. Therefore, the slope is

$$\frac{6 - 4}{4 - 3} = \frac{2}{1} = 2.$$

   (b) Use the point-slope formula. The equation of the line passing through the point $(3, 4)$ and having slope 2 is

$$y - 4 = 2(x - 3)$$

   or

$$y = 2x - 2.$$

2. The tangent line at $(4, 3)$ is, by definition, the line that best approximates the curve at $(4, 3)$. Since the "curve" in this case is itself a line, the curve and its tangent line at $(4, 3)$ (and at every other point) must be the same. Therefore, the equation is $y = \frac{1}{2}x + 1$.

1.3 The Derivative

Suppose that a curve is the graph of a function $f(x)$. It is usually possible to obtain a formula that gives the slope of the curve $y = f(x)$ at any point. This slope formula is called the *derivative* of $f(x)$ and is written $f'(x)$. For each value of x, $f'(x)$ gives the slope of the curve $y = f(x)$ at the point with first coordinate x.* (See Fig. 1.)

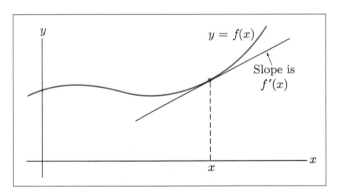

Figure 1. Definition of f'(x).

The process of computing $f'(x)$ for a given function $f(x)$ is called *differentiation.*

As we shall see later, the concept of a derivative occurs frequently in applications. In economics, the derivatives are often described by the adjective "marginal." For instance, if $C(x)$ is a cost function (the cost of producing x units of a commodity), then the derivative $C'(x)$ is called the *marginal cost function*. The derivative $P'(x)$ of a profit function $P(x)$ is called the *marginal profit function*; the derivative of a revenue function is called the *marginal revenue*

---

*As we shall see, there are curves that do not have tangent lines at every point. At values of x corresponding to such points, the derivative $f'(x)$ is not defined. For the sake of the current discussion, which is designed to develop an intuitive feeling for the derivative, let us assume that the graph of $f(x)$ has a tangent line for each x in the domain of f.

function; and so on. We shall discuss the economic meaning of these concepts in Section 1.8.

For the remainder of this section as well as the next few sections we will concentrate on calculating derivatives.

The case of a linear function $f(x) = mx + b$ is particularly simple. The graph of $y = mx + b$ is a straight line L of slope m. The tangent line to L (at any point) is just L itself, and so the slope of the graph is m at every point. (See Fig. 2.) In other words, the value of the derivative $f'(x)$ is always equal to m. We summarize this fact as follows:

If $f(x) = mx + b$, then we have

$$f'(x) = m. \tag{1}$$

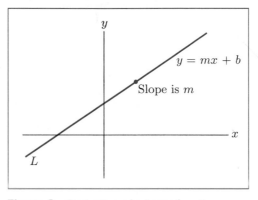

Figure 2. Derivative of a linear function.

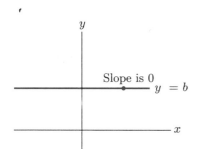

Figure 3. Derivative of a constant function.

Set $m = 0$ in equation (1). Then the function becomes $f(x) = b$, which has the value b for each value of x. The graph is a horizontal line of slope 0, so $f'(x) = 0$ for all x. (See Fig. 3.) Thus we have

The derivative of a constant function $f(x) = b$ is zero; that is,

$$f'(x) = 0. \tag{2}$$

Next, consider the function $f(x) = x^2$. As we stated in Section 1.2 (and will prove at the end of this section), the slope of the graph of $y = x^2$ at the point (x, y) is equal to $2x$. That is, the value of the derivative $f'(x)$ is $2x$:

If $f(x) = x^2$, then its derivative is the function $2x$. That is,

$$f'(x) = 2x. \tag{3}$$

In Exercises 29–34 of Section 1.2 we made use of the fact that the slope of the graph of $y = x^3$ at the point (x, y) is $3x^2$. This can be restated in terms of derivatives as follows:

If $f(x) = x^3$, then the derivative is $3x^2$. That is,

$$f'(x) = 3x^2. \tag{4}$$

We should, at this stage at least, keep the geometric meaning of these formulas clearly in mind. Figure 4 shows the graphs of x^2 and x^3 together with the interpretations of formulas (3) and (4) in terms of slope.

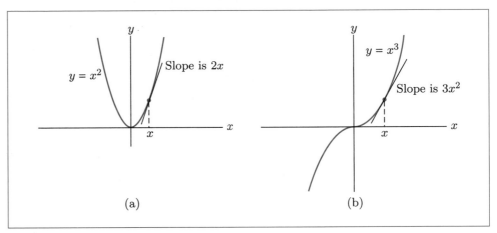

(a)   (b)

Figure 4. Derivatives of x^2 and x^3.

One of the reasons calculus is so useful is that it provides general techniques that can be easily used to determine derivatives. One such general rule, which contains formulas (3) and (4) as special cases, is the so-called power rule.

Power Rule Let r be any number and let $f(x) = x^r$. Then $f'(x) = rx^{r-1}$.

Indeed, if $r = 2$, then $f(x) = x^2$ and $f'(x) = 2x^{2-1} = 2x$, which is formula (3). If $r = 3$, then $f(x) = x^3$ and $f'(x) = 3x^{3-1} = 3x^2$, which is (4). We shall prove the power rule in Chapter 4. Until then, we shall use it to calculate derivatives.

▶ Example 1 Let $f(x) = \sqrt{x}$. What is $f'(x)$?

Solution Recall that $\sqrt{x} = x^{1/2}$. We may apply the power rule with $r = \frac{1}{2}$.

$$f(x) = x^{1/2}$$
$$f'(x) = \tfrac{1}{2}x^{1/2-1} = \tfrac{1}{2}x^{-1/2}$$
$$= \frac{1}{2} \cdot \frac{1}{x^{1/2}} = \frac{1}{2\sqrt{x}}.$$

◆

Another important special case of the power rule occurs for $r = -1$, corresponding to $f(x) = x^{-1}$. In this case, $f'(x) = (-1)x^{-1-1} = -x^{-2}$. However, since $x^{-1} = 1/x$ and $x^{-2} = 1/x^2$, the power rule for $r = -1$ may also be written as follows:*

If $f(x) = \dfrac{1}{x}$, then $f'(x) = -\dfrac{1}{x^2}$ $(x \neq 0)$.   (5)

---

*The formula gives $f'(x)$ for $x \neq 0$. The derivative of $f(x)$ is not defined at $x = 0$ since $f(x)$ itself is not defined there.

▶ Example 2   Find the slope of the curve $y = 1/x$ at $\left(2, \frac{1}{2}\right)$.

Solution   Set $f(x) = 1/x$. The point $\left(2, \frac{1}{2}\right)$ corresponds to $x = 2$, so in order to find the slope at this point, we compute $f'(2)$. From formula (5) we find that

$$f'(2) = -\frac{1}{2^2} = -\frac{1}{4}.$$

Thus the slope of $y = 1/x$ at the point $\left(2, \frac{1}{2}\right)$ is $-\frac{1}{4}$. (See Fig. 5.)   ◆

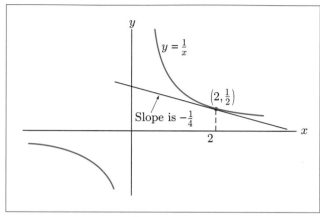

Figure 5. Derivative of $\frac{1}{x}$.

Warning   Do not confuse $f'(2)$, the value of the derivative at 2, with $f(2)$, the value of the y-coordinate at the point on the graph at which $x = 2$. In Example 2 we have $f'(2) = -\frac{1}{4}$, whereas $f(2) = \frac{1}{2}$. The number $f'(2)$ gives the *slope* of the graph at $x = 2$; the number $f(2)$ gives the *height* of the graph at $x = 2$.

Notation   The operation of forming a derivative $f'(x)$ from a function $f(x)$ is also indicated by the symbol $\dfrac{d}{dx}$ (read "the derivative with respect to x"). Thus

$$\frac{d}{dx} f(x) = f'(x).$$

For example,

$$\frac{d}{dx}(x^6) = 6x^5, \qquad \frac{d}{dx}(x^{5/3}) = \frac{5}{3} x^{2/3}, \qquad \frac{d}{dx}\left(\frac{1}{x}\right) = -\frac{1}{x^2}.$$

When working with an equation of the form $y = f(x)$, we often write $\dfrac{dy}{dx}$ as a symbol for the derivative $f'(x)$. For example, if $y = x^6$, we may write

$$\frac{dy}{dx} = 6x^5.$$

The Secant-Line Calculation of the Derivative

So far, we have said nothing about how to derive differentiation formulas such as (3), (4), or (5). Let us remedy that omission now. The derivative gives the slope of the tangent line, so we must describe a procedure for calculating that slope.*

---

*The following discussion is designed to develop geometric intuition for the derivative. It will also provide the basis for a formal definition of the derivative in terms of limits in Section 1.4.

The fundamental idea for calculating the slope of the tangent line at a point P is to approximate the tangent line very closely by *secant lines*. A secant line at P is a straight line passing through P and a nearby point Q on the curve. (See Fig. 6.) By taking Q very close to P, we can make the slope of the secant line approximate the slope of the tangent line to any desired degree of accuracy. Let us see what this amounts to in terms of calculations.

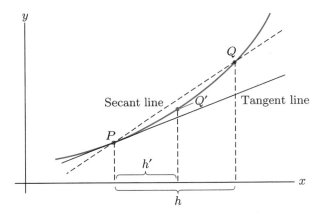

Figure 6. A secant line approximation to a tangent line.

Suppose that the point P is $(x, f(x))$. Suppose also that Q is h horizontal units away from P. Then Q has x-coordinate $x+h$ and y-coordinate $f(x+h)$. The slope of the secant line through the points $P = (x, f(x))$ and $Q = (x+h, f(x+h))$ is simply

$$[\text{slope of secant line}] = \frac{f(x+h) - f(x)}{(x+h) - x} = \frac{f(x+h) - f(x)}{h}.$$

(See Fig. 7.)

In order to move Q close to P along the curve, we let h approach zero. Then the secant line approaches the tangent line, and so

$$[\text{slope of secant line}] \quad \text{approaches} \quad [\text{slope of tangent line}];$$

that is,

$$\frac{f(x+h) - f(x)}{h} \quad \text{approaches} \quad f'(x).$$

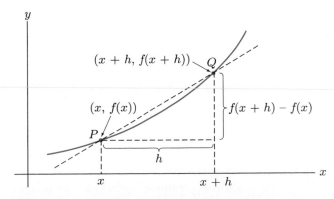

Figure 7. Computing the slope of a secant line.

Since we can make the secant line as close to the tangent line as we wish by taking h sufficiently small, the quantity $[f(x+h)-f(x)]/h$ can be made to approximate $f'(x)$ to any desired degree of accuracy. Thus we arrive at the following method to compute the derivative $f'(x)$.

To calculate $f'(x)$:

1. First calculate the quotient $\dfrac{f(x+h)-f(x)}{h}$ for $h \neq 0$.

2. Then let h approach zero.

3. The quantity $\dfrac{f(x+h)-f(x)}{h}$ will approach $f'(x)$.

The quotient in Step 1 turns out to be so important, we give it a name: the *difference quotient*.

Let us use this method to verify differentiation formulas (3) and (5), which, as we saw, were special cases of the power rule for $r = 2$ and $r = -1$, respectively.

Verification of the Power Rule for r = 2

Here $f(x) = x^2$, so the slope of the secant line is

$$\frac{f(x+h)-f(x)}{h} = \frac{(x+h)^2 - x^2}{h}.$$

However, by multiplying out we have $(x+h)^2 = x^2 + 2xh + h^2$, so that

$$\frac{f(x+h)-f(x)}{h} = \frac{x^2 + 2xh + h^2 - x^2}{h} = \frac{(2x+h)h}{h} = 2x + h.$$

As h approaches zero (i.e., as the secant line approaches the tangent line), the quantity $2x + h$ approaches $2x$. Thus we have

$$f'(x) = 2x,$$

which is formula (3).

Verification of the Power Rule for r = -1

Here $f(x) = x^{-1} = 1/x$, so that the slope of the secant line is

$$\frac{f(x+h)-f(x)}{h} = \frac{1}{h}\left[\frac{1}{x+h} - \frac{1}{x}\right] = \frac{1}{h}\left[\frac{x-(x+h)}{(x+h)x}\right]$$

$$= \frac{1}{h}\left[\frac{-h}{(x+h)x}\right] = -\frac{1}{x(x+h)}.$$

As h approaches zero, $-\dfrac{1}{x(x+h)}$ approaches $-\dfrac{1}{x^2}$. Hence

$$f'(x) = -\frac{1}{x^2},$$

which is formula (5).

Similar arguments can be used to verify the power rule for other values of r. The cases $r = 3$ and $r = \frac{1}{2}$ are outlined in Exercises 55 and 56, respectively.

Derivative Calculation in Delta Notation   As x changes from $x = a$ to $x = a + h$, then the change in x is given by

$$\Delta x = (a + h) - a = h$$

and the change in y by

$$\Delta y = f(a + h) - f(a).$$

Comparing the last two equations with our procedure for computing the derivative, we can reformulate that procedure using delta notation:

$f'(a)$ is obtained by computing $\dfrac{\Delta y}{\Delta x}$ and allowing Δx to approach 0.

The quotient $\dfrac{\Delta y}{\Delta x}$ gives the rate of change of $f(x)$ over the interval from $x = a$ to $x = a + h = a + \Delta x$. As Δx approaches 0, the quotient gives the rate of change over successively narrower intervals and eventually gives the rate of change of $f(x)$ at the point $x = a$ itself. That is, we have

$f'(a)$ measures the instantaneous rate of change of $f(x)$ when $x = a$. $\hspace{1cm}$ (6)

In Chapter 0, we introduced the rule of three, which says that all mathematical principles should be considered from three viewpoints: graphical, numerical, and analytical. Our discussion of derivatives has so far concentrated on the graphical (slope) and the analytical (formulas for derivatives). Let us now turn to a numerical discussion of the derivative and consider functions derived from real-world data. Consider the data on personal computer shipments in the U.S (1987–1995) shown in Fig. 8.

$x = year$ with 1 corresponding to 1987 $y = number$ of personal computers shipped in year x		
Year	x	y
1987	1	8391351
1988	2	9615739
1989	3	9329970
1990	4	9848593
1991	5	10903000
1992	6	12544374
1993	7	14775000
1994	8	18605000
1995	9	22582900

Figure 8. Number of personal computer shipments in U.S. (1987–1995).

$x = year\ with\ 1\ corresponding\ to\ 1987$
$y = number\ of\ personal\ computers\ shipped\ in\ year\ x$

Year	x	y	Δx	Δy	$\dfrac{\Delta y}{\Delta x}$
1987	1	8391351			
1988	2	9615739	1	1224388	1224388
1989	3	9329970	1	−285769	−285769
1990	4	9848593	1	518623	518623
1991	5	10903000	1	1054407	1054407
1992	6	12544374	1	1641374	1641374
1993	7	14775000	1	2230626	2230626
1994	8	18605000	1	3830000	3830000
1995	9	22582900	1	3977900	3977900

Figure 9. Number of personal computer shipments in U.S. (1987–1995).

In Fig. 8, note that y is a function of x, say $y = f(x)$. Although the function values are given only for certain values of x, the function $y = f(x)$ is defined for all positive values of x. For example, $f(1.5)$ equals the number of personal computers shipped in the 12-month period July 1, 1987—June 30, 1988. We can numerically analyze the year-to-year changes in the number of personal computers shipped by computing Δx and Δy for each pair of consecutive table entries, as shown in Fig. 9.

The final column in Fig. 9 gives the rate at which personal computers were changing at various times. For example, the first entry in the final column is 1,224,388. Its interpretation is that between $x = 1$ and $x = 2$ (a period of one year), the number of personal computer shipments was changing at a rate of 1,224,388 computers per year. The second entry in the final column gives the rate of change in the number of personal computers between $x = 2$ and $x = 3$, and so forth.

The data in Fig. 8 only provides information on an annual basis. However, personal computer shipments vary considerably throughout the year. Suppose we wish to determine the rate at which shipments were changing between July 1, 1995 and August 1, 1995. We would need to collect monthly data to make such a computation. Moreover, in this case, the time between consecutive observations would be $\Delta x = \dfrac{1}{12}$; the ratio $\dfrac{\Delta y}{\Delta x}$ would give the rate of change of y per year during the interval from July 1, 1995 to August 1, 1995.

Suppose we wish to determine the rate at which shipments were changing between July 1, 1995 and July 2, 1995. We would need to collect daily data to make such a computation. Moreover, in this case, the time between consecutive observations would be $\Delta x = \dfrac{1}{365}$; the ratio $\dfrac{\Delta y}{\Delta x}$ would give the rate of change of y per year during the interval from July 1, 1995 to July 2, 1995.

The rates of change just considered are often called *average rates of change* since they measure the rate of change as the independent variable ranges over an interval (a year, a week, or a day). By making the length of the interval very short, we can get an instantaneous "snapshot" of how y is changing at a particular value of x. By equation (6), this instantaneous rate of change is just the derivative $f'(a)$.

Note that the quotient $\dfrac{\Delta y}{\Delta x}$ may be used to approximate the derivative, as the following example shows.

▶ Example 3 Estimate the rate of change of personal computer shipments on January 1, 1992.

Solution As our estimate, we may take $\dfrac{\Delta y}{\Delta x}$ when $x = 5$. From Fig. 8, we see that on January 1, 1992, the number of personal computer shipments in the United States was increasing at a rate of 1,054,407 per year. ◆

The technique of using $\dfrac{\Delta y}{\Delta x}$ to approximate the derivative at a particular point is called *numerical differentiation*.

Incorporating Technology Graphing calculators have a numerical derivative function (sometimes called **nDeriv** or **nDer**) that gives approximate derivatives by essentially using a small value of h. Some calculators (such as the TI-85 and TI-86) have a function (called **der1**) that computes exact derivatives. In Fig. 10 the value of $f'(2)$ is first approximated and then calculated exactly for $f(x) = 1/x$. In Fig. 11, $\mathbf{Y_2}$ is specified as the derivative of $f(x)$. In Fig. 12, $\mathbf{Y_2}$ is graphed and TRACE is used to obtain the value of $f'(2)$.

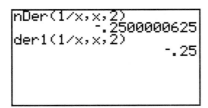

Figure 10. Derivatives on the TI-86.

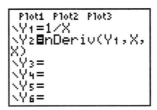

Figure 11. Derivative function on the TI-83.

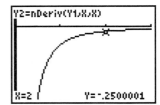

Figure 12. Graph of the function Y_2 from Fig. 9.

Practice Problems 1.3

1. Consider the curve $y = f(x)$ in Fig. 13.
   (a) Find $f(5)$. (b) Find $f'(5)$.

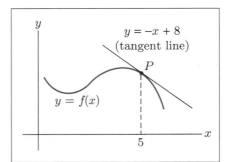

Figure 13.

2. Let $f(x) = 1/x^4$.
   (a) Find its derivative. (b) Find $f'(2)$.

▶ Exercises 1.3

Use (1), (2), and the power rule to find the derivatives of the following functions.

1. $f(x) = 2x - 5$

2. $f(x) = 3 - \frac{1}{2}x$

3. $f(x) = x^8$

4. $f(x) = x^{75}$

5. $f(x) = x^{5/2}$

6. $f(x) = x^{4/3}$

7. $f(x) = \sqrt[3]{x}$

8. $f(x) = x^{3/4}$

9. $f(x) = x^{-2}$

10. $f(x) = 5$

11. $f(x) = x^{-1/4}$

12. $f(x) = x^{-3}$

13. $f(x) = \frac{3}{4}$

14. $f(x) = 1/\sqrt[3]{x}$

15. $f(x) = 1/x^3$

16. $f(x) = 1/x^5$

In Exercises 17–24, find the derivative of $f(x)$ at the designated value of x.

17. $f(x) = x^6$ at $x = -2$

18. $f(x) = x^3$ at $x = \frac{1}{4}$

19. $f(x) = \dfrac{1}{x}$ at $x = 3$

20. $f(x) = 5x$ at $x = 2$

21. $f(x) = 4 - x$ at $x = 5$

22. $f(x) = x^{2/3}$ at $x = 1$

23. $f(x) = x^{3/2}$ at $x = 9$

24. $f(x) = \dfrac{1}{x^2}$ at $x = 2$

25. Find the slope of the curve $y = x^4$ at $x = 3$.

26. Find the slope of the curve $y = x^5$ at $x = -2$.

27. Find the slope of the curve $y = \sqrt{x}$ at $x = 9$.

28. Find the slope of the curve $y = x^{-3}$ at $x = 3$.

29. If $f(x) = x^2$, compute $f(-5)$ and $f'(-5)$.

30. If $f(x) = x + 6$, compute $f(3)$ and $f'(3)$.

31. If $f(x) = 1/x^5$, compute $f(2)$ and $f'(2)$.

32. If $f(x) = 1/x^2$, compute $f(5)$ and $f'(5)$.

33. If $f(x) = x^{4/3}$, compute $f(8)$ and $f'(8)$.

34. If $f(x) = x^{3/2}$, compute $f(16)$ and $f'(16)$.

35. Find the slope of the tangent line to the curve $y = x^3$ at the point $(4, 64)$, and write the equation of this line.

36. Find the slope of the tangent line to the curve $y = \sqrt{x}$ at the point $(25, 5)$, and write the equation of this line.

In Exercises 37–44, find the indicated derivative.

37. $\dfrac{d}{dx}\left(x^8\right)$

38. $\dfrac{d}{dx}\left(x^{-3}\right)$

39. $\dfrac{d}{dx}\left(x^{3/4}\right)$

40. $\dfrac{d}{dx}\left(x^{-1/3}\right)$

41. $\dfrac{dy}{dx}$ if $y = 1$

42. $\dfrac{dy}{dx}$ if $y = x^{-4}$

43. $\dfrac{dy}{dx}$ if $y = x^{1/5}$

44. $\dfrac{dy}{dx}$ if $y = \dfrac{x - 1}{3}$

45. Consider the curve $y = f(x)$ in Fig. 14. Find $f(6)$ and $f'(6)$.

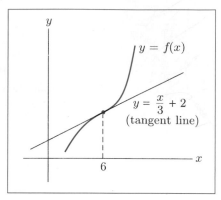

Figure 14.

46. Consider the curve $y = f(x)$ in Fig. 15. Find $f(1)$ and $f'(1)$.

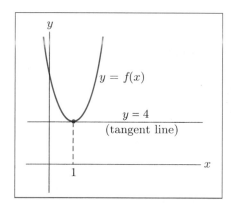

Figure 15.

47. In Fig. 16 the straight line $y = \frac{1}{4}x + b$ is tangent to the graph of $f(x) = \sqrt{x}$. Find the values of a and b.

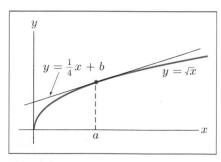

Figure 16.

48. In Fig. 17 the straight line is tangent to the graph of $f(x) = 1/x$. Find the value of a.

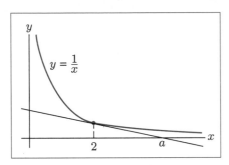

Figure 17.

49. Consider the curve $y = f(x)$ in Fig. 18. Find a and $f(a)$. Estimate $f'(a)$.

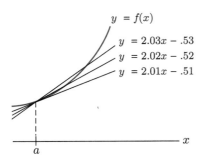

Figure 18.

50. Consider the curve $y = f(x)$ in Fig. 19. Estimate $f'(1)$.

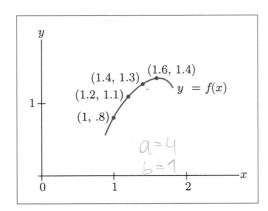

Figure 19.

51. As h approaches 0, what value is approached by

$$\frac{(3+h)^2 - (3)^2}{h}?$$

[*Hint*: This quotient is a secant-line approximation for $f(x) = x^2$.]

52. As h approaches 0, what value is approached by

$$\frac{\sqrt{4+h} - \sqrt{4}}{h}?$$

[*Hint*: This quotient is a secant-line approximation for $f(x) = \sqrt{x}$.]

53. Given that $f(x+h) - f(x) = 2xh + 5h + h^2$, use the secant-line calculation to find $f'(x)$.

54. Given that $g(x+h) - g(x) = 8xh + h + 4h^2$, use the secant-line calculation to find $g'(x)$.

55. Use an argument like those used to verify formulas (3) and (5) to show that the derivative of $f(x) = x^3$ is $3x^2$. [*Hint*: Recall that $(x+h)^3 = x^3 + 3x^2h + 3xh^2 + h^3$.]

56. Use an argument like those used to verify formulas (3) and (5) to show that the derivative of $f(x) = \sqrt{x}$ is $1/(2\sqrt{x})$. [*Hint*: After forming the expression for the slope of a secant line, eliminate the square roots from the numerator by multiplying both the numerator and denominator by the quantity $\sqrt{x+h} + \sqrt{x}$.]

57. In Fig. 20, find the equation of the tangent line to $f(x)$ at the point A.

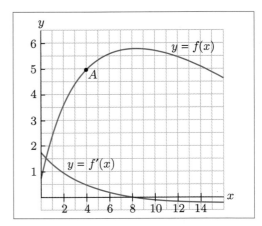

Figure 20.

58. In Fig. 21, find the equation of the tangent line to $f(x)$ at the point P.

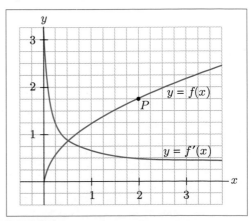

Figure 21.

Technology Exercises

In Exercises 59–64, use a derivative routine to obtain the value of the derivative. Give the value to 5 decimal places.

59. $f'(0)$, where $f(x) = 2^x$

60. $f'(1)$, where $f(x) = \dfrac{1}{1+x^2}$

61. $f'(1)$, where $f(x) = \sqrt{1+x^2}$

62. $f'(3)$, where $f(x) = \sqrt{25-x^2}$

63. $f'(2)$, where $f(x) = \dfrac{x}{1+x}$

64. $f'(0)$, where $f(x) = 10^{1+x}$

In Exercises 65 and 66, let Y_1 be the specified function and use a derivative routine to set Y_2 as its derivative. For instance, you might use $Y_2 = \texttt{nDeriv}(Y_1, X, X)$ or $\texttt{y2} = \texttt{der1}(\texttt{y1}, \texttt{x}, \texttt{x})$. Then graph Y_2 in the specified window and use TRACE to obtain the value of the derivative of Y_1 at $x = 2$.

65. $f(x) = 3x^2 - 5$, $[0, 4]$ by $[-5, 40]$

66. $f(x) = \sqrt{2x}$, $[0, 4]$ by $[-.5, 3]$

In Exercises 67–70, use paper and pencil to find the equation of the tangent line to the graph of the function at the designated point. Then graph both the function and the line to confirm it is indeed the sought-after tangent line.

67. $f(x) = \sqrt{x}$, $(9, 3)$ **68.** $f(x) = \dfrac{1}{x}$, $(.5, 2)$

69. $f(x) = \dfrac{1}{x^2}$, $(.5, 4)$ **70.** $f(x) = x^3$, $(1, 1)$

In Exercises 71–74, $g(x)$ is the tangent line to the graph of $f(x)$ at $x = a$. Graph $f(x)$ and $g(x)$ and determine the value of a.

71. $f(x) = \dfrac{5}{x}$, $g(x) = 5 - 1.25x$

72. $f(x) = \sqrt{8x}$, $g(x) = x + 2$

73. $f(x) = x^3 - 12x^2 + 46x - 50$, $g(x) = 14 - 2x$

74. $f(x) = (x-3)^3 + 4$, $g(x) = 4$

Suppose that $y = x^2 - x + 4$. Calculate Δy if

75. x increases from 1 to 1.1.

76. x decreases from 1 to .99.

77. x increases by .02 from the value 3.

78. x decreases by .1 from the value 3.

Calculate the average rate of change of y per unit change in x, as x changes from -1 to -1.01, provided that

79. $y = 3x - 2$ **80.** $y = -3x^2 - 1$

81. $y = \dfrac{1}{x}$ $(x \neq 0)$ **82.** $y = x^3$

Let $y = -x^2 + 3x - 4$. Suppose that x increases from 2 to $2 + \Delta x$. Calculate $\dfrac{\Delta y}{\Delta x}$ in case

83. $\Delta x = .1$ **84.** $\Delta x = .01$

85. $\Delta x = .001$ **86.** $\Delta x = .00005$

87. General Δx

88. Use the result of the preceding exercise to determine the value of $\dfrac{dy}{dx}\Big|_{x=2}$.

Exercises 89–92 refer to Table 1.

89. Make a table of $\dfrac{\Delta y}{\Delta x}$ for x varying from one data item to the next. (A spreadsheet or a graphing calculator would be helpful here.)

90. For which values of x is $\dfrac{\Delta y}{\Delta x}$ negative? Interpret this in terms of the data.

91. Determine the rate at which the Dow Jones Industrial Average was changing on Dec. 31, 1990.

92. Estimate $\dfrac{dy}{dx}\Big|_{x=7}$ and give an interpretation of this number.

| **Solutions to Practice Problems 1.3** | **1.** (a) The number $f(5)$ is the y-coordinate of the point P. Since the tangent line passes through P, the coordinates of P satisfy the equation $y = -x + 8$. Since its x-coordinate is 5, its y-coordinate is $-5 + 8 = 3$. Therefore, $f(5) = 3$. |

      (b) The number $f'(5)$ is the slope of the tangent line at P, which is readily seen to be -1.

2. (a) The function $1/x^4$ can be written as the power function x^{-4}. Here $r = -4$. Therefore,

$$f'(x) = (-4)x^{(-4)-1} = -4x^{-5} = \frac{-4}{x^5}.$$

      (b) $f'(2) = -4/2^5 = -4/32 = -\frac{1}{8}$.

| Table 1 | Dow Jones Industrial Average on Dec. 31 |

x = year with 0 corresponding to 1970
y = closing Dow Jones Industrial Average on Dec. 31 of year x

Year	x	y	Year	x	y	Year	x	y
1970	0	$838.92	1980	10	$963.99	1990	20	$2,633.66
1971	1	$890.20	1981	11	$875.00	1991	21	$3,168.83
1972	2	$1,020.02	1982	12	$1,046.54	1992	22	$3,301.11
1973	3	$850.86	1983	13	$1,258.64	1993	23	$3,754.09
1974	4	$616.24	1984	14	$1,211.57	1994	24	$3,834.44
1975	5	$852.41	1985	15	$1,546.47	1995	25	$5,117.12
1976	6	$1,004.65	1986	16	$1,895.95	1996	26	$6,448.27
1977	7	$831.17	1987	17	$1,938.83	1997	27	$7,908.25
1978	8	$805.01	1988	18	$2,168.57			
1979	9	$838.74	1989	19	$2,753.20			

1.4 Limits and the Derivative

The notion of a limit is one of the fundamental ideas of calculus. Indeed, any "theoretical" development of calculus rests on an extensive use of the theory of limits. Even in this book, where we have adopted an intuitive viewpoint, limit arguments are used occasionally (although in an informal way). In this section we give a brief introduction to limits and their role in calculus. As we shall see, the limit concept will allow us to define the notion of a derivative independently of our geometric reasoning.

Actually, we have already considered a limit in our discussion of the derivative, although we did not use the term "limit." Using the geometric reasoning of the previous section, we have the following procedure for calculating the derivative of a function $f(x)$ at $x = a$. First calculate the *difference quotient*

$$\frac{f(a + h) - f(a)}{h},$$

where h is a nonzero number. Next, allow h to approach zero by allowing it to assume both positive and negative numbers arbitrarily close to zero but different from zero. In symbols, we write $h \to 0$. The values of the difference quotient then approach the value of the derivative $f'(a)$. We say that the number $f'(a)$ is the *limit* of the difference quotient as h approaches zero, and in symbols we write

$$f'(a) = \lim_{h \to 0} \frac{f(a + h) - f(a)}{h}. \tag{1}$$

As a numerical example of formula (1), consider the case of the derivative of $f(x) = x^2$ at $x = 2$. The difference quotient in this case has the form

$$\frac{f(2 + h) - f(2)}{h} = \frac{(2 + h)^2 - 2^2}{h}.$$

Table 1	Values of a Difference Quotient as h Approaches 0		
h	$\dfrac{(2+h)^2 - 2^2}{h}$	h	$\dfrac{(2+h)^2 - 2^2}{h}$
1	$\dfrac{(2+1)^2 - 2^2}{1} = 5$	-1	$\dfrac{(2+(-1))^2 - 2^2}{-1} = 3$
.1	$\dfrac{(2+.1)^2 - 2^2}{.1} = 4.10$	$-.1$	$\dfrac{(2+(-.1))^2 - 2^2}{-.1} = 3.9$
.01	$\dfrac{(2+.01)^2 - 2^2}{.01} = 4.01$	$-.01$	$\dfrac{(2+(-.01))^2 - 2^2}{-.01} = 3.99$
.001	$\dfrac{(2+.001)^2 - 2^2}{.001} = 4.001$	$-.001$	$\dfrac{(2+(-.001))^2 - 2^2}{-.001} = 3.999$
.0001	$\dfrac{(2+.0001)^2 - 2^2}{.0001} = 4.0001$	$-.0001$	$\dfrac{(2+(-.0001))^2 - 2^2}{-.0001} = 3.9999$

Table 1 gives some typical values of this difference quotient for progressively smaller values of h, both positive and negative. It is clear that the values of the difference quotient are approaching 4 as $h \to 0$. In other words, 4 is the *limit* of the difference quotient as $h \to 0$. Thus

$$\lim_{h \to 0} \frac{(2+h)^2 - 2^2}{h} = 4.$$

Since the values of the difference quotient approach the derivative $f'(2)$, we conclude that $f'(2) = 4$.

Our discussion of the derivative has been based on an intuitive geometric concept of the tangent line. However, the limit on the right in (1) may be considered independently of its geometric interpretation. In fact, we may use (1) to *define* $f'(a)$. We say that f is *differentiable* at $x = a$ if

$$\frac{f(a+h) - f(a)}{h}$$

approaches some number as $h \to 0$, and we denote this limiting number by $f'(a)$. If the difference quotient

$$\frac{f(a+h) - f(a)}{h}$$

does not approach any specific number as $h \to 0$, we say that f is *nondifferentiable* at $x = a$. Essentially all the functions in this text are differentiable at all points in their domain. A few exceptions are described in Section 1.5.

To better understand the limit concept used to define the derivative, it will be helpful to look at limits in a more general setting. If we let

$$g(h) = \frac{(2+h)^2 - 2^2}{h},$$

then Table 1 gives values of $g(h)$ as $h \to 0$. These values obviously approach 4. We may express this by writing

$$\lim_{h \to 0} g(h) = 4.$$

(b) $\lim\limits_{x\to 2} g(x)$ does not exist. As x approaches 2 from the right, $g(x)$ approaches 2. However, as x approaches 2 from the left, $g(x)$ approaches 1. In order for a limit to exist, the function must approach the *same* value from each direction.

(c) $\lim\limits_{x\to 2} g(x)$ does not exist. As x approaches 2, the values of $g(x)$ become larger and larger and do not approach a fixed number. ◆

The following limit theorems, which we cite without proof, allow us to reduce the computation of limits for combinations of functions to computations of limits involving the constituent functions.

Limit Theorems Suppose that $\lim\limits_{x\to a} f(x)$ and $\lim\limits_{x\to a} g(x)$ both exist. Then we have the following results.

(I) If k is a constant, then $\lim\limits_{x\to a} k \cdot f(x) = k \cdot \lim\limits_{x\to a} f(x)$.

(II) If r is a positive constant, then $\lim\limits_{x\to a} [f(x)]^r = \left[\lim\limits_{x\to a} f(x)\right]^r$.

(III) $\lim\limits_{x\to a} [f(x) + g(x)] = \lim\limits_{x\to a} f(x) + \lim\limits_{x\to a} g(x)$.

(IV) $\lim\limits_{x\to a} [f(x) - g(x)] = \lim\limits_{x\to a} f(x) - \lim\limits_{x\to a} g(x)$.

(V) $\lim\limits_{x\to a} [f(x) \cdot g(x)] = \left[\lim\limits_{x\to a} f(x)\right] \cdot \left[\lim\limits_{x\to a} g(x)\right]$.

(VI) If $\lim\limits_{x\to a} g(x) \neq 0$, then $\lim\limits_{x\to a} \dfrac{f(x)}{g(x)} = \dfrac{\lim\limits_{x\to a} f(x)}{\lim\limits_{x\to a} g(x)}$.

▶ **Example 3** Use the limit theorems to compute the following limits.

(a) $\lim\limits_{x\to 2} x^3$ (b) $\lim\limits_{x\to 2} 5x^3$ (c) $\lim\limits_{x\to 2} (5x^3 - 15)$

(d) $\lim\limits_{x\to 2} \sqrt{5x^3 - 15}$ (e) $\lim\limits_{x\to 2} \left(\sqrt{5x^3 - 15}/x^5\right)$

Solution (a) Since $\lim\limits_{x\to 2} x = 2$, we have by Limit Theorem II that

$$\lim_{x\to 2} x^3 = \left(\lim_{x\to 2} x\right)^3 = 2^3 = 8.$$

(b) $\lim\limits_{x\to 2} 5x^3 = 5 \lim\limits_{x\to 2} x^3$ (Limit Theorem I with $k = 5$)

$\qquad\qquad = 5 \cdot 8$ [by part (a)]

$\qquad\qquad = 40.$

(c) $\lim\limits_{x\to 2} (5x^3 - 15) = \lim\limits_{x\to 2} 5x^3 - \lim\limits_{x\to 2} 15$ (Limit Theorem IV).

Note that $\lim\limits_{x\to 2} 15 = 15$. This is because the constant function $g(x) = 15$ always has the value 15, and so its limit as x approaches *any* number is 15. By part (b), $\lim\limits_{x\to 2} 5x^3 = 40$. Thus

$$\lim_{x\to 2} (5x^3 - 15) = 40 - 15 = 25.$$

The preceding discussion suggests the following definition: Let $g(x)$ be a function, a a number. We say that the number L *is the limit of* $g(x)$ *as* x *approaches* a provided that $g(x)$ can be made arbitrarily close to L by taking x sufficiently close (but not equal) to a. In this case we write

$$\lim_{x \to a} g(x) = L.$$

If, as x approaches a, the values $g(x)$ do *not* approach a specific number, then we say that the limit of $g(x)$ as x approaches a *does not exist*. Let us give some further examples of limits.

▶ **Example 1**   Determine $\lim_{x \to 2} (3x - 5)$.

Solution   Let us make a table of values of x approaching 2 and the corresponding values of $3x - 5$:

x	$3x - 5$	x	$3x - 5$
2.1	1.3	1.9	.7
2.01	1.03	1.99	.97
2.001	1.003	1.999	.997
2.0001	1.0003	1.9999	.9997

As x approaches 2, we see that $3x - 5$ approaches 1. In terms of our notation,

$$\lim_{x \to 2} (3x - 5) = 1. \qquad \blacklozenge$$

▶ **Example 2**   For each of the functions in Fig. 1, determine if $\lim_{x \to 2} g(x)$ exists. (The circles drawn on the graphs are meant to represent breaks in the graph, indicating that the functions under consideration are not defined at $x = 2$.)

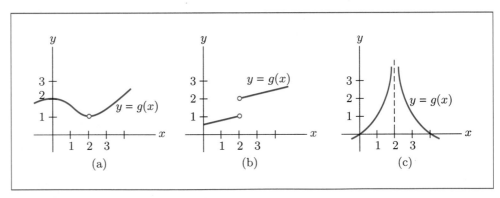

Figure 1. Functions with an undefined point.

Solution   (a) $\lim_{x \to 2} g(x) = 1$. We can see that as x gets closer and closer to 2, the values of $g(x)$ get closer and closer to 1. This is true for values of x to both the right and the left of 2.

| Table 1 | Values of a Difference Quotient as h Approaches 0 |

h	$\dfrac{(2+h)^2 - 2^2}{h}$	h	$\dfrac{(2+h)^2 - 2^2}{h}$
1	$\dfrac{(2+1)^2 - 2^2}{1} = 5$	-1	$\dfrac{(2+(-1))^2 - 2^2}{-1} = 3$
.1	$\dfrac{(2+.1)^2 - 2^2}{.1} = 4.10$	$-.1$	$\dfrac{(2+(-.1))^2 - 2^2}{-.1} = 3.9$
.01	$\dfrac{(2+.01)^2 - 2^2}{.01} = 4.01$	$-.01$	$\dfrac{(2+(-.01))^2 - 2^2}{-.01} = 3.99$
.001	$\dfrac{(2+.001)^2 - 2^2}{.001} = 4.001$	$-.001$	$\dfrac{(2+(-.001))^2 - 2^2}{-.001} = 3.999$
.0001	$\dfrac{(2+.0001)^2 - 2^2}{.0001} = 4.0001$	$-.0001$	$\dfrac{(2+(-.0001))^2 - 2^2}{-.0001} = 3.9999$

Table 1 gives some typical values of this difference quotient for progressively smaller values of h, both positive and negative. It is clear that the values of the difference quotient are approaching 4 as $h \to 0$. In other words, 4 is the *limit* of the difference quotient as $h \to 0$. Thus

$$\lim_{h \to 0} \frac{(2+h)^2 - 2^2}{h} = 4.$$

Since the values of the difference quotient approach the derivative $f'(2)$, we conclude that $f'(2) = 4$.

Our discussion of the derivative has been based on an intuitive geometric concept of the tangent line. However, the limit on the right in (1) may be considered independently of its geometric interpretation. In fact, we may use (1) to *define* $f'(a)$. We say that f is *differentiable* at $x = a$ if

$$\frac{f(a+h) - f(a)}{h}$$

approaches some number as $h \to 0$, and we denote this limiting number by $f'(a)$. If the difference quotient

$$\frac{f(a+h) - f(a)}{h}$$

does not approach any specific number as $h \to 0$, we say that f is *nondifferentiable* at $x = a$. Essentially all the functions in this text are differentiable at all points in their domain. A few exceptions are described in Section 1.5.

To better understand the limit concept used to define the derivative, it will be helpful to look at limits in a more general setting. If we let

$$g(h) = \frac{(2+h)^2 - 2^2}{h},$$

then Table 1 gives values of $g(h)$ as $h \to 0$. These values obviously approach 4. We may express this by writing

$$\lim_{h \to 0} g(h) = 4.$$

Table 1	Dow Jones Industrial Average on Dec. 31

x = year with 0 corresponding to 1970
y = closing Dow Jones Industrial Average on Dec. 31 of year x

Year	x	y	Year	x	y	Year	x	y
1970	0	$838.92	1980	10	$963.99	1990	20	$2,633.66
1971	1	$890.20	1981	11	$875.00	1991	21	$3,168.83
1972	2	$1,020.02	1982	12	$1,046.54	1992	22	$3,301.11
1973	3	$850.86	1983	13	$1,258.64	1993	23	$3,754.09
1974	4	$616.24	1984	14	$1,211.57	1994	24	$3,834.44
1975	5	$852.41	1985	15	$1,546.47	1995	25	$5,117.12
1976	6	$1,004.65	1986	16	$1,895.95	1996	26	$6,448.27
1977	7	$831.17	1987	17	$1,938.83	1997	27	$7,908.25
1978	8	$805.01	1988	18	$2,168.57			
1979	9	$838.74	1989	19	$2,753.20			

1.4 Limits and the Derivative

The notion of a limit is one of the fundamental ideas of calculus. Indeed, any "theoretical" development of calculus rests on an extensive use of the theory of limits. Even in this book, where we have adopted an intuitive viewpoint, limit arguments are used occasionally (although in an informal way). In this section we give a brief introduction to limits and their role in calculus. As we shall see, the limit concept will allow us to define the notion of a derivative independently of our geometric reasoning.

Actually, we have already considered a limit in our discussion of the derivative, although we did not use the term "limit." Using the geometric reasoning of the previous section, we have the following procedure for calculating the derivative of a function $f(x)$ at $x = a$. First calculate the *difference quotient*

$$\frac{f(a+h) - f(a)}{h},$$

where h is a nonzero number. Next, allow h to approach zero by allowing it to assume both positive and negative numbers arbitrarily close to zero but different from zero. In symbols, we write $h \to 0$. The values of the difference quotient then approach the value of the derivative $f'(a)$. We say that the number $f'(a)$ is the *limit* of the difference quotient as h approaches zero, and in symbols we write

$$f'(a) = \lim_{h \to 0} \frac{f(a+h) - f(a)}{h}. \tag{1}$$

As a numerical example of formula (1), consider the case of the derivative of $f(x) = x^2$ at $x = 2$. The difference quotient in this case has the form

$$\frac{f(2+h) - f(2)}{h} = \frac{(2+h)^2 - 2^2}{h}.$$

(d) $\displaystyle\lim_{x\to 2} \sqrt{5x^3 - 15} = \lim_{x\to 2}(5x^3 - 15)^{1/2}$

$$= \left[\lim_{x\to 2}(5x^3 - 15)\right]^{1/2} \quad \begin{bmatrix}\text{Limit Theorem II with} \\ r = \frac{1}{2},\ f(x) = 5x^3 - 15\end{bmatrix}$$

$$= 25^{1/2} \qquad\qquad \left[\text{by part (c)}\right]$$

$$= 5.$$

(e) The limit of the denominator is $\displaystyle\lim_{x\to 2} x^5$, which is $2^5 = 32$, a nonzero number. So by Limit Theorem VI we have

$$\lim_{x\to 2} \frac{\sqrt{5x^3 - 15}}{x^5} = \frac{\displaystyle\lim_{x\to 2}\sqrt{5x^3 - 15}}{\displaystyle\lim_{x\to 2} x^5}$$

$$= \frac{5}{32} \quad \left[\text{by part (d)}\right]. \qquad\qquad \blacklozenge$$

The following facts, which may be deduced by repeated applications of the various limit theorems, are extremely handy in evaluating limits.

Limit of a Polynomial Function   Let $p(x)$ be a polynomial function, a any number. Then

$$\lim_{x\to a} p(x) = p(a).$$

Limit of a Rational Function   Let $r(x) = p(x)/q(x)$ be a rational function, where $p(x)$ and $q(x)$ are polynomials. Let a be a number such that $q(a) \neq 0$. Then

$$\lim_{x\to a} r(x) = r(a).$$

In other words, to determine a limit of a polynomial or a rational function, simply evaluate the function at $x = a$, provided, of course, that the function is defined at $x = a$. For instance, we can rework the solution to Example 3(c) as follows:

$$\lim_{x\to 2}(5x^3 - 15) = 5(2)^3 - 15 = 25.$$

Many situations require algebraic simplifications before the limit theorems can be applied.

▶ **Example 4**   Compute the following limits.

(a) $\displaystyle\lim_{x\to 3} \frac{x^2 - 9}{x - 3}$ (b) $\displaystyle\lim_{x\to 0} \frac{\sqrt{x+4} - 2}{x}$

Solution   (a) The function $\dfrac{x^2 - 9}{x - 3}$ is not defined when $x = 3$, since

$$\frac{3^2 - 9}{3 - 3} = \frac{0}{0},$$

which is undefined. That causes no difficulty, since the limit as x approaches 3 depends only on the values of x *near* 3 and excludes considerations of the value at $x = 3$ itself. To evaluate the limit, note that $x^2 - 9 = (x-3)(x+3)$. So for $x \neq 3$,

$$\frac{x^2 - 9}{x - 3} = \frac{(x-3)(x+3)}{x-3} = x + 3.$$

As x approaches 3, $x + 3$ approaches 6. Therefore,

$$\lim_{x \to 3} \frac{x^2 - 9}{x - 3} = 6.$$

(b) Since the denominator approaches zero when taking the limit, we may not apply Limit Theorem VI directly. However, if we first apply an algebraic trick, the limit may be evaluated. Multiply numerator and denominator by $\sqrt{x + 4} + 2$.

$$\frac{\sqrt{x+4} - 2}{x} \cdot \frac{\sqrt{x+4} + 2}{\sqrt{x+4} + 2} = \frac{(x+4) - 4}{x(\sqrt{x+4} + 2)}$$

$$= \frac{x}{x(\sqrt{x+4} + 2)}$$

$$= \frac{1}{\sqrt{x+4} + 2}$$

Thus

$$\lim_{x \to 0} \frac{\sqrt{x+4} - 2}{x} = \lim_{x \to 0} \frac{1}{\sqrt{x+4} + 2}$$

$$= \frac{\displaystyle\lim_{x \to 0} 1}{\displaystyle\lim_{x \to 0} (\sqrt{x+4} + 2)} \qquad \text{(Limit Theorem VI)}$$

$$= \frac{1}{4}. \qquad\qquad\qquad\qquad\qquad\qquad \blacklozenge$$

The basic differentiation rules are obtained from the limit definition of the derivative. The next example (and Exercises 29–40) illustrates this process. There are three main steps.

Using Limits to Calculate a Derivative

1. Write the difference quotient $\dfrac{f(a+h) - f(a)}{h}$.

2. Simplify the difference quotient.

3. Find the limit as $h \to 0$.

▶ **Example 5** Use limits to compute the derivative $f'(5)$ for the following functions.

(a) $f(x) = 15 - x^2$ 　　　　　　　　(b) $f(x) = \dfrac{1}{2x - 3}$

Solution 　In each case, we must calculate $\displaystyle\lim_{h \to 0} \dfrac{f(5+h) - f(5)}{h}$.

(a) $\dfrac{f(5+h)-f(5)}{h} = \dfrac{\left[15-(5+h)^2\right]-\left(15-5^2\right)}{h}$ (*step 1*)

$$= \dfrac{15-\left(25+10h+h^2\right)-\left(15-25\right)}{h} \qquad (\textit{step 2})$$

$$= \dfrac{-10h-h^2}{h} = -10-h.$$

Therefore, $f'(5) = \lim\limits_{h\to 0}(-10-h) = -10.$ (*step 3*)

(b) $\dfrac{f(5+h)-f(5)}{h} = \dfrac{\dfrac{1}{2(5+h)-3} - \dfrac{1}{2(5)-3}}{h}$ (*step 1*)

$$= \dfrac{\dfrac{1}{7+2h} - \dfrac{1}{7}}{h} = \dfrac{\dfrac{7-(7+2h)}{(7+2h)7}}{h} \qquad (\textit{step 2})$$

$$= \dfrac{-2h}{(7+2h)7\cdot h} = \dfrac{-2}{(7+2h)7} = \dfrac{-2}{49+14h}.$$

$$f'(5) = \lim_{h\to 0}\dfrac{-2}{49+14h} = -\dfrac{2}{49} \qquad (\textit{step 3})$$

◆

Remark   When computing the limit in Example 5, we considered only values of h near zero (and not $h = 0$ itself). Therefore, we were freely able to divide both numerator and denominator by h.

Infinity and Limits   Consider the function $f(x)$ whose graph is sketched in Fig. 2. As x grows large the value of $f(x)$ approaches 2. In this circumstance, we say that 2 is the *limit of* $f(x)$ *as* x *approaches infinity*. Infinity is denoted

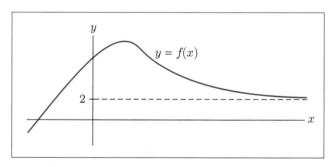

Figure 2.   Function with a limit as x approaches infinity.

by the symbol ∞. The preceding limit statement is expressed in the following notation:

$$\lim_{x\to\infty} f(x) = 2.$$

In a similar vein, consider the function whose graph is sketched in Fig. 3. As x grows large in the negative direction, the value of $f(x)$ approaches 0. In this circumstance, we say that 0 is the *limit of* $f(x)$ *as* x *approaches minus infinity*. In symbols,

$$\lim_{x\to-\infty} f(x) = 0.$$

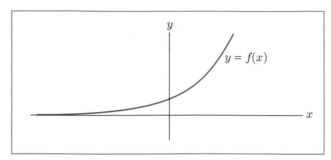

Figure 3. Function with a limit as x approaches minus infinity.

▶ Example 6   Calculate the following limits.

(a) $\displaystyle\lim_{x\to\infty}\frac{1}{x^2+1}$ (b) $\displaystyle\lim_{x\to\infty}\frac{6x-1}{2x+1}$

Solution   (a) As x increases without bound, so does $x^2 + 1$. Therefore, $1/(x^2+1)$ approaches zero as x approaches ∞.

(b) Both $6x - 1$ and $2x + 1$ increase without bound as x does. To determine the limit of their quotient, we employ an algebraic trick. Divide both numerator and denominator by x to obtain

$$\lim_{x\to\infty}\frac{6x-1}{2x+1}=\lim_{x\to\infty}\frac{6-\dfrac{1}{x}}{2+\dfrac{1}{x}}.$$

As x increases without bound, $1/x$ approaches zero, so $6 - (1/x)$ approaches 6 and $2 + (1/x)$ approaches 2. Thus the desired limit is $6/2 = 3$. ◆

Incorporating Technology   Figure 4 shows the graph of the function

$$\frac{x^2-9}{x-3}$$

from Example 4(a) with the window setting $[0,6]$ *by* $[-2,9]$. (Notice that 3 is in the middle of the x-range.) If you look closely, you will see a little blank space at the point where $x = 3$. However, if you examine function values near $x = 3$, you will be convinced that

$$\lim_{x\to3}\frac{x^2-9}{x-3}=6.$$

Figure 5(a) shows the graph of the function

$$\frac{6x-1}{2x+1}$$

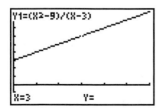

Figure 4.

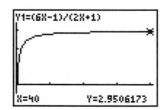

Figure 5(a)

X	Y1
1000	2.998
2000	2.999
3000	2.9993
4000	2.9995
5000	2.9996
6000	2.9997
7000	2.9997

Y1=2.99971430612

Figure 5(b)

from Example 6(b) with the window $[0, 41]$ *by* $[-1, 4]$, and Fig. 5(b) shows the function values for increasing values of x. These figures provide convincing evidence that

$$\lim_{x \to \infty} \frac{6x - 1}{2x + 1} = 3.$$

**Practice Problems
1.4**

Determine which of the following limits exist. Compute the limits that exist.

1. $\lim_{x \to 6} \dfrac{x^2 - 4x - 12}{x - 6}$

2. $\lim_{x \to 6} \dfrac{4x + 12}{x - 6}$

▶ Exercises 1.4

For each of the following functions $g(x)$, determine whether or not $\lim_{x \to 3} g(x)$ exists. If so, give the limit.

1.

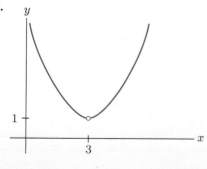

2.

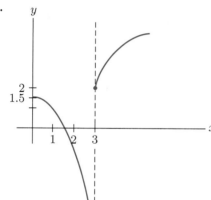

3.

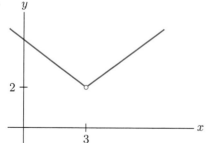

4.

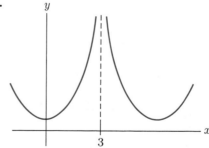

5.

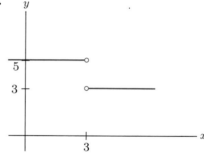

6.

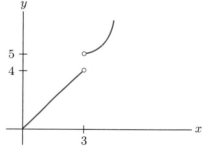

Determine which of the following limits exist. Compute the limits that exist.

7. $\lim_{x \to 1} (1 - 6x)$

8. $\lim\limits_{x \to 2} \dfrac{x}{x-2}$

9. $\lim\limits_{x \to 3} \sqrt{x^2 + 16}$

10. $\lim\limits_{x \to 4} \left(x^3 - 7 \right)$

11. $\lim\limits_{x \to 5} \dfrac{x^2 + 1}{5 - x}$

12. $\lim\limits_{x \to 6} \left(\sqrt{6x} + 3x - \dfrac{1}{x} \right) \left(x^2 - 4 \right)$

13. $\lim\limits_{x \to 7} \left(x + \sqrt{x - 6} \right) \left(x^2 - 2x + 1 \right)$

14. $\lim\limits_{x \to 8} \dfrac{\sqrt{5x - 4} - 1}{3x^2 + 2}$

15. $\lim\limits_{x \to 9} \dfrac{\sqrt{x^2 - 5x - 36}}{8 - 3x}$

16. $\lim\limits_{x \to 10} \left(2x^2 - 15x - 50 \right)^{20}$

17. $\lim\limits_{x \to 0} \dfrac{x^2 + 3x}{x}$

18. $\lim\limits_{x \to 1} \dfrac{x^2 - 1}{x - 1}$

19. $\lim\limits_{x \to 2} \dfrac{-2x^2 + 4x}{x - 2}$

20. $\lim\limits_{x \to 3} \dfrac{x^2 - x - 6}{x - 3}$

21. $\lim\limits_{x \to 4} \dfrac{x^2 - 16}{4 - x}$

22. $\lim\limits_{x \to 5} \dfrac{2x - 10}{x^2 - 25}$

23. $\lim\limits_{x \to 6} \dfrac{x^2 - 6x}{x^2 - 5x - 6}$

24. $\lim\limits_{x \to 7} \dfrac{x^3 - 2x^2 + 3x}{x^2}$

25. $\lim\limits_{x \to 8} \dfrac{x^2 + 64}{x - 8}$

26. $\lim\limits_{x \to 9} \dfrac{1}{(x - 9)^2}$

27. $\lim\limits_{x \to 0} \dfrac{-2}{\sqrt{x + 16} + 7}$

28. $\lim\limits_{x \to 0} \dfrac{4x}{x(x^2 + 3x + 5)}$

Use limits to compute the following derivatives.

29. $f'(3)$ where $f(x) = x^2 + 1$

30. $f'(2)$ where $f(x) = x^3$

31. $f'(0)$ where $f(x) = x^3 + 3x + 1$

32. $f'(0)$ where $f(x) = x^2 + 2x + 2$

33. $f'(3)$ where $f(x) = \dfrac{1}{2x + 5}$

34. $f'(4)$ where $f(x) = \sqrt{2x - 1}$

35. $f'(2)$ where $f(x) = \sqrt{5 - x}$

36. $f'(3)$ where $f(x) = \dfrac{1}{7 - 2x}$

37. $f'(0)$ where $f(x) = \sqrt{1 - x^2}$

38. $f'(2)$ where $f(x) = (5x - 4)^2$

39. $f'(0)$ where $f(x) = (x + 1)^3$

40. $f'(0)$ where $f(x) = \sqrt{x^2 + x + 1}$

Each limit in Exercises 41–46 is a definition of $f'(a)$. Determine the function $f(x)$ and the value of a.

41. $\lim\limits_{h \to 0} \dfrac{(9 + h)^{1/2} - 3}{h}$

42. $\lim\limits_{h \to 0} \dfrac{(2 + h)^3 - 8}{h}$

43. $\lim\limits_{h \to 0} \dfrac{\dfrac{1}{10 + h} - .1}{h}$

44. $\lim\limits_{h \to 0} \dfrac{(64 + h)^{1/3} - 4}{h}$

45. $\lim\limits_{h \to 0} \dfrac{(3(1 + h)^2 + 4) - 7}{h}$

46. $\lim\limits_{h \to 0} \dfrac{(1 + h)^{-1/2} - 1}{h}$

Compute the following limits.

47. $\lim\limits_{x \to \infty} \dfrac{1}{x^2}$

48. $\lim\limits_{x \to -\infty} \dfrac{1}{x^2}$

49. $\lim\limits_{x \to \infty} \dfrac{1}{x - 8}$

50. $\lim\limits_{x \to \infty} \dfrac{1}{3x + 5}$

51. $\lim\limits_{x \to \infty} \dfrac{2x + 1}{x + 2}$

52. $\lim\limits_{x \to \infty} \dfrac{x^2 + x}{x^2 - 1}$

Technology Exercises

Examine the graph of the function and evaluate the function at large values of x to guess the value of the limit.

53. $\lim\limits_{x \to \infty} \sqrt{25 + x} - \sqrt{x}$

54. $\lim\limits_{x \to \infty} \dfrac{x^2}{2^x}$

55. $\lim\limits_{x \to \infty} \dfrac{x^2 - 2x + 3}{2x^2 + 1}$

56. $\lim\limits_{x \to \infty} \dfrac{-8x^2 + 1}{x^2 + 1}$

Solutions to Practice Problems 1.4

1. The function under consideration is a rational function. Since the denominator has value 0 at $x = 6$, we cannot immediately determine the limit by just evaluating the function at $x = 6$. Also, $\lim\limits_{x \to 6} (x - 6) = 0$. Since the function in the denominator has the limit zero, we cannot apply Limit Theorem VI. However, since the definition of limit considers only values of x different from 6, the quotient can be simplified by factoring and canceling.

$$\frac{x^2 - 4x - 12}{x - 6} = \frac{(x + 2)(x - 6)}{(x - 6)} = x + 2 \qquad \text{for } x \neq 6.$$

Now $\lim\limits_{x \to 6} (x + 2) = 8$. Therefore,

$$\lim\limits_{x \to 6} \frac{x^2 - 4x - 12}{x - 6} = 8.$$

2. No limit exists. It is easily seen that $\lim\limits_{x \to 6}(4x + 12) = 36$ and $\lim\limits_{x \to 6}(x - 6) = 0$. As x approaches 6, the denominator gets very small and the numerator approaches 36. For example, if $x = 6.00001$, then the numerator is 36.00004 and the denominator is .00001. The quotient is 3,600,004. As x approaches 6 even more closely, the quotient gets arbitrarily large and cannot possibly approach a limit.

1.5 Differentiability and Continuity

In the preceding section we defined differentiability of $f(x)$ at $x = a$ in terms of a limit. If this limit does not exist, then we say that $f(x)$ is *nondifferentiable* at $x = a$. Geometrically, the nondifferentiability of $f(x)$ at $x = a$ can manifest itself in several different ways. First of all, the graph of $f(x)$ could have no tangent line at $x = a$. Second, the graph could have a vertical tangent line at $x = a$. (Recall that slope is not defined for vertical lines.) Some of the various geometric possibilities are illustrated in Fig. 1.

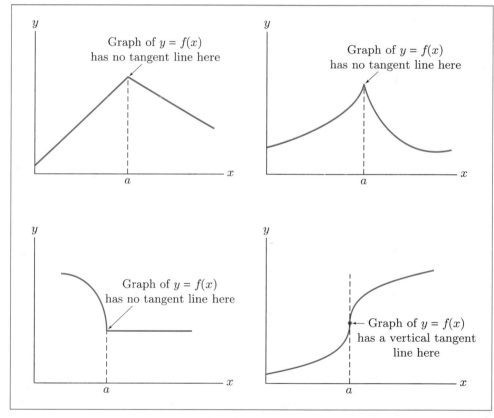

Figure 1. Functions that are nondifferentiable at x = a.

The following example illustrates how nondifferentiable functions can arise in practice.

▶ Example 1   A railroad company charges $10 per mile to haul a boxcar up to 200 miles and $8 per mile for each mile exceeding 200. In addition, the railroad charges a $1000 handling charge per boxcar. Graph the cost of sending a boxcar x miles.

Solution   If x is at most 200 miles, then the cost $C(x)$ is given by $C(x) = 1000 + 10x$ dollars. The cost for 200 miles is $C(200) = 1000 + 2000 = 3000$ dollars. If x

exceeds 200 miles, then the total cost will be

$$C(x) = \underbrace{3000}_{\substack{\text{cost of first} \\ \text{200 miles}}} + \underbrace{8(x - 200)}_{\substack{\text{cost of miles in} \\ \text{excess of 200}}} = 1400 + 8x.$$

Thus

$$C(x) = \begin{cases} 1000 + 10x, & \text{for } 0 < x \le 200, \\ 1400 + 8x, & \text{for } x > 200. \end{cases}$$

The graph of $C(x)$ is sketched in Fig. 2. Note that $C(x)$ is not differentiable at $x = 200$. ◆

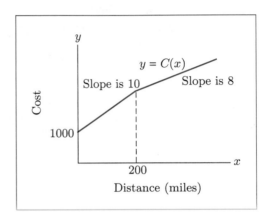

Figure 2. Cost of hauling a boxcar.

Closely related to the concept of differentiability is that of continuity. We say that a function $f(x)$ is *continuous* at $x = a$ provided that, roughly speaking, its graph has no breaks (or gaps) as it passes through the point $(a, f(a))$. That is, $f(x)$ is continuous at $x = a$ provided that we can draw the graph through $(a, f(a))$ without lifting our pencil from the paper. The functions whose graphs are drawn in Figs. 1 and 2 are continuous for all values of x. By contrast, however, the function whose graph is drawn in Fig. 3(a) is not continuous (we say it is *discontinuous*) at $x = 1$ and $x = 2$, since the graph has breaks there. Similarly, the function whose graph is drawn in Fig. 3(b) is discontinuous at $x = 2$.

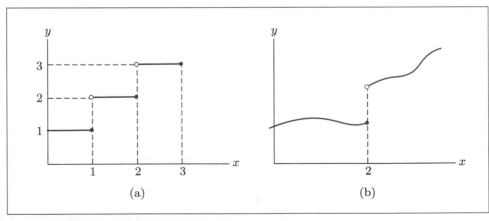

Figure 3. Functions with discontinuities.

Discontinuous functions can occur in applications, as the following example shows.

▶ Example 2   Suppose that a manufacturing plant is capable of producing 15,000 units in one shift of 8 hours. For each shift worked, there is a fixed cost of $2000 (for light, heat, etc.). Suppose that the variable cost (the cost of labor and raw materials) is $2 per unit. Graph the cost $C(x)$ of manufacturing x units.

Solution   If $x \leq 15{,}000$, a single shift will suffice, so that

$$C(x) = 2000 + 2x, \qquad 0 \leq x \leq 15{,}000.$$

If x is between 15,000 and 30,000, one extra shift will be required, and

$$C(x) = 4000 + 2x, \qquad 15{,}000 < x \leq 30{,}000.$$

If x is between 30,000 and 45,000, the plant will need to work three shifts, and

$$C(x) = 6000 + 2x, \qquad 30{,}000 < x \leq 45{,}000.$$

The graph of $C(x)$ for $0 \leq x \leq 45{,}000$ is drawn in Fig. 4. Note that the graph has breaks at two points.   ◆

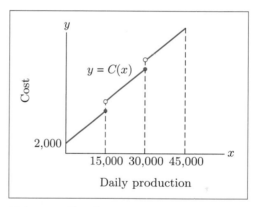

Figure 4. Cost function of a manufacturing plant.

The relationship between differentiability and continuity is this:

Theorem I   If $F'(x) = 0$ for all x, then $F(x) = C$ for some constant C.

Note, however, that the converse statement is definitely false: A function may be continuous at $x = a$ but still not be differentiable there. The functions whose graphs are drawn in Fig. 1 provide examples of this phenomenon.

Just as with differentiability, the notion of continuity can be phrased in terms of limits. In order for $f(x)$ to be continuous at $x = a$, the values of $f(x)$ for all x near a must be close to $f(a)$ (otherwise, the graph would have a break at $x = a$). In fact, the closer x is to a, the closer $f(x)$ must be to $f(a)$ (again, in order to avoid a break in the graph). In terms of limits, we must therefore have

$$\lim_{x \to a} f(x) = f(a).$$

Conversely, an intuitive argument shows that if the limit relation above holds, then the graph of $y = f(x)$ has no break at $x = a$.

Limit Definition of Continuity   A function $f(x)$ is continuous at $x = a$ provided the following limit relation holds:

$$\lim_{x \to a} f(x) = f(a). \tag{1}$$

In order for (1) to hold, three conditions must be fulfilled.

1. $f(x)$ must be defined at $x = a$.
2. $\lim_{x \to a} f(x)$ must exist.
3. The limit $\lim_{x \to a} f(x)$ must have the value $f(a)$.

A function will fail to be continuous at $x = a$ when any one of these conditions fails to hold. The various possibilities are illustrated in the next example.

▶ **Example 3**   Determine whether the functions whose graphs are drawn in Fig. 5 are continuous at $x = 3$. Use the limit definition.

Solution   (a) Here $\lim_{x \to 3} f(x) = 2$. However, $f(3) = 4$. So

$$\lim_{x \to 3} f(x) \neq f(3)$$

and $f(x)$ is not continuous at $x = 3$. (Geometrically, this is clear. The graph has a break at $x = 3$.)

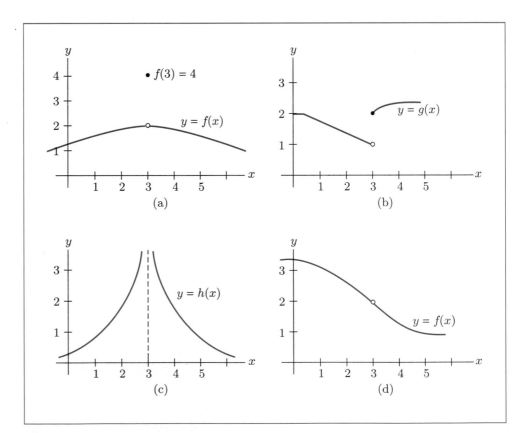

Figure 5.

(b) $\lim\limits_{x\to 3} g(x)$ does not exist, so $g(x)$ is not continuous at $x = 3$.

(c) $\lim\limits_{x\to 3} h(x)$ does not exist, so $h(x)$ is not continuous at $x = 3$.

(d) $f(x)$ is not defined at $x = 3$, so $f(x)$ is not continuous at $x = 3$.   ◆

Using our result on the limit of a polynomial function (Section 1.4), we see that

$$p(x) = a_0 + a_1 x + \cdots + a_n x^n, \qquad a_0, \ldots, a_n \text{ constants,}$$

is continuous for all x. Similarly, a rational function

$$\frac{p(x)}{q(x)}, \qquad p(x),\, q(x) \text{ polynomials,}$$

is continuous at all x for which $q(x) \neq 0$.

**Practice Problems
1.5**

Let $f(x) = \begin{cases} \dfrac{x^2 - x - 6}{x - 3} & \text{for } x \neq 3 \\ 4 & \text{for } x = 3. \end{cases}$

1. Is $f(x)$ continuous at $x = 3$?

2. Is $f(x)$ differentiable at $x = 3$?

▸ Exercises 1.5

Is the function whose graph is drawn in Fig. 6 continuous at the following values of x?

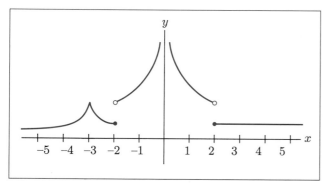

Figure 6.

1. $x = 0$
2. $x = -3$
3. $x = 3$
4. $x = .001$
5. $x = -2$
6. $x = 2$

Is the function whose graph is drawn in Fig. 6 differentiable at the following values of x?

7. $x = 0$
8. $x = -3$
9. $x = 3$
10. $x = .001$

11. $x = -2$
12. $x = 2$

Determine whether each of the following functions is continuous and/or differentiable at $x = 1$.

13. $f(x) = x^2$

14. $f(x) = \dfrac{1}{x}$

15. $f(x) = \begin{cases} x + 2 & \text{for } -1 \le x \le 1 \\ 3x & \text{for } 1 < x \le 5 \end{cases}$

16. $f(x) = \begin{cases} x^3 & \text{for } 0 \le x < 1 \\ x & \text{for } 1 \le x \le 2 \end{cases}$

17. $f(x) = \begin{cases} 2x - 1 & \text{for } 0 \le x \le 1 \\ 1 & \text{for } 1 < x \end{cases}$

18. $f(x) = \begin{cases} x & \text{for } x \neq 1 \\ 2 & \text{for } x = 1 \end{cases}$

19. $f(x) = \begin{cases} \dfrac{1}{x - 1} & \text{for } x \neq 1 \\ 0 & \text{for } x = 1 \end{cases}$

20. $f(x) = \begin{cases} x - 1 & \text{for } 0 \le x < 1 \\ 1 & \text{for } x = 1 \\ 2x - 2 & \text{for } x > 1 \end{cases}$

The functions in Exercises 21–26 are defined for all x except for one value of x. If possible, define $f(x)$ at the exceptional point in a way that makes $f(x)$ continuous for all x.

21. $f(x) = \dfrac{x^2 - 7x + 10}{x - 5},\ x \neq 5$

22. $f(x) = \dfrac{x^2 + x - 12}{x + 4},\ x \neq -4$

23. $f(x) = \dfrac{x^3 - 5x^2 + 4}{x^2},\ x \neq 0$

24. $f(x) = \dfrac{x^2 + 25}{x - 5},\ x \neq 5$

25. $f(x) = \dfrac{(6 + x)^2 - 36}{x},\ x \neq 0$

26. $f(x) = \dfrac{\sqrt{9 + x} - \sqrt{9}}{x},\ x \neq 0$

Solutions to Practice Problems 1.5

1. The function $f(x)$ is defined at $x = 3$, namely, $f(3) = 4$. When computing $\lim\limits_{x \to 3} f(x)$, we exclude consideration of $x = 3$; therefore, we can simplify the expression for $f(x)$ as follows:

$$f(x) = \frac{x^2 - x - 6}{x - 3} = \frac{(x - 3)(x + 2)}{x - 3} = x + 2$$

$$\lim_{x \to 3} f(x) = \lim_{x \to 3} (x + 2) = 5.$$

Since $\lim\limits_{x \to 3} f(x) = 5 \neq 4 = f(3)$, $f(x)$ is not continuous at $x = 3$.

2. There is no need to compute any limits in order to answer this question. By Theorem 1, since $f(x)$ is not continuous at $x = 3$, it cannot possibly be differentiable there.

1.6 Some Rules for Differentiation

Three additional rules of differentiation greatly extend the number of functions that we can differentiate.

1. Constant-Multiple Rule

$$\frac{d}{dx}[k \cdot f(x)] = k \cdot \frac{d}{dx}[f(x)], \quad k \text{ a constant}$$

2. Sum Rule

$$\frac{d}{dx}[f(x) + g(x)] = \frac{d}{dx}[f(x)] + \frac{d}{dx}[g(x)]$$

3. General Power Rule

$$\frac{d}{dx}\left([g(x)]^r\right) = r \cdot [g(x)]^{r-1} \cdot \frac{d}{dx}[g(x)]$$

We shall discuss these rules and then prove the first two.

The Constant-Multiple Rule Starting with a function $f(x)$, we can multiply it by a constant number k in order to obtain a new function $k \cdot f(x)$. For instance, if $f(x) = x^2 - 4x + 1$ and $k = 2$, then

$$2f(x) = 2(x^2 - 4x + 1) = 2x^2 - 8x + 2.$$

The constant-multiple rule says that the derivative of the new function $k \cdot f(x)$ is just k times the derivative of the original function.* In other words, when faced with the differentiation of a constant times a function, simply carry along the constant and differentiate the function.

▶ **Example 1** Calculate.

(a) $\dfrac{d}{dx}\left(2x^5\right)$ (b) $\dfrac{d}{dx}\left(\dfrac{x^3}{4}\right)$ (c) $\dfrac{d}{dx}\left(-\dfrac{3}{x}\right)$ (d) $\dfrac{d}{dx}\left(5\sqrt{x}\right)$

Solution (a) With $k = 2$ and $f(x) = x^5$, we have

$$\frac{d}{dx}\left(2 \cdot x^5\right) = 2 \cdot \frac{d}{dx}\left(x^5\right) = 2\left(5x^4\right) = 10x^4.$$

(b) Write $\dfrac{x^3}{4}$ in the form $\dfrac{1}{4} \cdot x^3$. Then

$$\frac{d}{dx}\left(\frac{x^3}{4}\right) = \frac{1}{4} \cdot \frac{d}{dx}\left(x^3\right) = \frac{1}{4}\left(3x^2\right) = \frac{3}{4}x^2.$$

(c) Write $-\dfrac{3}{x}$ in the form $(-3) \cdot \dfrac{1}{x}$. Then

$$\frac{d}{dx}\left(-\frac{3}{x}\right) = (-3) \cdot \frac{d}{dx}\left(\frac{1}{x}\right) = (-3) \cdot \frac{-1}{x^2} = \frac{3}{x^2}.$$

(d) $\dfrac{d}{dx}\left(5\sqrt{x}\right) = 5\dfrac{d}{dx}\left(\sqrt{x}\right) = 5\dfrac{d}{dx}\left(x^{1/2}\right) = \dfrac{5}{2}x^{-1/2}.$

This answer may also be written in the form $\dfrac{5}{2\sqrt{x}}$. ◆

The Sum Rule

To differentiate a sum of functions, differentiate each function individually and add the derivatives together.† Another way of saying this is "the derivative of a sum of functions is the sum of the derivatives."

▶ **Example 2** Find each of the following.

(a) $\dfrac{d}{dx}\left(x^3 + 5x\right)$ (b) $\dfrac{d}{dx}\left(x^4 - \dfrac{3}{x^2}\right)$ (c) $\dfrac{d}{dx}\left(2x^7 - x^5 + 8\right)$

Solution (a) Let $f(x) = x^3$ and $g(x) = 5x$. Then

$$\frac{d}{dx}\left(x^3 + 5x\right) = \frac{d}{dx}\left(x^3\right) + \frac{d}{dx}\left(5x\right) = 3x^2 + 5.$$

(b) The sum rule applies to differences as well as sums (see Exercise 46). Indeed,

---

*More precisely, the constant-multiple rule asserts that if $f(x)$ is differentiable at $x = a$, then so is the function $k \cdot f(x)$, and the derivative of $k \cdot f(x)$ at $x = a$ may be computed using the given formula.

†More precisely, the sum rule asserts that if both $f(x)$ and $g(x)$ are differentiable at $x = a$, then so is $f(x) + g(x)$, and the derivative (at $x = a$) of the sum is then the sum of the derivatives (at $x = a$).

by the sum rule,

$$\frac{d}{dx}\left(x^4 - \frac{3}{x^2}\right) = \frac{d}{dx}(x^4) + \frac{d}{dx}\left(-\frac{3}{x^2}\right) \qquad \text{(sum rule)}$$

$$= \frac{d}{dx}(x^4) - 3\frac{d}{dx}(x^{-2}) \qquad \text{(constant-multiple rule)}$$

$$= 4x^3 - 3(-2x^{-3})$$

$$= 4x^3 + 6x^{-3}.$$

After some practice, one usually omits most or all of the intermediate steps and simply writes

$$\frac{d}{dx}\left(x^4 - \frac{3}{x^2}\right) = 4x^3 + 6x^{-3}.$$

(c) We apply the sum rule repeatedly and use the fact that the derivative of a constant function is 0:

$$\frac{d}{dx}(2x^7 - x^5 + 8) = \frac{d}{dx}(2x^7) - \frac{d}{dx}(x^5) + \frac{d}{dx}(8)$$

$$= 2(7x^6) - 5x^4 + 0$$

$$= 14x^6 - 5x^4. \qquad \blacklozenge$$

Remark   The differentiation of a function *plus* a constant is different from the differentiation of a constant *times* a function. Figure 1 shows the graphs of $f(x)$, $f(x) + 2$, and $2 \cdot f(x)$, where $f(x) = x^3 - \frac{3}{2}x^2$. For each x, the graphs of $f(x)$ and $f(x) + 2$ have the same slope. In contrast, for each x, the slope of the graph of $2 \cdot f(x)$ is twice the slope of the graph of $f(x)$. Upon differentiation, an added constant disappears, whereas a constant that multiplies a function is carried along.

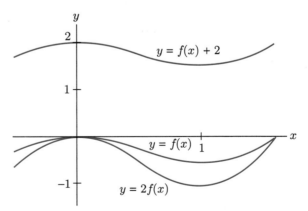

Figure 1. Two effects of a constant on the graph of f(x).

The General Power Rule   Frequently, we will encounter expressions of the form $[g(x)]^r$—for instance, $(x^3 + 5)^2$, where $g(x) = x^3 + 5$ and $r = 2$. The general power rule says that, to differentiate $[g(x)]^r$, we must first treat $g(x)$ as if it were simply an x, form $r[g(x)]^{r-1}$, and then multiply it by a "correction factor"

$g'(x).^*$ Thus

$$\frac{d}{dx}(x^3+5)^2 = 2(x^3+5)^1 \cdot \frac{d}{dx}(x^3+5)$$
$$= 2(x^3+5) \cdot (3x^2)$$
$$= 6x^2(x^3+5).$$

In this special case it is easy to verify that the general power rule gives the correct answer. We first expand $(x^3+5)^2$ and then differentiate.

$$(x^3+5)^2 = (x^3+5)(x^3+5) = x^6 + 10x^3 + 25$$

From the constant-multiple rule and the sum rule, we have

$$\frac{d}{dx}(x^3+5)^2 = \frac{d}{dx}(x^6 + 10x^3 + 25)$$
$$= 6x^5 + 30x^2 + 0$$
$$= 6x^2(x^3+5).$$

The two methods give the same answer.

Note that if we set $g(x) = x$ in the general power rule, we recover the power rule. So the general power rule contains the power rule as a special case.

▶ **Example 3**   Differentiate $\sqrt{1-x^2}$.

Solution
$$\frac{d}{dx}(\sqrt{1-x^2}) = \frac{d}{dx}[(1-x^2)^{1/2}] = \frac{1}{2}(1-x^2)^{-1/2} \cdot \frac{d}{dx}(1-x^2)$$
$$= \frac{1}{2}(1-x^2)^{-1/2} \cdot (-2x)$$
$$= \frac{-x}{(1-x^2)^{1/2}} = \frac{-x}{\sqrt{1-x^2}} \qquad \blacklozenge$$

▶ **Example 4**   Differentiate $y = \dfrac{1}{x^3+4x}$.

Solution
$$y = \frac{1}{x^3+4x} = (x^3+4x)^{-1}$$
$$\frac{dy}{dx} = (-1)(x^3+4x)^{-2} \cdot \frac{d}{dx}(x^3+4x)$$
$$= \frac{-1}{(x^3+4x)^2}(3x^2+4)$$
$$= -\frac{3x^2+4}{(x^3+4x)^2} \qquad \blacklozenge$$

Proofs of the Constant-Multiple and Sum Rules   Let us verify both rules when x has the value a. Recall that if $f(x)$ is differentiable at $x = a$, then its derivative is the limit

$$\lim_{h \to 0} \frac{f(a+h) - f(a)}{h}.$$

---

*More precisely, the general power rule asserts that if $g(x)$ is differentiable at $x = a$ and if $[g(x)]^{r-1}$ is defined at $x = a$, then $[g(x)]^r$ is also differentiable at $x = a$ and its derivative is given by the formula stated.

Constant-Multiple Rule   We assume that $f(x)$ is differentiable at $x = a$. We must prove that $k \cdot f(x)$ is differentiable at $x = a$ and that its derivative is $k \cdot f'(x)$. This amounts to showing that the limit

$$\lim_{h \to 0} \frac{k \cdot f(a + h) - k \cdot f(a)}{h}$$

exists and has the value $k \cdot f'(a)$. However,

$$\lim_{h \to 0} \frac{k \cdot f(a + h) - k \cdot f(a)}{h}$$

$$= \lim_{h \to 0} k \left[\frac{f(a + h) - f(a)}{h} \right]$$

$$= k \cdot \lim_{h \to 0} \frac{f(a + h) - f(a)}{h} \qquad \text{(by Limit Theorem I)}$$

$$= k \cdot f'(a) \qquad \text{[since } f(x) \text{ is differentiable at } x = a\text{]},$$

which is what we desired to show.

Sum Rule   We assume that both $f(x)$ and $g(x)$ are differentiable at $x = a$. We must prove that $f(x) + g(x)$ is differentiable at $x = a$ and that its derivative is $f'(a) + g'(a)$. That is, we must show that the limit

$$\lim_{h \to 0} \frac{[f(a + h) + g(a + h)] - [f(a) + g(a)]}{h}$$

exists and equals $f'(a) + g'(a)$. Using Limit Theorem III and the fact that $f(x)$ and $g(x)$ are differentiable at $x = a$, we have

$$\lim_{h \to 0} \frac{[f(a + h) + g(a + h)] - [f(a) + g(a)]}{h}$$

$$= \lim_{h \to 0} \left[\frac{f(a + h) - f(a)}{h} + \frac{g(a + h) - g(a)}{h} \right]$$

$$= \lim_{h \to 0} \frac{f(a + h) - f(a)}{h} + \lim_{h \to 0} \frac{g(a + h) - g(a)}{h}$$

$$= f'(a) + g'(a).$$

The general power rule will be proven as a special case of the chain rule in Chapter 3.

Practice Problems 1.6

1. Find the derivative $\dfrac{d}{dx}(x)$.

2. Differentiate the function $y = \dfrac{x + (x^5 + 1)^{10}}{3}$.

▶ Exercises 1.6

Differentiate.

1. $y = x^3 + x^2$

2. $y = x^2 + \dfrac{1}{x}$

3. $y = x^2 + 3x - 1$

4. $y = x^3 + 2x + 5$

5. $f(x) = x^5 + \dfrac{1}{x}$

6. $f(x) = x^8 - x$

7. $f(x) = x^4 + x^3 + x$

8. $f(x) = x^5 + x^2 - x$

9. $y = 3x^2$

10. $y = 2x^3$

11. $y = x^3 + 7x^2$

12. $y = -2x$

13. $y = \dfrac{4}{x^2}$

14. $y = 2\sqrt{x}$

15. $y = 3x - \dfrac{1}{x}$

16. $y = -x^2 + 3x + 1$

17. $f(x) = \frac{1}{3}x^3 - \frac{1}{2}x^2$

18. $f(x) = 100x^{100}$

19. $f(x) = -\dfrac{1}{5x^5}$

20. $f(x) = x^2 - \dfrac{1}{x^2}$

21. $f(x) = 1 - \sqrt{x}$

22. $f(x) = -3x^2 + 7$

23. $f(x) = (3x + 1)^{10}$

24. $f(x) = \dfrac{1}{x^2 + x + 1}$

25. $f(x) = 5\sqrt{3x^3 + x}$

26. $y = \dfrac{1}{(x^2 - 7)^5}$

27. $y = (2x^2 - x + 4)^6$

28. $y = \sqrt{-2x + 1}$

29. $y = \dfrac{x}{3} + \dfrac{3}{x}$

30. $y = \dfrac{2x - 1}{5}$

31. $y = \dfrac{2}{1 - 5x}$

32. $y = \dfrac{4}{3\sqrt{x}}$

33. $y = \dfrac{1}{1 - x^4}$

34. $y = \left(x^3 + \dfrac{x}{2} + 1\right)^5$

35. $f(x) = \dfrac{4}{\sqrt{x^2 + x}}$

36. $f(x) = \dfrac{6}{x^2 + 2x + 5}$

37. $f(x) = \left(\dfrac{\sqrt{x}}{2} + 1\right)^{3/2}$

38. $f(x) = \left(4 - \dfrac{2}{x}\right)^3$

Find the slope of the graph of $y = f(x)$ at the designated point.

39. $f(x) = 3x^2 - 2x + 1$, $(1, 2)$

40. $f(x) = x^{10} + 1 + \sqrt{1 - x}$, $(0, 2)$

41. Find the slope of the tangent line to the curve $y = x^3 + 3x - 8$ at $(2, 6)$.

42. Write the equation of the tangent line to the curve $y = x^3 + 3x - 8$ at $(2, 6)$.

43. Find the slope of the tangent line to the curve $y = (x^2 - 15)^6$ at $x = 4$. Then write the equation of this tangent line.

44. Find the equation of the tangent line to the curve $y = \dfrac{8}{x^2 + x + 2}$ at $x = 2$.

45. Differentiate the function $f(x) = (3x^2 + x - 2)^2$ in two ways.

(a) Use the general power rule.

(b) Multiply $3x^2 + x - 2$ by itself and then differentiate the resulting polynomial.

46. Using the sum rule and the constant-multiple rule, show that for any functions $f(x)$ and $g(x)$

$$\frac{d}{dx}[f(x) - g(x)] = \frac{d}{dx}f(x) - \frac{d}{dx}g(x).$$

47. Figure 2 contains the curves $y = f(x)$ and $y = g(x)$ and the tangent line to $y = f(x)$ at $x = 1$, with $g(x) = 3 \cdot f(x)$. Find $g(1)$ and $g'(1)$.

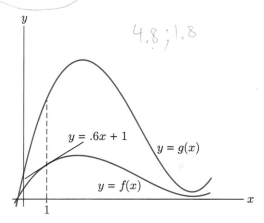

4.8 ; 1.8

Figure 2. Graphs of f(x) and g(x) = 3 f(x).

48. Figure 3 contains the curves $y = f(x)$, $y = g(x)$, and $y = h(x)$ and the tangent lines to $y = f(x)$ and $y = g(x)$ at $x = 1$, with $h(x) = f(x) + g(x)$. Find $h(1)$ and $h'(1)$.

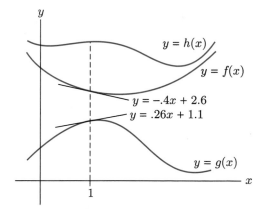

Figure 3. Graphs of f(x), g(x), and h(x) = f(x) + g(x).

49. Suppose $f(5) = 2$, $f'(5) = 3$, $g(5) = 4$, and $g'(5) = 1$. Find $h(5)$ and $h'(5)$, where $h(x) = 3f(x) + 2g(x)$.

50. Suppose $g(3) = 2$ and $g'(3) = 4$. Find $f(3)$ and $f'(3)$, where $f(x) = 2 \cdot [g(x)]^3$.

51. Suppose $g(1) = 4$ and $g'(1) = 3$. Find $f(1)$ and $f'(1)$, where $f(x) = 5 \cdot \sqrt{g(x)}$.

52. The tangent line to the curve $y = x^3 - 6x^2 - 34x - 9$ has slope 2 at two points on the curve. Find the two points.

53. The tangent line to the curve $y = \frac{1}{3}x^3 - 4x^2 + 18x + 22$

is parallel to the line $6x - 2y = 1$ at two points on the curve. Find the two points.

54. In Fig. 4 the straight line is tangent to the parabola. Find the value of b.

55. In Fig. 5 the straight line is tangent to the graph of $f(x)$. Find $f(4)$ and $f'(4)$.

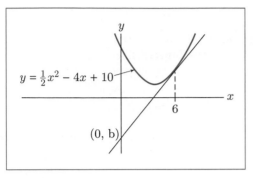

Figure 4.

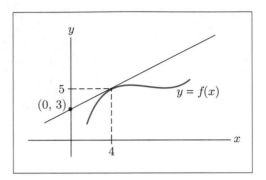

Figure 5.

1. The problem asks for the derivative of the function $y = x$, a straight line of slope 1. Therefore, $\dfrac{d}{dx}(x) = 1$. The result can also be obtained from the power rule with $r = 1$. If $f(x) = x^1$, then

$$\frac{d}{dx}(f(x)) = 1 \cdot x^{1-1} = x^0 = 1.$$

See Fig. 6.

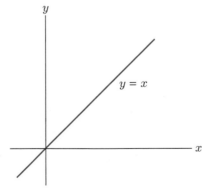

Figure 6.

Solutions to
Practice Problems
1.6 (Continued)

2. All three rules are required to differentiate this function.

$$\frac{dy}{dx} = \frac{d}{dx}\frac{1}{3}\cdot[x + (x^5 + 1)^{10}]$$

$$= \frac{1}{3}\frac{d}{dx}[x + (x^5 + 1)^{10}] \qquad \text{(constant-multiple rule)}$$

$$= \frac{1}{3}\left[\frac{d}{dx}(x) + \frac{d}{dx}(x^5 + 1)^{10}\right] \qquad \text{(sum rule)}$$

$$= \tfrac{1}{3}[1 + 10(x^5 + 1)^9\cdot(5x^4)] \qquad \text{(general power rule)}$$

$$= \tfrac{1}{3}[1 + 50x^4(x^5 + 1)^9]$$

1.7 More About Derivatives

In many applications it is convenient to use variables other than x and y. One might, for instance, study the function $f(t) = t^2$ instead of writing $f(x) = x^2$. In this case, the notation for the derivative involves t rather than x, but the concept of the derivative as a slope formula is unaffected. (See Fig. 1.) When the independent variable is t instead of x, we write $\frac{d}{dt}$ in place of $\frac{d}{dx}$. For instance,

$$\frac{d}{dt}(t^3) = 3t^2, \qquad \frac{d}{dt}(2t^2 + 3t) = 4t + 3.$$

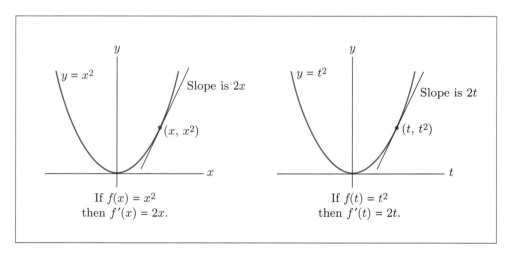

Figure 1. The same function but different variables.

Recall that if y is a function of x, say $y = f(x)$, then we may write $\frac{dy}{dx}$ in place of $f'(x)$. We sometimes call $\frac{dy}{dx}$ "the derivative of y with respect to x." Similarly, if v is a function of t, then the derivative of v with respect to t is written as $\frac{dv}{dt}$. For example, if $v = 4t^2$, then $\frac{dv}{dt} = 8t$.

Of course, other letters can be used to denote variables. The formulas

$$\frac{d}{dP}(P^3) = 3P^2, \qquad \frac{d}{ds}(s^3) = 3s^2, \qquad \frac{d}{dz}(z^3) = 3z^2$$

all express the same basic fact that the slope formula for the cubic curve $y = x^3$ is given by $3x^2$.

▶ Example 1   Compute.

(a) $\dfrac{ds}{dp}$ if $s = 3(p^2 + 5p + 1)^{10}$ 　　　　(b) $\dfrac{d}{dt}(at^2 + St^{-1} + S^2)$

Solution   (a) $\dfrac{d}{dp}3(p^2 + 5p + 1)^{10} = 30(p^2 + 5p + 1)^9 \cdot \dfrac{d}{dp}(p^2 + 5p + 1)$

$$= 30(p^2 + 5p + 1)^9(2p + 5)$$

(b) Although the expression $at^2 + St^{-1} + S^2$ contains several letters, the notation $\dfrac{d}{dt}$ indicates that for purposes of calculating the derivative with respect to t, all letters except t are to be considered as constants. Hence

$$\frac{d}{dt}(at^2 + St^{-1} + S^2) = \frac{d}{dt}(at^2) + \frac{d}{dt}(St^{-1}) + \frac{d}{dt}(S^2)$$

$$= a \cdot \frac{d}{dt}(t^2) + S \cdot \frac{d}{dt}(t^{-1}) + 0$$

$$= 2at - St^{-2}.$$

$$\left[\text{The derivative } \frac{d}{dt}(S^2) \text{ is zero because } S^2 \text{ is a constant.} \right]　　◆$$

The Second Derivative   When we differentiate a function $f(x)$, we obtain a new function $f'(x)$ that is a formula for the slope of the curve $y = f(x)$. If we differentiate the function $f'(x)$, we obtain what is called the *second derivative* of $f(x)$, denoted by $f''(x)$. That is,

$$\frac{d}{dx}f'(x) = f''(x).$$

▶ Example 2   Find the second derivatives of the following functions.

(a) $f(x) = x^3 + (1/x)$ 　　(b) $f(x) = 2x + 1$ 　　(c) $f(t) = t^{1/2} + t^{-1/2}$

Solution   (a) $f(x) = x^3 + (1/x) = x^3 + x^{-1}$

$f'(x) = 3x^2 - x^{-2}$

$f''(x) = 6x + 2x^{-3}$

(b) $f(x) = 2x + 1$

$f'(x) = 2$ (a constant function whose value is 2)

$f''(x) = 0$ (The derivative of a constant function is zero.)

(c) $f(t) = t^{1/2} + t^{-1/2}$

$f'(t) = \frac{1}{2}t^{-1/2} - \frac{1}{2}t^{-3/2}$

$f''(t) = -\frac{1}{4}t^{-3/2} + \frac{3}{4}t^{-5/2}$　　　　　　　　◆

The first derivative of a function $f(x)$ gives the slope of the graph of $f(x)$ at any point. The second derivative of $f(x)$ gives important additional information about the shape of the curve near any point. We shall examine this subject carefully in the next chapter.

Other Notation for Derivatives Unfortunately, the process of differentiation does not have a standardized notation. Consequently, it is important to become familiar with alternative terminology.

If y is a function of x, say $y = f(x)$, then we may denote the first and second derivatives of this function in several ways.

Prime Notation	$\dfrac{d}{dx}$ Notation
$f'(x)$	$\dfrac{d}{dx} f(x)$
y'	$\dfrac{dy}{dx}$
$f''(x)$	$\dfrac{d^2}{dx^2} f(x)$
y''	$\dfrac{d^2 y}{dx^2}$

The notation $\dfrac{d^2}{dx^2}$ is purely symbolic. It reminds us that the second derivative is obtained by differentiating $\dfrac{d}{dx} f(x)$; that is,

$$f'(x) = \frac{d}{dx} f(x)$$

$$f''(x) = \frac{d}{dx} \left[\frac{d}{dx} f(x) \right].$$

If we evaluate the derivative $f'(x)$ at a specific value of x, say $x = a$, we get a number $f'(a)$ that gives the slope of the curve $y = f(x)$ at the point $(a, f(a))$. Another way of writing $f'(a)$ is

$$\frac{dy}{dx}\bigg|_{x=a}.$$

If we have a second derivative $f''(x)$, then its value when $x = a$ is written

$$f''(a) \quad \text{or} \quad \frac{d^2 y}{dx^2}\bigg|_{x=a}.$$

▶ **Example 3** If $y = x^4 - 5x^3 + 7$, find $\dfrac{d^2 y}{dx^2}\bigg|_{x=3}$.

Solution

$$\frac{dy}{dx} = \frac{d}{dx}(x^4 - 5x^3 + 7) = 4x^3 - 15x^2$$

$$\frac{d^2 y}{dx^2} = \frac{d}{dx}(4x^3 - 15x^2) = 12x^2 - 30x$$

$$\frac{d^2 y}{dx^2}\bigg|_{x=3} = 12(3)^2 - 30(3) = 108 - 90 = 18$$

◆

▶ Example 4   If $s = t^3 - 2t^2 + 3t$, find

$$\left.\frac{ds}{dt}\right|_{t=-2} \quad \text{and} \quad \left.\frac{d^2 s}{dt^2}\right|_{t=-2}.$$

Solution

$$\frac{ds}{dt} = \frac{d}{dt}(t^3 - 2t^2 + 3t) = 3t^2 - 4t + 3$$

$$\left.\frac{ds}{dt}\right|_{t=-2} = 3(-2)^2 - 4(-2) + 3 = 12 + 8 + 3 = 23$$

To find the value of the second derivative at $t = -2$, we must first differentiate $\frac{ds}{dt}$.

$$\frac{d^2 s}{dt^2} = \frac{d}{dt}(3t^2 - 4t + 3) = 6t - 4$$

$$\left.\frac{d^2 s}{dt^2}\right|_{t=-2} = 6(-2) - 4 = -12 - 4 = -16 \qquad \blacklozenge$$

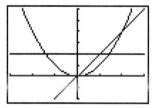

Figure 2. Graph of x^2 and its first two derivatives.

Incorporating Technology

Although functions can be specified (and differentiated) in graphing calculators with letters other than X, only functions in X can be graphed. Therefore, we will always use X as the variable.

Figure 2 contains the graphs of $f(x) = x^2$ and its first and second derivatives. These three graphs can be obtained with any of the function editor settings in Figs. 3(a), 3(b), and 3(c).

```
Plot1 Plot2 Plot3
\Y1▤X²
\Y2▤2X
\Y3▤2
\Y4=
\Y5=
\Y6=
\Y7=
```

Figure 3(a) TI-83

```
Plot1 Plot2 Plot3
\Y1▤X²
\Y2▤nDeriv(Y1,X,
X)
\Y3▤nDeriv(Y2,X,
X)
\Y4=
\Y5=
```

Figure 3(b) TI-83

```
Plot1 Plot2 Plot3
\y1▤x²
\y2▤der1(y1,x,x)
\y3▤der2(y1,x,x)

Y(X)▤ WIND ZOOM TRACE GRAPH
  x    y    INSf DELf SELCT▶
```

Figure 3(c) TI-86

Practice Problems 1.7	1. Let $f(t) = t + (1/t)$. Find $f''(2)$.
	2. Differentiate $g(r) = 2\pi rh$.

▶ Exercises 1.7

Find the first derivatives.

1. $f(t) = (t^2 + 1)^5$

2. $f(P) = P^4 - P^3 + 4P^2 - P$

3. $v = \sqrt{2t - 1}$

4. $g(z) = (z^3 - z + 1)^2$

5. $y = (T^3 + 5T)^{2/3}$

6. $s = \sqrt{t} + \dfrac{1}{\sqrt{t}}$

7. Find $\dfrac{d}{dP}(3P^2 - \frac{1}{2}P + 1)$.

8. Find $\dfrac{d}{dz}(\sqrt{z^2 - 1})$.

9. Find $\dfrac{d}{dt}(a^2t^2 + b^2t + c^2)$.

10. Find $\dfrac{d}{dx}(x^3 + t^3)$.

Find the first and second derivatives.

11. $f(x) = \frac{1}{2}x^2 - 7x + 2$ **12.** $y = \dfrac{1}{x^2} + 1$

13. $y = \sqrt{x}$ **14.** $f(t) = t^{100} + t + 1$

15. $f(r) = \pi hr^2 + 2\pi r$ **16.** $v = t^{3/2} + t$

17. $g(x) = 2 - 5x$ **18.** $V(r) = \frac{4}{3}\pi r^3$

19. $f(P) = (3P + 1)^5$ **20.** $u = \dfrac{t^6}{30} - \dfrac{t^4}{12}$

Compute the following.

21. $\dfrac{d}{dx}(2x^2 - 3)\Big|_{x=5}$ **22.** $\dfrac{d}{dt}(1 - 2t - 3t^2)\Big|_{t=-1}$

23. $\dfrac{d}{dz}(z^2 - 4)^3\Big|_{z=1}$ **24.** $\dfrac{d}{dT}\left(\dfrac{1}{3T + 1}\right)\Big|_{T=2}$

25. $\dfrac{d^2}{dx^2}(3x^3 - x^2 + 7x - 1)\Big|_{x=2}$

26. $\dfrac{d}{dt}\left(\dfrac{dv}{dt}\right)$, where $v = 2t^{-3}$

27. $\dfrac{d}{dP}\left(\dfrac{dy}{dP}\right)$, where $y = \dfrac{k}{2P - 1}$

28. $\dfrac{d^2V}{dr^2}\Big|_{r=2}$, where $V = ar^3$

29. $f'(3)$ and $f''(3)$, when $f(x) = \sqrt{10 - 2x}$

30. $g'(2)$ and $g''(2)$, when $g(T) = (3T - 5)^{10}$

31. Suppose a company finds that the revenue R generated by spending x dollars on advertising is given by $R = 1000 + 80x - .02x^2$, for $0 \le x \le 2000$. Find $\dfrac{dR}{dx}\Big|_{x=1500}$.

32. A supermarket finds that its average daily volume of business V (in thousands of dollars) and the number of hours t the store is open for business each day are approximately related by the formula

$$V = 20\left(1 - \dfrac{100}{100 + t^2}\right), \qquad 0 \le t \le 24.$$

Find $\dfrac{dV}{dt}\Big|_{t=10}$.

33. The *third derivative* of a function $f(x)$ is the derivative of the second derivative $f''(x)$ and is denoted by $f'''(x)$. Compute $f'''(x)$ for the following functions.
   (a) $f(x) = x^5 - x^4 + 3x$ (b) $f(x) = 4x^{5/2}$

34. Compute the third derivatives of the following functions.
   (a) $f(t) = t^{10}$ (b) $f(z) = \dfrac{1}{z + 5}$

Technology Exercises

In Exercises 35 and 36, simultaneously graph the functions $f(x)$, $f'(x)$, and $f''(x)$ with the specified window setting. *Note*: Since we have not yet learned how to differentiate the given functions, you must use your graphing utility's differentiation command to define the derivatives.

35. $f(x) = \dfrac{x}{1 + x^2}$, $[-4, 4]$ by $[-2, 2]$

36. $f(x) = x\sqrt{1 + x^2}$, $[-2, 2]$ by $[-4, 4]$

In Exercises 37 and 38, evaluate $\dfrac{d^2y}{dx^2}\Big|_{x=3}$ in the home screen.

37. $y = x^x$ **38.** $y = 2^x$

Solutions to Practice Problems 1.7	

1. $f(t) = t + t^{-1}$

$f'(t) = 1 + (-1)t^{(-1)-1} = 1 - t^{-2}$

$f''(t) = -(-2)t^{(-2)-1} = 2t^{-3} = \dfrac{2}{t^3}$

Therefore, $f''(2) = \dfrac{2}{2^3} = \dfrac{1}{4}$. [*Note*: It is essential first to compute the function $f''(t)$ and *then* to evaluate the function at $t = 2$.]

2. The expression $2\pi rh$ contains two numbers, 2 and π, and two letters, r and h. The notation $g(r)$ tells us that the expression $2\pi rh$ is to be regarded as a function of r. Therefore, h—and hence $2\pi h$—is to be treated as a constant, and differentiation is done with respect to the variable r. That is,

$$g(r) = (2\pi h)r$$
$$g'(r) = 2\pi h.$$

1.8 The Derivative as a Rate of Change

An important interpretation of the slope of a function at a point is as a rate of change. In this section, we examine this interpretation and discuss some of the applications in which it proves useful. The first step is to understand what is meant by "average rate of change" of a function $f(x)$.

Consider a function $y = f(x)$ defined on the interval $a \leq x \leq b$. The average rate of change of $f(x)$ over this interval is the change in $f(x)$ divided by the length of the interval. That is,

$$\begin{bmatrix} \text{average rate of change of } f(x) \\ \text{over the interval } a \leq x \leq b \end{bmatrix} = \frac{\Delta y}{\Delta x} = \frac{f(b) - f(a)}{b - a}.$$

▶ **Example 1** Suppose that $f(x) = x^2$. Calculate the average rate of change of $f(x)$ over the following intervals:

(a) $1 \leq x \leq 2$ (b) $1 \leq x \leq 1.1$ (c) $1 \leq x \leq 1.01$

Solution   (a) The average rate of change over the interval $1 \leq x \leq 2$ is

$$\frac{2^2 - 1^2}{2 - 1} = \frac{3}{1} = 3.$$

(b) The average rate of change over the interval $1 \leq x \leq 1.1$ is

$$\frac{1.1^2 - 1^2}{1.1 - 1} = \frac{.21}{.1} = 2.1.$$

(c) The average rate of change over the interval $1 \leq x \leq 1.01$ is

$$\frac{1.01^2 - 1^2}{1.01 - 1} = \frac{.0201}{.01} = 2.01. \qquad \blacklozenge$$

In the special case where b is $a + h$, the value of $b - a$ is $(a + h) - a$ or h, and the average rate of change of the function over the interval is the familiar difference quotient

$$\frac{f(a + h) - f(a)}{h}.$$

Geometrically, this quotient is the slope of the secant line in Fig. 1. Recall that as h approaches 0, the slope of the secant line approaches the slope of the tangent line. Thus the average rate of change approaches $f'(a)$. For this reason, $f'(a)$ is called the (*instantaneous*) *rate of change* of $f(x)$ exactly at the point where $x = a$. From now on, unless we explicitly use the word "average" when we refer to the rate of change of a function, we mean the "instantaneous" rate of change.

The derivative $f'(a)$ measures the rate of change of $f(x)$ at $x = a$.

▶ **Example 2** Consider the function $f(x) = x^2$ of Example 1. Calculate the rate of change of $f(x)$ at $x = 1$.

Solution   The rate of change of $f(x)$ at $x = 1$ is equal to $f'(1)$. We have

$$f'(x) = 2x$$
$$f'(1) = 2 \cdot 1 = 2.$$

That is, the rate of change is 2 units per unit change in x. Notice how the average rates of change in Example 1 approach the (instantaneous) rate of change as the intervals beginning at $x = 1$ shrink.   ◆

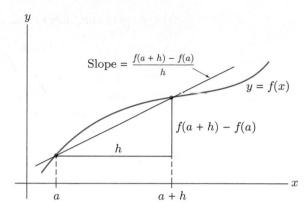

Figure 1. Average rate of change is the slope of the secant line.

▶ Example 3   The function $f(t)$ in Fig. 2 gives the population of the United States t years after the beginning of 1800. The figure also shows the tangent line through the point $(40, 17)$.

(a) What was the average rate of growth of the United States population from 1840 to 1870?

(b) How fast was the population growing in 1840?

(c) Was the population growing faster in 1810 or 1880?

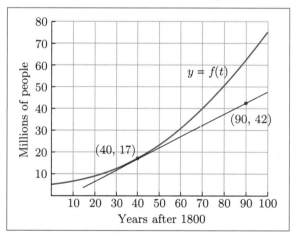

Figure 2. United States population from 1800 to 1900.

Solution   (a) Since 1870 is 70 years after 1800 and 1840 is 40 years after 1800, the question asks for the average rate of change of $f(t)$ over the interval $40 \leq t \leq 70$. This value is

$$\frac{f(70) - f(40)}{70 - 40} = \frac{40 - 17}{30} = \frac{23}{30} \approx .77.$$

Therefore, from 1840 to 1870, the population grew at the average rate of about .77 million or 770,000 people per year.

(b) The rate of growth of $f(t)$ at $t = 40$ is $f'(40)$; that is, the slope of the tangent line at $t = 40$. Since $(40, 17)$ and $(90, 42)$ are two points on the tangent line, the slope of the tangent line is

$$\frac{42-17}{90-40} = \frac{25}{50} = .5.$$

Therefore, in 1840 the population was growing at the rate of .5 million or 500,000 people per year.

(c) The graph is clearly steeper in 1880 than in 1810. Therefore, the population was growing faster in 1880 than in 1810. ◆

Velocity and Acceleration   An everyday illustration of rate of change is given by the velocity of a moving object. Suppose that we are driving a car along a straight road and at each time t we let $s(t)$ be our position on the road, measured from some convenient reference point. See Fig. 3, where distances are positive to the right of the car. For the moment we shall assume that we are proceeding in only the positive direction along the road.

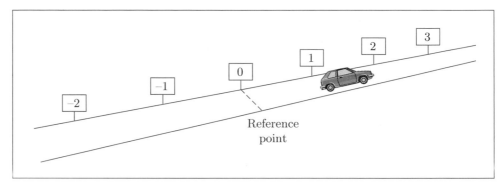

Figure 3. Position of a car traveling on a straight road.

At any instant, the car's speedometer tells us how fast we are moving—that is, how fast our position $s(t)$ is changing. To show how the speedometer reading is related to our calculus concept of a derivative, let us examine what is happening at a specific time, say $t = 2$. Consider a short time interval of duration h from $t = 2$ to $t = 2 + h$. Our car will move from position $s(2)$ to position $s(2 + h)$, a distance of $s(2 + h) - s(2)$. Thus the *average velocity from $t = 2$ to $t = 2 + h$* is

$$\frac{[\text{distance traveled}]}{[\text{time elapsed}]} = \frac{s(2+h) - s(2)}{h}. \tag{1}$$

If the car is traveling at a steady speed during this time period, then the speedometer reading will equal the average velocity in (1).

From our discussion in Section 1.3 the ratio (1) approaches the derivative $s'(2)$ as h approaches zero. For this reason we call $s'(2)$ *the (instantaneous) velocity at $t = 2$*. This number will agree with the speedometer reading at $t = 2$ because when h is very small, the car's speed will be nearly steady over the time interval from $t = 2$ to $t = 2 + h$, and so the average velocity over this time interval will be nearly the same as the speedometer reading at $t = 2$.

The reasoning used for $t = 2$ holds for an arbitrary t as well. Thus the following definition makes sense:

If $s(t)$ denotes the position function of an object moving in a straight line, then the velocity $v(t)$ of the object at time t is given by

$$v(t) = s'(t).$$

In our discussion we assumed that the car moved in the positive direction. If the car moves in the opposite direction, the ratio (1) and the limiting value $s'(2)$ will be negative. So we interpret negative velocity as movement in the negative direction along the road.

The derivative of the velocity function $v(t)$ is called the acceleration function and is often written as $a(t)$:

$$a(t) = v'(t).$$

Because $v'(t)$ measures the rate of change of the velocity $v(t)$, this use of the word "acceleration" agrees with our common usage in connection with automobiles. Note that since $v(t) = s'(t)$, the acceleration is actually the second derivative of the position function $s(t)$,

$$a(t) = s''(t).$$

▶ Example 4   When a ball is thrown straight up into the air, its position may be measured as the vertical distance from the ground. Regard "up" as the positive direction, and let $s(t)$ be the height of the ball in feet after t seconds. Suppose that $s(t) = -16t^2 + 128t + 5$. Figure 4 contains a graph of the function $s(t)$. *Note:* The graph does *not* show the path of the ball. The ball is rising *straight* up and then falling *straight* down.

(a) What is the velocity after 2 seconds?

(b) What is the acceleration after 2 seconds?

(c) At what time is the velocity -32 feet per second? (The negative sign indicates that the ball's height is decreasing; that is, the ball is falling.)

(d) When is the ball at a height of 117 feet?

Solution   (a) The velocity is the rate of change of the height function, so

$$v(t) = s'(t) = -32t + 128.$$

The velocity when $t = 2$ is $v(2) = -32(2) + 128 = 64$ feet per second.

(b) $a(t) = v'(t) = -32$. The acceleration is -32 feet per second per second for all t. This constant acceleration is due to the downward (and therefore negative) force of gravity.

(c) Since the velocity is given and the time is unknown, we set $v(t) = -32$ and solve for t:

$$-32t + 128 = -32$$
$$-32t = -160$$
$$t = 5.$$

The velocity is -32 feet per second when t is 5 seconds.

(d) The question here involves the height function, not the velocity. Since the height is given and the time is unknown, we set $s(t) = 117$ and solve for t:

$$-16t^2 + 128t + 5 = 117$$
$$-16(t^2 - 8t + 7) = 0$$
$$-16(t - 1)(t - 7) = 0.$$

The ball is at a height of 117 feet once on the way up (when $t = 1$ second) and once on the way down (when $t = 7$ seconds).   ◆

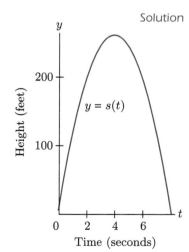

Figure 4. Height of ball in the air.

Approximating the Change in a Function   Consider the function $f(x)$ near $x = a$. As we saw in previous sections, for small values of h

$$\frac{f(a+h) - f(a)}{h} \approx f'(a).$$

That is, for small values of h, the average rate of change over a small interval is approximately equal to the instantaneous change at an endpoint of the interval. Multiplying both sides of this approximation by h, we have

$$f(a+h) - f(a) \approx f'(a) \cdot h \tag{2}$$

If x changes from a to $a + h$, then the change in the value of the function $f(x)$ is approximately $f'(a)$ times the change h in the value of x. This result holds for both positive and negative values of h. In applications, the right side of (2) is calculated and used to estimate the left side.

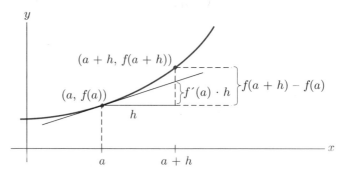

Figure 5. Change in y along the tangent line and along the graph of y = f(x).

Figure 5 contains a geometric interpretation of (2). Given the small change h in x, the quantity $f'(a) \cdot h$ gives the corresponding change in y *along the tangent line* at $(a, f(a))$. In contrast, the quantity $f(a + h) - f(a)$ gives the change in y *along the curve* $y = f(x)$. When h is small, $f'(a) \cdot h$ is a good approximation to the change in $f(x)$.

▶ **Example 5**   Let the production function $p(x)$ give the number of units of goods produced when employing x units of labor. Suppose 5000 units of labor are currently employed, $p(5000) = 300$, and $p'(5000) = 2$.

(a) Interpret $p(5000) = 300$.

(b) Interpret $p'(5000) = 2$.

(c) Estimate the number of *additional* units of goods produced when x is increased from 5000 to $5000\frac{1}{2}$ units of labor.

(d) Estimate the *change* in the number of units of goods produced when x is decreased from 5000 to 4999 units of labor.

Solution   (a) When 5000 units of labor are employed, 300 units of goods will be produced.

(b) If 5000 units of labor are currently employed and we consider adding more labor, productivity will increase at approximately the rate of 2 units of goods for each additional unit of labor.

(c) Here $h = \frac{1}{2}$. By (2), the change in $p(x)$ will be approximately

$$p'(5000) \cdot \tfrac{1}{2} = 2 \cdot \tfrac{1}{2} = 1.$$

About one additional unit will be produced. Therefore, about 301 units of goods will be produced when $5000\frac{1}{2}$ units of labor are employed.

(d) Here $h = -1$, since the amount of labor is reduced. The change in $p(x)$ will be approximately

$$p'(5000) \cdot (-1) = 2 \cdot (-1) = -2.$$

About two fewer (that is, 298 units) of goods will be produced. ◆

The Marginal Concept in Economics Recall from Section 1.3 that economists often use the adjective *marginal* to denote a derivative. For example, if $C(x)$ is a cost function (the cost of producing x units of a commodity), then the value of the derivative $C'(a)$ is called the marginal cost at production level a. The number $C'(a)$ gives the rate at which costs are increasing with respect to the level of production, when the production is currently at level a. If we apply (2) to the cost function and take h to be 1 unit, then

$$C(a + 1) - C(a) \approx C'(a) \cdot 1 = C'(a)$$

$$C'(a) \approx C(a + 1) - C(a). \tag{3}$$

The quantity $C(a + 1) - C(a)$ is the amount the cost rises when the production level is increased from a units to $a + 1$ units. See Fig. 6. Economists interpret (3) by saying that the marginal cost is approximately the increase in cost incurred when the production level is raised by one unit. Other marginal quantities in economics have analogous interpretations. For instance, if $P(x)$ is the profit at production (or sales) level x, the marginal profit $P'(a)$ is approximately the change in profit earned when the production level is raised by one unit from a to $a + 1$.

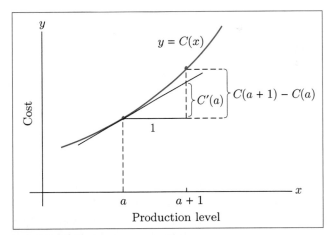

Figure 6. Approximating the change in a cost function.

▶ Example 6 Suppose the cost of producing x items is $C(x) = .005x^3 - .5x^2 + 28x + 300$ dollars, and daily production is 50 items.

(a) What is the extra cost of increasing daily production from 50 to 51 items?

(b) What is the marginal cost when $x = 50$?

Solution   (a) The change in cost when daily production is raised from 50 to 51 items is $C(51) - C(50)$, which equals

$$[.005(51)^3 - .5(51)^2 + 28(51) + 300] - [.005(50)^3 - .5(50)^2 + 28(50) + 300]$$
$$= 1090.76 - 1075.00$$
$$= 15.76.$$

(b) The marginal cost at production level 50 is $C'(50)$.

$$C'(x) = .015x^2 - x + 28$$
$$C'(50) = 15.5$$

Notice that 15.5 is close to the actual cost in (a) of increasing production by one item.   ◆

Units for Rate of Change   Table 1 shows the units appearing in several examples from this section. In general,

$$[\text{unit of measure for } f'(x)]$$
$$= [\text{unit of measure for } f(x)] \text{ per } [\text{unit of measure for } x].$$

Table 1   **Units of measure**

Example	Unit for $f(t)$ or $f(x)$	Unit for t or x	Unit for $f'(t)$ or $f'(x)$
U.S. population	millions of people	year	millions of people per year
Ball in air	feet	second	feet per second
Cost function	dollars	item	dollars per item

**Practice Problems
1.8**

Let $f(t)$ be the temperature (in degrees Celsius) of a liquid at time t (in hours). The rate of temperature change at time a has the value $f'(a)$. Listed next are typical questions about $f(t)$ and its slope at various points. Match each question with the proper method of solution.

Questions:

1. What is the temperature of the liquid after 6 hours (that is, when $t = 6$)?
2. When is the temperature rising at the rate of 6 degrees per hour?
3. By how many degrees did the temperature rise during the first 6 hours?
4. When is the liquid's temperature only 6 degrees?
5. How fast is the temperature of the liquid changing after 6 hours?
6. What is the average rate of increase in the temperature during the first 6 hours?

Methods of Solution:
   (a) Compute $f(6)$.
   (b) Set $f(t) = 6$ and solve for t.
   (c) Compute $[f(6) - f(0)]/6$.
   (d) Compute $f'(6)$.
   (e) Find a value of a for which $f'(a) = 6$.
   (f) Compute $f(6) - f(0)$.

▶ Exercises 1.8

1. Suppose that $f(x) = 4x^2$.
   (a) What is the average rate of change of $f(x)$ over each of the intervals 1 to 2, 1 to 1.5, and 1 to 1.1?
   (b) What is the (instantaneous) rate of change of $f(x)$ when $x = 1$?

2. Suppose that $f(x) = -6/x$.
   (a) What is the average rate of change of $f(x)$ over each of the intervals 1 to 2, 1 to 1.5, and 1 to 1.2?
   (b) What is the (instantaneous) rate of change of $f(x)$ when $x = 1$?

3. Suppose that $f(t) = t^2 + 3t - 7$.
   (a) What is the average rate of change of $f(t)$ over the interval 5 to 6?
   (b) What is the (instantaneous) rate of change of $f(t)$ when $t = 5$?

4. Suppose that $f(t) = 3t + 2 - \dfrac{12}{t}$.
   (a) What is the average rate of change of $f(t)$ over the interval 2 to 3?
   (b) What is the (instantaneous) rate of change of $f(t)$ when $t = 2$?

5. Refer to Fig. 7, where $f(t)$ is the percentage yield (interest rate) on a 3-month T-bill (U.S. Treasury bill) t years after January 1, 1980.
   (a) What was the average rate of change in the yield from January 1, 1981, to January 1, 1985?
   (b) How fast was the percentage yield rising on January 1, 1989?
   (c) Was the percentage yield rising faster on January 1, 1980, or January 1, 1989?

6. Refer to Fig. 8, where $f(t)$ is the mean (average) size of a farm (in acres) t years after January 1, 1900.
   (a) What was the average rate of growth in the mean size of farms from 1920 to 1960?
   (b) How fast was the mean farm size rising on January 1, 1950?
   (c) Was the mean farm size rising faster on January 1, 1960, or January 1, 1980?

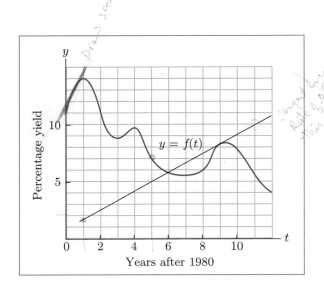

Figure 7. Percentage yield on 3-month T-bill (1980–1992).

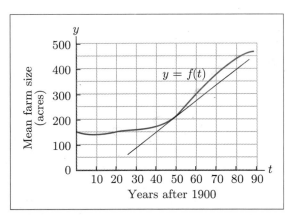

Figure 8. Mean size of farms from 1900 to 1990.

7. Suppose that the weight of a cancerous tumor after t weeks is $W(t) = .1t^2$ grams. What is the average rate of growth of the tumor during the fifth week, that is, from time $t = 4$ to $t = 5$? What is the (instantaneous) rate of growth of the tumor at time $t = 4$?

8. After an advertising campaign, the sales of a product often increase and then decrease. Suppose that t days after the end of the advertising, the daily sales are $f(t) = -3t^2 + 32t + 100$ units. What is the average rate of growth in sales during the fourth day, that is, from time $t = 3$ to $t = 4$? At what (instantaneous) rate are the sales changing when $t = 2$?

9. An analysis of the daily output of a factory assembly line shows that about $60t + t^2 - \frac{1}{12}t^3$ units are produced after t hours of work, $0 \leq t \leq 8$. What is the rate of production (in units per hour) when $t = 2$?

10. Liquid is pouring into a large vat. After t hours, there are $5t - \sqrt{t}$ gallons in the vat. At what rate is the liquid flowing into the vat (in gallons per hour) when $t = 4$?

11. An object moving in a straight line travels $s(t)$ kilometers in t hours, where $s(t) = 2t^2 + 4t$.

    (a) What is the object's velocity when $t = 6$?

    (b) How far has the object traveled in 6 hours?

    (c) When is the object traveling at the rate of 6 kilometers per hour?

12. Suppose that the position of a car at time t (in hours) is given by $s(t) = 50t - 7/(t+1)$, where the position is measured in kilometers. Find the velocity and acceleration of the car at $t = 0$.

13. A toy rocket fired straight up into the air has height $s(t) = 160t - 16t^2$ feet after t seconds.

    (a) What is the rocket's initial velocity (when $t = 0$)?

    (b) What is the velocity after 2 seconds?

    (c) What is the acceleration when $t = 3$?

    (d) At what time will the rocket hit the ground?

    (e) At what velocity will the rocket be traveling just as it smashes into the ground?

14. A helicopter is rising straight up in the air. Its distance from the ground t seconds after takeoff is $s(t)$ feet, where $s(t) = t^2 + t$.

    (a) How long will it take for the helicopter to rise 20 feet?

    (b) Find the velocity and the acceleration of the helicopter when it is 20 feet above the ground.

15. Let $s(t)$ be the height (in feet) after t seconds of a ball thrown straight up into the air. Match each question with the proper solution.

Questions:

   A. What will be the velocity of the ball after 3 seconds?

   B. When will the velocity be 3 feet per second?

   C. What is the average velocity during the first 3 seconds?

   D. When will the ball be 3 feet above the ground?

   E. When will the ball hit the ground?

   F. How high will the ball be after 3 seconds?

   G. How far did the ball travel during the first 3 seconds?

Solutions:

   a. Set $s(t) = 0$ and solve for t.

   b. Compute $s'(3)$.

   c. Compute $s(3)$.

   d. Set $s'(t) = 3$ and solve for t.

   e. Find a value of a for which $s(a) = 3$.

   f. Compute $[s(3) - s(0)]/3$.

   g. Compute $s(3) - s(0)$.

16. Table 2 gives a car's trip-meter reading (in miles) at 1 hour into a trip and at several nearby times. What is the average speed during the time interval from 1 to 1.05 hours? Estimate the speed at time 1 hour into the trip.

Table 2	Trip-meter Readings at Several Times
Time	**Trip Meter**
.96	43.2
.97	43.7
.98	44.2
.99	44.6
1.00	45.0
1.01	45.4
1.02	45.8
1.03	46.3
1.04	46.8
1.05	47.4

17. A particle is moving in a straight line in such a way that its position at time t (in seconds) is $s(t) = t^2 + 3t + 2$ feet to the right of a reference point, for $t \geq 0$.

    (a) What is the velocity of the object when the time is 6 seconds?

    (b) Is the object moving toward the reference point when $t = 6$? Explain your answer.

    (c) What is the object's velocity when the object is 6 feet from the reference point?

18. A car is traveling from New York to Boston and is partway between the two cities. Let $s(t)$ be the distance from New York during the next minute. Match each behavior with the corresponding graph of $s(t)$ in Fig. 9.

    (a) The car travels at a steady speed.

(b) The car is stopped.

(c) The car is backing up.

(d) The car is accelerating.

(e) The car is decelerating.

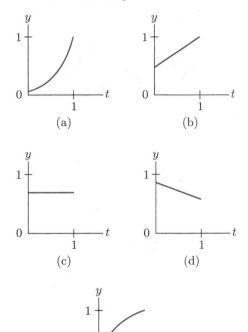

(a)

(b)

(c)

(d)

(e)

Figure 9. Possible graphs of s(t).

19. Suppose $f(100) = 5000$ and $f'(100) = 10$. Estimate each of the following.

(a) $f(101)$ (b) $f(100.5)$ (c) $f(99)$

(d) $f(98)$ (e) $f(99.75)$

20. Suppose $f(25) = 10$ and $f'(25) = -2$. Estimate each of the following.

(a) $f(27)$ (b) $f(26)$ (c) $f(25.25)$

(d) $f(24)$ (e) $f(23.5)$

21. Let $f(t)$ be the temperature of a cup of coffee, t minutes after it has been poured. Interpret $f(4) = 120$ and $f'(4) = -5$. Estimate the temperature of the coffee after 4 minutes and 6 seconds, that is, after 4.1 minutes.

22. Suppose 5 mg of a drug is injected into the bloodstream. Let $f(t)$ be the amount present in the bloodstream after t hours. Interpret $f(3) = 2$ and $f'(3) = -.5$. Estimate the number of milligrams of the drug in the bloodstream after $3\frac{1}{2}$ hours.

23. Price affects sales. Let $f(p)$ be the number of cars sold when the price is p dollars per car. Interpret the statements $f(10,000) = 200,000$ and $f'(10,000) = -3$.

24. Advertising affects sales. Let $f(x)$ be the number of toys sold when x dollars are spent on advertising. Interpret the statements $f(100,000) = 3,000,000$ and $f'(100,000) = 30$.

25. Let $C(x)$ be the cost (in dollars) of manufacturing x bicycles per day in a certain factory. Interpret $C(50) = 5000$ and $C'(50) = 45$. Estimate the cost of manufacturing 52 bicycles per day.

26. Let $C(x)$ be the cost (in dollars) of manufacturing x radios. Interpret the statements $C(2000) = 50,000$ and $C'(2000) = 10$. Estimate the cost of manufacturing 1998 radios.

27. Let $P(x)$ be the profit (in dollars) from manufacturing and selling x luxury cars. Interpret $P(100) = 90,000$ and $P'(100) = 1200$. Estimate the profit from manufacturing and selling 99 cars.

28. Let $P(x)$ be the profit from producing (and selling) x units of goods. Match each question with the proper solution.

Questions:

  A. What is the profit from producing 1000 units of goods?

  B. At what level of production will the marginal profit be 1000 dollars?

  C. What is the marginal profit from producing 1000 units of goods?

  D. For what level of production will the profit be 1000 dollars?

Solutions:

  a. Compute $P'(1000)$.

  b. Find a value of a for which $P'(a) = 1000$.

  c. Set $P(x) = 1000$ and solve for x.

  d. Compute $P(1000)$.

29. Suppose the revenue from producing (and selling) x units of a product is given by $R(x) = 3x - .01x^2$ dollars.

(a) Find the marginal revenue at a production level of 20.

(b) Find the production levels where the revenue is $200.

30. In an 8-second test run, a vehicle accelerates for several seconds and then decelerates. The function $s(t)$ gives the number of feet traveled after t seconds and is graphed in Fig. 10.

(a) How far has the vehicle traveled after 3.5 seconds?

(b) What is the velocity after 2 seconds?

(c) What is the acceleration after 1 second?

(d) When will the vehicle have traveled 120 feet?

(e) When, during the second part of the test run, will the vehicle be traveling at the rate of 20 feet per second?

(f) What is the greatest velocity? At what time is this greatest velocity reached? How far has the vehicle traveled at this time?

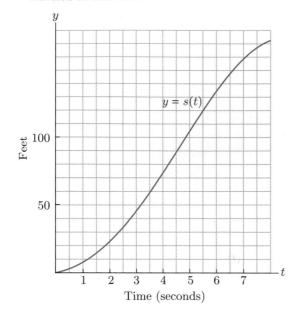

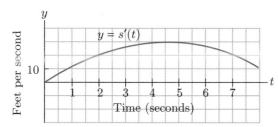

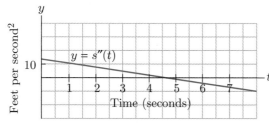

Figure 10.

31. National health expenditures (in billions of dollars) from 1980 to 1998 are given by the function $f(t)$ in Fig. 11.

(a) How much money was spent in 1987?

(b) Approximately how fast were expenditures rising in 1987?

(c) When did expenditures reach one trillion dollars?

(d) When were expenditures rising at the rate of $100 billion per year?

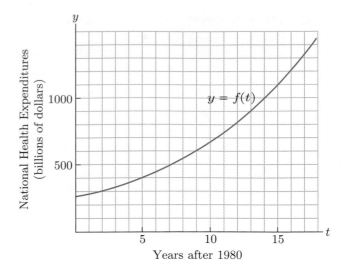

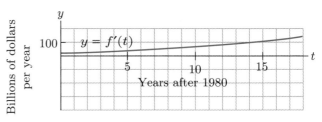

Figure 11.

Technology Exercises

32. Consider the cost function of Example 6.

(a) Graph $C(x)$ in the window $[0, 60]$ by $[-300, 1260]$.

(b) For what level of production will the cost be $535?

(c) For what level of production will the marginal cost be $14?

33. In a psychology experiment,[*] people improved their ability to recognize common verbal and semantic information with practice. Their judgment time after t days of practice was $f(t) = .36 + .77(t - .5)^{-.36}$ seconds.

(a) Display the graphs of $f(t)$ and $f'(t)$ in the window $[.5, 6]$ by $[-3, 3]$. Use these graphs to answer the following questions.

(b) What was the judgment time after 4 days of practice?

(c) After how many days of practice was the judgment time about .8 seconds?

(d) After 4 days of practice, at what rate was judgment time changing with respect to days of practice?

(e) After how many days was judgment time changing at the rate of $-.08$ seconds per day of practice?

34. A ball thrown straight up into the air has height $s(t) = 102t - 16t^2$ feet after t seconds.

---

[*]John R. Anderson, "Automaticity and the ACT Theory," *American Journal of Psychology*, 105:2 (Summer 1992), 165–180.

(a) Display the graphs of $s(t)$ and $s'(t)$ in the window $[0, 7]$ by $[-100, 200]$. Use these graphs to answer the remaining questions.

(b) How high is the ball after 2 seconds?

(c) When, during descent, is the height 110 feet?

(d) What is the velocity after 6 seconds?

(e) When is the velocity 70 feet per second?

(f) How fast is the ball traveling when it hits the ground?

Solutions to Practice Problems 1.8

1. Method (a). The question involves $f(t)$, the temperature at time t. Since the time is given, compute $f(6)$.

2. Method (e). The question involves $f'(t)$, the rate of change of temperature. The interrogative "when" indicates that the time is unknown. Set $f'(t) = 6$ and solve for t.

3. Method (f). The question asks for the change in the value of the function from time 0 to time 6, $f(6) - f(0)$.

4. Method (b). The question involves $f(t)$, and the time is unknown. Set $f(t) = 6$ and solve for t.

5. Method (d). The question involves $f'(t)$, and the time is given. Compute $f'(6)$.

6. Method (c). The question asks for the average rate of change of the function during the time interval 0 to 6, $[f(6) - f(0)]/6$.

Review of Fundamental Concepts of Chapter 1

1. Define the slope of a line and give a physical description.

2. What is the point-slope form of the equation of a line?

3. Describe how to find an equation for a line when you know the coordinates of two points on the graph of a line.

4. What can you say about the slopes of parallel lines? Perpendicular lines?

5. Give a physical description of what is meant by the slope of $f(x)$ at the point $(2, f(2))$.

6. What does $f'(2)$ represent?

7. Explain why the derivative of a constant function is 0.

8. State the power rule, constant-multiple rule, and sum rule, and give an example of each.

9. Explain how to calculate $f'(2)$ as the limit of slopes of secant lines through the point $(2, f(2))$.

10. In your own words, explain the meaning of $\lim_{x \to 2} f(x) = 3$. Give an example of a function with this property.

11. Give the limit definition of $f'(2)$, that is, the slope of $f(x)$ at the point $(2, f(2))$.

12. In your own words, explain the meaning of $\lim_{x \to \infty} f(x) = 3$. Give an example of such a function $f(x)$. Do the same for $\lim_{x \to -\infty} f(x) = 3$.

13. In your own words, explain the meaning of "$f(x)$ is continuous at $x = 2$." Give an example of a function $f(x)$ that is not continuous at $x = 2$.

14. In your own words, explain the meaning of "$f(x)$ is differentiable at $x = 2$." Give an example of a function $f(x)$ that is not differentiable at $x = 2$.

15. State the general power rule and give an example.

16. Give two different notations for the first derivative of $f(x)$ at $x = 2$. Second derivative.

17. What is meant by the average rate of change of a function over an interval?

18. How is an (instantaneous) rate of change related to average rates of change?

19. Explain the relationship between derivatives and velocity and acceleration.

20. What expression involving a derivative gives an approximation to $f(a + h) - f(a)$?

21. Describe marginal cost in your own words.

22. How do you determine the proper units for a rate of change? Give an example.

▶ Chapter 1 Supplementary Exercises

Find the equation and sketch the graph of the following lines.

1. With slope -2, y-intercept $(0, 3)$.

2. With slope $\frac{3}{4}$, y-intercept $(0, -1)$.

3. Through $(2, 0)$, with slope 5.

4. Through $(1, 4)$, with slope $-\frac{1}{3}$.

5. Parallel to $y = -2x$, passing through $(3, 5)$.

6. Parallel to $-2x + 3y = 6$, passing through $(0, 1)$.

7. Through $(-1, 4)$ and $(3, 7)$.

8. Through $(2, 1)$ and $(5, 1)$.

9. Perpendicular to $y = 3x + 4$ and passing through $(1, 2)$.

10. Perpendicular to $3x + 4y = 5$ and passing through $(6, 7)$.

11. Horizontal with height 3 units above the x-axis.

12. Vertical and 4 units to the right of the y-axis.

13. The y-axis.

14. The x-axis.

Differentiate.

15. $y = x^7 + x^3$

16. $y = 5x^8$

17. $y = 6\sqrt{x}$

18. $y = x^7 + 3x^5 + 1$

19. $y = \dfrac{3}{x}$

20. $y = x^4 - \dfrac{4}{x}$

21. $y = (3x^2 - 1)^8$

22. $y = \frac{3}{4}x^{4/3} + \frac{4}{3}x^{3/4}$

23. $y = \dfrac{1}{5x - 1}$

24. $y = (x^3 + x^2 + 1)^5$

25. $y = \sqrt{x^2 + 1}$

26. $y = \dfrac{5}{7x^2 + 1}$

27. $f(x) = 1/\sqrt[4]{x}$

28. $f(x) = (2x + 1)^3$

29. $f(x) = 5$

30. $f(x) = \dfrac{5x}{2} - \dfrac{2}{5x}$

31. $f(x) = \left[x^5 - (x - 1)^5\right]^{10}$

32. $f(t) = t^{10} - 10t^9$

33. $g(t) = 3\sqrt{t} - \dfrac{3}{\sqrt{t}}$

34. $h(t) = 3\sqrt{2}$

35. $f(t) = \dfrac{2}{t - 3t^3}$

36. $g(P) = 4P^{.7}$

37. $h(x) = \frac{3}{2}x^{3/2} - 6x^{2/3}$

38. $f(x) = \sqrt{x + \sqrt{x}}$

39. If $f(t) = 3t^3 - 2t^2$, find $f'(2)$.

40. If $V(r) = 15\pi r^2$, find $V'\left(\frac{1}{3}\right)$.

41. If $g(u) = 3u - 1$, find $g(5)$ and $g'(5)$.

42. If $h(x) = -\frac{1}{2}$, find $h(-2)$ and $h'(-2)$.

43. If $f(x) = x^{5/2}$, what is $f''(4)$?

44. If $g(t) = \frac{1}{4}(2t - 7)^4$, what is $g''(3)$?

45. Find the slope of the graph of $y = (3x - 1)^3 - 4(3x - 1)^2$ at $x = 0$.

46. Find the slope of the graph of $y = (4 - x)^5$ at $x = 5$.

Compute.

47. $\dfrac{d}{dx}(x^4 - 2x^2)$

48. $\dfrac{d}{dt}(t^{5/2} + 2t^{3/2} - t^{1/2})$

49. $\dfrac{d}{dP}(\sqrt{1 - 3P})$

50. $\dfrac{d}{dn}(n^{-5})$

51. $\dfrac{d}{dz}(z^3 - 4z^2 + z - 3)\Big|_{z=-2}$

52. $\dfrac{d}{dx}(4x - 10)^5\Big|_{x=3}$

53. $\dfrac{d^2}{dx^2}(5x + 1)^4$

54. $\dfrac{d^2}{dt^2}(2\sqrt{t})$

55. $\dfrac{d^2}{dt^2}(t^3 + 2t^2 - t)\Big|_{t=-1}$

56. $\dfrac{d^2}{dP^2}(3P + 2)\Big|_{P=4}$

57. $\dfrac{d^2y}{dx^2}$, where $y = 4x^{3/2}$

58. $\dfrac{d}{dt}\left(\dfrac{dy}{dt}\right)$, where $y = \dfrac{1}{3t}$

59. What is the slope of the graph of $f(x) = x^3 - 4x^2 + 6$ at $x = 2$? Write the equation of the line tangent to the graph of $f(x)$ at $x = 2$.

60. What is the slope of the curve $y = 1/(3x - 5)$ at $x = 1$? Write the equation of the line tangent to this curve at $x = 1$.

61. Find the equation of the tangent line to the curve $y = x^2$ at the point $\left(\frac{3}{2}, \frac{9}{4}\right)$. Sketch the graph of $y = x^2$ and sketch the tangent line at $\left(\frac{3}{2}, \frac{9}{4}\right)$.

62. Find the equation of the tangent line to the curve $y = x^2$ at the point $(-2, 4)$. Sketch the graph of $y = x^2$ and sketch the tangent line at $(-2, 4)$.

63. Determine the equation of the tangent line to the curve $y = 3x^3 - 5x^2 + x + 3$ at $x = 1$.

64. Determine the equation of the tangent line to the curve $y = (2x^2 - 3x)^3$ at $x = 2$.

65. In Fig. 1 the straight line has slope -1 and is tangent to the graph of $f(x)$. Find $f(2)$ and $f'(2)$.

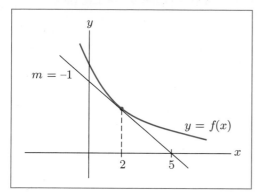

Figure 1.

66. In Fig. 2 the straight line is tangent to the graph of $f(x) = x^3$. Find the value of a.

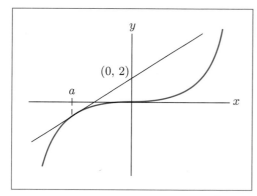

Figure 2.

67. A helicopter is rising at a rate of 32 feet per second. At a height of 128 feet the pilot drops a pair of binoculars. After t seconds, the binoculars have height $s(t) = -16t^2 + 32t + 128$ feet from the ground. How fast will they be falling when they hit the ground?

68. Each day, the total output of a coal mine after t hours of operation is approximately $40t + t^2 - \frac{1}{15}t^3$ tons, $0 \le t \le 12$. What is the rate of output (in tons of coal per hour) at $t = 5$ hours?

Exercises 69–72 refer to Fig. 3, where $s(t)$ is the number of feet traveled by a person after t seconds of walking along a straight path.

69. How far has the person traveled after 6 seconds?

70. What is the person's average velocity from time $t = 1$ to $t = 4$?

71. What is the person's velocity at time $t = 3$?

72. Without calculating velocities, determine whether the person is traveling faster at $t = 5$ or at $t = 6$.

73. A manufacturer estimates that the hourly cost of producing x units of a product on an assembly line is $C(x) = .1x^3 - 6x^2 + 136x + 200$ dollars.

(a) Compute $C(21) - C(20)$, the extra cost of raising the production from 20 to 21 units.

(b) Find the marginal cost when the production level is 20 units.

74. The number of people riding the subway daily from Silver Spring to Metro Center is a function $f(x)$ of the fare, x cents. Suppose $f(235) = 4600$ and $f'(235) = -100$. Approximate the daily number of riders for each of the following costs:

(a) 237 cents   (b) 234 cents

(c) 240 cents   (d) 232 cents

75. Let $h(t)$ be a boy's height (in inches) after t years. If $h'(12) = 1.5$, how much will his height increase (approximately) between ages 12 and $12\frac{1}{2}$?

76. If you deposit \$100 into a savings account at the end of each month for 2 years, the balance will be a function $f(r)$ of the interest rate, $r\%$. At 7% interest (compounded monthly), $f(7) = 2568.10$ and $f'(7) = 25.06$. Approximately how much additional money would you earn if the bank paid $7\frac{1}{2}\%$ interest?

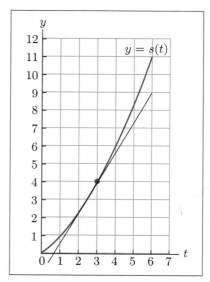

Figure 3. Walker's progress.

Determine whether the following limits exist. If so, compute the limit.

77. $\lim\limits_{x \to 2} \dfrac{x^2 - 4}{x - 2}$ **78.** $\lim\limits_{x \to 3} \dfrac{1}{x^2 - 4x + 3}$

79. $\lim\limits_{x \to 4} \dfrac{x - 4}{x^2 - 8x + 16}$ **80.** $\lim\limits_{x \to 5} \dfrac{x - 5}{x^2 - 7x + 2}$

Use limits to compute the following derivatives.

81. $f'(5)$, where $f(x) = 1/(2x)$.

82. $f'(3)$, where $f(x) = x^2 - 2x + 1$.

83. What geometric interpretation may be given to $\dfrac{(3+h)^2 - 3^2}{h}$ in connection with the graph of $f(x) = x^2$?

84. As h approaches 0, what value is approached by $\dfrac{\dfrac{1}{2+h} - \dfrac{1}{2}}{h}$?

Chapter Project

The following table gives the total number of pieces of mail and pieces of first class mail (in millions) handled by the U.S. Post Office in the years 1990–1997.

Table 1	U.S. Postal Service—Pieces of mail and pieces of first class mail (in millions), 1990–1997	

Source: U.S. Postal Service, Annual Report of the Postmaster General

Year	All Mail	First Class Mail
1990	166,301	89,270
1992	166,443	90,781
1993	171,220	92,169
1994	178,039	95,333
1995	180,734	96,296
1996	183,440	98,216
1997	190,888	99,660

Let t measure time in years, with $t = 0$ corresponding to 1990. Let

$f(t) =$ amount of mail in 12 months preceding time t

$g(t) =$ amount of first class mail in 12 months preceding time t.

The following functions give approximate models for $f(t)$ and $g(t)$:

$$f(t) = 373t^2 + 1020t + 165,378$$
$$g(t) = 72t^2 + 1081t + 88,935.$$

(a) Plot the data in the table and the graphs of the functions on a single coordinate system.

(b) Use the foregoing models to estimate the rate at which the number of pieces of mail is increasing for each of the years 1990–1997.

(c) Use the models to estimate the rate at which the number of pieces of first class mail is increasing for each of the years 1990–1997.

(d) Compare your answers to parts (b) and (c) to determine the relationship between the rates of increase of all mail versus first class mail for each of the years 1990–1997.

Applications of the Derivative

2

Calculus techniques can be applied to a wide variety of problems in real life, such as the description of the growth of a yeast culture in Chapter 1. We consider many examples in this chapter. In each case we construct a function as a "mathematical model" of some problem and then analyze the function and its derivatives in order to gain information about the original problem. Our principal method for analyzing a function will be to sketch its graph. For this reason we devote the first part of the chapter to curve sketching.

2.1   Describing Graphs of Functions

Let's examine the graph of a typical function, such as the one shown in Fig. 1, and introduce some terminology to describe its behavior. First observe that the graph is either rising or falling, depending on whether we look at it from left to right or from right to left. To avoid confusion, we shall always follow the accepted practice of reading a graph from left to right.

Let's now examine the behavior of a function $f(x)$ in an interval throughout which it is defined. We say that a function $f(x)$ is *increasing in the interval* if the graph continuously rises as x goes from from left to right through the interval.

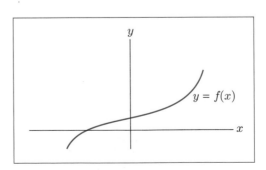

Figure 1. An increasing function.

That is, whenever x_1 and x_2 are in the interval with $x_1 < x_2$, then we have $f(x_1) < f(x_2)$. We say that $f(x)$ is increasing at $x = c$ provided that $f(x)$ is increasing in some open interval on the x-axis that contains the point c.

We say that a function $f(x)$ is *decreasing in an interval* provided that the graph continuously falls as x goes from left to right through the interval. That is, whenever x_1 and x_2 are in the interval with $x_1 < x_2$, then we have $f(x_1) > f(x_2)$. We say that $f(x)$ is decreasing at $x = c$ provided that $f(x)$ is decreasing in some open interval that contains the point c. Figure 2 shows graphs that are increasing and decreasing at $x = c$. Observe in Fig. 2(d) that when $f(c)$ is negative and $f(x)$ is decreasing, the values of $f(x)$ become *more* negative. When $f(c)$ is negative and $f(x)$ is increasing, as in Fig. 2(e), the values of $f(x)$ become *less* negative.

Extreme Points   A *relative extreme point* of a function is a point at which its graph changes from increasing to decreasing, or vice versa. We distinguish the two possibilities in an obvious way. A *relative maximum point* is a point at which the graph changes from increasing to decreasing; a *relative minimum point* is a point at which the graph changes from decreasing to increasing. (See Fig. 3.) The adjective "relative" in these definitions indicates that a point is maximal or minimal only relative to nearby points on the graph.

The *maximum value* (or *absolute maximum value*) of a function is the largest value that the function assumes on its domain. The *minimum value* (or *absolute minimum value*) of a function is the smallest value that the function assumes on its domain. Functions may or may not have maximum or minimum values. (See Fig. 4.) However, it can be shown that a continuous function whose domain is an interval of the form $a \le x \le b$ has both a maximum and a minimum value.

Maximum values and minimum values of functions usually occur at relative maximum points and relative minimum points, as in Fig. 4(a). However, they

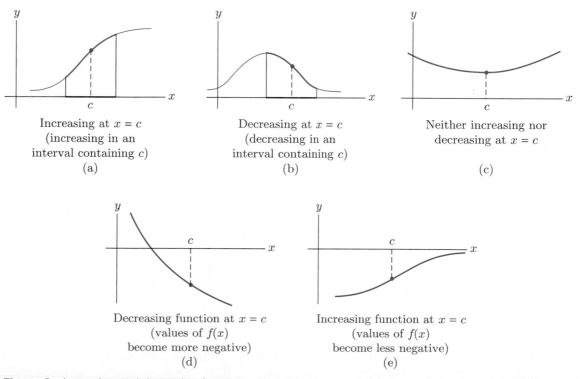

Figure 2. Increasing and decreasing functions at x = c.

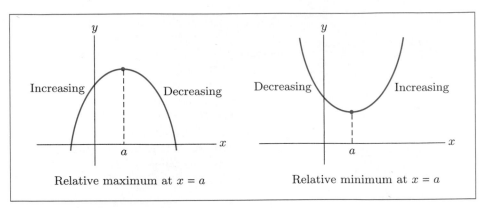

Figure 3.   Relative extreme points.

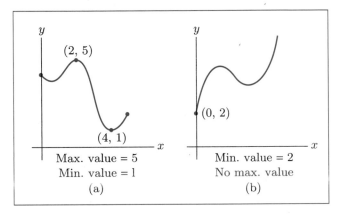

Figure 4.

can occur at endpoints of the domain, as in Fig. 4(b). If so, we say that the function has an *endpoint extreme value* (or *endpoint extremum*).

Relative maximum points and endpoint maximum points are higher than any nearby points. The maximum value of a function is the y-coordinate of the highest point on its graph. (The highest point is called the *absolute maximum point*.) Similar considerations apply to minima.

▶ **Example 1**   When a drug is injected intramuscularly (into a muscle), the concentration of the drug in the veins has the time-concentration curve shown in Fig. 5. Describe this graph, using the terms introduced above.

Solution   Initially (when $t = 0$), there is no drug in the veins. When the drug is injected into the muscle, it begins to diffuse into the bloodstream. The concentration in the veins increases until it reaches its maximum value at $t = 2$. After this time the concentration begins to decrease, as the body's metabolic processes remove the drug from the blood. Eventually the drug concentration decreases to a level so small that, for all practical purposes, it is zero.   ◆

Changing Slope   An important but subtle feature of a graph is the way the graph's slope *changes* (as we look from left to right). The graphs in Fig. 6 are both increasing, but there is a fundamental difference in the way they are increasing. Graph I, which describes the U.S. gross federal debt per person, is

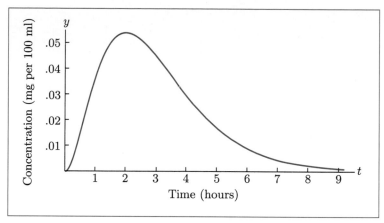

Figure 5. A drug time-concentration curve.

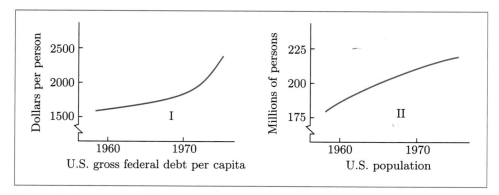

Figure 6. Increasing and decreasing slopes.

steeper for 1970 than for 1960. That is, the *slope* of graph I *increases* as we move from left to right. A newspaper description of graph I might read,

> U.S. gross federal debt per capita rose at an increasing rate during the decade from 1960 to 1970.

In contrast, the *slope* of graph II *decreases* as we move from left to right. Although the U.S. population is rising each year, the rate of increase declines throughout the decade from 1960 to 1970. That is, the slope becomes less positive. The media might say,

> During the 1960s, U.S. population rose at a decreasing rate.

▶ **Example 2** The daily number of hours of sunlight in Washington, D.C., increased from 9.45 hours on December 21 to 12 hours on March 21 and then increased to 14.9 hours on June 21. From December 22 to March 21, the daily increase was greater than the previous daily increase and from March 22 to June 21, the daily increase was less than the previous daily increase. Draw a possible graph of the number of hours of daylight as a function of time.

Solution Let $f(t)$ be the number of hours of daylight t months after December 21. See Fig. 7. The first part of the graph, December 21 to March 21, is increasing at an increasing rate. The second part of the graph, March 21 to June 21, is increasing at a decreasing rate. ◆

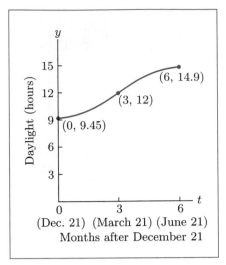

Figure 7. Hours of daylight in Washington, D.C.

Warning Recall that when a negative quantity decreases, it becomes more negative. (Think about the temperature outside when it is below zero and the temperature is falling.) So if the slope of a graph is negative and the slope is decreasing, then the slope is becoming more negative, as in Fig. 8. This technical use of the term "decreasing" runs counter to our intuition, because in popular discourse, decrease often means to become smaller in size.

It is true that the curve in Fig. 9 is becoming "less steep" in a nontechnical sense (since steepness, if it were defined, would probably refer to the magnitude of the slope). However, the slope of the curve in Fig. 9 is increasing because it is becoming less negative. The popular press would probably describe the curve in Fig. 9 as decreasing at a decreasing rate, since the rate of fall tends to taper off. Since this terminology is potentially confusing, we shall not use it.

Concavity The U.S. debt and population graphs in Fig. 6 may also be described in geometric terms: Graph I opens up and lies above its tangent line at each point, whereas graph II opens down and lies below its tangent line at each point (Fig. 10).

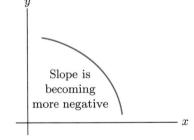

Figure 8. Slope is decreasing.

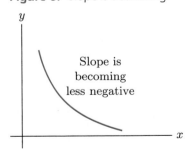

Figure 9. Slope is increasing.

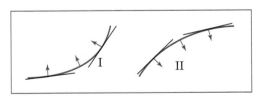

Figure 10. Relationship between concavity and tangent lines.

We say that a function $f(x)$ is *concave up* at $x = a$ if there is an open interval on the x-axis containing a throughout which the graph of $f(x)$ lies above its tangent line. Equivalently, $f(x)$ is concave up at $x = a$ if the slope of the graph increases as we move from left to right through $(a, f(a))$. Graph I is an example of a function that is concave up at each point.

Similarly, we say that a function $f(x)$ is *concave down* at $x = a$ if there is an open interval on the x-axis containing a throughout which the graph of $f(x)$ lies below its tangent line. Equivalently, $f(x)$ is concave down at $x = a$ if the slope of the graph decreases as we move from left to right through $(a, f(a))$. Graph II is concave down at each point.

An *inflection point* is a point on the graph of a function at which the function is continuous and at which the graph changes from concave up to concave down, or vice versa. At such a point, the graph crosses its tangent line (Fig. 11). (The continuity condition means that the graph cannot break at an inflection point.)

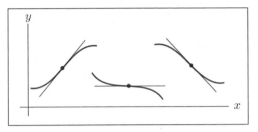

Figure 11. Inflection points.

▶ **Example 3**   Use the terms defined earlier to describe the graph shown in Fig. 12.

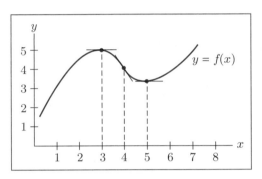

Figure 12.

Solution   (a) For $x < 3$, $f(x)$ is increasing and concave down.

(b) Relative maximum point at $x = 3$.

(c) For $3 < x < 4$, $f(x)$ is decreasing and concave down.

(d) Inflection point at $x = 4$.

(e) For $4 < x < 5$, $f(x)$ is decreasing and concave up.

(f) Relative minimum point at $x = 5$.

(g) For $x > 5$, $f(x)$ is increasing and concave up.   ◆

Intercepts, Undefined Points, and Asymptotes   A point at which a graph crosses the y-axis is called a *y-intercept*, and a point at which it crosses the x-axis is called an *x-intercept*. A function can have at most one y-intercept. Otherwise its graph would violate the vertical line test for a function. Note, however, that a function can have any number of x-intercepts (possibly none). The x-coordinate of an x-intercept is sometimes called a "zero" of the function, since the function has the value zero there. (See Fig. 13.)

Some functions are not defined for all values of x. For instance, $f(x) = 1/x$ is not defined for $x = 0$, and $f(x) = \sqrt{x}$ is not defined for $x < 0$. (See Fig. 14.) Many functions that arise in applications are defined only for $x \geq 0$. A properly drawn graph should leave no doubt as to the values of x for which the function is defined.

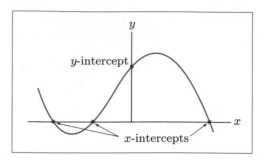

Figure 13. Intercepts of a graph.

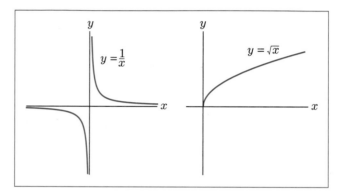

Figure 14. Graphs with undefined points.

Graphs in applied problems sometimes straighten out and approach some straight line as x gets large (Fig. 15). Such a straight line is called an *asymptote* of the curve. The most common asymptotes are horizontal as in (a) and (b) of Fig. 15. In Example 1, the t-axis is an asymptote of the drug time-concentration curve.

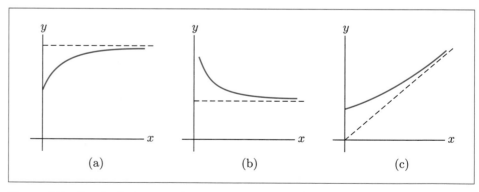

Figure 15. Graphs that approach asymptotes as x gets large.

The horizontal asymptotes of a graph may be determined by calculating the limits

$$\lim_{x \to \infty} f(x) \quad \text{and} \quad \lim_{x \to -\infty} f(x).$$

If either limit exists, then the value of the limit determines a horizontal asymptote.

Occasionally, a graph will approach a vertical line as x approaches some fixed value, as in Fig. 16. Such a line is a *vertical asymptote*. Most often, we expect a vertical asymptote at a value x that would result in division by zero in the definition of $f(x)$. For example, $f(x) = 1/(x-3)$ has a vertical asymptote $x = 3$.

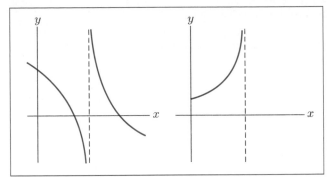

Figure 16. Examples of vertical asymptotes.

We now have six categories for describing the graph of a function.

1. Intervals in which the function is increasing (resp. decreasing), relative maximum points, relative minimum points

2. Maximum value, minimum value

3. Intervals in which the function is concave up (resp. concave down), inflection points

4. x-intercepts, y-intercept

5. Undefined points

6. Asymptotes

For us, the first three categories will be the most important. However, the last three categories should not be forgotten.

Practice Problems 2.1

1. Does the slope of the curve in Fig. 17 increase or decrease as x increases?

2. At what value of x is the slope of the curve in Fig. 18 minimized?

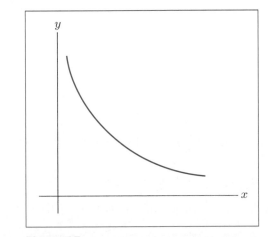

Figure 17.

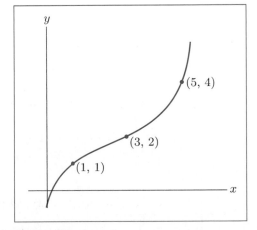

Figure 18.

▶ Exercises 2.1

Exercises 1–4 refer to graphs (a)–(f) in Fig. 19.

1. Which functions are increasing for all x?

2. Which functions are decreasing for all x?

3. Which functions have the property that the slope always increases as x increases?

4. Which functions have the property that the slope always decreases as x increases?

reading from left → right

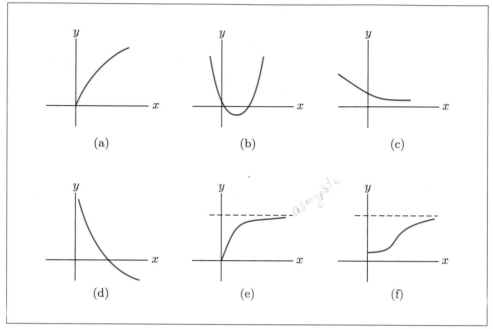

asymptote

(a)   (b)   (c)

(d)   (e)   (f)

Figure 19.

Describe each of the following graphs. Your description should include each of the six categories mentioned previously.

5.

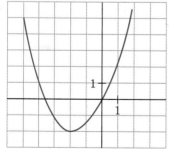

6.

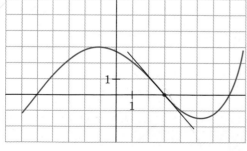

7.

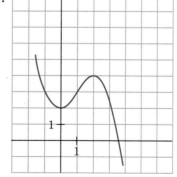

8.

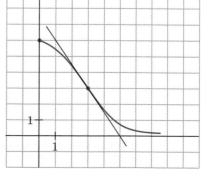

9.

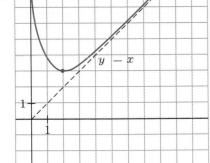

10.

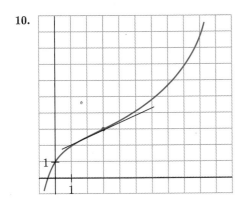

11.

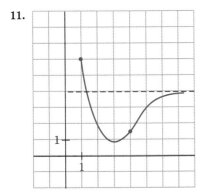

12.

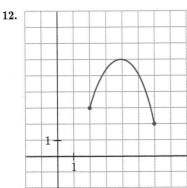

13. Describe the way the *slope* changes as you move along the graph (from left to right) in Exercise 5.

14. Describe the way the *slope* changes on the graph in Exercise 6.

15. Describe the way the *slope* changes on the graph in Exercise 8.

16. Describe the way the *slope* changes on the graph in Exercise 10.

Exercises 17 and 18 refer to the graph in Fig. 20.

17. (a) At which labeled points is the function increasing?

(b) At which labeled points is the graph concave up?

(c) Which labeled point has the most positive slope?

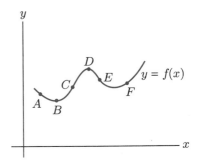

Figure 20.

18. (a) At which labeled points is the function decreasing?

(b) At which labeled points is the graph concave down?

(c) Which labeled point has the most negative slope (that is, negative and with the greatest magnitude)? *dropping most steeply closest to −∞*

In Exercises 19–22, draw the graph of a function $y = f(x)$ with the stated properties.

19. Both the function and the slope increase as x increases.

20. The function increases and the slope decreases as x increases.

21. The function decreases and the slope increases as x increases. [*Note:* The slope is negative but becomes less negative.]

22. Both the function and the slope decrease as x increases. [*Note:* The slope is negative and becomes more negative.]

23. The annual world consumption of oil rises each year. Furthermore, the amount of the annual *increase* in oil consumption is also rising each year. Sketch a graph that could represent the annual world consumption of oil.

24. In certain professions the average annual income has been rising at an increasing rate. Let $f(T)$ denote the average annual income at year T for persons in one of these professions and sketch a graph that could represent $f(T)$.

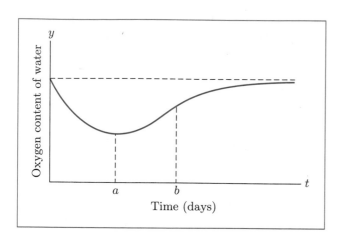

Figure 21. A lake's recovery from pollution.

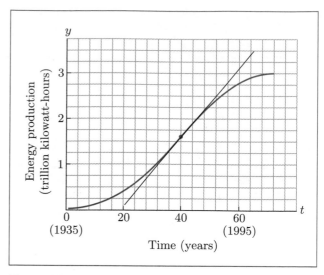

Figure 22. U.S. electrical energy production.

25. At noon a child's temperature is 101° and is rising at an increasing rate. At 1 P.M. the child is given medicine. After 2 P.M. the temperature is still increasing but at a decreasing rate. The temperature reaches a peak of 103° at 3 P.M. and decreases to 100° by 5 P.M. Draw a possible graph of the function $T(t)$, the child's temperature at time t.

26. The number of parking tickets given out each year in the District of Columbia increased from 114,000 in 1950 to 1,500,000 in 1990. Also, each year's increase was greater than the increase for the previous year. Draw a possible graph of the annual number of parking tickets as a function of time t. [*Note:* If $f(t)$ is the yearly rate t years after 1950, then $f(0) = 114$ and $f(40) = 1500$ thousand tickets.]

27. Let $C(x)$ denote the total cost of manufacturing x units of some product. Then $C(x)$ is an increasing function for all x. For small values of x, the rate of increase of $C(x)$ decreases. (This is because of the savings that are possible with "mass production.") Eventually, however, for large values of x, the cost $C(x)$ increases at an increasing rate. (This happens when production facilities are strained and become less efficient.) Sketch a graph that could represent $C(x)$.

28. One method of determining the level of blood flow through the brain requires the person to inhale air containing a fixed concentration of N_2O, nitrous oxide. During the first minute, the concentration of N_2O in the jugular vein grows at an increasing rate to a level of .25%. Thereafter it grows at a decreasing rate and reaches a concentration of about 4% after 10 minutes. Draw a possible graph of the concentration of N_2O in the vein as a function of time.

29. Suppose that some organic waste products are dumped into a lake at time $t = 0$, and suppose that the oxygen content of the lake at time t is given by the graph in

Fig. 21. Describe the graph in physical terms. Indicate the significance of the inflection point at $t = b$.

30. Figure 22 gives the U.S. electrical energy production in trillion kilowatt-hours from 1935 ($t = 0$) to 1995 ($t = 60$) with projections. In what year was the level of production growing at the greatest rate?

31. Figure 23 gives the number of U.S. farms in millions from 1920 ($t = 20$) to 1990 ($t = 90$). In what year was the number of farms decreasing most rapidly?

32. Figure 24 shows the graph of the consumer price index for the years 1976 ($t = 0$) through 1992 ($t = 16$). This index measures how much a basket of commodities that cost $100 in the beginning of 1976 would cost at any given time. In what year was the rate of increase of the index greatest? The least?

33. Let $s(t)$ be the distance (in feet) traveled by a parachutist after t seconds from the time of opening the chute, and suppose that $s(t)$ has the line $y = -15t + 10$ as an asymptote. What does this imply about the velocity of the parachutist? [*Note:* Distance traveled downward is given a negative value.]

34. Let $P(t)$ be the population of a bacteria culture after t days and suppose that $P(t)$ has the line $y = 25,000,000$ as an asymptote. What does this imply about the size of the population?

In Exercises 35–38, sketch the graph of a function having the given properties.

35. Defined for $0 \le x \le 10$; relative maximum point at $x = 3$; absolute maximum value at $x = 10$.

36. Relative maximum points at $x = 1$ and $x = 5$; relative minimum point at $x = 3$; inflection points at $x = 2$ and $x = 4$.

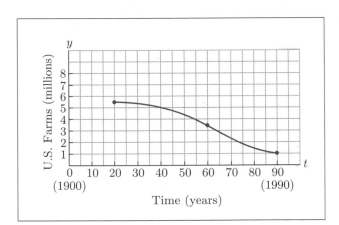

Figure 23. Number of U.S. farms.

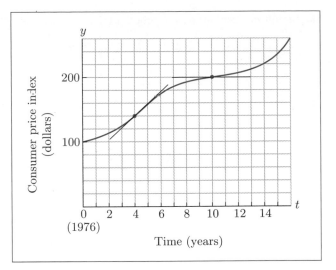

Figure 24. Consumer price index.

37. Defined and increasing for all $x \geq 0$; inflection point at $x = 5$; asymptotic to the line $y = (3/4)x + 5$.

38. Defined for $x \geq 0$; absolute minimum value at $x = 0$; relative maximum point at $x = 4$; asymptotic to the line $y = (x/2) + 1$.

39. Consider a smooth curve with no undefined points.

(a) If it has two relative maximum points, must it have a relative minimum point?

(b) If it has two relative extreme points, must it have an inflection point?

40. Suppose the function $f(x)$ has a relative minimum at $x = a$ and a relative maximum at $x = b$. Must $f(a)$ be less than $f(b)$?

The difference between the pressure inside the lungs and the pressure surrounding the lungs is called the *transmural pressure* (or transmural pressure gradient). Figure 25 shows for three different persons how the volume of the lungs is related to the transmural pressure, based on static measurements taken while there is no air flowing through the mouth. (The functional reserve capacity mentioned on the vertical axis is the volume of air in the lungs at the end of a normal expiration.) The rate of change in lung volume with respect to transmural pressure is called the *lung compliance.*

41. If the lungs are less flexible than normal, an increase in pressure will cause a smaller change in lung volume than in a normal lung. In this case, is the compliance relatively high or low?

42. Most lung diseases cause a decrease in lung compliance. However, the compliance of a person with emphysema is higher than normal. Which curve (I or II) in the figure could correspond to a person with emphysema?

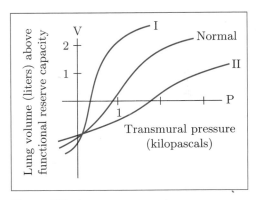

Figure 25. Lung pressure-volume curves.

Technology Exercises

43. Graph the function

$$f(x) = \frac{1}{x^3 - 2x^2 + x - 2}$$

in the window $[0, 4]$ *by* $[-15, 15]$. For what value of x does $f(x)$ have a vertical asymptote?

44. The graph of the function

$$f(x) = \frac{2x^2 - 1}{.5x^2 + 6}$$

has a horizontal asymptote of the form $y = c$. Estimate the value of c by graphing $f(x)$ in the window $[0, 50]$ *by* $[-1, 6]$.

45. Simultaneously graph the functions

$$y = \frac{1}{x} + x \quad \text{and} \quad y = x$$

in the window $[-6, 6]$ *by* $[-6, 6]$. Convince yourself that the straight line is an asymptote of the first curve. How far apart are the two graphs when $x = 6$?

46. Simultaneously graph the functions

$$y = \frac{1}{x} + x + 2 \quad \text{and} \quad y = x + 2$$

in the window $[-6, 6]$ *by* $[-4, 8]$. Convince yourself that the straight line is an asymptote of the first curve. How far apart are the two graphs when $x = 6$?

Solutions to Practice Problems 2.1

1. The curve is concave up, so the slope increases. Even though the curve itself is decreasing, the slope becomes less negative as we move from left to right.

2. At $x = 3$. We have drawn in tangent lines at various points (Fig. 26). Note that as we move from left to right, the slopes decrease steadily until the point $(3, 2)$, at which time they start to increase. This is consistent with the fact that the graph is concave down (hence, slopes are decreasing) to the left of $(3, 2)$ and concave up (hence, slopes are increasing) to the right of $(3, 2)$. Extreme values of slopes always occur at inflection points.

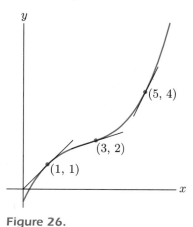

Figure 26.

2.2   The First and Second Derivative Rules

We shall now show how properties of the graph of a function $f(x)$ are determined by properties of the derivatives, $f'(x)$ and $f''(x)$. These relationships will provide the key to the curve-sketching and optimization discussed in the rest of the chapter.*

We begin with a discussion of the first derivative of a function $f(x)$. Suppose that for some value of x, say $x = a$, the derivative $f'(a)$ is positive. Then the tangent line at $(a, f(a))$ has positive slope and is a rising line (moving from left to right, of course). Since the graph of $f(x)$ near $(a, f(a))$ resembles its tangent line, the function must be increasing at $x = a$. Similarly, when $f'(a) < 0$, the function is decreasing at $x = a$. (See Fig. 1.)

Thus we have the following useful result.

First Derivative Rule   If $f'(a) > 0$, then $f(x)$ is increasing at $x = a$. If $f'(a) < 0$, then $f(x)$ is decreasing at $x = a$.

---

*Throughout this chapter, we shall assume that we are dealing with functions that are not "too badly behaved." More precisely, it suffices to assume that all of our functions have continuous first and second derivatives in the interval(s) (in x) where we are considering their graphs.

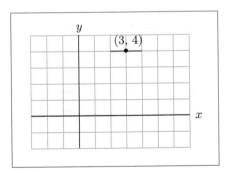

Figure 1. *Illustration of the first derivative rule.*

In other words, a function is increasing whenever the value of its derivative is positive; a function is decreasing whenever the value of its derivative is negative. The first derivative rule says nothing about the case when the derivative of a function is zero. If $f'(a) = 0$, the function might be increasing or decreasing or have a relative extreme point at $x = a$.

▶ **Example 1**   Sketch the graph of a function $f(x)$ that has all the following properties.

(a) $f(3) = 4$
(b) $f'(x) > 0$ for $x < 3$, $f'(3) = 0$, and $f'(x) < 0$ for $x > 3$

Solution   The only specific point on the graph is $(3, 4)$ [property (a)]. We plot this point and then use the fact that $f'(3) = 0$ to sketch the tangent line at $x = 3$ (Fig. 2).

From property (b) and the first derivative rule, we know that $f(x)$ must be increasing for x less than 3 and decreasing for x greater than 3. A graph with these properties might look like the curve in Fig. 3.   ◆

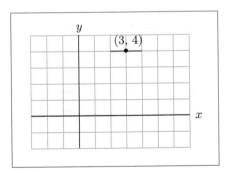

Figure 2.

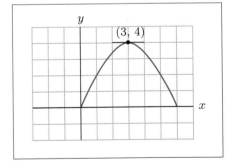

Figure 3.

The second derivative of a function $f(x)$ gives useful information about the concavity of the graph of $f(x)$. Suppose that $f''(a)$ is negative. Then since $f''(x)$ is the derivative of $f'(x)$, we conclude that $f'(x)$ has a negative derivative at $x = a$. In this case, $f'(x)$ must be a decreasing function at $x = a$; that is, the slope of the graph of $f(x)$ is decreasing as we move from left to right on the graph near $(a, f(a))$. (See Fig. 4.) This means that the graph of $f(x)$ is concave down at $x = a$. A similar analysis shows that if $f''(a)$ is positive, then $f(x)$ is concave up at $x = a$. Thus we have the following rule.

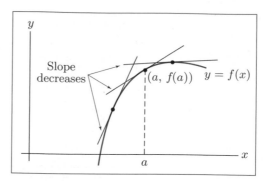

Figure 4. Illustration of the second derivative rule.

Second Derivative Rule   If $f''(a) > 0$, then $f(x)$ is concave up at $x = a$. If $f''(a) < 0$, then $f(x)$ is concave down at $x = a$.

When $f''(a) = 0$, the second derivative rule gives no information. In this case, the function might be concave up, concave down, or neither at $x = a$.

The following chart shows how a graph may combine the properties of increasing, decreasing, concave up, and concave down.

Conditions on the Derivatives	Description of $f(x)$ at $x = a$	Graph of $y = f(x)$ Near $x = a$
1. $f'(a)$ positive $f''(a)$ positive	$f(x)$ increasing $f(x)$ concave up	
2. $f'(a)$ positive $f''(a)$ negative	$f(x)$ increasing $f(x)$ concave down	
3. $f'(a)$ negative $f''(a)$ positive	$f(x)$ decreasing $f(x)$ concave up	
4. $f'(a)$ negative $f''(a)$ negative	$f(x)$ decreasing $f(x)$ concave down	

▶ **Example 2**   Sketch the graph of a function $f(x)$ with all the following properties.

(a) $(2, 3)$, $(4, 5)$, and $(6, 7)$ are on the graph.
(b) $f'(6) = 0$ and $f'(2) = 0$.
(c) $f''(x) > 0$ for $x < 4$, $f''(4) = 0$, and $f''(x) < 0$ for $x > 4$.

Solution   First we plot the three points from property (a) and then sketch two tangent lines, using the information from property (b). (See Fig. 5). From property (c) and the second derivative rule, we know that $f(x)$ is concave up for $x < 4$. In particular, $f(x)$ is concave up at $(2, 3)$. Also $f(x)$ is concave down for $x > 4$,

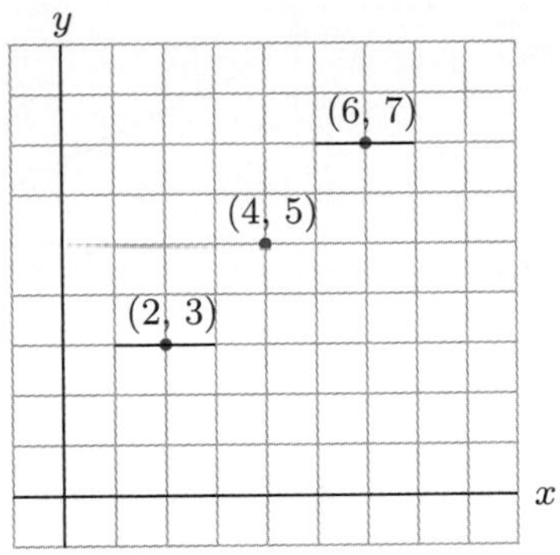

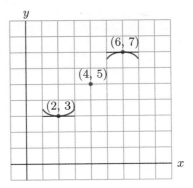

Figure 5.

Figure 6.

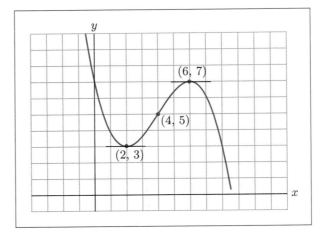

Figure 7.

in particular, at $(6, 7)$. Note that $f(x)$ must have an inflection point at $x = 4$ because the concavity changes there. We now sketch small portions of the curve near $(2, 3)$ and $(6, 7)$. (See Fig. 6.) We can now complete the sketch (Fig. 7), taking care to make the curve concave up for $x < 4$ and concave down for $x > 4$. ◆

Connections Between the Graphs of f(x) and f'(x)　Think of the derivative of $f(x)$ as a "slope-function" for $f(x)$. The "y-values" on the graph of $y = f'(x)$ are the *slopes* of the corresponding points on the original graph $y = f(x)$. This important connection is illustrated in the next three examples.

▶ Example 3　The function $f(x) = 8x - x^2$ is graphed in Fig. 8 along with the slope at several points. How is the slope changing on the graph? Compare the slopes on the graph with the y-coordinates of the points on the graph of $f'(x)$ in Fig. 9.

Solution　The slopes are decreasing (as we move from left to right). That is, $f'(x)$ is a decreasing function. Observe that the y-values of $f'(x)$ decrease to zero at $x = 4$ and then continue to decrease for x-values greater than 4. ◆

　　The graph in Fig. 8 is the shape of a typical revenue curve for a manufacturer. In this case, the graph of $f'(x)$ in Fig. 9 would be the *marginal revenue curve*. The graph in the next example has the shape of a typical cost curve. Its derivative produces a *marginal cost curve*.

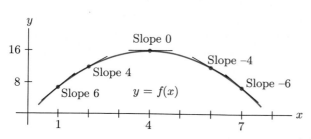

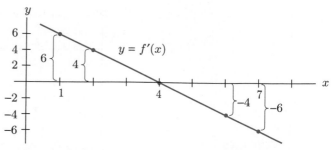

Figure 8. Graph of f(x) = 8x - x².

Figure 9. Graph of the derivative of the function in Fig. 8.

▶ **Example 4**   The function $\frac{1}{3}x^3 - 4x^2 + 18x + 10$ is graphed in Fig. 10. The slope decreases at first and then increases. Use the graph of $f'(x)$ to verify that the slope in Fig. 10 is minimum at the inflection point, where $x = 4$.

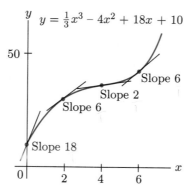

Figure 10. Graph of f(x).

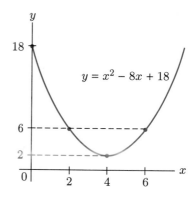

Figure 11. Graph of f'(x).

Solution   Several slopes are marked on the graph of $f(x)$. These values are y-coordinates of points on the graph of the derivative, $f'(x) = x^2 - 8x + 18$. Observe in Fig. 11 that the y-values on the graph of $f'(x)$ decrease at first and then begin to increase. The minimum value of $f'(x)$ occurs at $x = 4$.   ◆

▶ **Example 5**   Figure 12 shows the graph of $y = f'(x)$, the derivative of a function $f(x)$.

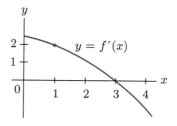

Figure 12.

(a) What is the slope of the graph of $f(x)$ when $x = 1$?
(b) Describe how the values of $f'(x)$ change on the interval $1 \le x \le 2$.
(c) Describe the shape of the graph of $f(x)$ on the interval $1 \le x \le 2$.
(d) For what values of x does the graph of $f(x)$ have a horizontal tangent line?
(e) Explain why $f(x)$ has a relative maximum at $x = 3$.

Solution (a) Since $f'(1)$ is 2, $f(x)$ has slope 2 when $x = 1$.

(b) The values of $f'(x)$ are positive and decreasing as x increases from 1 to 2.

(c) On the interval $1 \leq x \leq 2$, the slope of the graph of $f(x)$ is positive and is decreasing as x increases. Therefore, the graph of $f(x)$ is increasing and concave down.

(d) The graph of $f(x)$ has a horizontal tangent line when the slope is 0—that is, when $f'(x)$ is 0. This occurs at $x = 3$.

(e) Since $f'(x)$ is positive to the left of $x = 3$ and negative to the right of $x = 3$, the graph of $f(x)$ changes from increasing to decreasing at $x = 3$. Therefore, $f(x)$ has a relative maximum at $x = 3$. ◆

Figure 13 shows the graph of a function $f(x)$ whose derivative has the shape shown in Fig. 12. Reread Example 5 and its solution while referring to Fig. 13.

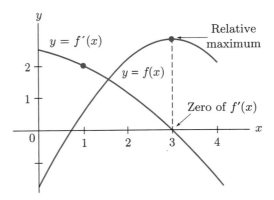

Figure 13.

1. Make a good sketch of the function $f(x)$ near the point where $x = 2$, given that $f(2) = 5$, $f'(2) = 1$, and $f''(2) = -3$.

2. The graph of $f(x) = x^3$ is shown in Fig. 14.

   (a) Is the function increasing at $x = 0$?

   (b) Compute $f'(0)$.

   (c) Reconcile your answers to parts (a) and (b) with the first derivative rule.

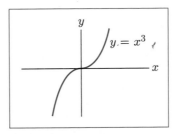

Figure 14.

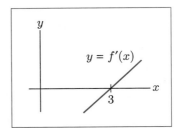

Figure 15.

3. The graph of $y = f'(x)$ is shown in Fig. 15. Explain why $f(x)$ must have a relative minimum point at $x = 3$.

▶ Exercises 2.2

Exercises 1–4 refer to the functions whose graphs are given in Fig. 16.

1. Which functions *are increasing* have a positive first derivative for all x?

2. Which functions *are decreasing* have a negative first derivative for all x?

3. Which functions *are concave up* have a positive second derivative for all x?

4. Which functions *are concave down* have a negative second derivative for all x?

5. Which one of the graphs in Fig. 17 could represent a function $f(x)$ for which $f(a) > 0$, $f'(a) = 0$, and $f''(a) < 0$?

6. Which one of the graphs in Fig. 17 could represent a function $f(x)$ for which $f(a) = 0$, $f'(a) < 0$, and $f''(a) > 0$?

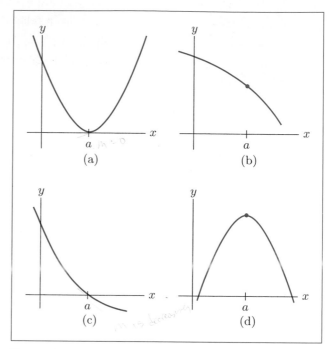

Figure 17.

In Exercises 7–12, sketch the graph of a function that has the properties described.

7. $f(2) = 1$; $f'(2) = 0$; concave up for all x.

8. $f(-1) = 0$; $f'(x) < 0$ for $x < -1$, $f'(-1) = 0$ and $f'(x) > 0$ for $x > -1$.

9. $f(3) = 5$; $f'(x) > 0$ for $x < 3$, $f'(3) = 0$ and $f'(x) > 0$ for $x > 3$.

10. $(-2, -1)$ and $(2, 5)$ are on the graph; $f'(-2) = 0$ and $f'(2) = 0$; $f''(x) > 0$ for $x < 0$, $f''(0) = 0$, $f''(x) < 0$ for $x > 0$.

11. $(0, 6)$, $(2, 3)$, and $(4, 0)$ are on the graph; $f'(0) = 0$ and $f'(4) = 0$; $f''(x) < 0$ for $x < 2$, $f''(2) = 0$, $f''(x) > 0$ for $x > 2$.

12. $f(x)$ defined only for $x \geq 0$; $(0, 0)$ and $(5, 6)$ are on the graph; $f'(x) > 0$ for $x \geq 0$; $f''(x) < 0$ for $x < 5$, $f''(5) = 0$, $f''(x) > 0$ for $x > 5$.

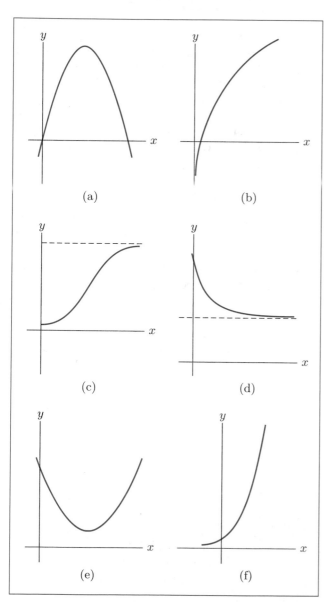

Figure 16.

In Exercises 13–18, use the given information to make a good sketch of the function $f(x)$ near $x = 3$.

13. $f(3) = 4$, $f'(3) = -\frac{1}{2}$, $f''(3) = 5$

14. $f(3) = -2$, $f'(3) = 0$, $f''(3) = 1$

15. $f(3) = 1$, $f'(3) = 0$, inflection point at $x = 3$, $f'(x) > 0$ for $x > 3$

16. $f(3) = 4$, $f'(3) = -\frac{3}{2}$, $f''(3) = -2$

17. $f(3) = -2$, $f'(3) = 2$, $f''(3) = 3$

18. $f(3) = 3$, $f'(3) = 1$, inflection point at $x = 3$, $f''(x) < 0$ for $x > 3$

19. Refer to the graph in Fig. 18. Fill in each entry of the grid with POS, NEG, or 0.

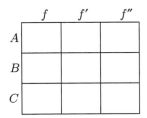

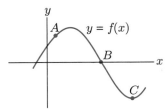

Figure 18.

20. The first and second derivatives of the function $f(x)$ have the values given in Table 1.

(a) Find the x-coordinates of all relative extreme points.

(b) Find the x-coordinates of all inflection points.

Table 1	Values of the First Two Derivatives of a Function	
x	$f'(x)$	$f''(x)$
$0 \le x < 2$	Positive	Negative
2	0	Negative
$2 < x < 3$	Negative	Negative
3	Negative	0
$3 < x < 4$	Negative	Positive
4	0	0
$4 < x \le 6$	Negative	Negative

21. Suppose that Fig. 19 contains the graph of $y = s(t)$, the distance traveled by a car after t hours. Is the car going faster at $t = 1$ or $t = 2$?

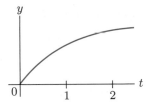

Figure 19.

22. Suppose that Fig. 19 contains the graph of $y = v(t)$, the velocity of a car after t hours. Is the car going faster at $t = 1$ or $t = 2$?

23. Refer to Fig. 20.

(a) Looking at the graph of $f'(x)$, determine whether $f(x)$ is increasing or decreasing at $x = 9$. Look at the graph of $f(x)$ to confirm your answer.

(b) Looking at the values of $f'(x)$ for $1 \le x < 2$ and $2 < x \le 3$, explain why the graph of $f(x)$ must have a relative maximum at $x = 2$. What are the coordinates of the relative maximum point?

(c) Looking at the values of $f'(x)$ for x close to 10, explain why the graph of $f(x)$ has a relative minimum at $x = 10$.

(d) Looking at the graph of $f''(x)$, determine whether $f(x)$ is concave up or concave down at $x = 2$. Look at the graph of $f(x)$ to confirm your answer.

(e) Looking at the graph of $f''(x)$, determine where $f(x)$ has an inflection point. Look at the graph of $f(x)$ to confirm your answer. What are the coordinates of the inflection point?

(f) Find the x-coordinate of the point on the graph of $f(x)$ at which $f(x)$ is increasing at the rate of 6 units per unit change in x.

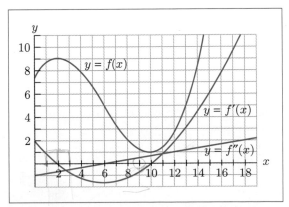

Figure 20.

24. In Fig. 21, the t-axis represents time in minutes.

  (a) What is $f(2)$?

  (b) Solve $f(t) = 1$.

  (c) When does $f(t)$ attain its greatest value?

  (d) When does $f(t)$ attain its least value?

  (e) What is the rate of change of $f(t)$ at $t = 7.5$?

  (f) When is $f(t)$ decreasing at the rate of 1 unit per minute? That is, when is the rate of change equal to -1?

  (g) When is $f(t)$ decreasing at the greatest rate?

  (h) When is $f(t)$ increasing at the greatest rate?

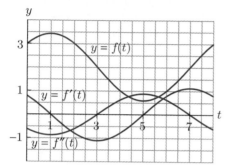

Figure 21.

Exercises 25–36 refer to Fig. 22, which contains the graph of $f'(x)$, the derivative of the function $f(x)$.

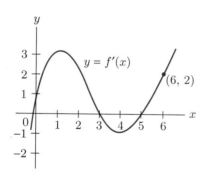

Figure 22.

25. Explain why $f(x)$ must be increasing at $x = 6$.

26. Explain why $f(x)$ must be decreasing at $x = 4$.

27. Explain why $f(x)$ has a relative maximum at $x = 3$.

28. Explain why $f(x)$ has a relative minimum at $x = 5$.

29. Explain why $f(x)$ must be concave up at $x = 0$.

30. Explain why $f(x)$ must be concave down at $x = 2$.

31. Explain why $f(x)$ has an inflection point at $x = 1$.

32. Explain why $f(x)$ has an inflection point at $x = 4$.

33. If $f(6) = 3$, what is the equation of the tangent line to the graph of $y = f(x)$ at $x = 6$?

34. If $f(6) = 8$, what is an approximate value of $f(6.5)$?

35. If $f(0) = 3$, what is an approximate value of $f(.25)$?

36. If $f(0) = 3$, what is the equation of the tangent line to the graph of $y = f(x)$ at $x = 0$?

37. Suppose melting snow causes a river to overflow its banks and $h(t)$ is the number of inches of water on Main Street t hours after the melting begins.

  (a) If $h'(100) = \frac{1}{3}$, by approximately how much will the water level change during the next half-hour?

  (b) Which of the following two conditions are the better news?

    (i) $h(100) = 3$, $h'(100) = 2$, $h''(100) = -5$

    (ii) $h(100) = 3$, $h'(100) = -2$, $h''(100) = 5$

38. Suppose $T(t)$ is the temperature on a hot summer day at time t hours.

  (a) If $T'(10) = 4$, by approximately how much will the temperature rise from 10:00 to 10:45?

  (b) Which of the following two conditions are the better news?

    (i) $T(10) = 95$, $T'(10) = 4$, $T''(10) = -3$

    (ii) $T(10) = 95$, $T'(10) = -4$, $T''(10) = 3$

39. By looking at the first derivative, decide which of the curves in Fig. 23 could *not* be the graph of $f(x) = (3x^2 + 1)^4$ for $x \geq 0$.

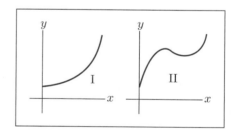

Figure 23.

40. By looking at the first derivative, decide which of the curves in Fig. 23 could *not* be the graph of $f(x) = x^3 - 9x^2 + 24x + 1$ for $x \geq 0$. [*Hint:* Factor the formula for $f'(x)$.]

41. By looking at the second derivative, decide which of the curves in Fig. 24 could be the graph of $f(x) = x^{5/2}$.

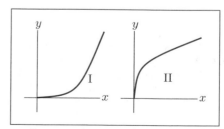

Figure 24.

42. Match each observation (a)–(e) with a conclusion (A)–(E).

Observations:

   (a) The point $(3, 4)$ is on the graph of $f'(x)$.

   (b) The point $(3, 4)$ is on the graph of $f(x)$.

   (c) The point $(3, 4)$ is on the graph of $f''(x)$.

   (d) The point $(3, 0)$ is on the graph of $f'(x)$ and the point $(3, 4)$ is on the graph of $f''(x)$.

   (e) The point $(3, 0)$ is on the graph of $f'(x)$ and the point $(3, -4)$ is on the graph of $f''(x)$.

Conclusions:

   (A) $f(x)$ has a relative minimum point at $x = 3$.

   (B) When $x = 3$, the graph of $f(x)$ is concave up.

   (C) When $x = 3$, the tangent line to the graph of $y = f(x)$ has slope 4.

   (D) When $x = 3$, the value of $f(x)$ is 4.

   (E) $f(x)$ has a relative maximum point at $x = 3$.

43. The number of farms in the United States t years after 1925 is $f(t)$ million, where f is the function graphed in Fig. 25(a). (The graphs of $f'(t)$ and $f''(t)$ are shown in Fig. 25(b).)

   (a) Approximately how many farms were there in 1990?

   (b) At what rate was the number of farms declining in 1990?

   (c) In what year were there about 6 million farms?

   (d) When was the number of farms declining at the rate of 60,000 farms per year?

   (e) When was the number of farms declining fastest?

44. After a drug is taken orally, the amount of the drug in the bloodstream after t hours is $f(t)$ units. Figure 26 shows partial graphs of $f'(t)$ and $f''(t)$.

   (a) Is the amount of the drug in the bloodstream increasing or decreasing at $t = 5$ hours?

   (b) Is the graph of $f(t)$ concave up or concave down at $t = 5$ hours?

   (c) When is the level of the drug in the bloodstream decreasing the fastest?

   (d) At what time is the greatest level of drug in the bloodstream reached?

   (e) When is the level of the drug in the bloodstream decreasing at the rate of 3 units per hour?

Technology Exercises

In Exercises 45 and 46, display the graph of the *derivative* of $f(x)$ in the specified window. Then use the graph of $f'(x)$ to determine the approximate values of x at which the graph of $f(x)$ has relative extreme points and inflection points. Then check your conclusions by displaying the graph of $f(x)$.

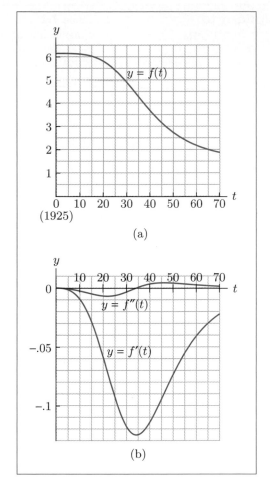

(a)

(b)

Figure 25.

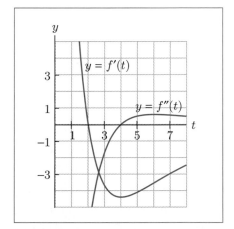

Figure 26.

45. $3x^5 - 20x^3 - 120x$; $[-4, 4]$ *by* $[-325, 325]$

46. $x^4 - x^2$; $[-1.5, 1.5]$ *by* $[-.75, 1]$

1. Since $f(2) = 5$, the point $(2, 5)$ is on the graph [Fig. 27(a)]. Since $f'(2) = 1$, the tangent line at the point $(2, 5)$ has slope 1. Draw in the tangent line [Fig. 27(b)]. Near the point $(2, 5)$ the graph looks approximately like the tangent line. Since $f''(2) = -3$, a negative number, the graph is concave down at the point $(2, 5)$. Now we are ready to sketch the graph [Fig. 27(c)].

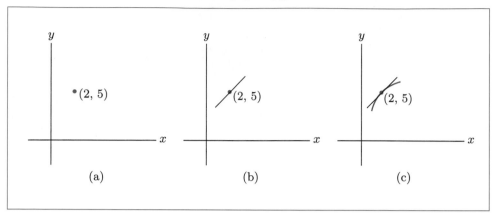

Figure 27.

2. **(a)** Yes. The graph is steadily increasing as we pass through the point $(0, 0)$.

　　(b) Since $f'(x) = 3x^2$, $f'(0) = 3 \cdot 0^2 = 0$.

　　(c) There is no contradiction here. The first derivative rule says that if the derivative is positive, the function is increasing. However, it does not say that this is the only condition under which a function is increasing. As we have just seen, sometimes we can have the first derivative zero and the function still increasing.

3. Since $f'(x)$, the derivative of $f(x)$, is negative to the left of $x = 3$ and positive to the right of $x = 3$, $f(x)$ is decreasing to the left of $x = 3$ and increasing to the right of $x = 3$. Therefore, by the definition of a relative minimum point, $f(x)$ has a relative minimum point at $x = 3$.

2.3 Curve Sketching (Introduction)

In this section and the next we develop our ability to sketch the graphs of functions. There are two important reasons for doing so. First, a geometric "picture" of a function is often easier to comprehend than its abstract formula. Second, the material in this section will provide a foundation for the applications in Sections 2.5 through 2.7.

　　A "sketch" of the graph of a function $f(x)$ should convey the general shape of the graph—it should show where $f(x)$ is defined and where it is increasing and decreasing and it should indicate, insofar as possible, where $f(x)$ is concave up and concave down. In addition, one or more key points should be accurately located on the graph. These points usually include extreme points, inflection points, and x- and y-intercepts. Other features of a graph may be important, too, but we shall discuss them as they arise in examples and applications.

　　Our general approach to curve sketching will involve four main steps:

　　1. Starting with $f(x)$, we compute $f'(x)$ and $f''(x)$.

2. Next, we locate all relative maximum and relative minimum points and make a partial sketch.

3. We study the concavity of $f(x)$ and locate all inflection points.

4. We consider other properties of the graph, such as the intercepts, and complete the sketch.

The first step was the main subject of the preceding chapter. We discuss the second and third steps in this section and then present several completely worked examples that combine all four steps in the next section.

Locating Relative Extreme Points   The tangent line at a relative maximum or a relative minimum point of a function $f(x)$ has zero slope; that is, the derivative is zero there. Thus we may state the following useful rule.

> Look for possible relative extreme points of
> $f(x)$ by setting $f'(x) = 0$ and solving for x.   (1)

Suppose that $f'(a) = 0$. Then $x = a$ is a candidate for a relative extreme point of $f(x)$. There are several ways to determine if $f(x)$ has a relative maximum or a relative minimum (or neither) at $x = a$. The method that works in most cases is described in our first two examples.

▶ **Example 1**   The graph of the quadratic function $f(x) = \frac{1}{4}x^2 - x + 2$ is a parabola and so has one relative extreme point. Find it and sketch the graph.

Solution   We begin by computing the first and second derivatives of $f(x)$.

$$f(x) = \tfrac{1}{4}x^2 - x + 2$$

$$f'(x) = \tfrac{1}{2}x - 1$$

$$f''(x) = \tfrac{1}{2}$$

Setting $f'(x) = 0$, we have $\frac{1}{2}x - 1 = 0$, so that $x = 2$. Thus, $f'(2) = 0$. Geometrically, this means that the graph of $f(x)$ will have a horizontal tangent line at the point where $x = 2$. To plot this point, we substitute the value 2 for x in the original expression for $f(x)$.

$$f(2) = \tfrac{1}{4}(2)^2 - (2) + 2 = 1$$

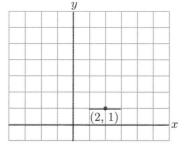

Figure 1.

Figure 1 shows the point $(2, 1)$ together with the horizontal tangent line. Is $(2, 1)$ a relative extreme point? In order to decide, we look at $f''(x)$. Since $f''(x) = \frac{1}{2}$, which is positive, the graph of $f(x)$ is concave up at $x = 2$. So a partial sketch of the graph near $(2, 1)$ should look something like Fig. 2.

We see that $(2, 1)$ is a relative minimum point. In fact, it is the only relative extreme point, for there is no other place where the tangent line is horizontal. Since the graph has no other "turning points," it must be decreasing before it gets to $(2, 1)$ and then increasing to the right of $(2, 1)$. Note that since $f''(x)$ is positive (and equal to $\frac{1}{2}$) for all x, the graph is concave up at each point. A completed sketch is given in Fig. 3.   ◆

▶ **Example 2**   Locate all possible relative extreme points on the graph of the function $f(x) = x^3 - 3x^2 + 5$. Check the concavity at these points and use this information to sketch the graph of $f(x)$.

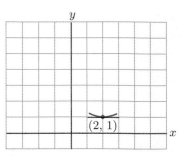

Figure 2.

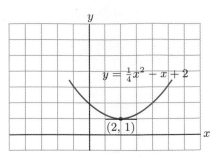

$$y = \tfrac{1}{4}x^2 - x + 2$$

Figure 3.

Solution   We have

$$f(x) = x^3 - 3x^2 + 5$$
$$f'(x) = 3x^2 - 6x$$
$$f''(x) = 6x - 6.$$

The easiest way to find those values of x for which $f'(x)$ is zero is to factor the expression for $f'(x)$:

$$3x^2 - 6x = 3x(x-2).$$

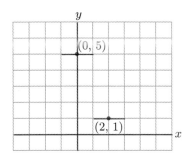

Figure 4.

From this factorization it is clear that $f'(x)$ will be zero if and only if $x = 0$ or $x = 2$. In other words, the graph will have horizontal tangent lines when $x = 0$ and $x = 2$, and nowhere else.

To plot the points on the graph where $x = 0$ and $x = 2$, we substitute these values back into the original expression for $f(x)$. That is, we compute

$$f(0) = (0)^3 - 3(0)^2 + 5 = 5$$
$$f(2) = (2)^3 - 3(2)^2 + 5 = 1.$$

Figure 4 shows the points $(0,5)$ and $(2,1)$, along with the corresponding tangent lines.

Next, we check the concavity of the graph at these points by evaluating $f''(x)$ at $x = 0$ and $x = 2$:

$$f''(0) = 6(0) - 6 = -6$$
$$f''(2) = 6(2) - 6 = +6.$$

Since $f''(0)$ is negative, the graph is concave down at $x = 0$; since $f''(2)$ is positive, the graph is concave up at $x = 2$. A partial sketch of the graph is given in Fig. 5.

It is clear from Fig. 5 that $(0,5)$ is a relative maximum point and $(2,1)$ is a relative minimum point. Since they are the only turning points, the graph must be increasing before it gets to $(0,5)$, decreasing from $(0,5)$ to $(2,1)$, and then increasing again to the right of $(2,1)$. A sketch incorporating these properties appears in Fig. 6.   ◆

The facts that we used to sketch Fig. 6 could equally well be used to produce the graph in Fig. 7. Which graph really corresponds to $f(x) = x^3 - 3x^2 + 5$? The answer will be clear when we find the inflection points on the graph of $f(x)$.

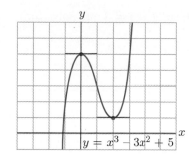

Figure 5.

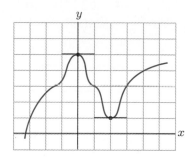

Figure 6.   **Figure 7.**

Locating Inflection Points   An inflection point of a function $f(x)$ can occur only at a value of x for which $f''(x)$ is zero, because the curve is concave up where $f''(x)$ is positive and concave down where $f''(x)$ is negative. Thus we have the following test.

> Look for possible inflection points by setting
> $f''(x) = 0$ and solving for x.   (2)

Once we have a value of x where the second derivative is zero, say at $x = b$, we must check the concavity of $f(x)$ at nearby points to see if the concavity really changes at $x = b$.

▶ **Example 3**   Find the inflection points of the function $f(x) = x^3 - 3x^2 + 5$ and explain why the graph in Fig. 6 has the correct shape.

Solution   From Example 2 we have $f''(x) = 6x - 6 = 6(x - 1)$. Clearly, $f''(x) = 0$ if and only if $x = 1$. We will want to plot the corresponding point on the graph, so we compute

$$f(1) = (1)^3 - 3(1)^2 + 5 = 3.$$

Therefore, the only possible inflection point is $(1, 3)$.

Now look back at Fig. 5, where we indicated the concavity of the graph at the relative extreme points. Since $f(x)$ is concave down at $(0, 5)$ and concave up at $(2, 1)$, the concavity must reverse somewhere between these points. Hence $(1, 3)$ must be an inflection point. Furthermore, since the concavity of $f(x)$ reverses nowhere else, the concavity at all points to the left of $(1, 3)$ must be the same (i.e., concave down). Similarly, the concavity of all points to the right of $(1, 3)$ must be the same (i.e., concave up). Thus the graph in Fig. 6 has the correct shape. The graph in Fig. 7 has too many "wiggles," caused by frequent changes in concavity; that is, there are too many inflection points. A correct sketch showing the one inflection point at $(1, 3)$ is given in Fig. 8.   ◆

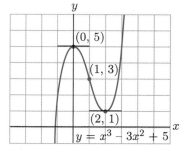

Figure 8.

▶ **Example 4**   Sketch the graph of $y = -\frac{1}{3}x^3 + 3x^2 - 5x$.

Solution   Let

$$f(x) = -\frac{1}{3}x^3 + 3x^2 - 5x.$$

Then

$$f'(x) = -x^2 + 6x - 5$$
$$f''(x) = -2x + 6.$$

We set $f'(x) = 0$ and solve for x:

$$-(x^2 - 6x + 5) = 0$$
$$-(x - 1)(x - 5) = 0$$
$$x = 1 \quad \text{or} \quad x = 5.$$

Substituting these values of x back into $f(x)$, we find that

$$f(1) = -\tfrac{1}{3}(1)^3 + 3(1)^2 - 5(1) = -\tfrac{7}{3}$$
$$f(5) = -\tfrac{1}{3}(5)^3 + 3(5)^2 - 5(5) = \tfrac{25}{3}.$$

The information we have so far is given in Fig. 9(a). The sketch in Fig. 9(b) is obtained by computing

$$f''(1) = -2(1) + 6 = 4$$
$$f''(5) = -2(5) + 6 = -4.$$

The curve is concave up at $x = 1$ because $f''(1)$ is positive, and the curve is concave down at $x = 5$ because $f''(5)$ is negative.

Since the concavity reverses somewhere between $x = 0$ and $x = 5$, there must be at least one inflection point. If we set $f''(x) = 0$, we find that

$$-2x + 6 = 0$$
$$x = 3.$$

So the inflection point must occur at $x = 3$. In order to plot the inflection point, we compute

$$f(3) = -\tfrac{1}{3}(3)^3 + 3(3)^2 - 5(3) = 3.$$

The final sketch of the graph is given in Fig. 10. ◆

The argument in Example 4 that there must be an inflection point because concavity reverses is valid whenever $f(x)$ is a polynomial. However, it does not always apply to a function whose graph has a break in it. For example, the

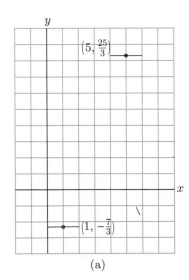

(a)

Figure 9.

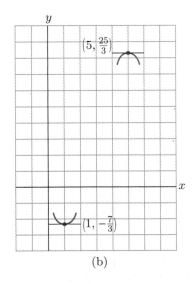

(b)

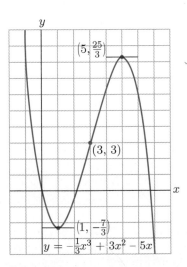

Figure 10.

function $f(x) = 1/x$ is concave down at $x = -1$ and concave up at $x = 1$, but there is no inflection point in between.

A summary of curve-sketching techniques appears at the end of the next section. You may find steps 1, 2, and 3 helpful when working the exercises.

**Practice Problems
2.3**

1. Which of the curves in Fig. 11 could possibly be the graph of a function of the form $f(x) = ax^2 + bx + c$, where $a \neq 0$?

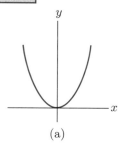

(a)

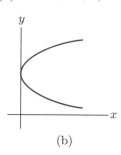

(b)

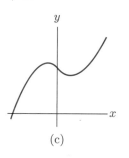

(c)

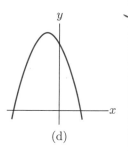

(d)

Figure 11.

2. Which of the curves in Fig. 12 could possibly be the graph of a function of the form $f(x) = ax^3 + bx^2 + cx + d$, where $a \neq 0$?

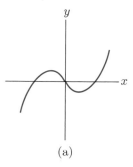

(a)

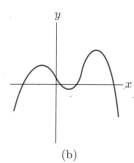

(b)

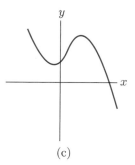

(c)

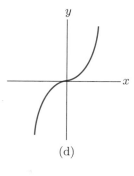

(d)

Figure 12.

▶ Exercises 2.3

Each of the graphs of the functions in Exercises 1–8 has one relative extreme point. Plot this point and check the concavity there. Using only this information, sketch the graph. [As you work the problems, observe that if $f(x) = ax^2 + bx + c$, then $f(x)$ has a relative minimum point when $a > 0$ and a relative maximum point when $a < 0$.]

1. $f(x) = 2x^2 - 8$ **2.** $f(x) = 3x^2 + 6x - 5$

3. $f(x) = \frac{1}{2}x^2 + x - 4$ **4.** $f(x) = -\frac{1}{2}x^2 + x - 4$

5. $f(x) = 1 + 6x - x^2$ **6.** $f(x) = 1 + x + x^2$

7. $f(x) = -x^2 - 8x - 10$ **8.** $f(x) = -3x^2 + 18x - 20$

Each of the graphs of the functions in Exercises 9–16 has one relative maximum and one relative minimum point. Plot these two points and check the concavity there. Using only this information, sketch the graph.

9. $f(x) = x^3 + 6x^2 + 9x$ **10.** $f(x) = \frac{1}{9}x^3 - x^2$

11. $f(x) = x^3 - 12x$ **12.** $f(x) = -\frac{1}{3}x^3 + 9x - 2$

13. $f(x) = -\frac{1}{9}x^3 + x^2 + 9x$

14. $f(x) = 2x^3 - 15x^2 + 36x - 24$

15. $f(x) = -\frac{1}{3}x^3 + 2x^2 - 12$

16. $f(x) = \frac{1}{3}x^3 + 2x^2 - 5x + \frac{8}{3}$

Sketch the following curves, indicating all relative extreme points and inflection points.

17. $y = x^3 - 3x + 2$ **18.** $y = x^3 - 6x^2 + 9x + 3$

19. $y = 1 + 3x^2 - x^3$ **20.** $y = -x^3 + 12x - 4$

21. $y = \frac{1}{3}x^3 - x^2 - 3x + 5$ **22.** $y = x^3 + \frac{3}{2}x^2 - 6x + 4$

23. $y = 2x^3 - 3x^2 - 36x + 20$

24. $y = 11 + 9x - 3x^2 - x^3$

25. Let a, b, c be fixed numbers with $a \neq 0$ and let $f(x) = ax^2 + bx + c$. Is it possible for the graph of $f(x)$ to have an inflection point? Explain your answer.

26. Let a, b, c, d be fixed numbers with $a \neq 0$ and let $f(x) = ax^3 + bx^2 + cx + d$. Is it possible for the graph of $f(x)$ to have more than one inflection point? Explain your answer.

The graph of each function in Exercises 27–32 has one relative extreme point. Find it (giving both x- and y-coordinates) and determine if it is a relative maximum or a relative minimum point. Do not include a sketch of the graph of the function.

27. $f(x) = \frac{1}{4}x^2 - 2x + 7$

28. $f(x) = 5 - 12x - 2x^2$

29. $g(x) = 3 + 4x - 2x^2$

30. $g(x) = x^2 + 10x + 10$

31. $f(x) = 5x^2 + x - 3$

32. $f(x) = 30x^2 - 1800x + 29{,}000$

In Exercises 33 and 34, determine which function is the derivative of the other.

33.

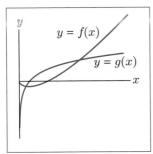

34.

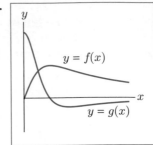

35. Consider the graph of $g(x)$ in Fig. 13.

(a) If $g(x)$ is the first derivative of $f(x)$, what is the nature of $f(x)$ when $x = 2$?

(b) If $g(x)$ is the second derivative of $f(x)$, what is the nature of $f(x)$ when $x = 2$?

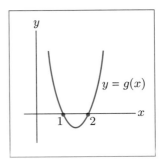

Figure 13.

36. The population (in millions) of the United States (excluding Alaska and Hawaii) t years after 1800 is given by the function $f(t)$ in Fig. 14(a). The graphs of $f'(t)$ and $f''(t)$ are shown in Figs. 14(b) and 14(c).

(a) What was the population in 1925?

(b) Approximately when was the population 25 million?

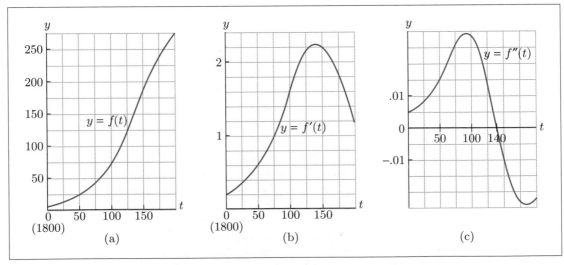

Figure 14. Population (in millions) of the United States from 1800 to 1998.

(c) How fast was the population growing in 1950?

(d) When during the last 50 years was the population growing at the rate of 1.8 million people per year?

(e) In what year was the population growing at the greatest rate?

37. The function $f(t)$ shown in Fig. 15 gives the percentage of homes (with TV sets) that also had a VCR in year t, where $t = 0$ corresponds to 1980.

(a) In what year did 10% of the homes have a VCR?

(b) What percentage of homes had a VCR in 1990?

(c) When was the percentage of homes with TVs that also had VCRs growing at the greatest rate? Approximately what percentage of homes with TVs also had VCRs at that time?

Technology Exercises

38. Draw the graph of $f(x) = \frac{1}{6}x^3 - x^2 + 3x + 3$ in the window $[-2, 6]$ by $[-10, 20]$. It has an inflection point at $x = 2$, but no relative extreme points. Enlarge the window a few times to convince yourself that there are no relative extreme points anywhere. What does this tell you about $f'(x)$?

39. Draw the graph of $f(x) = \frac{1}{6}x^3 - \frac{5}{2}x^2 + 13x - 20$ in the window $[0, 10]$ by $[-20, 30]$. Algebraically determine the coordinates of the inflection point. Zoom in and zoom out to convince yourself that there are no relative extreme points anywhere.

40. Draw the graph of

$$f(x) = 2x + \frac{18}{x} - 10$$

in the window $[0, 16]$ by $[0, 16]$. In what ways is this graph like the graph of a parabola that opens upward? In what ways is it different?

41. Draw the graph of

$$f(x) = 3x + \frac{75}{x} - 25$$

in the window $[0, 25]$ by $[0, 50]$. Use the trace feature of the calculator or computer to estimate the coordinates of the relative minimum point. Then determine the coordinates algebraically. Convince yourself both graphically (with the calculator or computer) and algebraically that this function has no inflection points.

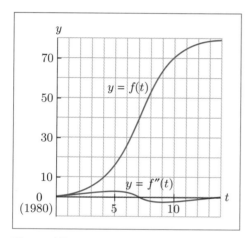

Figure 15. Percentage of homes with TVs that also had VCRs, from 1980 to 1994.

Solutions to Practice Problems 2.3	

1. Answer: (a) and (d). Curve (b) has the shape of a parabola, but it is not the graph of any function, since vertical lines cross it twice. Curve (c) has two relative extreme points, but the derivative of $f(x)$ is a linear function, which could not be zero for two different values of x.

2. Answer: (a), (c), (d). Curve (b) has three relative extreme points, but the derivative of $f(x)$ is a quadratic function which could not be zero for three different values of x.

2.4 Curve Sketching (Conclusion)

In Section 2.3 we discussed the main techniques for curve sketching. Here we add a few finishing touches and examine some slightly more complicated curves.

The more points we plot on a graph, the more accurate the graph becomes. This statement is true even for the simple quadratic and cubic curves in Section 2.3. Of course, the most important points on a curve are the relative extreme points and the inflection points. In addition, the x- and y-intercepts often have some intrinsic interest in an applied problem. The y-intercept is $(0, f(0))$. To find the x-intercepts on the graph of $f(x)$, we must find those values of x for which $f(x) = 0$. Since this can be a difficult (or impossible) problem, we shall find x-intercepts only when they are easy to find or when a problem specifically requires us to find them.

When $f(x)$ is a quadratic function, as in Example 1, we can easily compute the x-intercepts (if they exist) either by factoring the expression for $f(x)$ or by using the quadratic formula.

▶ **Example 1** Sketch the graph of $y = \frac{1}{2}x^2 - 4x + 7$.

Solution Let

$$f(x) = \tfrac{1}{2}x^2 - 4x + 7.$$

Then

$$f'(x) = x - 4$$
$$f''(x) = 1.$$

Since $f'(x) = 0$ only when $x = 4$ and since $f''(4)$ is positive, $f(x)$ must have a relative minimum point at $x = 4$. The relative minimum point is $(4, f(4)) = (4, -1)$.

The y-intercept is $(0, f(0)) = (0, 7)$. To find the x-intercepts, we set $f(x) = 0$ and solve for x:

$$\tfrac{1}{2}x^2 - 4x + 7 = 0.$$

The expression for $f(x)$ is not easily factored, so we use the quadratic formula to solve the equation:

$$x = \frac{-(-4) \pm \sqrt{(-4)^2 - 4(\frac{1}{2})(7)}}{2(\frac{1}{2})} = 4 \pm \sqrt{2}.$$

The x-intercepts are $(4 - \sqrt{2}, 0)$ and $(4 + \sqrt{2}, 0)$. To plot these points we use the approximation $\sqrt{2} \approx 1.4$. (See Fig. 1.) ◆

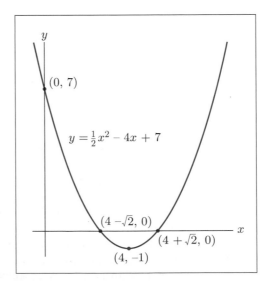

Figure 1.

▶ **Example 2**   Sketch the graph of $f(x) = \frac{1}{6}x^3 - \frac{3}{2}x^2 + 5x + 1$.

Solution

$$f(x) = \tfrac{1}{6}x^3 - \tfrac{3}{2}x^2 + 5x + 1$$
$$f'(x) = \tfrac{1}{2}x^2 - 3x + 5$$
$$f''(x) = x - 3$$

Let us set $f'(x) = 0$ and try to solve for x:

$$\tfrac{1}{2}x^2 - 3x + 5 = 0. \tag{1}$$

If we apply the quadratic formula with $a = \frac{1}{2}$, $b = -3$, and $c = 5$, we see that $b^2 - 4ac$ is negative, and so there is no solution to (1). In other words, $f'(x)$ is never zero. Thus the graph cannot have relative extreme points. If we evaluate $f'(x)$ at some x, say $x = 0$, we see that the first derivative is positive, and so $f(x)$ is increasing there. Since the graph of $f(x)$ is a smooth curve with no relative extreme points and no breaks, $f(x)$ must be increasing for all x. (If a function were increasing at $x = a$ and decreasing at $x = b$, then it would have a relative extreme point between a and b.)

Now let us check the concavity.

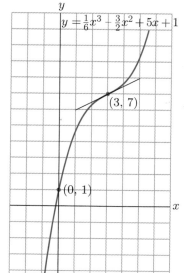

	$f''(x) = x - 3$	Graph of $f(x)$
$x < 3$	Negative	Concave down
$x = 3$	Zero	Concavity reverses
$x > 3$	Positive	Concave up

The inflection point is $(3, f(3)) = (3, 7)$. The y-intercept is $(0, f(0)) = (0, 1)$. We omit the x-intercept because it is difficult to solve the cubic equation $\frac{1}{6}x^3 - \frac{3}{2}x^2 + 5x + 1 = 0$.

The quality of our sketch of the curve will be improved if we first sketch the tangent line at the inflection point. To do this, we need to know the slope of the graph at $(3, 7)$:

$$f'(3) = \tfrac{1}{2}(3)^2 - 3(3) + 5 = \tfrac{1}{2}.$$

We draw a line through $(3, 7)$ with slope $\frac{1}{2}$ and then complete the sketch as shown in Fig. 2. ◆

Figure 2.

The methods used so far will work for most of the functions we shall study. Occasionally, however, $f'(x)$ and $f''(x)$ are both zero at some value of x, say $x = a$, and we cannot tell from the second derivative if the function has a relative minimum or a relative maximum at $x = a$ or neither. This exceptional situation can be handled using the following observation: As we move from left to right in the vicinity of a relative maximum, the slope decreases and changes sign from positive to negative. In the vicinity of a relative minimum, the slope increases and changes sign from negative to positive.

The following example shows how this observation may be used in curve sketching.

▶ **Example 3**   Sketch the graph of $f(x) = (x-2)^4 - 1$.

Solution

$$f(x) = (x-2)^4 - 1$$
$$f'(x) = 4(x-2)^3$$
$$f''(x) = 12(x-2)^2.$$

Clearly, $f'(x) = 0$ only if $x = 2$. So the curve has a horizontal tangent at $(2, f(2)) = (2, -1)$. Since $f''(2) = 0$, the second derivative rule cannot be used to determine whether the point $(2, -1)$ is a relative maximum, a relative minimum, or neither. However, note that

$$f'(x) = 4(x-2)^3 \quad \begin{cases} \text{negative} & \text{if } x < 2 \\ \text{positive} & \text{if } x > 2 \end{cases}$$

since the cube of a negative number is negative and the cube of a positive number is positive. Therefore, as x goes from left to right in the vicinity of 2, the first derivative goes from negative to positive. By the preceding observation, this means that the point $(2, -1)$ is a relative minimum.

The y-intercept is $(0, f(0)) = (0, 15)$. To find the x-intercepts, we set $f(x) = 0$ and solve for x:

$$(x-2)^4 - 1 = 0$$
$$(x-2)^4 = 1$$
$$x - 2 = 1 \quad \text{or} \quad x - 2 = -1$$
$$x = 3 \quad \text{or} \quad x = 1.$$

(See Fig. 3.)

◆

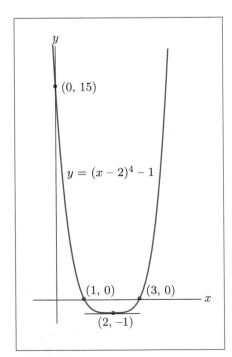

Figure 3.

A Graph with Asymptotes Graphs similar to the one in the next example will arise in several applications later in this chapter.

▶ **Example 4** Sketch the graph of $f(x) = x + (1/x)$, for $x > 0$.

Solution

$$f(x) = x + \frac{1}{x}$$

$$f'(x) = 1 - \frac{1}{x^2}$$

$$f''(x) = \frac{2}{x^3}$$

We set $f'(x) = 0$ and solve for x:

$$1 - \frac{1}{x^2} = 0$$

$$1 = \frac{1}{x^2}$$

$$x^2 = 1$$

$$x = 1.$$

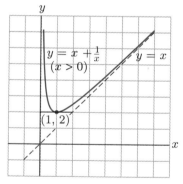

Figure 4.

(We exclude the case $x = -1$ because we are only considering positive values of x.) The graph has a horizontal tangent at $(1, f(1)) = (1, 2)$. Now, $f''(1) = 2 > 0$, and so the graph is concave up at $x = 1$ and $(1, 2)$ is a relative minimum point. In fact, $f''(x) = (2/x^3) > 0$ for all positive x, and therefore the graph is concave up at all points.

Before sketching the graph, notice that as x approaches zero [a point at which $f(x)$ is not defined], the term $1/x$ in the formula for $f(x)$ is dominant. That is, this term becomes arbitrarily large, whereas the term x contributes a diminishing proportion to the function value as x approaches 0. Thus $f(x)$ has the y-axis as an asymptote. For large values of x, the term x is dominant. The value of $f(x)$ is only slightly larger than x since the term $1/x$ has decreasing significance as x becomes arbitrarily large; that is, the graph of $f(x)$ is slightly above the graph of $y = x$. As x increases, the graph of $f(x)$ has the line $y = x$ as an asymptote. (See Fig. 4.) ◆

Summary of Curve-Sketching Techniques

1. Compute $f'(x)$ and $f''(x)$.

2. Find all relative extreme points.

  (a) Set $f'(x) = 0$ and solve for x. Suppose that $x = a$ is a solution. Substitute $x = a$ into $f(x)$ to find $f(a)$, plot the point $(a, f(a))$, and draw a small horizontal tangent line through the point. Compute $f''(a)$.

  (i) If $f''(a) > 0$, draw a small concave up arc with $(a, f(a))$ as its lowest point. The curve has a relative minimum at $x = a$.

  (ii) If $f''(a) < 0$, draw a small concave down arc with $(a, f(a))$ as its peak. The curve has a relative maximum at $x = a$.

  (iii) If $f''(a) = 0$, examine $f'(x)$ to the left and right of $x = a$ in order to determine if the function changes from increasing to decreasing, or vice versa. If a relative extreme point is indicated, draw an appropriate arc as in parts (i) and (ii).

  (b) Repeat the preceding steps for each of the solutions to $f'(x) = 0$.

3. Find all the inflection points of $f(x)$.

   (a) Set $f''(x) = 0$ and solve for x. Suppose that $x = b$ is a solution. Compute $f(b)$ and plot the point $(b, f(b))$.

   (b) Test the concavity of $f(x)$ to the right and left of b. If the concavity changes at $x = b$, then $(b, f(b))$ is an inflection point.

4. Consider other properties of the function and complete the sketch.

   (a) If $f(x)$ is defined at $x = 0$, the y-intercept is $(0, f(0))$.

   (b) Does the partial sketch suggest that there are x-intercepts? If so, they are found by setting $f(x) = 0$ and solving for x. (Solve only in easy cases or when a problem essentially requires you to calculate the x-intercepts.)

   (c) Observe where $f(x)$ is defined. Sometimes the function is given only for restricted values of x. Sometimes the formula for $f(x)$ is meaningless for certain values of x.

   (d) Look for possible asymptotes.

       (i) Examine the formula for $f(x)$. If some terms become insignificant as x gets large and if the rest of the formula gives the equation of a straight line, then that straight line is an asymptote.

       (ii) Suppose that there is some point a such that $f(x)$ is defined for x near a but not at a (e.g., $1/x$ at $x = 0$). If $f(x)$ gets arbitrarily large (in the positive or negative sense) as x approaches a, then the vertical line $x = a$ is an asymptote for the graph.

   (e) Complete the sketch.

| **Practice Problems 2.4** | Determine whether each of the following functions has an asymptote as x gets large. If so, give the equation of the straight line which is the asymptote. |

1. $f(x) = \dfrac{3}{x} - 2x + 1$ 2. $f(x) = \sqrt{x} + x$ 3. $f(x) = \dfrac{1}{2x}$

▶ Exercises 2.4

Find the x-intercepts of the following curves.

1. $y = x^2 - 3x + 1$

2. $y = x^2 + 5x + 5$

3. $y = 2x^2 + 5x + 2$

4. $y = 4 - 2x - x^2$

5. $y = 4x - 4x^2 - 1$

6. $y = 3x^2 + 7x + 2$

7. Show that the function $f(x) = \frac{1}{3}x^3 - 2x^2 + 5x$ has no relative extreme points.

8. Show that the function $f(x) = \frac{1}{3}x^3 - x^2 + 5x - 3$ is always increasing.

Sketch the graphs of the following functions.

9. $f(x) = x^3 - 6x^2 + 12x - 6$

10. $f(x) = -x^3$

11. $f(x) = x^3 + 3x + 1$

12. $f(x) = 4 - x - x^3$

13. $f(x) = 5 - 13x + 6x^2 - x^3$

14. $f(x) = 2x^3 + x - 2$

15. $f(x) = \frac{4}{3}x^3 - 2x^2 + x$

16. $f(x) = 1 - 4x - 3x^2 - x^3$

17. $f(x) = 1 - 3x + 3x^2 - x^3$

18. $f(x) = x^3 - 6x^2 + 12x - 5$

19. $f(x) = x^4 - 6x^2$

20. $f(x) = 1 + 6x^2 - 3x^4$

21. $f(x) = (x - 3)^4$

22. $f(x) = (x + 2)^4 - 1$

Sketch the graphs of the following functions for $x > 0$.

23. $y = \dfrac{1}{x} + \dfrac{1}{4}x$

24. $y = \dfrac{1}{x} + 9x$

25. $y = \dfrac{9}{x} + x + 1$

26. $y = \dfrac{12}{x} + 3x + 1$

27. $y = \dfrac{2}{x} + \dfrac{x}{2} + 2$

28. $y = \dfrac{x}{3} + \dfrac{12}{x} - 1$

29. $y = 6\sqrt{x} - x$

30. $y = 21x + \dfrac{8400}{x}$

In Exercises 31 and 32, determine which function is the derivative of the other.

31.

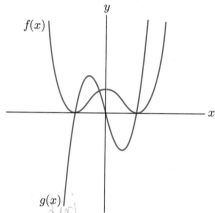

32.

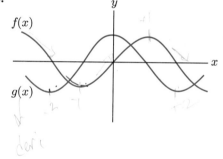

Technology Exercises

33. In a medical experiment,[*] the body weight of a baby rat in the control group after t days was $f(t) = 4.96 + .48t + .17t^2 - .0048t^3$ grams.

   (a) Graph $f(t)$ in the window $[0, 20]$ by $[-12, 50]$.

   (b) Approximately how much did the rat weigh after 7 days?

   (c) Approximately when did the rat's weight reach 27 grams?

   (d) Approximately how fast was the rat gaining weight after 4 days?

   (e) Approximately when was the rat gaining weight at the rate of 2 grams per day?

   (f) Approximately when was the rat gaining weight at the fastest rate?

34. The canopy height (in meters) of the tropical bunch-grass elephantmillet[†] t days after mowing (for $t \geq 32$) is $f(t) = -3.14 + .142t - .0016t^2 + .0000079t^3 - .0000000133t^4$.

   (a) Graph $f(t)$ in the window $[32, 250]$ by $[-1.2, 4.5]$.

   (b) How tall was the canopy after 100 days?

   (c) When was the canopy 2 meters high?

   (d) How fast was the canopy growing after 80 days?

   (e) When was the canopy growing at the rate of .02 meters per day?

   (f) Approximately when was the canopy growing slowest?

   (g) Approximately when was the canopy growing fastest?

| **Solutions to Practice Problems 2.4** | Functions with asymptotes as x gets large have the form $f(x) = g(x) + mx + b$, where $g(x)$ approaches zero as x gets large. The function $g(x)$ often looks like c/x or $c/(ax+d)$. The asymptote will be the straight line $y = mx + b$. |

  1. Here $g(x)$ is $3/x$ and the asymptote is $y = -2x + 1$.

  2. This function has no asymptote as x gets large. Of course, it can be written as $g(x) + mx + b$, where $m = 1$ and $b = 0$. However, $g(x) = \sqrt{x}$ does not approach zero as x gets large.

  3. Here $g(x)$ is $\dfrac{1}{2x}$ and the asymptote is $y = 0$. That is, the function has the x-axis as an asymptote.

---

[*]Johnson, Wogenrich, Hsi, Skipper, and Greenberg, "Growth Retardation During the Sucking Period in Expanded Litters of Rats; Observations of Growth Patterns and Protein Turnover," *Growth, Development and Aging*, 55(1991), 263–273.

[†]Woodward and Prine, "Crop Quality and Utilization," *Crop Science*, 33(1993), 818–824.

2.5 Optimization Problems

One of the most important applications of the derivative concept is to "optimization" problems, in which some quantity must be maximized or minimized. Examples of such problems abound in many areas of life. An airline must decide how many daily flights to schedule between two cities in order to maximize its profits. A doctor wants to find the minimum amount of a drug that will produce a desired response in one of her patients. A manufacturer needs to determine how often to replace certain equipment in order to minimize maintenance and replacement costs.

Our purpose in this section is to illustrate how calculus can be used to solve optimization problems. In each example we will find or construct a function that provides a "mathematical model" for the problem. Then, by sketching the graph of this function, we will be able to determine the answer to the original optimization problem by locating the highest or lowest point on the graph. The y-coordinate of this point will be the maximum value or minimum value of the function.

The first two examples are quite simple because the functions to be studied are given explicitly.

⟩ **Example 1** Find the minimum value of the function $f(x) = 2x^3 - 15x^2 + 24x + 19$ for $x \geq 0$.

Solution Using the curve-sketching techniques from Section 2.3, we obtain the graph in Fig. 1. As part of that process, we compute the derivatives:

$$f'(x) = 6x^2 - 30x + 24$$
$$f''(x) = 12x - 30.$$

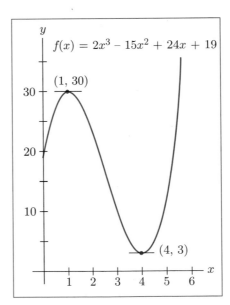

Figure 1.

The x-coordinate of the minimum point satisfies the equation

$$f'(x) = 0$$
$$6x^2 - 30x + 24 = 0$$
$$6(x - 4)(x - 1) = 0$$
$$x = 1, 4.$$

The corresponding points on the curve are

$$(1, f(1)) = (1, 30)$$
$$(4, f(4)) = (4, 3).$$

Applying the second derivative test, we see that

$$f''(1) = 12 \cdot 1 - 30 = -18 < 0 \quad \text{(maximum)}$$
$$f''(4) = 12 \cdot 4 - 30 = +18 > 0 \quad \text{(minimum)}.$$

That is, the point $(1, 30)$ is a relative maximum and the point $(4, 3)$ is a relative minimum. The lowest point on the graph is $(4, 3)$. The minimum *value* of the function $f(x)$ is the y-coordinate of this point—namely, 3. ◆

▶ **Example 2**   Suppose that a ball is thrown straight up into the air and its height after t seconds is $4 + 48t - 16t^2$ feet. Determine how long it will take for the ball to reach its maximum height and determine the maximum height.

Solution   Consider the function $f(t) = 4 + 48t - 16t^2$. For each value of t, $f(t)$ is the height of the ball at time t. We want to find the value of t for which $f(t)$ is the greatest. Using the techniques of Section 2.3, we sketch the graph of $f(t)$. (See Fig. 2.)

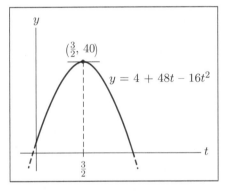

Figure 2.

Note that we may neglect the portions of the graph corresponding to points for which either $t < 0$ or $f(t) < 0$. [A negative value of $f(t)$ would correspond to the ball being underneath the ground.] The t-coordinate giving the maximum height is the solution of the equation:

$$f'(t) = 48 - 32t = 0$$
$$t = \tfrac{3}{2}.$$

Since

$$f''(t) = -32$$
$$f''(\tfrac{3}{2}) = -32 < 0,$$

we see that $t = \tfrac{3}{2}$ is the location of a relative maximum. So $f(t)$ is greatest when $t = \tfrac{3}{2}$. At this value of t, the ball attains a height of 40 feet. [Note that the curve in Fig. 2 is the graph of $f(t)$, *not* a picture of the physical path of the ball.]

ANSWER

The ball reaches its maximum height of 40 feet in 1.5 seconds. ◆

▶ Example 3   A person wants to plant a rectangular garden along one side of a house, with a picket fence on the other three sides of the garden. Find the dimensions of the largest garden that can be enclosed using 40 feet of fencing.

Solution   The first step is to make a simple diagram and assign letters to the quantities that may vary. Let us denote the dimensions of the rectangular garden by w and x (Fig. 3). The phrase "largest garden" indicates that we must maximize the

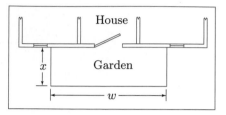

Figure 3.

area, A, of the garden. In terms of the variables w and x,

$$A = wx. \tag{1}$$

The fencing on three sides must total 40 running feet; that is,

$$2x + w = 40. \tag{2}$$

We now solve equation (2) for w in terms of x:

$$w = 40 - 2x. \tag{3}$$

Substituting this expression for w into equation (1), we have

$$A = (40 - 2x)x = 40x - 2x^2. \tag{4}$$

We now have a formula for the area A that depends on just one variable, and so we may graph A as a function of x. From the statement of the problem, the value of $2x$ can be at most 40, so that the domain of the function consists of x in the interval $(0, 20)$.

Using curve-sketching techniques, we obtain the graph in Fig. 4. The x-coordinate of the maximum point can be found from the equation

$$A'(x) = 0$$
$$40 - 4x = 0$$
$$x = 10.$$

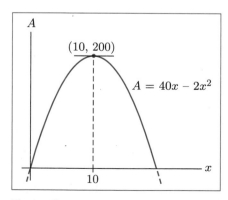

Figure 4.

(The maximum area is 200 square feet, but this fact is not needed for the problem.) From equation (3) we find that when $x = 10$,

$$w = 40 - 2(10) = 20.$$

ANSWER

$w = 20$ feet, $x = 10$ feet. ◆

Equation (1) in Example 3 is called an *objective equation*. It expresses the quantity to be optimized (the area of the garden) in terms of the variables w and x. Equation (2) is called a *constraint equation* because it places a limit or constraint on the way x and w may vary.

▶ **Example 4**   The manager of a department store wants to build a 600-square-foot rectangular enclosure on the store's parking lot in order to display some equipment. Three sides of the enclosure will be built of redwood fencing, at a cost of $14 per running foot. The fourth side will be built of cement blocks, at a cost of $28 per running foot. Find the dimensions of the enclosure that will minimize the total cost of the building materials.

Solution   Let x be the length of the side built out of cement blocks and let y be the length of an adjacent side, as shown in Fig. 5. The phrase "minimize the total cost" tells us that the objective equation should be a formula giving the total cost of the building materials.

$$[\text{cost of redwood}] = [\text{length of redwood fencing}] \times [\text{cost per foot}]$$
$$= (x + 2y) \cdot 14 = 14x + 28y$$
$$[\text{cost of cement blocks}] = [\text{length of cement wall}] \times [\text{cost per foot}]$$
$$= x \cdot 28$$

If C denotes the total cost of the materials, then

$$C = (14x + 28y) + 28x$$
$$C = 42x + 28y \quad \text{(objective equation)}. \tag{5}$$

Since the area of the enclosure must be 600 square feet, the constraint equation is

$$xy = 600. \tag{6}$$

We simplify the objective equation by solving (6) for one of the variables, say y, and substituting into (5): Since $y = 600/x$,

$$C = 42x + 28 \left(\frac{600}{x} \right) = 42x + \frac{16,800}{x}.$$

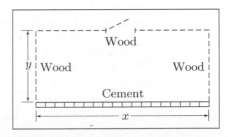

Figure 5. Rectangular enclosure.

We now have C as a function of the single variable x. From the context, we must have $x > 0$, since a length must be positive. However, to any positive value for x, there is a corresponding value for C. So the domain of C consists of all $x > 0$. We may now sketch the graph of C (Fig. 6). (A similar curve was sketched in Example 4 of Section 2.4.) The x-coordinate of the minimum point is a solution of:

$$C'(x) = 0$$
$$42 - \frac{16{,}800}{x^2} = 0$$
$$42x^2 = 16{,}800$$
$$x^2 = 400$$
$$x = 20.$$

The corresponding value of C is:

$$C(20) = 42(20) + \frac{16{,}800}{20} = \$1680.$$

That is, the minimum total cost of \$1680 occurs where $x = 20$. From equation (6) we find that the corresponding value of y is $\frac{600}{20} = 30$.

ANSWER

$x = 20$ feet, $y = 30$ feet. ◆

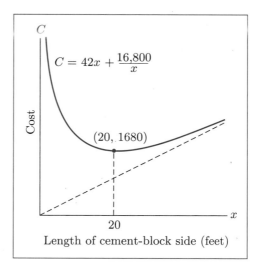

Figure 6.

▶ **Example 5** U.S. parcel post regulations state that packages must have length plus girth of no more than 84 inches. Find the dimensions of the cylindrical package of greatest volume that is mailable by parcel post.

Solution Let l be the length of the package and let r be the radius of the circular end. (See Fig. 7.) The phrase "greatest volume" tells us that the objective equation should express the volume of the package in terms of the dimensions l and r. Let V denote the volume. Then

$$V = [\text{area of base}] \cdot [\text{length}]$$
$$V = \pi r^2 l \quad (\text{objective equation}). \tag{7}$$

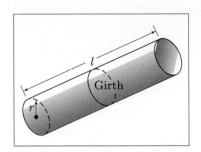

Figure 7. Cylindrical mailing package.

The girth equals the circumference of the end—that is, $2\pi r$. Since we want the package to be as large as possible, we must use the entire 84 inches allowable:

$$\text{length} + \text{girth} = 84$$
$$l + 2\pi r = 84 \quad \text{(constraint equation)}. \tag{8}$$

We now solve equation (8) for one of the variables, say $l = 84 - 2\pi r$. Substituting this expression into (7), we obtain

$$V = \pi r^2(84 - 2\pi r) = 84\pi r^2 - 2\pi^2 r^3. \tag{9}$$

Let $f(r) = 84\pi r^2 - 2\pi^2 r^3$. Then, for each value of r, $f(r)$ is the volume of the parcel with end radius r that meets the postal regulations. We want to find that value of r for which $f(r)$ is as large as possible.

Using curve-sketching techniques, we obtain the graph of $f(r)$ in Fig. 8. The domain excludes values of r that are negative and values of r for which the volume $f(r)$ is negative. Points corresponding to values of r not in the domain are shown with a dashed curve. We see that the volume is greatest when $r = 28/\pi$.

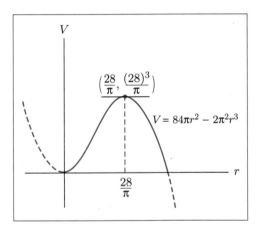

Figure 8.

From (8) we find that the corresponding value of l is

$$l = 84 - 2\pi r = 84 - 2\pi\left(\frac{28}{\pi}\right) = 84 - 56 = 28.$$

The girth when $r = 28/\pi$ is

$$2\pi r = 2\pi\left(\frac{28}{\pi}\right) = 56.$$

ANSWER

$l = 28$ inches, $r = 28/\pi$ inches.   ◆

Suggestions for Solving an Optimization Problem

1. Draw a picture, if possible.

2. Decide what quantity Q is to be maximized or minimized.

3. Assign letters to other quantities that may vary.

4. Determine the "objective equation" that expresses Q as a function of the variables assigned in step 3.

5. Find the "constraint equation" that relates the variables to each other and to any constants that are given in the problem.

6. Use the constraint equation to simplify the objective equation in such a way that Q becomes a function of only one variable. Determine the domain of this function.

7. Sketch the graph of the function obtained in step 6 and use this graph to solve the optimization problem.

Note Optimization problems often involve geometric formulas. The most common formulas are shown in Fig. 9.

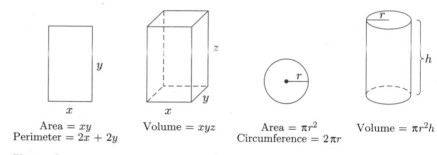

Area $= xy$
Perimeter $= 2x + 2y$

Volume $= xyz$

Area $= \pi r^2$
Circumference $= 2\pi r$

Volume $= \pi r^2 h$

Figure 9.

Practice Problems 2.5

1. A canvas wind shelter for the beach has a back, two square sides, and a top (Fig. 10). Suppose that 96 square feet of canvas are to be used. Find the dimensions of the shelter for which the space inside the shelter (i.e., the volume) will be maximized.

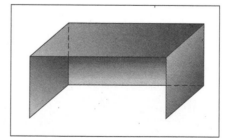

Figure 10. Wind shelter.

2. In Practice Problem 1, what are the objective equation and the constraint equation?

▶ Exercises 2.5

1. For what x does the function $g(x) = 10 + 40x - x^2$ have its maximum value?

2. Find the maximum value of the function $f(x) = 12x - x^2$ and give the value of x where this maximum occurs.

3. Find the minimum value of $f(t) = t^3 - 6t^2 + 40$, $t \geq 0$, and give the value of t where this minimum occurs.

4. For what t does the function $f(t) = t^2 - 24t$ have its minimum value?

5. Three hundred twenty dollars are available to fence in a rectangular garden. The fencing for the side of the garden facing the road costs \$6 per foot and the fencing for the other three sides costs \$2 per foot. [See Fig. 11(a).] Consider the problem of finding the dimensions of the largest possible garden.

   (a) Determine the objective and constraint equations.

   (b) Express the quantity to be maximized as a function of x.

   (c) Find the optimal values of x and y.

6. Figure 11(b) shows an open rectangular box with a square base. Consider the problem of finding the values of x and h for which the volume is 32 cubic feet and the total surface area of the box is minimal. (The surface area is the sum of the areas of the five faces of the box.)

   (a) Determine the objective and constraint equations.

   (b) Express the quantity to be minimized as a function of x.

   (c) Find the optimal values of x and h.

7. Postal requirements specify that parcels must have length plus girth at most 84 inches. Consider the problem of finding the dimensions of the square-ended rectangular package of greatest volume that is mailable.

   (a) Draw a square-ended rectangular box. Label each edge of the square end with the letter x and label the remaining dimension of the box with the letter h.

   (b) Express the length plus the girth in terms of x and h.

(c) Determine the objective and constraint equations.

(d) Express the quantity to be maximized as a function of x.

(e) Find the optimal values of x and h.

8. Consider the problem of finding the dimensions of the rectangular garden of area 100 square meters for which the amount of fencing needed to surround the garden is as small as possible.

   (a) Draw a picture of a rectangle and select appropriate letters for the dimensions.

   (b) Determine the objective and constraint equations.

   (c) Find the optimal values for the dimensions.

9. A rectangular garden of area 75 square feet is to be surrounded on three sides by a brick wall costing \$10 per foot and on one side by a fence costing \$5 per foot. Find the dimensions of the garden such that the cost of materials is minimized.

10. A closed rectangular box with square base and a volume of 12 cubic feet is to be constructed using two different types of materials. The top is made of a metal costing \$2 per square foot and the remainder of wood costing \$1 per square foot. Find the dimension of the box for which the cost of materials is minimized.

11. Find the dimensions of the closed rectangular box with square base and volume 8000 cubic centimeters that can be constructed with the least amount of material.

12. A canvas wind shelter for the beach has a back, two square sides, and a top. Find the dimensions for which the volume will be 250 cubic feet and that requires the least possible amount of canvas.

13. A farmer has \$1500 available to build an E-shaped fence along a straight river so as to create two identical rectangular pastures. (See Fig. 12.) The materials for the side parallel to the river cost \$6 per foot and the materials for the three sections perpendicular to the river cost \$5 per foot. Find the dimensions for which the total area is as large as possible.

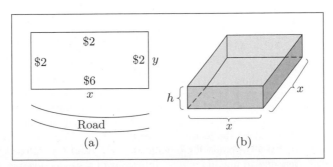

Figure 11.

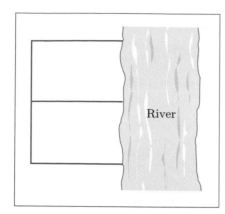

Figure 12. Rectangular pastures along a river.

14. Find the dimensions of the rectangular garden of greatest area that can be fenced off (all four sides) with 300 meters of fencing.

15. Find two positive numbers, x and y, whose sum is 100 and whose product is as large as possible.

16. Find two positive numbers, x and y, whose product is 100 and whose sum is as small as possible.

17. Figure 13(a) shows a Norman window, which consists of a rectangle capped by a semicircular region. Find the value of x such that the perimeter of the window will be 14 feet and the area of the window will be as large as possible.

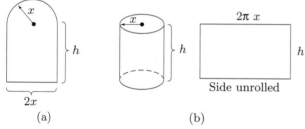

Figure 13.

18. A large soup can is to be designed so that the can will hold 16π cubic inches (about 28 ounces) of soup. [See Fig. 13(b).] Find the values of x and h for which the amount of metal needed is as small as possible.

19. In Example 3 one can solve the constraint equation (2) for x instead of w to get $x = 20 - \frac{1}{2}w$. Substituting this for x in (1), one has

$$A = xw = \left(20 - \tfrac{1}{2}w\right)w.$$

Sketch the graph of the equation $A = 20w - \frac{1}{2}w^2$ and show that the maximum occurs when $w = 20$ and $x = 10$.

20. A ship uses $5x^2$ dollars of fuel per hour when traveling at a speed of x miles per hour. The other expenses of operating the ship amount to \$2000 per hour. What speed minimizes the cost of a 500-mile trip? (*Hint*: Express cost in terms of speed and time. The constraint equation is *distance* = *speed* × *time*.)

21. Find the point on the graph of $y = \sqrt{x}$ that is closest to the point $(2, 0)$. See Fig. 14. [*Hint*: $\sqrt{(x-2)^2 + y^2}$ has its smallest value when $(x-2)^2 + y^2$ does. Therefore, just minimize the second expression.]

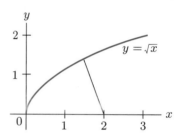

Figure 14. Shortest distance from a point to a curve.

Technology Exercises

22. Find the value of x for which the rectangle inscribed in the semicircle of radius 3 in Fig. 15 has the greatest area. (*Note*: The equation of the semicircle is $y = \sqrt{9 - x^2}$.)

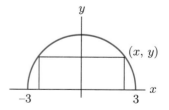

Figure 15.

Solutions to Practice Problems 2.5

1. Since the sides of the wind shelter are square, we may let x represent the length of each side of the square. The remaining dimension of the wind shelter can be denoted by the letter h. (See Fig. 16.)

    The volume of the shelter is x^2h, and this is to be maximized. Since we have learned to maximize only functions of a single variable, we must express h in terms

**Solutions to
Practice Problems
2.5 (Continued)**

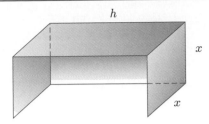

Figure 16.

of x. We must use the information that 96 feet of canvas are used—that is, $2x^2 + 2xh = 96$. (*Note*: The roof and the back each have an area xh, and each end has an area x^2.) We now solve this equation for h.

$$2x^2 + 2xh = 96$$

$$2xh = 96 - 2x^2$$

$$h = \frac{96}{2x} - \frac{2x^2}{2x} = \frac{48}{x} - x$$

The volume V is

$$x^2 h = x^2 \left(\frac{48}{x} - x \right) = 48x - x^3.$$

By sketching the graph of $V = 48x - x^3$, we see that V has a maximum value when $x = 4$. Then, $h = \frac{48}{4} - 4 = 12 - 4 = 8$. So each end of the shelter should be a 4-foot by 4-foot square and the top should be 8 feet long.

2. The objective equation is $V = x^2 h$, since it expresses the volume (the quantity to be maximized) in terms of the variables. The constraint equation is $2x^2 + 2xh = 96$, for it relates the variables to each other; that is, it can be used to express one of the variables in terms of the other.

2.6 Further Optimization Problems

In this section we apply the optimization techniques developed in the preceding section to some practical situations.

▶ Example 1   Suppose that, on a certain route, an airline carries 8000 passengers per month, each paying \$50. The airline wants to increase the fare. However, the market research department estimates that for each \$1 increase in fare, the airline will lose 100 passengers. Determine the price that maximizes the airline's revenue.

Solution   Since the problem calls for setting an optimum price, let x be the increased ticket price. The other variable is the number of passengers, which we can denote by n. The goal is to maximize revenue.

$$[\text{revenue}] = [\text{number of passengers}] \cdot [\text{price per ticket}] = n \cdot x$$

If R denotes revenue, then the objective equation is

$$R = nx.$$

Since the number of passengers n depends on the price x, the constraint equation can be derived by expressing n in terms of x. The number of passengers lost due to the fare increase is obtained by multiplying the number of dollars of fare increase, $x - 50$, by the number of passengers lost for each dollar of fare increase.

$$\begin{bmatrix} \text{number of} \\ \text{passengers} \end{bmatrix} = \begin{bmatrix} \text{original number} \\ \text{of passengers} \end{bmatrix} - \begin{bmatrix} \text{number of passengers lost} \\ \text{due to fare increase} \end{bmatrix}$$

$$n = 8000 - (x - 50) \cdot 100$$

$$= 13{,}000 - 100x$$

Therefore,

$$R = nx = (13{,}000 - 100x)x = 13{,}000x - 100x^2.$$

Figure 1 shows that revenue is maximized when the price per ticket is \$65. ◆

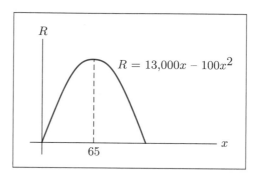

Figure 1. Revenue function for airline fares.

Inventory Control   When a firm regularly orders and stores supplies for later use or resale, it must decide on the size of each order. If it orders enough supplies to last an entire year, the business will incur heavy *carrying costs*. Such costs include insurance, storage costs, and cost of capital that is tied up in inventory. To reduce these carrying costs, the firm could order small quantities of the supplies at frequent intervals. However, such a policy increases the *ordering costs*. These might consist of minimum freight charges, the clerical costs of preparing the orders, and the costs of receiving and checking the orders when they arrive. Clearly, the firm must find an inventory ordering policy that lies between these two extremes.

The following example illustrates the use of calculus to minimize the firm's annual inventory cost, where

$$[\text{inventory cost}] = [\text{ordering cost}] + [\text{carrying cost}].$$

We assume that each order is the same size. The size of the order that minimizes the inventory cost is called the *economic order quantity*, commonly referred to in business as the EOQ.*

▶ Example 2   A supermarket manager wants to establish an optimal inventory policy for frozen orange juice. It is estimated that a total of 1200 cases will be sold at a steady rate during the next year. The manager plans to place several orders of the same size spaced equally throughout the year. Use the following data to determine the economic order quantity, that is, the order size that minimizes the total ordering and carrying cost.

---

*See James C. Van Horne, *Financial Management and Policy*, 6th ed. (Englewood Cliffs, N.J.: Prentice Hall, 1983), pp. 416–420.

1. The ordering cost for each delivery is $75.

2. It costs $8 to carry one case of orange juice in inventory for one year. (Carrying costs should be computed on the average inventory during the order-reorder period.)

Solution   Let x be the order quantity and r the number of orders placed during the year. The number of cases of orange juice in inventory declines steadily from x cases (each time a new order is filled) to 0 cases at the end of each order-reorder period. Figure 2 shows that the average number of cases in storage during the year is $x/2$. Since the carrying cost for one case is $8 per year, the cost for $x/2$ cases is $8 \cdot (x/2)$ dollars. Now

$$[\text{inventory cost}] = [\text{ordering cost}] + [\text{carrying cost}]$$
$$= 75r + 8 \cdot \frac{x}{2}$$
$$= 75r + 4x.$$

If C denotes the inventory cost, then the objective equation is

$$C = 75r + 4x.$$

Since there are r orders of x cases each, the total number of cases ordered during the year is $r \cdot x$. Therefore, the constraint equation is

$$r \cdot x = 1200.$$

The constraint equation says that $r = 1200/x$. Substitution into the objective equation yields

$$C = \frac{90{,}000}{x} + 4x.$$

Figure 3 is the graph of C as a function of x, for $x > 0$. The minimum point of the graph occurs where the first derivative equals zero. We find this point using

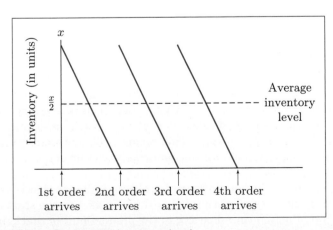

Figure 2. Average inventory level.

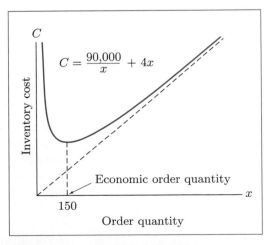

Figure 3. Cost function for inventory problem.

the calculation

$$C'(x) = -\frac{90{,}000}{x^2} + 4 = 0$$
$$4x^2 = 90{,}000$$
$$x^2 = 22{,}500$$
$$x = 150 \quad \text{(ignore the negative root since } x > 0\text{)}.$$

The total cost is at a minimum when $x = 150$. Therefore, the optimum inventory policy is to order 150 cases at a time and to place $1200/150 = 8$ orders during the year. ◆

▶ **Example 3** What should the inventory policy of Example 2 be if sales of frozen orange juice increase fourfold (i.e., 4800 cases are sold each year), but all other conditions are the same?

Solution The only change in our previous solution is in the constraint equation, which now becomes

$$r \cdot x = 4800.$$

The objective equation is, as before,

$$C = 75r + 4x.$$

Since $r = 4800/x$,

$$C = 75 \cdot \frac{4800}{x} + 4x = \frac{360{,}000}{x} + 4x.$$

Now

$$C' = -\frac{360{,}000}{x^2} + 4.$$

Setting $C' = 0$ yields

$$\frac{360{,}000}{x^2} = 4$$

$$90{,}000 = x^2$$

$$x = 300.$$

Therefore, the economic order quantity is 300 cases. ◆

Notice that although the sales increased by a factor of 4, the economic order quantity increased by only a factor of 2 ($= \sqrt{4}$). In general, a store's inventory of an item should be proportional to the square root of the expected sales. (See Exercise 9 for a derivation of this result.) Many stores tend to keep their average inventories at a fixed percentage of sales. For example, each order may contain enough goods to last for 4 or 5 weeks. This policy is likely to create excessive inventories of high-volume items and uncomfortably low inventories of slower-moving items.

Manufacturers have an inventory-control problem similar to that of retailers. They have the carrying costs of storing finished products and the startup costs of setting up each production run. The size of the production run that minimizes the sum of these two costs is called the *economic lot size*. See Exercises 6 and 7.

▶ Example 4 — Doctors studying coughing problems need to know the most effective condition for clearing the lungs and trachea (windpipe); that is, what radius of the trachea gives the greatest air velocity. Using the following principles of fluid flow, determine this optimum value. (See Fig. 4.) Let

r_0 = normal radius of the trachea,

r = radius during a cough,

P = increase in air pressure in the trachea during cough,

v = velocity of air through trachea during cough

1. $r_0 - r = aP$, for some positive constant a. (Experiment has shown that, during coughing, the decrease in the radius of the trachea is nearly proportional to the increase in the air pressure.)
2. $v = b \cdot P \cdot \pi r^2$, for some positive constant b. (The theory of fluid flow requires that the velocity of the air forced through the trachea is proportional to the product of the increase in the air pressure and the area of a cross section of the trachea.)

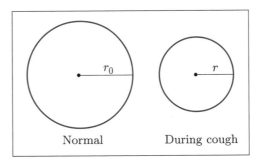

Normal   During cough

Figure 4. Contraction of the windpipe during coughing.

Solution — In this problem the constraint equation **(1)** and the objective equation **(2)** are given directly. Solving equation **(1)** for P and substituting this result into equation **(2)**, we have

$$v = b \left(\frac{r_0 - r}{a} \right) \pi r^2 = k(r_0 - r)r^2,$$

where $k = b\pi/a$. To find the radius at which the velocity v is a maximum, we first compute the derivatives with respect to r:

$$v = k(r_0 r^2 - r^3)$$

$$\frac{dv}{dr} = k(2r_0 r - 3r^2) = kr(2r_0 - 3r)$$

$$\frac{d^2 v}{dr^2} = k(2r_0 - 6r).$$

We see that $\dfrac{dv}{dr} = 0$ when $r = 0$ or when $2r_0 - 3r = 0$; that is, when $r = \frac{2}{3}r_0$.

It is easy to see that $\dfrac{d^2 v}{dr^2}$ is positive at $r = 0$ and is negative at $r = \frac{2}{3}r_0$. The graph of v as a function of r is drawn in Fig. 5. The air velocity is maximized at $r = \frac{2}{3}r_0$. ◆

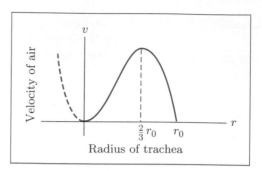

Figure 5. Graph for the windpipe problem.

When solving optimization problems, we look for the maximum or minimum point on a graph. From our discussion of curve sketching, we have seen that this point occurs either at a relative extreme point or at an endpoint of the domain of definition. In all our optimization problems so far, the maximum or minimum points were at relative extreme points. In the next example, the optimum point is an endpoint. This occurs often in applied problems.

▶ Example 5   A rancher has 204 meters of fencing from which to build two corrals: one square and the other rectangular with length that is twice the width. Find the dimensions that result in the greatest combined area.

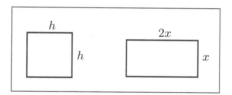

Figure 6. Two corrals.

Solution   Let x be the width of the rectangular corral and h be the length of each side of the square corral. (See Fig. 6.) Let A be the combined area. Then

$$A = [\text{area of square}] + [\text{area of rectangle}] = h^2 + 2x^2.$$

The constraint equation is

$$204 = [\text{perimeter of square}] + [\text{perimeter of rectangle}] = 4h + 6x.$$

Since the perimeter of the rectangle cannot exceed 204, we must have $0 \le 6x \le 204$, or $0 \le x \le 34$. Solving the constraint equation for h and substituting into the objective equation leads to the function graphed in Fig. 7. The graph reveals that the area is minimized when $x = 18$. However, the problem asks for the *maximum* possible area. From Fig. 7 we see that this occurs at the endpoint where $x = 0$. Therefore, the rancher should build only the square corral, with $h = 204/4 = 51$ meters. In this example, the objective function has an endpoint extremum; namely, the maximum value occurs at the endpoint where $x = 0$. ◆

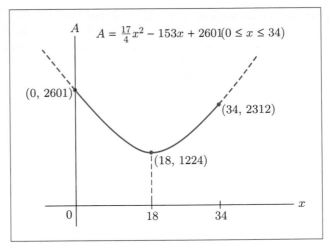

Figure 7. Combined area of the corrals.

Practice Problems 2.6

1. An apple orchard produces a profit of $40 a tree when planted with 1000 trees. Because of overcrowding, the profit per tree (for each tree in the orchard) is reduced by 2 cents for each additional tree planted. How many trees should be planted in order to maximize the total profit from the orchard?

2. In the inventory problem of Example 2, suppose that the sales of frozen orange juice increase ninefold; that is, 10,800 cases are sold each year. What is the new economic order quantity?

▶ Exercises 2.6

1. An artist is planning to sell signed prints of her latest work. If 50 prints are offered for sale, she can charge $400 each. However, if she makes more than 50 prints, she must lower the price of all the prints by $5 for each print in excess of the 50. How many prints should the artist make in order to maximize her revenue?

2. A swimming club offers memberships at the rate of $200, provided that a minimum of 100 people join. For each member in excess of 100, the membership fee will be reduced $1 per person (for each member). At most 160 memberships will be sold. How many memberships should the club try to sell in order to maximize its revenue?

3. In the planning of a sidewalk cafe, it is estimated that if there are 12 tables, the daily profit will be $10 per table. Because of overcrowding, for each additional table the profit per table (for every table in the cafe) will be reduced by $.50. How many tables should be provided to maximize the profit from the cafe?

4. A certain toll road averages 36,000 cars per day when charging $1 per car. A survey concludes that increasing the toll will result in 300 fewer cars for each cent of increase. What toll should be charged in order to maximize the revenue?

5. A California distributor of sporting equipment expects to sell 10,000 cases of tennis balls during the coming year at a steady rate. Yearly carrying costs (to be computed on the average number of cases in stock during the year) are $10 per case, and the cost of placing an order with the manufacturer is $80.

   (a) Find the inventory cost incurred if the distributor orders 500 cases at a time during the year.

   (b) Determine the economic order quantity, that is, the order quantity that minimizes the inventory cost.

6. The Great American Tire Co. expects to sell 600,000 tires of a particular size and grade during the next year. Sales tend to be roughly the same from month to month. Setting up each production run costs the company $15,000. Carrying costs, based on the average number of tires in storage, amount to $5 per year for one tire.

   (a) Determine the costs incurred if there are 10 production runs during the year.

   (b) Find the economic lot size (i.e., the production run size that minimizes the overall cost of producing the tires).

7. Foggy Optics, Inc. makes laboratory microscopes. Setting up each production run costs $2500. Insurance costs, based on the average number of microscopes in the warehouse, amount to $20 per microscope per year. Storage costs, based on the maximum number of microscopes in the warehouse, amount to $15 per microscope per year. Suppose that the company expects to sell 1600 microscopes at a fairly uniform rate throughout the year. Determine the number of production runs that will minimize the company's overall expenses.

8. A bookstore is attempting to determine the economic order quantity for a popular book. The store sells 8000 copies of this book a year. The store figures that it costs $40 to process each new order for books. The carrying cost (due primarily to interest payments) is $2 per book, to be figured on the maximum inventory during an order-reorder period. How many times a year should orders be placed?

9. A store manager wants to establish an optimal inventory policy for an item. Sales are expected to be at a steady rate and should total Q items sold during the year. Each time an order is placed, a cost of h dollars is incurred. Carrying costs for the year will be s dollars per item, to be figured on the average number of items in storage during the year. Show that the total inventory cost is minimized when each order calls for $\sqrt{2hQ/s}$ items.

10. Refer to the inventory problem of Example 2. Suppose that the distributor offers a discount of $1 per case for orders of 600 or more cases. Should the manager change the quantity ordered?

11. Starting with a 100-foot-long stone wall, a farmer would like to construct a rectangular enclosure by adding 400 feet of fencing as shown in Fig. 8(a). Find the values of x and w that result in the greatest possible area.

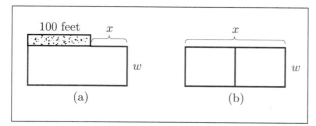

(a)   (b)

Figure 8. Rectangular enclosures.

12. Rework Exercise 11 for the case where only 200 feet of fencing is added to the stone wall.

13. A rectangular corral of 54 square meters is to be fenced off and then divided by a fence into two sections, as shown in Fig. 8(b). Find the dimensions of the corral so that the amount of fencing required is minimized.

14. Referring to Exercise 13, suppose that the cost of the fencing for the boundary is $5 per meter and the dividing fence costs $2 per meter. Find the dimensions of the corral that minimize the cost of the fencing.

15. A travel agency offers a boat tour of several Caribbean islands for 3 days and 2 nights. For a group of 12 people, the cost per person is $800. For each additional person above the 12-person minimum, the cost per person is reduced by $20 for each person in the group. The maximum tour group size is 25. What tour group size produces the greatest revenue for the travel agency?

16. Design an open rectangular box with square ends, having volume 36 cubic inches, that minimizes the amount of material required for construction.

17. A storage shed is to be built in the shape of a box with a square base. It is to have a volume of 150 cubic feet. The concrete for the base costs $4 per square foot, the material for the roof costs $2 per square foot, and the material for the sides costs $2.50 per square foot. Find the dimensions of the most economical shed.

18. A supermarket is to be designed as a rectangular building with a floor area of 12,000 square feet. The front of the building will be mostly glass and will cost $70 per running foot for materials. The other three walls will be constructed of brick and cement block, at a cost of $50 per running foot. Ignore all other costs (labor, cost of foundation and roof, etc.) and find the dimensions of the base of the building that will minimize the cost of the materials for the four walls of the building.

19. A certain airline requires that rectangular packages carried on an airplane by passengers be such that the sum of the three dimensions is at most 120 centimeters. Find the dimensions of the square-ended rectangular package of greatest volume that meets this requirement.

20. An athletic field [Fig. 9(a)] consists of a rectangular region with a semicircular region at each end. The perimeter will be used for a 440-yard track. Find the value of x for which the area of the rectangular region is as large as possible.

21. An open rectangular box is to be constructed by cutting square corners out of a 16- by 16-inch piece of cardboard and folding up the flaps. [See Fig. 9(b).] Find the value of x for which the volume of the box will be as large as possible.

22. A closed rectangular box is to be constructed with a base that is twice as long as it is wide. Suppose that the total surface area must be 27 square feet. Find the dimensions of the box that will maximize the volume.

23. Let $f(t)$ be the amount of oxygen (in suitable units) in a lake t days after sewage is dumped into the lake, and suppose that $f(t)$ is given approximately by

$$f(t) = 1 - \frac{10}{t + 10} + \frac{100}{(t + 10)^2}.$$

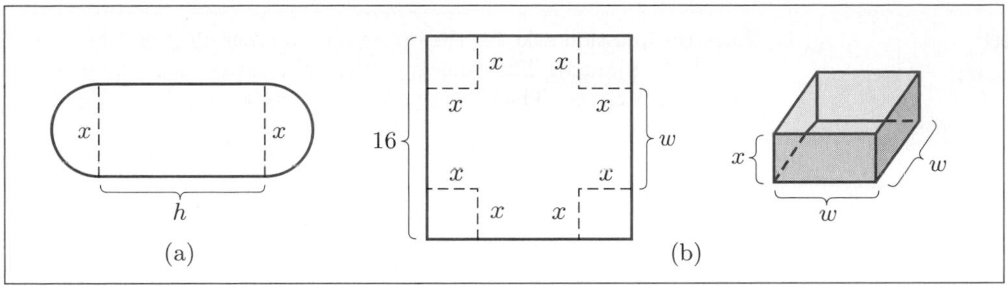

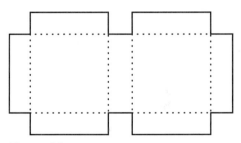

Figure 9.

At what time is the oxygen content increasing the fastest?

24. The daily output of a coal mine after t hours of operation is approximately $40t + t^2 - \frac{1}{15}t^3$ tons, $0 \le t \le 12$. Find the maximum rate of output (in tons of coal per hour).

25. Consider a parabolic arch whose shape may be represented by the graph of $y = 9 - x^2$, where the base of the arch lies on the x-axis from $x = -3$ to $x = 3$. Find the dimensions of the rectangular window of maximum area that can be constructed inside the arch.

26. Advertising for a certain product is terminated, and t weeks later the weekly sales are $f(t)$ cases, where

$$f(t) = 1000(t + 8)^{-1} - 4000(t + 8)^{-2}.$$

At what time is the weekly sales amount falling the fastest?

27. An open rectangular box of volume 400 cubic inches has a square base and a partition down the middle. See Fig. 10. Find the dimensions of the box for which the amount of material needed to construct the box is as small as possible.

Figure 10. Open rectangular box with dividing partition.

28. Suppose $f(x)$ is defined on the interval $0 \le x \le 5$ and $f'(x)$ is negative for all x. For what value of x will $f(x)$ have its greatest value?

Technology Exercises

29. A pizza box is formed from a 20 cm by 40 cm rectangular piece of cardboard by cutting out six squares of equal size, three from each long side of the rectangle, and then folding the cardboard in the obvious manner to create a box. See Fig. 11. Let x be the length of each side of the six squares. For what value of x will the box have greatest volume?

Figure 11.

30. Coffee consumption in the United States is greater on a per capita basis than anywhere else in the world. However, due to price fluctuations of coffee beans and worries over health effects of caffeine, coffee consumption has varied considerably over the years. According to data published in the *Wall Street Journal*, the number of cups $f(x)$ consumed daily per adult in year x (with 1955 corresponding to $x = 0$) is given by the mathematical model

$$f(x) = 2.77 + 0.0848x - 0.00832x^2 + 0.000144x^3.$$

(a) Graph $y = f(x)$ to show daily coffee consumption from 1955 through 1994.

(b) Use $f'(x)$ to determine the year in which coffee consumption was least during this period. What was the daily coffee consumption at that time?

(c) Use $f'(x)$ to determine the year in which coffee consumption was greatest during this period. What was the daily coffee consumption at that time?

(d) Use $f''(x)$ to determine the year in which coffee consumption was decreasing at the greatest rate.

<table>
<tr><td>

Solutions to Practice Problems 2.6

</td><td>

1. Since the question asks for the optimum "number of trees," let x be the number of trees to be planted. The other quantity that varies is "profit per tree." So let p be the profit per tree. The objective is to maximize total profit, call it T. Then

$$[\text{total profit}] = [\text{profit per tree}] \cdot [\text{number of trees}]$$
$$T = p \cdot x.$$

Since the profit per tree p depends on the number of trees planted, x, the constraint equation can be derived by expressing p in terms of x.

$$\begin{bmatrix} \text{profit} \\ \text{per tree} \end{bmatrix} = \begin{bmatrix} \text{original profit} \\ \text{per tree} \end{bmatrix} - \begin{bmatrix} \text{loss in profit (per tree)} \\ \text{due to increase} \end{bmatrix}$$

$$p = \quad 40 \quad - \quad (x - 1000)(.02)$$

$$= 60 - .02x$$

The loss in profit (per tree) due to the increase in the number of trees was obtained by multiplying $x - 1000$, the number of trees in excess of 1000, by the amount of money lost (per tree) for each excess tree. Therefore,

$$T = p \cdot x = (60 - .02x)x = 60x - .02x^2.$$

By computing first and second derivatives and sketching the graph, we easily find that total profit is maximized when $x = 1500$. Therefore, 1500 trees should be planted.

2. This problem can be solved in the same manner that Example 3 was solved. However, the comment made at the end of Example 3 indicates that the economic order quantity should increase by a factor of 3, since $3 = \sqrt{9}$. Therefore, the economic order quantity is $3 \cdot 150 = 450$ cases.

</td></tr>
</table>

2.7   Applications of Derivatives to Business and Economics

In recent years economic decision making has become more and more mathematically oriented. Faced with huge masses of statistical data, depending on hundreds or even thousands of different variables, business analysts and economists have increasingly turned to mathematical methods to help them describe what is happening, predict the effects of various policy alternatives, and choose reasonable courses of action from the myriad possibilities. Among the mathematical methods employed is derivatives. In this section we illustrate just a few of the many applications of derivatives to business and economics. All our applications will center around what economists call *the theory of the firm*. In other words, we study the activity of a business (or possibly a whole industry) and restrict our analysis to a time period during which background conditions (such as supplies of raw materials, wage rates, taxes) are fairly constant. We then show how derivatives can help the management of such a firm make vital production decisions.

Management, whether or not it knows calculus, utilizes many functions of the sort we have been considering. Examples of such functions are

$C(x) = $ cost of producing x units of the product,

$R(x) = $ revenue generated by selling x units of the product,

$P(x) = R(x) - C(x) = $ the profit (or loss) generated by producing and selling x units of the product.

Note that the functions $C(x)$, $R(x)$, $P(x)$ often are defined only for nonnegative integers—that is, for $x = 0, 1, 2, 3, \ldots$. The reason is that it does not make sense to speak about the cost of producing -1 cars or the revenue generated by selling 3.62 refrigerators. Thus each of the functions may give rise to a set of discrete points on a graph, as in Fig. 1(a). In studying these functions, however, economists usually draw a smooth curve through the points and assume that $C(x)$ is actually defined for all positive x. Of course, we must often interpret answers to problems in light of the fact that x is, in most cases, a nonnegative integer.

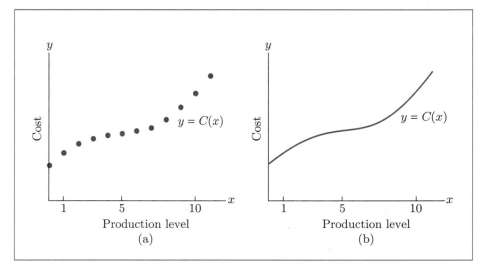

Figure 1. A cost function.

Cost Functions   If we assume that a cost function $C(x)$ has a smooth graph as in Fig. 1(b), we can use the tools of calculus to study it. A typical cost function is analyzed in Example 1.

▶ Example 1   Suppose that the cost function for a manufacturer is given by $C(x) = (10^{-6})x^3 - .003x^2 + 5x + 1000$ dollars.

(a) Describe the behavior of the marginal cost.

(b) Sketch the graph of $C(x)$.

Solution   The first two derivatives of $C(x)$ are given by

$$C'(x) = (3 \cdot 10^{-6})x^2 - .006x + 5$$
$$C''(x) = (6 \cdot 10^{-6})x - .006.$$

Let us sketch the marginal cost $C'(x)$ first. From the behavior of $C'(x)$, we will be able to graph $C(x)$. The marginal cost function $y = (3 \cdot 10^{-6})x^2 - .006x + 5$ has as its graph a parabola that opens upward. Since $y' = C''(x) = .000006(x - 1000)$, we see that the parabola has a horizontal tangent at $x = 1000$. So the minimum value of $C'(x)$ occurs at $x = 1000$. The corresponding y-coordinate is

$$(3 \cdot 10^{-6})(1000)^2 - .006 \cdot (1000) + 5 = 3 - 6 + 5 = 2.$$

The graph of $y = C'(x)$ is shown in Fig. 2. Consequently, at first the marginal cost decreases. It reaches a minimum of 2 at production level 1000 and increases thereafter. This answers part (a). Let us now graph $C(x)$. Since the graph shown

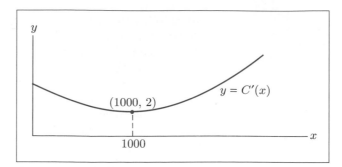

Figure 2. A marginal cost function.

in Fig. 2 is the graph of the derivative of $C(x)$, we see that $C'(x)$ is never zero, so there are no relative extreme points. Since $C'(x)$ is always positive, $C(x)$ is always increasing (as any cost curve should be). Moreover, since $C'(x)$ decreases for x less than 1000 and increases for x greater than 1000, we see that $C(x)$ is concave down for x less than 1000, is concave up for x greater than 1000, and has an inflection point at $x = 1000$. The graph of $C(x)$ is drawn in Fig. 3. Note that the inflection point of $C(x)$ occurs at the value of x for which marginal cost is a minimum. ◆

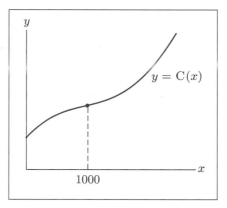

Figure 3. A cost function.

Actually, most marginal cost functions have the same general shape as the marginal cost curve of Example 1. For when x is small, production of additional units is subject to economies of production, which lowers unit costs. Thus, for x small, marginal cost decreases. However, increased production eventually leads to overtime, use of less efficient, older plants, and competition for scarce raw materials. As a result, the cost of additional units will increase for very large x. So we see that $C'(x)$ initially decreases and then increases.

Revenue Functions In general, a business is concerned not only with its costs, but also with its revenues. Recall that if $R(x)$ is the revenue received from the sale of x units of some commodity, then the derivative $R'(x)$ is called the *marginal revenue*. Economists use this to measure the rate of increase in revenue per unit increase in sales.

If x units of a product are sold at a price p per unit, then the total revenue $R(x)$ is given by

$$R(x) = x \cdot p.$$

If a firm is small and is in competition with many other companies, its sales have little effect on the market price. Then since the price is constant as far as the one firm is concerned, the marginal revenue $R'(x)$ equals the price p [that is, $R'(x)$ is the amount that the firm receives from the sale of one additional unit]. In this case, the revenue function will have a graph as in Fig. 4.

An interesting problem arises when a single firm is the only supplier of a certain product or service—that is, when the firm has a monopoly. Consumers will buy large amounts of the commodity if the price per unit is low and less if the price is raised. For each quantity x, let $f(x)$ be the highest price per unit that can be set in order to sell all x units to customers. Since selling greater quantities requires a lowering of the price, $f(x)$ will be a decreasing function. Figure 5 shows a typical "demand curve" that relates the quantity demanded, x, to the price $p = f(x)$.

The *demand equation* $p = f(x)$ determines the total revenue function. If the firm wants to sell x units, the highest price it can set is $f(x)$ dollars per unit, and so the total revenue from the sale of x units is

$$R(x) = x \cdot p = x \cdot f(x). \tag{1}$$

The concept of a demand curve applies to an entire industry (with many producers) as well as to a single monopolistic firm. In this case, many producers offer the same product for sale. If x denotes the total output of the industry, then $f(x)$ is the market price per unit of output and $x \cdot f(x)$ is the total revenue earned from the sale of the x units.

▶ Example 2   The demand equation for a certain product is $p = 6 - \frac{1}{2}x$ dollars. Find the level of production that results in maximum revenue.

Solution   In this case, the revenue function $R(x)$ is

$$R(x) = x \cdot p = x \left(6 - \tfrac{1}{2}x\right) = 6x - \tfrac{1}{2}x^2$$

dollars. The marginal revenue is given by

$$R'(x) = 6 - x.$$

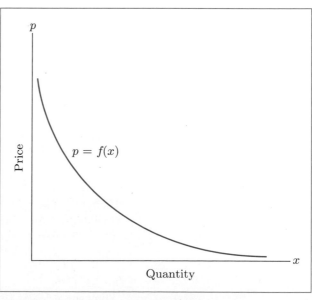

Figure 5. A demand curve.

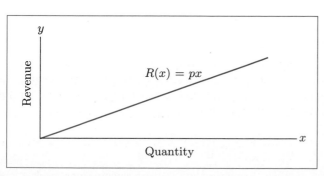

Figure 4. A revenue curve.

The graph of $R(x)$ is a parabola that opens downward (Fig. 6). It has a horizontal tangent precisely at those x for which $R'(x) = 0$—that is, for those x at which marginal revenue is 0. The only such x is $x = 6$. The corresponding value of revenue is

$$R(6) = 6 \cdot 6 - \tfrac{1}{2}(6)^2 = 18 \text{ dollars.}$$

Thus the rate of production resulting in maximum revenue is $x = 6$, which results in total revenue of 18 dollars.   ◆

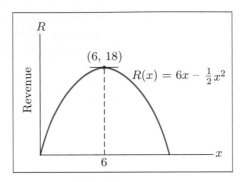

Figure 6. Maximizing revenue.

▶ Example 3   The WMA Bus Lines offers sightseeing tours of Washington, D.C. One of the tours, priced at \$7 per person, had an average demand of about 1000 customers per week. When the price was lowered to \$6, the weekly demand jumped to about 1200 customers. Assuming that the demand equation is linear, find the tour price that should be charged per person in order to maximize the total revenue each week.

Solution   First we must find the demand equation. Let x be the number of customers per week and let p be the price of a tour ticket. Then $(x, p) = (1000, 7)$ and $(x, p) = (1200, 6)$ are on the demand curve (Fig. 7). Using the point-slope formula for the line through these two points, we have

$$p - 7 = \frac{7 - 6}{1000 - 1200} \cdot (x - 1000) = -\frac{1}{200}(x - 1000) = -\frac{1}{200}x + 5,$$

so

$$p = 12 - \frac{1}{200}x. \tag{2}$$

From equation (1) we obtain the revenue function

$$R(x) = x\left(12 - \frac{1}{200}x\right) = 12x - \frac{1}{200}x^2.$$

The marginal revenue function is

$$R'(x) = 12 - \frac{1}{100}x = -\frac{1}{100}(x - 1200).$$

Using $R(x)$ and $R'(x)$, we can sketch the graph of $R(x)$ (Fig. 8). The maximum revenue occurs when the marginal revenue is zero—that is, when $x = 1200$. The price corresponding to this number of customers is found from the demand equation (2),

$$p = 12 - \frac{1}{200}(1200) = 6 \text{ dollars.}$$

Thus the price of \$6 is most likely to bring the greatest revenue per week.   ◆

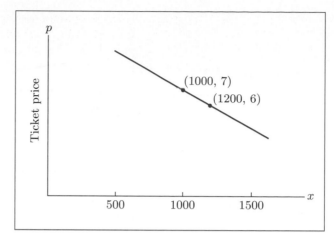

Figure 7. A demand curve.

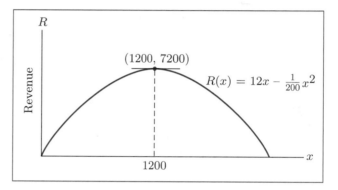

Figure 8. Maximizing revenue.

Profit Functions Once we know the cost function $C(x)$ and the revenue function $R(x)$, we can compute the profit function $P(x)$ from

$$P(x) = R(x) - C(x).$$

▶ Example 4 Suppose that the demand equation for a monopolist is $p = 100 - .01x$ and the cost function is $C(x) = 50x + 10{,}000$. Find the value of x that maximizes the profit and determine the corresponding price and total profit for this level of production. (See Fig. 9.)

Solution The total revenue function is

$$R(x) = x \cdot p = x(100 - .01x) = 100x - .01x^2.$$

Hence the profit function is

$$\begin{aligned} P(x) &= R(x) - C(x) \\ &= 100x - .01x^2 - (50x + 10{,}000) \\ &= -.01x^2 + 50x - 10{,}000. \end{aligned}$$

The graph of this function is a parabola that opens downward. (See Fig. 10.) Its highest point will be where the curve has zero slope—that is, where the marginal

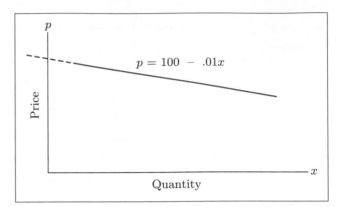

Figure 9. A demand curve.

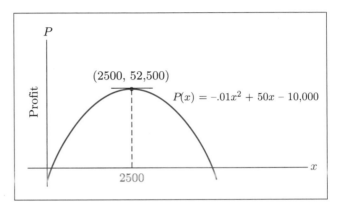

Figure 10. Maximizing profit.

profit $P'(x)$ is zero. Now

$$P'(x) = -.02x + 50 = -.02(x - 2500).$$

So $P'(x) = 0$ when $x = 2500$. The profit for this level of production is

$$P(2500) = -.01(2500)^2 + 50(2500) - 10,000 = 52,500 \text{ dollars.}$$

Finally, we return to the demand equation to find the highest price that can be charged per unit to sell all 2500 units:

$$p = 100 - .01(2500) = 100 - 25 = 75 \text{ dollars.}$$

ANSWER

Produce 2500 units and sell them at \$75 per unit. The profit will be \$52,500. ◆

▶ **Example 5**   Rework Example 4 under the condition that the government imposes an excise tax of \$10 per unit.

Solution   For each unit sold, the manufacturer will have to pay \$10 to the government. In other words, $10x$ dollars are added to the cost of producing and selling x units. The cost function is now

$$C(x) = (50x + 10,000) + 10x = 60x + 10,000.$$

The demand equation is unchanged by this tax, so the revenue is still

$$R(x) = 100x - .01x^2.$$

Proceeding as before, we have

$$P(x) = R(x) - C(x)$$
$$= 100x - .01x^2 - (60x + 10,000)$$
$$= -.01x^2 + 40x - 10,000.$$
$$P'(x) = -.02x + 40 = -.02(x - 2000).$$

The graph of $P(x)$ is still a parabola that opens downward, and the highest point is where $P'(x) = 0$—that is, where $x = 2000$. (See Fig. 11.) The corresponding profit is

$$P(2000) = -.01(2000)^2 + 40(2000) - 10,000 = 30,000 \text{ dollars.}$$

From the demand equation, $p = 100 - .01x$, we find the price that corresponds to $x = 2000$:

$$p = 100 - .01(2000) = 80 \text{ dollars.}$$

ANSWER

Produce 2000 units and sell them at $80 per unit. The profit will be $30,000. ◆

Notice in Example 5 that the optimal price is raised from $75 to $80. If the monopolist wishes to maximize profits, he or she should pass only half the $10 tax on to the customer. The monopolist cannot avoid the fact that profits will be substantially lowered by the imposition of the tax. This is one reason why industries lobby against taxation.

Setting Production Levels Suppose that a firm has cost function $C(x)$ and revenue function $R(x)$. In a free-enterprise economy the firm will set production x in such a way as to maximize the profit function

$$P(x) = R(x) - C(x).$$

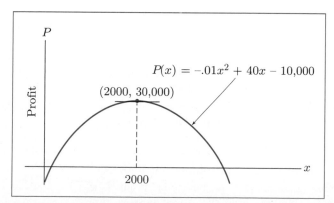

Figure 11. Profit after an excise tax.

We have seen that if $P(x)$ has a maximum at $x = a$, then $P'(a) = 0$. In other words, since $P'(x) = R'(x) - C'(x)$,

$$R'(a) - C'(a) = 0$$
$$R'(a) = C'(a).$$

Thus profit is maximized at a production level for which marginal revenue equals marginal cost. (See Fig. 12.)

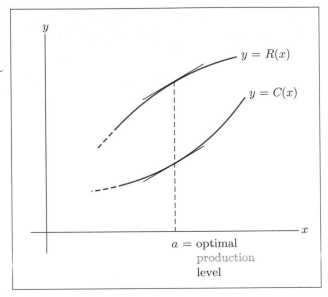

Figure 12.

Practice Problems 2.7

1. Rework Example 4 by finding the production level at which marginal revenue equals marginal cost.

2. Rework Example 4 under the condition that the fixed cost is increased from $10,000 to $15,000.

▶ Exercises 2.7

1. Given the cost function $C(x) = x^3 - 6x^2 + 13x + 15$, find the minimum marginal cost.

2. Suppose that a total cost function is $C(x) = .0001x^3 - .06x^2 + 12x + 100$. Is the marginal cost increasing, decreasing, or not changing at $x = 100$? Find the minimum marginal cost.

3. The revenue function for a one-product firm is

$$R(x) = 200 - \frac{1600}{x + 8} - x.$$

Find the value of x that results in maximum revenue.

4. The revenue function for a particular product is $R(x) = x(4 - .0001x)$. Find the largest possible revenue.

5. A one-product firm estimates that its daily total cost function (in suitable units) is $C(x) = x^3 - 6x^2 + 13x + 15$ and its total revenue function is $R(x) = 28x$. Find the value of x that maximizes the daily profit.

6. A small tie shop sells ties for $3.50 each. The daily cost function is estimated to be $C(x)$ dollars, where x is the number of ties sold on a typical day and $C(x) = .0006x^3 - .03x^2 + 2x + 20$. Find the value of x that will maximize the store's daily profit.

7. The demand equation for a certain commodity is

$$p = \tfrac{1}{12}x^2 - 10x + 300,$$

$0 \le x \le 60$. Find the value of x and the corresponding price p that maximize the revenue.

8. The demand equation for a product is $p = 2 - .001x$. Find the value of x and the corresponding price p that maximize the revenue.

9. Some years ago it was estimated that the demand for steel approximately satisfied the equation $p = 256 - 50x$, and the total cost of producing x units of steel was $C(x) = 182 + 56x$. (The quantity x was measured in millions of tons and the price and total cost were measured in millions of dollars.) Determine the level of production and the corresponding price that maximize the profits.

10. Consider a rectangle in the xy-plane, with corners at $(0,0)$, $(a,0)$, $(0,b)$, and (a,b). Suppose that (a,b) lies on the graph of the equation $y = 30 - x$. Find a and b such that the area of the rectangle is maximized. What economic interpretations can be given to your answer if the equation $y = 30 - x$ represents a demand curve and y is the price corresponding to the demand x?

11. Until recently hamburgers at the city sports arena cost $2 each. The food concessionaire sold an average of 10,000 hamburgers on a game night. When the price was raised to $2.40, hamburger sales dropped off to an average of 8000 per night.

(a) Assuming a linear demand curve, find the price of a hamburger that will maximize the nightly hamburger revenue.

(b) Suppose that the concessionaire has fixed costs of $1000 per night and the variable cost is $.60 per hamburger. Find the price of a hamburger that will maximize the nightly hamburger profit.

12. The average ticket price for a concert at the opera house was $50. The average attendance was 4000. When the ticket price was raised to $52, attendance declined to an average of 3800 persons per performance. What should the ticket price be in order to maximize the revenue for the opera house? (Assume a linear demand curve.)

13. The monthly demand equation for an electric utility company is estimated to be

$$p = 60 - (10^{-5})x,$$

where p is measured in dollars and x is measured in thousands of kilowatt-hours. The utility has fixed costs of $7,000,000 per month and variable costs of $30 per 1000 kilowatt-hours of electricity generated, so that the cost function is

$$C(x) = 7 \cdot 10^6 + 30x.$$

(a) Find the value of x and the corresponding price for 1000 kilowatt-hours that maximize the utility's profit.

(b) Suppose that rising fuel costs increase the utility's variable costs from $30 to $40 so that its new cost function is

$$C_1(x) = 7 \cdot 10^6 + 40x.$$

Should the utility pass all this increase of $10 per thousand kilowatt-hours on to consumers? Explain your answer.

14. The demand equation for a monopolist is $p = 200 - 3x$, and the cost function is

$$C(x) = 75 + 80x - x^2, \qquad 0 \le x \le 40.$$

(a) Determine the value of x and the corresponding price that maximize the profit.

(b) Suppose that the government imposes a tax on the monopolist of $4 per unit quantity produced. Determine the new price that maximizes the profit.

(c) Suppose that the government imposes a tax of T dollars per unit quantity produced so that the new cost function is

$$C(x) = 75 + (80 + T)x - x^2, \qquad 0 \le x \le 40.$$

Determine the new value of x that maximizes the monopolist's profit as a function of T. Assuming that the monopolist cuts back production to this level, express the tax revenues received by the government as a function of T. Finally, determine the value of T that will maximize the tax revenue received by the government.

15. A savings and loan association estimates that the amount of money on deposit will be 1,000,000 times the percentage rate of interest. For instance, a 4% interest rate will generate $4,000,000 in deposits. Suppose the savings and loan association can loan all the money it takes in at 10% interest. What interest rate on deposits generates the greatest profit?

16. Let $P(x)$ be the annual profit for a certain product, where x is the amount of money spent on advertising. See Fig. 13.

(a) Interpret $P(0)$.

(b) Describe how the marginal profit changes as the amount of money spent on advertising increases.

(c) Explain the economic significance of the inflection point.

17. The revenue for a manufacturer is $R(x)$ thousand dollars, where x is the number of units of goods produced (and sold) and R and R' are the functions given in Figs. 14(a) and 14(b).

(a) What is the revenue from producing 40 units of goods?

(b) What is the marginal revenue when 17.5 units of goods are produced?

(c) At what level of production is the revenue $45,000?

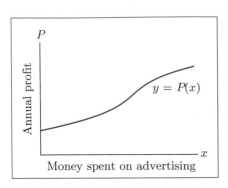

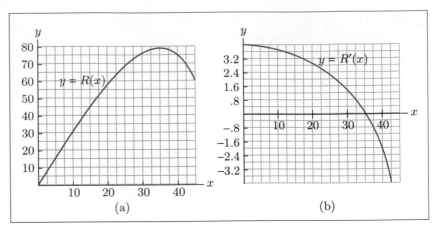

Figure 13. Profit as a function of advertising.

Figure 14. Revenue function and its first derivative.

(d) At what level(s) of production is the marginal revenue $800?

(e) At what level of production is the revenue greatest?

18. The cost function for a manufacturer is $C(x)$ dollars, where x is the number of units of goods produced and C, C', and C'' are the functions given in Fig. 15.

(a) What is the cost of manufacturing 60 units of goods?

(b) What is the marginal cost when 40 units of goods are manufactured?

(c) At what level of production is the cost $1200?

(d) At what level(s) of production is the marginal cost $22.50?

(e) At what level of production does the marginal cost have the least value? What is the marginal cost at this level of production?

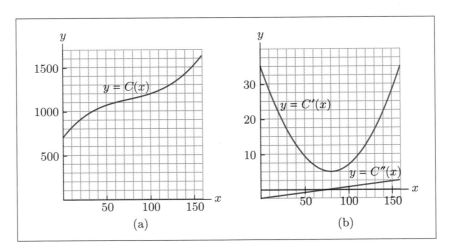

Figure 15. Cost function and its derivatives.

1. The revenue function is $R(x) = 100x - .01x^2$, so the marginal revenue function is $R'(x) = 100 - .02x$. The cost function is $C(x) = 50x + 10{,}000$, so the marginal cost function is $C'(x) = 50$. Let us now equate the two marginal functions and solve for x.

$$R'(x) = C'(x)$$
$$100 - .02x = 50$$
$$-.02x = -50$$
$$x = \frac{-50}{-.02} = \frac{5000}{2} = 2500$$

Of course, we obtain the same level of production as before.

2. If the fixed cost is increased from \$10,000 to \$15,000, the new cost function will be $C(x) = 50x + 15{,}000$ but the marginal cost function will still be $C'(x) = 50$. Therefore, the solution will be the same: 2500 units should be produced and sold at \$75 per unit. (Increases in fixed costs should not necessarily be passed on to the consumer if the objective is to maximize the profit.)

Review of Fundamental Concepts of Chapter 2

1. State as many terms used to describe graphs of functions as you can recall.

2. What is the difference between having a relative maximum at $x = 2$ and having an absolute maximum at $x = 2$?

3. Give three characterizations of what it means for the graph of $f(x)$ to be concave up at $x = 2$. Concave down.

4. What does it mean to say that the graph of $f(x)$ has an inflection point at $x = 2$?

5. What is the difference between an x-intercept and a zero of a function?

6. How do you determine the y-intercept of a function?

7. What is an asymptote? Give an example.

8. State the first derivative rule. Second derivative rule.

9. Give two connections between the graphs of $f(x)$ and $f'(x)$.

10. Outline a method for locating the relative extreme points of a function.

11. Outline a method for locating the inflection points of a function.

12. Outline a procedure for sketching the graph of a function.

13. What is an objective equation?

14. What is a constraint equation?

15. Outline the procedure for solving an optimization problem.

16. How are the cost, revenue, and profit functions related?

▶ Chapter 2 Supplementary Exercises

1. Figure 1 contains the graph of $f'(x)$, the derivative of $f(x)$. Use the graph to answer the following questions about the graph of $f(x)$.

   (a) For what values of x is the graph of $f(x)$ increasing? Decreasing?

   (b) For what values of x is the graph of $f(x)$ concave up? Concave down?

2. Figure 2 shows the graph of the function $f(x)$ and its tangent line at $x = 3$. Find $f(3)$, $f'(3)$, and $f''(3)$.

In Exercises 3–6, draw the graph of a function $f(x)$ for which the function and its first derivative have the stated property for all x.

3. $f(x)$ and $f'(x)$ increasing.

4. $f(x)$ and $f'(x)$ decreasing.

5. $f(x)$ increasing and $f'(x)$ decreasing.

6. $f(x)$ decreasing and $f'(x)$ increasing.

Exercises 7–12 refer to the graph in Fig. 3. List the labeled values of x at which the derivative has the stated property.

7. $f'(x)$ is positive.

8. $f'(x)$ is negative.

9. $f''(x)$ is positive.

10. $f''(x)$ is negative.

11. $f'(x)$ is maximized.

12. $f'(x)$ is minimized.

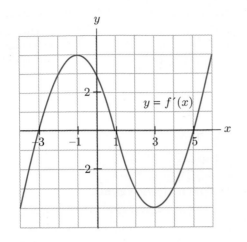

Figure 1.

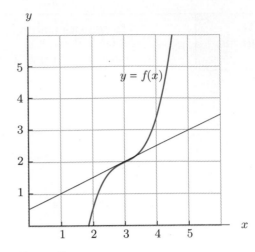

Figure 2.

Properties of various functions are described next. In each case draw some conclusion about the graph of the function.

13. $f(1) = 2$, $f'(1) > 0$

14. $g(1) = 5$, $g'(1) = -1$

15. $h'(3) = 4$, $h''(3) = 1$

16. $F'(2) = -1$, $F''(2) < 0$

17. $G(10) = 2$, $G'(10) = 0$, $G''(10) > 0$

18. $f(4) = -2$, $f'(4) > 0$, $f''(4) = -1$

19. $g(5) = -1$, $g'(5) = -2$, $g''(5) = 0$

20. $H(0) = 0$, $H'(0) = 0$, $H''(0) = 1$

21. In Figs. 4(a) and 4(b), the t-axis represents time in hours.

   (a) When is $f(t) = 1$?

   (b) Find $f(5)$.

   (c) When is $f(t)$ changing at the rate of $-.08$ units per hour?

   (d) How fast is $f(t)$ changing after 8 hours?

22. U.S. electrical energy production (in trillions of kilowatt-hours) in year t (with 1900 corresponding to $t = 0$) is given by $f(t)$, where f and its derivatives are graphed in Figs. 5(a) and 5(b).

   (a) How much electrical energy was produced in 1950?

   (b) How fast was energy production rising in 1950?

   (c) When did energy production reach 3000 trillion kilowatt-hours?

   (d) When was the level of energy production rising at the rate of 10 trillion kilowatt-hours per year?

   (e) When was energy production growing at the greatest rate? What was the level of production at that time?

Sketch the following parabolas. Include their x- and y-intercepts.

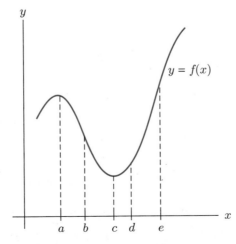

Figure 3.

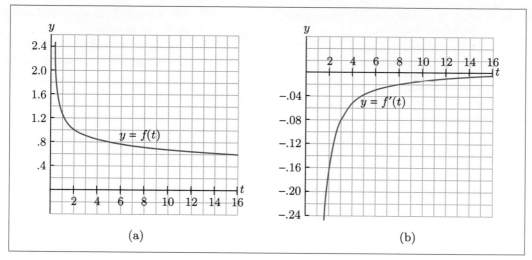

Figure 4.

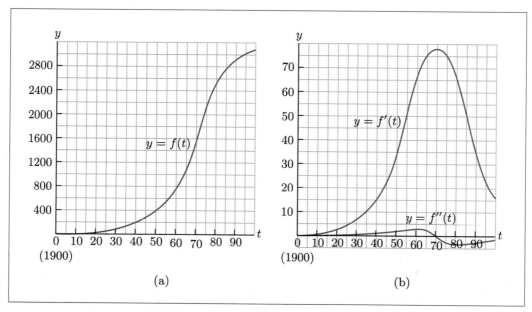

Figure 5. U.S. electrical energy production.

23. $y = 3 - x^2$

24. $y = 7 + 6x - x^2$

25. $y = x^2 + 3x - 10$

26. $y = 4 + 3x - x^2$

27. $y = -2x^2 + 10x - 10$

28. $y = x^2 - 9x + 19$

29. $y = x^2 + 3x + 2$

30. $y = -x^2 + 8x - 13$

31. $y = -x^2 + 20x - 90$

32. $y = 2x^2 + x - 1$

Sketch the following curves.

33. $y = 2x^3 + 3x^2 + 1$

34. $y = x^3 - \frac{3}{2}x^2 - 6x$

35. $y = x^3 - 3x^2 + 3x - 2$

36. $y = 100 + 36x - 6x^2 - x^3$

37. $y = \frac{11}{3} + 3x - x^2 - \frac{1}{3}x^3$

38. $y = x^3 - 3x^2 - 9x + 7$

39. $y = -\frac{1}{3}x^3 - 2x^2 - 5x$

40. $y = x^3 - 6x^2 - 15x + 50$

41. $y = x^4 - 2x^2$

42. $y = x^4 - 4x^3$

43. $y = \frac{x}{5} + \frac{20}{x} + 3 \quad (x > 0)$

44. $y = \frac{1}{2x} + 2x + 1 \quad (x > 0)$

45. Let $f(x) = (x^2 + 2)^{3/2}$. Show that the graph of $f(x)$ has a possible relative extreme point at $x = 0$.

46. Show that the function $f(x) = (2x^2 + 3)^{3/2}$ is decreasing for $x < 0$ and increasing for $x > 0$.

47. Let $f(x)$ be a function whose *derivative* is

$$f'(x) = \frac{1}{1 + x^2}.$$

Note that $f'(x)$ is always positive. Show that the graph of $f(x)$ definitely has an inflection point at $x = 0$.

48. Let $f(x)$ be a function whose *derivative* is

$$f'(x) = \sqrt{5x^2 + 1}.$$

Show that the graph of $f(x)$ definitely has an inflection point at $x = 0$.

49. Suppose a car is traveling on a straight road and $s(t)$ is the distance traveled after t hours. Match each set of information about $s(t)$ and its derivatives with the corresponding description of the car's motion.
   Information:
   A. $s(t)$ is a constant function.
   B. $s'(t)$ is a positive constant function.
   C. $s'(t)$ is positive at $t = a$.
   D. $s'(t)$ is negative at $t = a$.
   E. $s'(t)$ and $s''(t)$ are positive at $t = a$.
   F. $s'(t)$ is positive and $s''(t)$ is negative at $t = a$.
   Descriptions:
   a. The car is moving forward and speeding up at time a.
   b. The car is backing up at time a.
   c. The car is standing still.
   d. The car is moving forward but slowing down at time a.
   e. The car is moving forward at a steady rate.
   f. The car is moving forward at time a.

50. The water level in a reservoir varies during the year. Let $h(t)$ be the depth (in feet) of the water at time t days, where $t = 0$ at the beginning of the year. Match each set of information about $h(t)$ and its derivatives with the corresponding description of the reservoir's activity.
   Information:
   A. $h(t)$ has the value 50 for $1 \leq t \leq 2$.
   B. $h'(t)$ has the value .5 for $1 \leq t \leq 2$.
   C. $h'(t)$ is positive at $t = a$.
   D. $h'(t)$ is negative at $t = a$.
   E. $h'(t)$ and $h''(t)$ are positive at $t = a$.
   F. $h'(t)$ is positive and $h''(t)$ is negative at $t = a$.
   Descriptions:
   a. The water level is rising at an increasing rate at time a.
   b. The water level is receding at time a.
   c. The water level is constant at 50 feet on January 2.
   d. At time a the water level is rising, but the rate of increase is slowing down.

   e. On January 2 the water rose steadily at a rate of .5 feet per day.
   f. The water level is rising at time a.

51. Let $f(x)$ be the number of people living within x miles of the center of New York City.
   (a) What does $f(10 + h) - f(10)$ represent?
   (b) Explain why $f'(10)$ cannot be negative.

52. For what x does the function $f(x) = \frac{1}{4}x^2 - x + 2$, $0 \leq x \leq 8$, have its maximum value?

53. Find the maximum value of the function $f(x) = 2 - 6x - x^2$, $0 \leq x \leq 5$, and give the value of x where this maximum occurs.

54. Find the minimum value of the function $g(t) = t^2 - 6t + 9$, $1 \leq t \leq 6$.

55. An open rectangular box is to be 4 feet long and have a volume of 200 cubic feet. Find the dimensions for which the amount of material needed to construct the box is as small as possible.

56. A closed rectangular box with a square base is to be constructed using two different types of wood. The top is made of wood costing $3 per square foot and the remainder is made of wood costing $1 per square foot. Suppose that $48 is available to spend. Find the dimensions of the box of greatest volume that can be constructed.

57. A long rectangular sheet of metal 30 inches wide is to be made into a gutter by turning up strips vertically along the two sides. How many inches should be turned up on each side in order to maximize the amount of water that the gutter can carry?

58. A small orchard yields 25 bushels of fruit per tree when planted with 40 trees. Because of overcrowding, the yield per tree (for each tree in the orchard) is reduced by $\frac{1}{2}$ bushel for each additional tree that is planted. How many trees should be planted in order to maximize the total yield of the orchard?

59. A publishing company sells 400,000 copies of a certain book each year. Ordering the entire amount printed at the beginning of the year ties up valuable storage space and capital. However, running off the copies in several partial runs throughout the year results in added costs for setting up each printing run. Setting up each production run costs $1000. The carrying costs, figured on the average number of books in storage, are 50¢ per book. Find the economic lot size, that is, the production run size that minimizes the total setting up and carrying costs.

60. A poster is to have an area of 125 square inches. The printed material is to be surrounded by a margin of 3 inches at the top and margins of 2 inches at the bottom and sides. Find the dimensions of the poster that maximize the area of the printed material.

61. Suppose that the demand equation for a monopolist is $p = 150 - .02x$ and the cost function is $C(x) = 10x + 300$. Find the value of x that maximizes the profit.

Chapter Project

The following table gives the amount of retail store credit card debt, total credit card debt, and the number of retail store credit cards, for various years. The values for years 2000 and 2005 are projections. Suppose that t represents time with $t = 0$ corresponding to 1990.

Table 1   **Credit Card—Holders, Numbers, Spending, and Debt**

Source: Statistical Abstracts of the United States, U.S. Department of Commerce, 1999.

	Retail store credit card debt (billion \$)	Total credit card debt (billion \$)	Total retail store credit cards (millions of cards)
1990	51	243	459
1995	78	460	588
1996	86	522	602
1997	88	560	614
2000	99	677	652
2005	129	951	720

Let $f(t)$, $g(t)$, and $h(t)$ be functions defined by columns 2, 3, and 4 of the table, respectively. Using regression analysis, we can determine models $f_1(t)$, $g_1(t)$, and $h_1(t)$, where

$$f_1(t) = .026t^3 - .572t^2 + 8.03t + 50.8$$
$$g_1(t) = .105t^3 - 1.98t^2 + 53.2t + 242$$
$$h_1(t) = .09t^3 - 2.6t^2 + 36.2t + 459.$$

(a) Compute the first and second derivatives of each of the model functions.

(b) Determine where the graphs of each of the model functions is concave up (respectively, concave down).

(c) For each model function, determine when it is increasing the fastest.

(d) Determine a model for the proportion of credit card debt coming from retail store cards as a function of time. Graph this model and describe its behavior.

(e) Determine a model for the amount of debt per card as a function of time. Graph this model and describe its behavior.

CHAPTER 3

Techniques of Differentiation

e have seen that the derivative is useful in many applications. However, our ability to differentiate functions is somewhat limited. For example, we cannot yet readily differentiate

$$(x^2 - 1)^4(x^2 + 1)^5, \qquad \frac{x^3}{(x^2 + 1)^4}.$$

In this chapter we develop differentiation techniques that apply to functions like those just given. Two new rules are the *product rule* and the *quotient rule*. In Section 3.2 we extend the general power rule into a powerful formula called the *chain rule*.

3.1 The Product and Quotient Rules

We observed in our discussion of the sum rule for derivatives that the derivative of the sum of two differentiable functions is the sum of the derivatives. Unfortunately, however, the derivative of the product $f(x)g(x)$ is *not* the product of the derivatives. Rather, the derivative of a product is determined from the following rule.

Product Rule

$$\frac{d}{dx}[f(x)g(x)] = f(x)g'(x) + g(x)f'(x)$$

The derivative of the product of two functions is the first function times the derivative of the second plus the second function times the derivative of the first. At the end of the section we show why this statement is true.

▶ **Example 1** Show that the product rule works for the case $f(x) = x^2$, $g(x) = x^3$.

Solution Since $x^2 \cdot x^3 = x^5$, we know that

$$\frac{d}{dx}(x^2 \cdot x^3) = \frac{d}{dx}(x^5) = 5x^4.$$

214

On the other hand, using the product rule,

$$\frac{d}{dx}(x^2 \cdot x^3) = x^2 \cdot \frac{d}{dx}(x^3) + x^3 \cdot \frac{d}{dx}(x^2)$$

$$= x^2(3x^2) + x^3(2x)$$

$$= 3x^4 + 2x^4 = 5x^4.$$

Thus the product rule gives the correct answer. ◆

▶ **Example 2**   Differentiate the product $(2x^3 - 5x)(3x + 1)$.

Solution   Let $f(x) = 2x^3 - 5x$ and $g(x) = 3x + 1$. Then

$$\frac{d}{dx}\left[(2x^3 - 5x)(3x + 1)\right] = (2x^3 - 5x) \cdot \frac{d}{dx}(3x + 1) + (3x + 1) \cdot \frac{d}{dx}(2x^3 - 5x)$$

$$= (2x^3 - 5x)(3) + (3x + 1)(6x^2 - 5)$$

$$= 6x^3 - 15x + 18x^3 - 15x + 6x^2 - 5$$

$$= 24x^3 + 6x^2 - 30x - 5.$$ ◆

▶ **Example 3**   Apply the product rule to $y = g(x) \cdot g(x)$.

Solution

$$\frac{d}{dx}[g(x) \cdot g(x)] = g(x) \cdot g'(x) + g(x) \cdot g'(x)$$

$$= 2g(x)g'(x)$$

This answer is the same as that given by the general power rule:

$$\frac{d}{dx}[g(x) \cdot g(x)] = \frac{d}{dx}[g(x)]^2 = 2g(x)g'(x).$$ ◆

▶ **Example 4**   Find $\dfrac{dy}{dx}$, where $y = (x^2 - 1)^4(x^2 + 1)^5$.

Solution   Let $f(x) = (x^2 - 1)^4$, $g(x) = (x^2 + 1)^5$, and use the product rule. The general power rule is needed to compute $f'(x)$ and $g'(x)$.

$$\frac{dy}{dx} = (x^2 - 1)^4 \cdot \frac{d}{dx}(x^2 + 1)^5 + (x^2 + 1)^5 \cdot \frac{d}{dx}(x^2 - 1)^4$$

$$= (x^2 - 1)^4 \cdot 5(x^2 + 1)^4(2x) + (x^2 + 1)^5 \cdot 4(x^2 - 1)^3(2x) \qquad (1)$$

This form of $\dfrac{dy}{dx}$ is suitable for some purposes. For example, if we need to compute $\dfrac{dy}{dx}\bigg|_{x=2}$, it is easier just to substitute 2 for x than to simplify and then substitute. However, it is often helpful to simplify the formula for $\dfrac{dy}{dx}$, such as in the case when we need to find x such that $\dfrac{dy}{dx} = 0$.

To simplify the answer in (1), we shall write $\dfrac{dy}{dx}$ as a single product rather than as a sum of two products. The first step is to identify the common factors.

$$\frac{dy}{dx} = (x^2 - 1)^4 \cdot 5(x^2 + 1)^4 \, (2x) + (x^2 + 1)^5 \cdot 4(x^2 - 1)^3 \, (2x).$$

Both terms contain $2x$ and powers of $x^2 - 1$ and $x^2 + 1$. The most we can factor out of each term is $2x(x^2 - 1)^3(x^2 + 1)^4$. We obtain

$$\frac{dy}{dx} = 2x(x^2 - 1)^3(x^2 + 1)^4[5(x^2 - 1) + 4(x^2 + 1)].$$

Simplifying the right-most factor in this product, we have

$$\frac{dy}{dx} = 2x(x^2 - 1)^3(x^2 + 1)^4[9x^2 - 1]. \tag{2}$$

◆

The answers to the exercises for this section appear in two forms, similar to those in (1) and (2). The unsimplified answers will allow you to check if you have mastered the differentiation rules. In each case you should make an effort to transform your original answer into the simplified version. Examples 4 and 6 show how to do this.

The Quotient Rule   Another useful formula for differentiating functions is the quotient rule.

Quotient Rule

$$\frac{d}{dx}\left[\frac{f(x)}{g(x)}\right] = \frac{g(x)f'(x) - f(x)g'(x)}{[g(x)]^2}$$

One must be careful to remember the order of the terms in this formula because of the minus sign in the numerator.

▶ **Example 5**   Differentiate $\dfrac{x}{2x + 3}$.

Solution   Let $f(x) = x$ and $g(x) = 2x + 3$.

$$\frac{d}{dx}\left(\frac{x}{2x + 3}\right) = \frac{(2x + 3) \cdot \dfrac{d}{dx}(x) - (x) \cdot \dfrac{d}{dx}(2x + 3)}{(2x + 3)^2}$$

$$= \frac{(2x + 3) \cdot 1 - x \cdot 2}{(2x + 3)^2} = \frac{3}{(2x + 3)^2}$$

◆

▶ **Example 6**   Find $\dfrac{dy}{dx}$, where $y = \dfrac{x^3}{(x^2 + 1)^4}$.

Solution   Let $f(x) = x^3$ and $g(x) = (x^2 + 1)^4$.

$$\frac{d}{dx}\left(\frac{x^3}{(x^2+1)^4}\right) = \frac{(x^2+1)^4 \cdot \frac{d}{dx}(x^3) - (x^3)\cdot\frac{d}{dx}(x^2+1)^4}{[(x^2+1)^4]^2}$$

$$= \frac{(x^2+1)^4 \cdot 3x^2 - x^3 \cdot 4(x^2+1)^3(2x)}{(x^2+1)^8}$$

If a simplified form of $\dfrac{dy}{dx}$ is desired, we can divide the numerator and the denominator by the common factor $(x^2 + 1)^3$.

$$\frac{dy}{dx} = \frac{(x^2+1)\cdot 3x^2 - x^3 \cdot 4(2x)}{(x^2+1)^5}$$

$$= \frac{3x^4 + 3x^2 - 8x^4}{(x^2+1)^5}$$

$$= \frac{3x^2 - 5x^4}{(x^2+1)^5} = \frac{x^2(3-5x^2)}{(x^2+1)^5} \qquad \blacklozenge$$

▶ Example 7   Suppose that the total cost of manufacturing x units of a certain product is given by the function $C(x)$. Then the *average cost per unit*, AC, is defined by

$$AC = \frac{C(x)}{x}.$$

Recall that the *marginal cost*, MC, is defined by

$$MC = C'(x).$$

Show that at the level of production where the average cost is at a minimum, the average cost equals the marginal cost.

Solution   In practice, the marginal cost and average cost curves will have the general shapes shown in Fig. 1. So the minimum point on the average cost curve will occur when $\dfrac{d}{dx}(AC) = 0$. To compute the derivative, we need the quotient rule,

$$\frac{d}{dx}(AC) = \frac{d}{dx}\left(\frac{C(x)}{x}\right) = \frac{x\cdot C'(x) - C(x)}{x^2}.$$

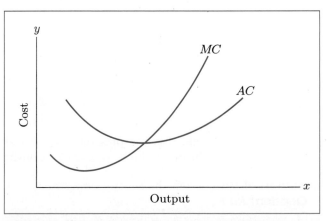

Figure 1. Marginal cost and average cost functions.

Setting the derivative equal to zero and multiplying by x^2, we obtain

$$0 = x \cdot C'(x) - C(x)$$

$$C(x) = x \cdot C'(x)$$

$$\frac{C(x)}{x} = C'(x)$$

$$AC = MC.$$

Thus when the output x is chosen so that the average cost is minimized, the average cost equals the marginal cost. ◆

Verification of the Product and Quotient Rules

Verification of the Product Rule

From our discussion of limits we compute the derivative of $f(x)g(x)$ at $x = a$ as the limit

$$\frac{d}{dx}[f(x)g(x)]\bigg|_{x=a} = \lim_{h \to 0} \frac{f(a+h)g(a+h) - f(a)g(a)}{h}.$$

Let us add and subtract the quantity $f(a)g(a + h)$ in the numerator. After factoring and applying Limit Theorem III, we obtain

$$\lim_{h \to 0} \frac{[f(a+h)g(a+h) - f(a)g(a+h)] + [f(a)g(a+h) - f(a)g(a)]}{h}$$

$$= \lim_{h \to 0} g(a+h) \cdot \frac{f(a+h) - f(a)}{h} + \lim_{h \to 0} f(a) \cdot \frac{g(a+h) - g(a)}{h}.$$

And this expression may be rewritten by Limit Theorem V as

$$\lim_{h \to 0} g(a+h) \cdot \lim_{h \to 0} \frac{f(a+h) - f(a)}{h} + \lim_{h \to 0} f(a) \cdot \lim_{h \to 0} \frac{g(a+h) - g(a)}{h}.$$

Note, however, that since $g(x)$ is differentiable at $x = a$, it is continuous there, so that $\lim_{h \to 0} g(a+h) = g(a)$. Therefore, the preceding expression equals

$$g(a)f'(a) + f(a)g'(a).$$

That is, we have proved that

$$\frac{d}{dx}[f(x)g(x)]\bigg|_{x=a} = g(a)f'(a) + f(a)g'(a),$$

which is the product rule. An alternative verification of the product rule not involving limit arguments is outlined in Exercise 60.

Verification of the Quotient Rule

From the general power rule, we know that

$$\frac{d}{dx}\left[\frac{1}{g(x)}\right] = \frac{d}{dx}[g(x)]^{-1} = (-1)[g(x)]^{-2} \cdot g'(x).$$

We can now derive the quotient rule from the product rule.

$$\frac{d}{dx}\left[\frac{f(x)}{g(x)}\right] = \frac{d}{dx}\left[\frac{1}{g(x)} \cdot f(x)\right]$$

$$= \frac{1}{g(x)} \cdot f'(x) + f(x) \cdot \frac{d}{dx}\left[\frac{1}{g(x)}\right]$$

$$= \frac{g(x)f'(x)}{[g(x)]^2} + f(x) \cdot (-1)[g(x)]^{-2} \cdot g'(x)$$

$$= \frac{g(x)f'(x) - f(x)g'(x)}{[g(x)]^2}$$

Incorporating Technology A quotient of continuous functions typically has a vertical asymptote at each zero of the denominator. Graphing calculators do not do well drawing graphs with vertical asymptotes. For example, the graph of

$$y = \frac{1}{x - 1}$$

has a vertical asymptote at $x = 1$, which is shown on the calculator screen as a spike. The problem is that the standard graphing mode on most calculators (CONNECTED mode on TI calculators) connects consecutive dots along the graph. There is no point over $x = 1$ since the function is not defined there. When the calculator connects the dots on either side of the asymptote, a spike results. To display a graph more accurately with an asymptote, most calculators have a DOT mode in which consecutive dots are displayed, but not connected with line segments. On the TI-83, the DOT mode is selected from the MODE screen. The DOT mode setting is retained until you change it.

Practice Problems 3.1

1. Consider the function $y = (\sqrt{x} + 1)x$.
   (a) Differentiate y by the product rule.
   (b) First multiply out the expression for y and then differentiate.

2. Differentiate $y = \dfrac{5}{x^4 - x^3 + 1}$.

▶ Exercises 3.1

Differentiate the functions in Exercises 1–30.

1. $(x + 1)(x^3 + 5x + 2)$

2. $(2x - 1)(x^2 - 3)$

3. $(3x^2 - x + 2)(2x^2 - 1)$

4. $(x^4 + 1)(3x + 5)$

5. $(2x - 7)(x - 1)^5$

6. $(x + 1)^2(x - 3)^3$

7. $(x^2 + 3)(x^2 - 3)^{10}$

8. $(x^2 - 4)(x^2 + 3)^6$

9. $\frac{1}{3}(4 - x)^3(4 + x)^3$

10. $\frac{1}{6}(3x + 1)^4(4x - 1)^3$

11. $\dfrac{4 - x}{4 + x}$

12. $\dfrac{x - 2}{x + 2}$

13. $\dfrac{x^2 - 1}{x^2 + 1}$

14. $\dfrac{1 + x^3}{1 - x^3}$

15. $\dfrac{1}{5x^2 + 2x + 5}$

16. $\dfrac{2x - 1}{x}$

17. $\dfrac{x^2 + 2x}{x + 1}$

18. $\dfrac{3x^2 - 2x}{x + 3}$

19. $\dfrac{3x^2 + 5x + 1}{3 - x^2}$

20. $\dfrac{1}{x^2 + 1}$

21. $\dfrac{x}{(x^2+1)^2}$

22. $\dfrac{x-1}{(2x+1)^2}$

23. $\dfrac{(x-1)^4}{(x+2)^2}$

24. $\dfrac{(x+3)^3}{(x+4)^2}$

25. $\dfrac{x^4-4x^2+3}{x}$

26. $(x^2+9)\left(x-\dfrac{3}{x}\right)$

27. $2\sqrt{x}(3x^2-1)^3$

28. $4\sqrt{x}(5x-1)^4$

29. $(x+3)\sqrt{2x-3}$

30. $(2x-5)\sqrt{4x-1}$

31. Find the equation of the tangent line to the curve $y=(x-2)^5(x+1)^2$ at the point $(3,16)$.

32. Find the equation of the tangent line to the curve $y=(x+1)/(x-1)$ at the point $(2,3)$.

33. Find all x such that $\dfrac{dy}{dx}=0$, where
$$y=(x^2-4)^3(2x^2+5)^5.$$

34. Find all x such that $\dfrac{dy}{dx}=0$, where
$$y=(3x-8)^4(2x-3)^3.$$

35. Find all x-coordinates of points (x,y) on the curve $y=(x-2)^5/(x-4)^3$ where the tangent line is horizontal.

36. Find all x-coordinates of points (x,y) on the curve $y=x^4/(x-1)$ where the tangent line is horizontal.

37. Find the point(s) on the graph of $y=(x^2+3x-1)/x$ where the slope is 5.

38. Find the point(s) on the graph of $y=(2x^4+1)(x-5)$ where the slope is 1.

39. An open rectangular box is 3 feet long and has a surface area of 16 square feet. Find the dimensions of the box for which the volume is as large as possible.

40. A closed rectangular box is to be constructed with one side 1 meter long. The material for the top costs $20 per square meter, and the material for the sides and bottom costs $10 per square meter. Find the dimensions of the box with the largest possible volume that can be built at a cost of $240 for materials.

41. A sugar refinery can produce x tons of sugar per week at a weekly cost of $.1x^2+5x+2250$ dollars. Find the level of production for which the average cost is at a minimum and show that the average cost equals the marginal cost at that level of production.

42. A cigar manufacturer produces x cases of cigars per day at a daily cost of $50x(x+200)/(x+100)$ dollars. Show that his cost increases and his average cost decreases as the output x increases.

43. Let $R(x)$ be the revenue received from the sale of x units of a product. The *average revenue per unit* is defined by $AR=R(x)/x$. Show that at the level of production where the average revenue is maximized, the average revenue equals the marginal revenue.

44. Let $s(t)$ be the number of miles a car travels in t hours. Then the average velocity during the first t hours is $\overline{v}(t)=s(t)/t$ miles per hour. Suppose that the average velocity is maximized at time t_0. Show that at this time the average velocity $\overline{v}(t_0)$ equals the instantaneous velocity $s'(t_0)$. [*Hint:* Compute the derivative of $\overline{v}(t)$.]

45. The width of a rectangle is increasing at a rate of 3 inches per second and its length is increasing at the rate of 4 inches per second. At what rate is the area of the rectangle increasing when its width is 5 inches and its length is 6 inches? [*Hint:* Let $W(t)$ and $L(t)$ be the widths and lengths, respectively, at time t.]

46. A manufacturer plans to decrease the amount of sulfur dioxide escaping from its smokestacks. The estimated cost-benefit function is
$$f(x)=\frac{3x}{105-x}, \qquad 0\le x\le100$$
where $f(x)$ is the cost in millions of dollars for eliminating x percent of the total sulfur dioxide. (See Fig. 2.) Find the value of x at which the rate of increase of the cost-benefit function is 1.4 million dollars per unit. (Each unit is 1 percentage point increase in pollutant removed.)

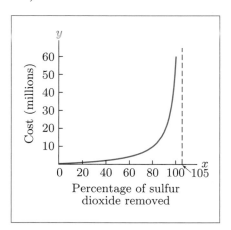

Figure 2. A cost-benefit function.

In Exercises 47 and 48, use the fact that at the beginning of 1998 the population of the United States was 268,924,000 people and growing at the rate of 1,856,000 people per year.

47. At the beginning of 1998, the annual per capita consumption of gasoline in the United States was 52.3 gallons and growing at the rate of .2 gallons per year. At what rate was the total annual consumption of gasoline in the United States increasing at that time? (*Hint:* [total annual consumption] = [population]·[annual per capita consumption].)

48. At the beginning of 1998, the annual consumption of ice cream in the United States was 12,582 million pints and growing at the rate of 212 million pints per year. At what rate was the annual per capita consumption of ice cream increasing at that time? (*Hint*: [annual per capita consumption] = $\frac{\text{annual consumption}}{\text{population size}}$.)

49. Figure 3 shows the graph of

$$y = \frac{10x}{1 + .25x^2}$$

for $x \geq 0$. Find the coordinates of the maximum point.

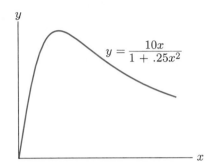

$$y = \frac{10x}{1 + .25x^2}$$

Figure 3.

50. Figure 4 shows the graph of

$$y = \frac{1}{2} + \frac{x^2 - 2x + 1}{x^2 - 2x + 2}$$

for $0 \leq x \leq 2$. Find the coordinates of the minimum point.

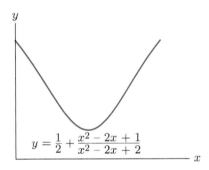

$$y = \frac{1}{2} + \frac{x^2 - 2x + 1}{x^2 - 2x + 2}$$

Figure 4.

51. Show that the derivative of $\dfrac{x^4}{x^2 + 1}$ is not $\dfrac{4x^3}{2x}$.

52. Show that the derivative of $(5x^3)(2x^4)$ is not $(15x^2)(8x^3)$.

53. Suppose that $f(x)$ is a function whose derivative is $f'(x) = \dfrac{1}{1 + x^2}$. Find the derivative of $\dfrac{f(x)}{1 + x^2}$.

54. Suppose that $f(x)$ and $g(x)$ are differentiable functions such that $f(1) = 2$, $f'(1) = 3$, $g(1) = 4$, and $g'(1) = 5$. Find $\dfrac{d}{dx}[f(x)g(x)]\Big|_{x=1}$.

55. Consider the functions of Exercise 54. Find $\dfrac{d}{dx}\left[\dfrac{f(x)}{g(x)}\right]\Big|_{x=1}$.

56. Suppose that $f(x)$ is a function whose derivative is $f'(x) = 1/x$. Find the derivative of $xf(x) - x$.

57. Let $f(x) = 1/x$ and $g(x) = x^3$.

(a) Show that the product rule yields the correct derivative of $(1/x)x^3 = x^2$.

(b) Compute the product $f'(x)g'(x)$ and note that it is *not* the derivative of $f(x)g(x)$.

58. The derivative of $(x^3 - 4x)/x$ is obviously $2x$ for $x \neq 0$, because $(x^3 - 4x)/x = x^2 - 4$ for $x \neq 0$. Verify that the quotient rule gives the same derivative.

59. Let $f(x)$, $g(x)$, and $h(x)$ be differentiable functions. Find a formula for the derivative of $f(x)g(x)h(x)$. [*Hint*: First differentiate $f(x)[g(x)h(x)]$.]

60. (*Alternative Verification of the Product Rule*) Apply the special case of the general power rule

$$\frac{d}{dx}[h(x)]^2 = 2h(x)h'(x)$$

and the identity

$$fg = \tfrac{1}{4}[(f + g)^2 - (f - g)^2]$$

to prove the product rule.

Technology Exercises

61. The relationship between the area of the pupil of the eye and the intensity of light was analyzed by B. H. Crawford.* Crawford concluded that the area of the pupil is

$$f(x) = \frac{160x^{-.4} + 94.8}{4x^{-.4} + 15.8} \qquad (0 \leq x \leq 37)$$

square millimeters when x units of light are entering the eye per unit time.

(a) Graph $f(x)$ and $f'(x)$ in the window $[0, 6]$ *by* $[-5, 20]$.

(b) How large is the pupil when 3 units of light are entering the eye per unit time?

(c) For what light intensity is the pupil size 11 square millimeters?

(d) When 3 units of light are entering the eye per unit time, what is the rate of change of pupil size with respect to a unit change in light intensity?

---

*The dependence of pupil size upon the external light stimulus under static and variable conditions; *Proceedings of the Royal Society Series B*, 121: 376–395, 1937.

Solutions to Practice Problems 3.1

1. **(a)** Apply the product rule to $y = (\sqrt{x} + 1)x$ with

$$f(x) = \sqrt{x} + 1 = x^{1/2} + 1$$
$$g(x) = x.$$

$$\frac{dy}{dx} = (x^{1/2} + 1) \cdot 1 + x \cdot \tfrac{1}{2}x^{-1/2}$$
$$= x^{1/2} + 1 + \tfrac{1}{2}x^{1/2}$$
$$= \tfrac{3}{2}\sqrt{x} + 1.$$

(b)
$$y = (\sqrt{x} + 1)x = (x^{1/2} + 1)x = x^{3/2} + x$$

$$\frac{dy}{dx} = \tfrac{3}{2}x^{1/2} + 1.$$

Comparing parts (a) and (b), we note that it is helpful to simplify the function before differentiating.

2. We may apply the quotient rule to $y = \dfrac{5}{x^4 - x^3 + 1}$.

$$\frac{dy}{dx} = \frac{(x^4 - x^3 + 1) \cdot 0 - 5 \cdot (4x^3 - 3x^2)}{(x^4 - x^3 + 1)^2} = \frac{-5x^2(4x - 3)}{(x^4 - x^3 + 1)^2}$$

However, it is slightly faster to use the general power rule, since $y = 5(x^4 - x^3 + 1)^{-1}$. Thus

$$\frac{dy}{dx} = -5(x^4 - x^3 + 1)^{-2}(4x^3 - 3x^2)$$

$$= -5x^2(x^4 - x^3 + 1)^{-2}(4x - 3).$$

3.2   The Chain Rule and the General Power Rule

In this section we show that the general power rule is a special case of a powerful differentiation technique called the chain rule. Applications of the chain rule appear throughout the text.

A useful way of combining functions $f(x)$ and $g(x)$ is to replace each occurrence of the variable x in $f(x)$ by the function $g(x)$. The resulting function is called the *composition* (or *composite*) of $f(x)$ and $g(x)$ and is denoted by $f(g(x))$.*

▶ **Example 1**   Let $f(x) = \dfrac{x - 1}{x + 1}$, $g(x) = x^3$. What is $f(g(x))$?

Solution   Replace each occurrence of x in $f(x)$ by $g(x)$ to obtain

$$f(g(x)) = \frac{g(x) - 1}{g(x) + 1} = \frac{x^3 - 1}{x^3 + 1}. \qquad \blacklozenge$$

Given a composite function $f(g(x))$, we may think of $f(x)$ as the "outside" function that acts on the values of the "inside" function $g(x)$. This point of view often helps us to recognize a composite function.

▶ **Example 2**   Write the following functions as composites of simpler functions.

---

*See Section 0.3 for additional information on function composition.

(a) $h(x) = (x^5 + 9x + 3)^8$
(b) $k(x) = \sqrt{4x^2 + 1}$

Solution   (a) $h(x) = f(g(x))$, where the outside function is the power function, $(\ldots)^8$, that is $f(x) = x^8$. Inside this power function is $g(x) = x^5 + 9x + 3$.

(b) $k(x) = f(g(x))$, where the outside function is the square root function, $f(x) = \sqrt{x}$, and the inside function is $g(x) = 4x^2 + 1$. ◆

A function of the form $[g(x)]^r$ is a composite $f(g(x))$, where the outside function is $f(x) = x^r$. We have already given a rule for differentiating this function—namely,

$$\frac{d}{dx}[g(x)]^r = r[g(x)]^{r-1}g'(x).$$

The *chain rule* has the same form, except that the outside function $f(x)$ can be *any* differentiable function.

The Chain Rule   To differentiate $f(g(x))$, first differentiate the outside function $f(x)$ and substitute $g(x)$ for x in the result. Then multiply by the derivative of the inside function $g(x)$. Symbolically,

$$\frac{d}{dx}f(g(x)) = f'(g(x))g'(x).$$

▶ **Example 3**   Use the chain rule to compute the derivative of $f(g(x))$, where $f(x) = x^8$ and $g(x) = x^5 + 9x + 3$.

Solution
$$f'(x) = 8x^7, \qquad g'(x) = 5x^4 + 9$$
$$f'(g(x)) = 8(x^5 + 9x + 3)^7$$

Finally, by the chain rule,

$$\frac{d}{dx}f(g(x)) = f'(g(x))g'(x) = 8(x^5 + 9x + 3)^7(5x^4 + 9).$$

Since in this example the outside function is a power function, the calculations are the same as when the general power rule is used to compute the derivative of $(x^5 + 9x + 3)^8$. However, the organization of the solution here emphasizes the notation of the chain rule. ◆

There is another way to write the chain rule. Given the function $y = f(g(x))$, set $u = g(x)$ so that $y = f(u)$. Then y may be regarded either as a function of u or, indirectly through u, as a function of x. With this notation, we have $\frac{du}{dx} = g'(x)$ and $\frac{dy}{du} = f'(u) = f'(g(x))$. Thus the chain rule says that

$$\frac{dy}{dx} = \frac{dy}{du}\frac{du}{dx}. \tag{1}$$

Although the derivative symbols in (1) are not really fractions, the apparent cancellation of the "*du*" symbols provides a mnemonic device for remembering this form of the chain rule.

Here is an illustration that makes (1) a plausible formula. Suppose that y, u, and x are three varying quantities, with y varying three times as fast as u and u varying two times as fast as x. It seems reasonable that y should vary six times as fast as x. That is, $\frac{dy}{dx} = \frac{dy}{du}\frac{du}{dx} = 3 \cdot 2 = 6$.

▶ Example 4   Find $\dfrac{dy}{dx}$ if $y = u^5 - 2u^3 + 8$ and $u = x^2 + 1$.

Solution   Since

$$\frac{dy}{du} = 5u^4 - 6u^2 \quad \text{and} \quad \frac{du}{dx} = 2x,$$

we have

$$\frac{dy}{dx} = (5u^4 - 6u^2) \cdot 2x.$$

It is usually desirable to express $\dfrac{dy}{dx}$ as a function of x alone, so we substitute $x^2 + 1$ for u to obtain

$$\frac{dy}{dx} = [5(x^2 + 1)^4 - 6(x^2 + 1)^2] \cdot 2x. \qquad \blacklozenge$$

You may have noticed another entirely mechanical way to work Example 4. Namely, first substitute $u = x^2 + 1$ into the original formula for y and obtain $y = (x^2 + 1)^5 - 2(x^2 + 1)^3 + 8$. Then $\dfrac{dy}{dx}$ is easily calculated by the sum rule, the general power rule, and the constant-multiple rule. The solution in Example 4, however, lays the foundation for applications of the chain rule in Section 3.3 and elsewhere.

In many situations involving composition of functions, the basic variable is time, t. It may happen that x is a function of t, say $x = g(t)$, and some other variable such as R is a function of x, say $R = f(x)$. Then $R = f(g(t))$, and the chain rule says that

$$\frac{dR}{dt} = \frac{dR}{dx}\frac{dx}{dt}.$$

▶ Example 5   A store sells ties for \$12 apiece. Let x be the number of ties sold in one day and let R be the revenue received from the sale of x ties, so that $R = 12x$. Suppose that daily sales are rising at the rate of four ties per day. How fast is the revenue rising?

Solution   Clearly, revenue is rising at the rate of \$48 per day, because each of the additional four ties brings in \$12. This intuitive conclusion also follows from the chain rule,

$$\frac{dR}{dt} = \frac{dR}{dx} \cdot \frac{dx}{dt}$$

$$\begin{bmatrix} \text{rate of change} \\ \text{of revenue with} \\ \text{respect to time} \end{bmatrix} = \begin{bmatrix} \text{rate of change} \\ \text{of revenue with} \\ \text{respect to sales} \end{bmatrix} \cdot \begin{bmatrix} \text{rate of change} \\ \text{of sales with} \\ \text{respect to time} \end{bmatrix}$$

$$\begin{bmatrix} \text{\$48 increase} \\ \text{per day} \end{bmatrix} = \begin{bmatrix} \text{\$12 increase} \\ \text{per add'l tie} \end{bmatrix} \cdot \begin{bmatrix} \text{four additional} \\ \text{ties per day} \end{bmatrix}.$$

Notice that $\dfrac{dR}{dx}$ is actually the marginal revenue, studied earlier. This example shows that the time rate of change of revenue, $\dfrac{dR}{dt}$, is the marginal revenue multiplied by the time rate of change of sales. $\qquad \blacklozenge$

Verification of the Chain Rule Suppose that $f(x)$ and $g(x)$ are differentiable, and let $x = a$ be a number in the domain of $f(g(x))$. Since every differentiable function is continuous, we have

$$\lim_{h \to 0} g(a + h) = g(a),$$

which implies that

$$\lim_{h \to 0} [g(a + h) - g(a)] = 0. \tag{2}$$

Now $g(a)$ is a number in the domain of f, and the limit definition of the derivative gives us

$$f'(g(a)) = \lim_{k \to 0} \frac{f(g(a) + k) - f(g(a))}{k}. \tag{3}$$

Let $k = g(a + h) - g(a)$. By equation (2), k approaches zero as h approaches zero. Also, $g(a + h) = g(a) + k$. Therefore, (3) may be rewritten in the form

$$f'(g(a)) = \lim_{h \to 0} \frac{f(g(a + h)) - f(g(a))}{g(a + h) - g(a)}. \tag{4}$$

[Strictly speaking, we must assume that the denominator in (4) is never zero. This assumption may be avoided by a somewhat different and more technical argument that we omit.] Finally, we show that the function $f(g(x))$ has a derivative at $x = a$. We use the limit definition of the derivative, Limit Theorem V, and equation (4).

$$\frac{d}{dx} f(g(x)) \Big|_{x=a} = \lim_{h \to 0} \frac{f(g(a + h)) - f(g(a))}{h}$$

$$= \lim_{h \to 0} \left[\frac{f(g(a + h)) - f(g(a))}{g(a + h) - g(a)} \cdot \frac{g(a + h) - g(a)}{h} \right]$$

$$= \lim_{h \to 0} \frac{f(g(a + h)) - f(g(a))}{g(a + h) - g(a)} \cdot \lim_{h \to 0} \frac{g(a + h) - g(a)}{h}$$

$$= f'(g(a)) \cdot g'(a)$$

Incorporating Technology Function combination with calculators is discussed in Appendices A–D as the last item in "How To Use Functions."

Practice Problems 3.2

Consider the function $h(x) = (2x^3 - 5)^5 + (2x^3 - 5)^4$.

1. Write $h(x)$ as a composite function, $f(g(x))$.
2. Compute $f'(x)$ and $f'(g(x))$.
3. Use the chain rule to differentiate $h(x)$.

▶ Exercises 3.2

Compute $f(g(x))$, where $f(x)$ and $g(x)$ are the following.

1. $f(x) = \dfrac{x}{x+1}$, $g(x) = x^3$

2. $f(x) = x\sqrt{x+1}$, $g(x) = x^4$

3. $f(x) = x^5 + 3x$, $g(x) = x^2 + 4$

4. $f(x) = \dfrac{x}{(x+1)^4}$, $g(x) = 5 - 3x$

Each of the following functions may be viewed as a composite function $f(g(x))$. Find $f(x)$ and $g(x)$.

5. $(x^3 + 8x - 2)^5$ **6.** $(9x^2 + 2x - 5)^7$

7. $\sqrt{4 - x^2}$ **8.** $(5x^2 + 1)^{-1/2}$

9. $\dfrac{1}{x^3 - 5x^2 + 1}$ **10.** $(4x - 3)^3 + \dfrac{1}{4x - 3}$

Differentiate the functions in Exercises 11–22 using one or more of the differentiation rules discussed thus far.

11. $(x^2 + 5)^{15}$ **12.** $(x^4 + x^2)^{10}$

13. $6x^2(x - 1)^3$ **14.** $5x^3(2 - x)^4$

15. $2(x^3 - 1)(3x^2 + 1)^4$ **16.** $2(2x - 1)^{5/4}(2x + 1)^{3/4}$

17. $\left(\dfrac{4}{1 - x}\right)^3$ **18.** $\dfrac{4x^2 + x}{\sqrt{x}}$

19. $\left(\dfrac{4x - 1}{3x + 1}\right)^3$ **20.** $\left(\dfrac{x}{x^2 + 1}\right)^2$

21. $\left(\dfrac{4 - x}{x^2}\right)^3$ **22.** $\left(\dfrac{1 - x^2}{x}\right)^3$

23. Sketch the graph of $y = 4x/(x + 1)^2$, $x > -1$.

24. Sketch the graph of $y = 2/(1 + x^2)$.

Compute $\dfrac{d}{dx} f(g(x))$, where $f(x)$ and $g(x)$ are the following.

25. $f(x) = x^5$, $g(x) = 6x - 1$

26. $f(x) = x^2$, $g(x) = x^3 + 1$

27. $f(x) = \dfrac{1}{x}$, $g(x) = 1 - x^2$

28. $f(x) = x^{10}$, $g(x) = x^2 + 3x$

29. $f(x) = x^4 - x^2$, $g(x) = x^2 - 4$ stop here

30. $f(x) = \dfrac{4}{x} + x^2$, $g(x) = 1 - x^4$

31. $f(x) = (x^3 + 1)^2$, $g(x) = x^2 + 5$

32. $f(x) = x(x - 2)^4$, $g(x) = x^3$

Compute $\dfrac{dy}{dx}$ using the chain rule in formula (1).

33. $y = u^{3/2}$, $u = 4x + 1$

34. $y = u^5$, $u = 2 + \sqrt{x}$

35. $y = \dfrac{4}{u^2} + \dfrac{u^2}{4}$, $u = x - 3x^2$

36. $y = \frac{1}{2}u^2 + 2u^{1/2}$, $u = 1 - 3x$

37. $y = u(u + 1)^5$, $u = x^2 + x$

38. $y = (u^2 + 1)^3$, $u = x - x^2$

39. $y = \dfrac{u - 1}{u + 1}$, $u = 2 + \sqrt{x}$

40. $y = \dfrac{u}{1 + u^2}$, $u = 5x + 1$

41. Find the equation of the line tangent to the graph of $y = 2x(x - 4)^6$ at the point $(5, 10)$.

42. Find the equation of the line tangent to the graph of $y = \dfrac{x}{\sqrt{2 - x^2}}$ at the point $(1, 1)$.

43. Find the x-coordinates of all points on the curve $y = (-x^2 + 4x - 3)^3$ with a horizontal tangent line.

44. The function $f(x) = \sqrt{x^2 - 6x + 10}$ has one relative minimum point for $x \geq 0$. Find it.

45. The length, x, of the edge of a cube is increasing.

   (a) Write the chain rule for $\dfrac{dV}{dt}$, the time rate of change of the volume of the cube.

   (b) For what value of x is $\dfrac{dV}{dt}$ equal to 12 times the rate of increase of x?

46. The weight of a male hognose snake is approximately $446x^3$ grams, where x is its length in meters.[*] Suppose a snake has length .4 meters and is growing at the rate of .2 meters per year. At what rate is the snake gaining weight?

47. Suppose that P, y, and t are variables, where P is a function of y and y is a function of t.

   (a) Write the derivative symbols for the following quantities: the rate of change of y with respect to t, the rate of change of P with respect to y, and the rate of change of P with respect to t. Select your answers from the following:

$$\frac{dP}{dy}, \quad \frac{dy}{dP}, \quad \frac{dy}{dt}, \quad \frac{dP}{dt}, \quad \frac{dt}{dP}, \quad \text{and} \quad \frac{dt}{dy}.$$

   (b) Write the chain rule for $\dfrac{dP}{dt}$.

---

[*] D. R. Platt, "Natural History of the Hognose Snakes *Heterodon Platyrhinos* and *Heterodon Nasicus*," University of Kansas Publications, *Museum of Natural History* 18:4 (1969), 253–420.

48. Suppose that Q, x, and y are variables, where Q is a function of x and x is a function of y. (Read this carefully.)

(a) Write the derivative symbols for the following quantities: the rate of change of x with respect to y, the rate of change of Q with respect to y, and the rate of change of Q with respect to x. Select your answers from the following:

$$\frac{dy}{dx}, \quad \frac{dx}{dy}, \quad \frac{dQ}{dx}, \quad \frac{dx}{dQ}, \quad \frac{dQ}{dy}, \quad \text{and} \quad \frac{dy}{dQ}.$$

(b) Write the chain rule for $\dfrac{dQ}{dy}$.

49. When a company produces and sells x thousand units per week its total weekly profit is P thousand dollars, where

$$P = \frac{200x}{100 + x^2}.$$

The production level at t weeks from the present is $x = 4 + 2t$.

(a) Find the marginal profit, $\dfrac{dP}{dx}$.

(b) Find the time rate of change of profit, $\dfrac{dP}{dt}$.

(c) How fast (with respect to time) are profits changing when $t = 8$?

50. The cost of manufacturing x cases of cereal is C dollars, where $C = 3x + 4\sqrt{x} + 2$. Weekly production at t weeks from the present is estimated to be $x = 6200 + 100t$ cases.

(a) Find the marginal cost, $\dfrac{dC}{dx}$.

(b) Find the time rate of change of cost, $\dfrac{dC}{dt}$.

(c) How fast (with respect to time) are costs rising when $t = 2$?

51. Ecologists estimate that when the population of a certain city is x thousand persons, the average level L of carbon monoxide in the air above the city will be L ppm (parts per million), where $L = 10 + .4x + .0001x^2$.

The population of the city is estimated to be $x = 752 + 23t + .5t^2$ thousand persons t years from the present.

(a) Find the rate of change of carbon monoxide with respect to the population of the city.

(b) Find the time rate of change of the population when $t = 2$.

(c) How fast (with respect to time) is the carbon monoxide level changing at time $t = 2$?

52. A manufacturer of microcomputers estimates that t months from now it will sell x thousand units of its main line of microcomputers per month, where $x = .05t^2 + 2t + 5$. Because of economies of scale, the profit P from manufacturing and selling x thousand units is estimated to be $P = .001x^2 + .1x - .25$ million dollars. Calculate the rate at which the profit will be increasing 5 months from now.

53. Suppose that $f(x)$ and $g(x)$ are differentiable functions. Find $g(x)$ if you know that

$$\frac{d}{dx}f(g(x)) = 3x^2 \cdot f'(x^3 + 1).$$

54. Suppose that $f(x)$ and $g(x)$ are differentiable functions. Find $g(x)$ if you know that $f'(x) = 1/x$ and

$$\frac{d}{dx}f(g(x)) = \frac{2x + 5}{x^2 + 5x - 4}.$$

55. Suppose that $f(x)$ and $g(x)$ are differentiable functions such that $f(1) = 2$, $f'(1) = 3$, $f'(5) = 4$, $g(1) = 5$, $g'(1) = 6$, $g'(2) = 7$, and $g'(5) = 8$. Find $\left.\dfrac{d}{dx}f(g(x))\right|_{x=1}$.

56. Consider the functions of Exercise 55. Find $\left.\dfrac{d}{dx}g(f(x))\right|_{x=1}$.

57. In an expression of the form $f(g(x))$, $f(x)$ is called the *outer* function and $g(x)$ is called the *inner* function. Give a written description of the chain rule using the words "inner" and "outer."

Solutions to Practice Problems 3.2	**1.** Let $f(x) = x^5 + x^4$ and $g(x) = 2x^3 - 5$.

2. $f'(x) = 5x^4 + 4x^3$, $f'(g(x)) = 5(2x^3 - 5)^4 + 4(2x^3 - 5)^3$.

3. We have $g'(x) = 6x^2$. Then, from the chain rule and the result of Problem 2, we have

$$h'(x) = f'(g(x))g'(x) = [5(2x^3 - 5)^4 + 4(2x^3 - 5)^3](6x^2).$$

3.3   Implicit Differentiation and Related Rates

This section presents two different applications of the chain rule. In each case we will need to differentiate one or more composite functions where the "inside" functions are not known explicitly.

Implicit Differentiation   In some applications, the variables are related by an equation rather than a function. In these cases, we can still determine the rate of change of one variable with respect to the other by the technique of implicit differentiation. As an illustration, consider the equation

$$x^2 + y^2 = 4. \tag{1}$$

The graph of this equation is the circle in Fig. 1. This graph obviously is not the graph of a function since, for instance, there are two points on the graph whose x-coordinate is 1. (Functions must satisfy the vertical line test. See Section 0.1.)

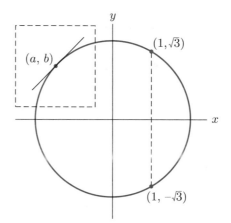

Figure 1. Graph of $x^2 + y^2 = 4$.

We denote the slope of the curve at the point $(1, \sqrt{3})$ by

$$\left. \frac{dy}{dx} \right|_{\substack{x = 1 \\ y = \sqrt{3}}}.$$

In general, the slope at the point (a, b) is denoted by

$$\left. \frac{dy}{dx} \right|_{\substack{x = a \\ y = b}}.$$

In a small vicinity of the point (a, b), the curve looks like the graph of a function. [That is, on this part of the curve, $y = g(x)$ for some function $g(x)$.] We say that this function is defined *implicitly* by the equation.* We obtain a formula for $\frac{dy}{dx}$ by differentiating both sides of the equation with respect to x while treating y as a function of x.

▶ Example 1   Consider the graph of the equation $x^2 + y^2 = 4$.

(a) Use implicit differentiation to compute $\frac{dy}{dx}$.

---

*Of course, the tangent line at the point (a, b) must not be vertical. In this section, we assume that the given equations implicitly determine differentiable functions.

(b) Find the slope of the graph at the points $(1, \sqrt{3})$ and $(1, -\sqrt{3})$.

Solution (a) The first term x^2 has derivative $2x$, as usual. We think of the second term y^2 as having the form $[g(x)]^2$. To differentiate we use the chain rule (specifically, the general power rule):

$$\frac{d}{dx}[g(x)]^2 = 2[g(x)]g'(x)$$

or, equivalently,

$$\frac{d}{dx}y^2 = 2y\frac{dy}{dx}.$$

On the right side of the original equation, the derivative of the constant function 4 is zero. Thus implicit differentiation of $x^2 + y^2 = 4$ yields

$$2x + 2y\frac{dy}{dx} = 0.$$

Solving for $\frac{dy}{dx}$, we have

$$2y\frac{dy}{dx} = -2x.$$

If $y \neq 0$, then

$$\frac{dy}{dx} = \frac{-2x}{2y} = -\frac{x}{y}.$$

Notice that this slope formula involves y as well as x. This reflects the fact that the slope of the circle at a point depends on the y-coordinate of the point as well as the x-coordinate.

(b) At the point $(1, \sqrt{3})$ the slope is

$$\frac{dy}{dx}\bigg|_{\substack{x=1 \\ y=\sqrt{3}}} = -\frac{x}{y}\bigg|_{\substack{x=1 \\ y=\sqrt{3}}} = -\frac{1}{\sqrt{3}}.$$

At the point $(1, -\sqrt{3})$ the slope is

$$\frac{dy}{dx}\bigg|_{\substack{x=1 \\ y=-\sqrt{3}}} = -\frac{x}{y}\bigg|_{\substack{x=1 \\ y=-\sqrt{3}}} = -\frac{1}{-\sqrt{3}} = \frac{1}{\sqrt{3}}.$$

(See Fig. 2.) The formula for $\frac{dy}{dx}$ gives the slope at every point on the graph of $x^2 + y^2 = 4$ except $(-2, 0)$ and $(2, 0)$. At these two points the tangent line is vertical and the slope of the curve is undefined. ◆

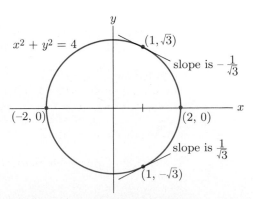

Figure 2. Slope of the tangent line.

The difficult step in Example 1(a) was to differentiate y^2 correctly. The derivative of y^2 *with respect to y* would be $2y$, by the ordinary power rule. But the derivative of y^2 *with respect to x* must be computed by the general power rule. In general,

$$\frac{d}{dx}y^r = ry^{r-1}\frac{dy}{dx}.$$

(2)

This rule is used to compute slope formulas in the next two examples.

▶ Example 2   Use implicit differentiation to calculate $\dfrac{dy}{dx}$ for the equation $x^2y^6 = 1$.

Solution   Differentiate each side of the equation $x^2y^6 = 1$ with respect to x. On the left side of the equation use the product rule and treat y as a function of x.

$$x^2\frac{d}{dx}(y^6) + y^6\frac{d}{dx}(x^2) = \frac{d}{dx}(1)$$

$$x^2 \cdot 6y^5\frac{dy}{dx} + y^6 \cdot 2x = 0$$

Solve for $\dfrac{dy}{dx}$ by moving the term not involving $\dfrac{dy}{dx}$ to the right side and dividing by the factor that multiplies $\dfrac{dy}{dx}$.

$$6x^2y^5\frac{dy}{dx} = -2xy^6$$

$$\frac{dy}{dx} = \frac{-2xy^6}{6x^2y^5} = -\frac{y}{3x}$$ ◆

▶ Example 3   Use implicit differentiation to calculate $\dfrac{dy}{dx}$ when y is related to x by the equation $x^2y + xy^3 - 3x = 5$.

Solution   Differentiate the equation term by term, taking care to differentiate x^2y and xy^3 by the product rule.

$$x^2\frac{d}{dx}(y) + y\frac{d}{dx}(x^2) + x\frac{d}{dx}(y^3) + y^3\frac{d}{dx}(x) - 3 = 0$$

$$x^2\frac{dy}{dx} + y \cdot 2x + x \cdot 3y^2\frac{dy}{dx} + y^3 \cdot 1 - 3 = 0$$

Solve for $\dfrac{dy}{dx}$ in terms of x and y.

$$x^2\frac{dy}{dx} + 3xy^2\frac{dy}{dx} = 3 - y^3 - 2xy$$

$$(x^2 + 3xy^2)\frac{dy}{dx} = 3 - y^3 - 2xy$$

$$\frac{dy}{dx} = \frac{3 - y^3 - 2xy}{x^2 + 3xy^2}$$ ◆

Reminder When a power of y is differentiated with respect to x, the result must include the factor $\dfrac{dy}{dx}$. When a power of x is differentiated, there is no factor $\dfrac{dy}{dx}$.

Here is the general procedure for implicit differentiation.

Finding $\dfrac{dy}{dx}$ by Implicit Differentiation

Differentiate each term of the equation *with respect to x*, treating y as a function of x.

Move all terms involving $\dfrac{dy}{dx}$ to the left side of the equation and move the other terms to the right side.

Factor out $\dfrac{dy}{dx}$ on the left side of the equation.

Divide both sides of the equation by the factor that multiplies $\dfrac{dy}{dx}$.

Equations that implicitly define functions arise frequently in economic models. The economic background for the equation in the next example is discussed in Section 7.1.

▶ **Example 4** Suppose that x and y represent the amounts of two basic inputs into a production process and the equation

$$60x^{3/4}y^{1/4} = 3240$$

describes all input amounts (x, y) for which the output of the process is 3240 units. (The graph of this equation is called a *production isoquant* or *constant product curve*.) See Fig. 3. Use implicit differentiation to calculate the slope of the graph at the point on the curve where $x = 81$, $y = 16$.

Solution We use the product rule and treat y as a function of x.

$$60x^{3/4}\frac{d}{dx}(y^{1/4}) + y^{1/4}\frac{d}{dx}(60x^{3/4}) = 0$$

$$60x^{3/4}\cdot\left(\frac{1}{4}\right)y^{-3/4}\frac{dy}{dx} + y^{1/4}\cdot 60\left(\frac{3}{4}\right)x^{-1/4} = 0$$

$$15x^{3/4}y^{-3/4}\frac{dy}{dx} = -45x^{-1/4}y^{1/4}$$

$$\frac{dy}{dx} = \frac{-45x^{-1/4}y^{1/4}}{15x^{3/4}y^{-3/4}} = \frac{-3y}{x}$$

When $x = 81$ and $y = 16$, we have

$$\left.\frac{dy}{dx}\right|_{\substack{x=81 \\ y=16}} = \frac{-3(16)}{81} = -\frac{16}{27}.$$

The number $-\frac{16}{27}$ is the slope of the production isoquant at the point $(81, 16)$. If the first input (corresponding to x) is increased by one small unit, then the second input (corresponding to y) must decrease by approximately $\frac{16}{27}$ unit in

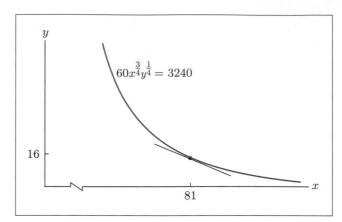

Figure 3. A production isoquant.

order to keep the production output unchanged [i.e., to keep (x, y) on the curve]. In economic terminology, the absolute value of $\dfrac{dy}{dx}$ is called the *marginal rate of substitution* of the first input for the second input. ◆

Related Rates In implicit differentiation we differentiate an equation involving x and y, with y treated as a function of x. However, in some applications where x and y are related by an equation, both variables are functions of a third variable t (which may represent time). Often the formulas for x and y as functions of t are not known. When we differentiate such an equation with respect to t, we derive a relationship between the rates of change $\dfrac{dy}{dt}$ and $\dfrac{dx}{dt}$. We say that these derivatives are *related rates*. The equation relating the rates may be used to find one of the rates when the other is known.

▶ Example 5 Suppose that x and y are both differentiable functions of t and are related by the equation

$$x^2 + 5y^2 = 36. \tag{3}$$

(a) Differentiate each term in the equation with respect to t and solve the resulting equation for $\dfrac{dy}{dt}$.

(b) Calculate $\dfrac{dy}{dt}$ at a time when $x = 4$, $y = 2$, and $\dfrac{dx}{dt} = 5$.

Solution (a) Since x is a function of t, the general power rule gives

$$\frac{d}{dt}(x^2) = 2x\frac{dx}{dt}.$$

A similar formula holds for the derivative of y^2. Differentiating each term in (3) with respect to t, we obtain

$$\frac{d}{dt}(x^2) + \frac{d}{dt}(5y^2) = \frac{d}{dt}(36)$$

$$2x\frac{dx}{dt} + 5\cdot 2y\frac{dy}{dt} = 0$$

$$10y\frac{dy}{dt} = -2x\frac{dx}{dt}$$

$$\frac{dy}{dt} = -\frac{x}{5y}\frac{dx}{dt}.$$

(b) When $x = 4$, $y = 2$, and $\dfrac{dx}{dt} = 5$,

$$\frac{dy}{dt} = -\frac{4}{5(2)}\cdot(5) = -2. \qquad\blacklozenge$$

There is a helpful graphical interpretation of the calculations in Example 5. Imagine a point that is moving clockwise along the graph of the equation $x^2 + 5y^2 = 36$. (See Fig. 4.) Suppose that when the point is at $(4, 2)$, the x-coordinate of the point is changing at the rate of 5 units per minute, so that $\dfrac{dx}{dt} = 5$. In Example 5(b) we found that $\dfrac{dy}{dt} = -2$. This means that the y-coordinate of the point is decreasing at the rate of 2 units per minute when the moving point reaches $(4, 2)$.

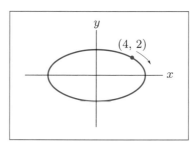

Figure 4. Graph of $x^2 + 5y^2 = 36$.

▶ Example 6   Suppose that x thousand units of a commodity can be sold weekly when the price is p dollars per unit, and suppose that x and p satisfy the demand equation

$$p + 2x + xp = 38.$$

See Fig. 5. How fast are weekly sales changing at a time when $x = 4$, $p = 6$, and the price is falling at the rate of \$.40 per week?

Solution   Assume that p and x are differentiable functions of t, and differentiate the demand equation with respect to t.

$$\frac{d}{dt}(p) + \frac{d}{dt}(2x) + \frac{d}{dt}(xp) = \frac{d}{dt}(38)$$

$$\frac{dp}{dt} + 2\frac{dx}{dt} + x\frac{dp}{dt} + p\frac{dx}{dt} = 0 \qquad (4)$$

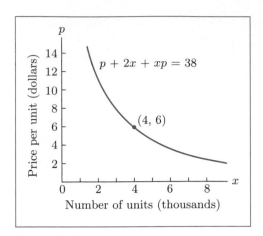

Figure 5. A demand curve.

We want to know $\dfrac{dx}{dt}$ at a time when $x = 4$, $p = 6$, and $\dfrac{dp}{dt} = -.40$. (The derivative $\dfrac{dp}{dt}$ is negative because the price is decreasing.) We could solve (4) for $\dfrac{dx}{dt}$ and then substitute the given values, but since we do not need a general formula for $\dfrac{dx}{dt}$, it is easier to substitute first and then solve.

$$-.40 + 2\frac{dx}{dt} + 4(-.40) + 6\frac{dx}{dt} = 0$$

$$8\frac{dx}{dt} = 2$$

$$\frac{dx}{dt} = .25$$

Thus sales are rising at the rate of .25 thousand units (i.e., 250 units) per week. ◆

Suggestions for Solving Related Rates Problems

1. Draw a picture, if possible.

2. Assign letters to quantities that vary, and identify one variable, say t, on which the other variables depend.

3. Find an equation that relates the variables to each other.

4. Differentiate the equation with respect to the independent variable t. Use the chain rule whenever appropriate.

5. Substitute all specified values for the variables and their derivatives.

6. Solve for the derivative that gives the unknown rate.

Incorporating Technology   The graph of an equation in x and y can be easily obtained when y can be expressed as one or more functions of x. For instance, the graph of $x^2 + y^2 = 4$ can be plotted by simultaneously graphing $y = \sqrt{4 - x^2}$ and $y = -\sqrt{4 - x^2}$.

Suppose that x and y are related by the equation $3y^2 - 3x^2 + y = 1$.

1. Use implicit differentiation to find a formula for the slope of the graph of the equation.

2. Suppose that the x and y in the preceding equation are both functions of t. Differentiate both sides of the equation with respect to t and find a formula for $\dfrac{dy}{dt}$ in terms of x, y, and $\dfrac{dx}{dt}$.

▶ Exercises 3.3

In Exercises 1–18, suppose that x and y are related by the given equation and use implicit differentiation to determine $\dfrac{dy}{dx}$.

1. $x^2 - y^2 = 1$

2. $x^3 + y^3 - 6 = 0$

3. $y^5 - 3x^2 = x$

4. $x^4 + (y+3)^4 = x^2$

5. $y^4 - x^4 = y^2 - x^2$

6. $x^3 + y^3 = x^2 + y^2$

7. $2x^3 + y = 2y^3 + x$

8. $x^4 + 4y = x - 4y^3$

9. $xy = 5$

10. $xy^3 = 2$

11. $x(y+2)^5 = 8$

12. $x^2 y^3 = 6$

13. $x^3 y^2 - 4x^2 = 1$

14. $(x+1)^2 (y-1)^2 = 1$

15. $x^3 + y^3 = x^3 y^3$

16. $x^2 + 4xy + 4y = 1$

17. $x^2 y + y^2 x = 3$

18. $x^3 y + xy^3 = 4$

Use implicit differentiation of the equations in Exercises 19–24 to determine the slope of the graph at the given point.

19. $4y^3 - x^2 = -5$; $x = 3$, $y = 1$

20. $y^2 = x^3 + 1$; $x = 2$, $y = -3$

21. $xy^3 = 2$; $x = -\frac{1}{4}$, $y = -2$

22. $\sqrt{x} + \sqrt{y} = 7$; $x = 9$, $y = 16$

23. $xy + y^3 = 14$; $x = 3$, $y = 2$

24. $y^2 = 3xy - 5$; $x = 2$, $y = 1$

25. Find the equation of the tangent line to the graph of $x^2 y^4 = 1$ at the point $(4, \frac{1}{2})$ and at the point $(4, -\frac{1}{2})$.

26. Find the equation of the tangent line to the graph of $x^4 y^2 = 144$ at the point $(2, 3)$ and at the point $(2, -3)$.

27. The graph of $x^4 + 2x^2 y^2 + y^4 = 4x^2 - 4y^2$ is the "lemniscate" in Fig. 6.

   (a) Find $\dfrac{dy}{dx}$ by implicit differentiation.

   (b) Find the slope of the tangent line to the lemniscate at $(\sqrt{6}/2, \sqrt{2}/2)$.

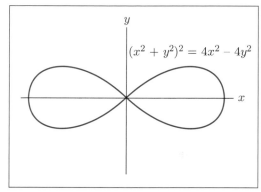

Figure 6. A lemniscate.

28. The graph of $x^4 + 2x^2 y^2 + y^4 = 9x^2 - 9y^2$ is a lemniscate similar to that in Fig. 6.

   (a) Find $\dfrac{dy}{dx}$ by implicit differentiation.

   (b) Find the slope of the tangent line to the lemniscate at $(\sqrt{5}, -1)$.

29. Suppose that x and y represent the amounts of two basic inputs for a production process and suppose that the equation

$$30x^{1/3} y^{2/3} = 1080$$

describes all input amounts where the output of the process is 1080 units. Find $\dfrac{dy}{dx}$ when $x = 16$, $y = 54$.

30. Suppose that x and y represent the amounts of two basic inputs for a production process and

$$10x^{1/2} y^{1/2} = 600.$$

Find $\dfrac{dy}{dx}$ when $x = 50$, $y = 72$.

In Exercises 31–36, suppose that x and y are both differentiable functions of t and are related by the given equation. Use implicit differentiation with respect to t to determine $\dfrac{dy}{dt}$ in terms of x, y, and $\dfrac{dx}{dt}$.

31. $x^4 + y^4 = 1$

32. $y^4 - x^2 = 1$

33. $3xy - 3x^2 = 4$

34. $y^2 = 8 + xy$

35. $x^2 + 2xy = y^3$

36. $x^2y^2 = 2y^3 + 1$

37. A point is moving along the graph of $x^2 - 4y^2 = 9$. When the point is at $(5, -2)$, its x-coordinate is increasing at the rate of 3 units per second. How fast is the y-coordinate changing at that moment?

38. A point is moving along the graph of $x^3y^2 = 200$. When the point is at $(2, 5)$, its x-coordinate is changing at the rate of -4 units per minute. How fast is the y-coordinate changing at that moment?

39. Suppose that the price p (in dollars) and the weekly sales x (in thousands of units) of a certain commodity satisfy the demand equation

$$2p^3 + x^2 = 4500.$$

Determine the rate at which sales are changing at a time when $x = 50$, $p = 10$, and the price is falling at the rate of \$.50 per week.

40. Suppose that the price p (in dollars) and the demand x (in thousands of units) of a commodity satisfy the demand equation

$$6p + x + xp = 94.$$

How fast is the demand changing at a time when $x = 4$, $p = 9$, and the price is rising at the rate of \$2 per week?

41. The monthly advertising revenue A and the monthly circulation x of a magazine are related approximately by the equation

$$A = 6\sqrt{x^2 - 400}, \qquad x \geq 20,$$

where A is given in thousands of dollars and x is measured in thousands of copies sold. At what rate is the advertising revenue changing if the current circulation is $x = 25$ thousand copies and the circulation is changing at the rate of 2 thousand copies per month? $\left(Hint\text{: Use the chain rule } \dfrac{dA}{dt} = \dfrac{dA}{dx}\dfrac{dx}{dt}.\right)$

42. Suppose that in Boston the wholesale price p of oranges (in dollars per crate) and the daily supply x (in thousands of crates) are related by the equation $px + 7x + 8p = 328$. If there are 4 thousand crates available today at a price of \$25 per crate, and if the supply is changing at the rate of $-.3$ thousand crates per day, at what rate is the price changing?

43. Under certain conditions (called adiabatic expansion) the pressure P and volume V of a gas such as oxygen satisfy the equation $P^5V^7 = k$, where k is a constant. Suppose that at some moment the volume of the gas is 4 liters, the pressure is 200 units, and the pressure is increasing at the rate of 5 units per second. Find the (time) rate at which the volume is changing.

44. The volume V of a spherical cancer tumor is given by $V = \pi x^3/6$, where x is the diameter of the tumor. A physician estimates that the diameter is growing at the rate of .4 millimeters per day, at a time when the diameter is already 10 millimeters. How fast is the volume of the tumor changing at that time?

45. Figure 7 shows a 10-foot ladder leaning against a wall.

    (a) Use the Pythagorean theorem to find an equation relating x and y.

    (b) Suppose that the foot of the ladder is being pulled along the ground at the rate of 3 feet per second. How fast is the top end of the ladder sliding down the wall at the time when the foot of the ladder is 8 feet from the wall? That is, what is $\dfrac{dy}{dt}$ at the time when $\dfrac{dx}{dt} = 3$ and $x = 8$?

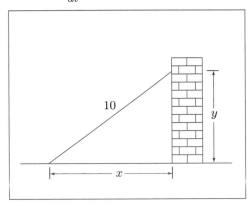

Figure 7.

46. An airplane flying 390 feet per second at an altitude of 5000 feet flew directly over an observer. Figure 8 shows the relationship of the airplane to the observer at a later time.

    (a) Find an equation relating x and y.

    (b) Find the value of x when y is 13,000.

    (c) How fast is the distance from the observer to the airplane changing at the time when the airplane is 13,000 feet from the observer? That is, what is $\dfrac{dy}{dt}$ at the time when $\dfrac{dx}{dt} = 390$ and $y = 13,000$?

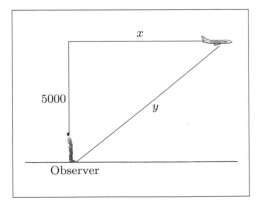

Figure 8.

47. A baseball diamond is a 90-foot by 90-foot square. See Fig. 9. A player runs from first to second base at the speed of 22 feet per second. How fast is the player's distance from third base changing when he is halfway between first and second base? (*Hint:* If x is the distance from the player to second base and y is his distance from third base, then $x^2 + 90^2 = y^2$.)

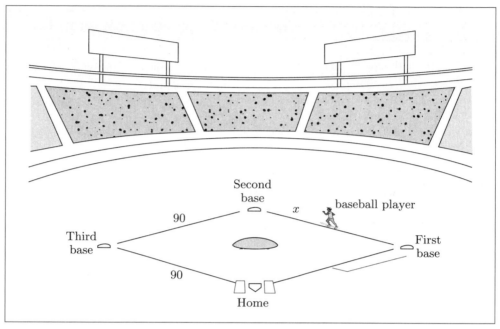

Figure 9. A baseball diamond.

Solutions to Practice Problems 3.3	

1.
$$\frac{d}{dx}(3y^2) - \frac{d}{dx}(3x^2) + \frac{d}{dx}(y) = \frac{d}{dx}(1)$$

$$6y\frac{dy}{dx} - 6x + \frac{dy}{dx} = 0$$

$$(6y + 1)\frac{dy}{dx} = 6x$$

$$\frac{dy}{dx} = \frac{6x}{6y + 1}$$

2. First, here is the solution without reference to Practice Problem 1.

$$\frac{d}{dt}(3y^2) - \frac{d}{dt}(3x^2) + \frac{d}{dt}(y) = \frac{d}{dt}(1)$$

$$6y\frac{dy}{dt} - 6x\frac{dx}{dt} + \frac{dy}{dt} = 0$$

$$(6y + 1)\frac{dy}{dt} = 6x\frac{dx}{dt}$$

$$\frac{dy}{dt} = \frac{6x}{6y + 1}\frac{dx}{dt}$$

<table>
<tr><td>

Solutions to Practice Problems 3.3 (Continued)

</td><td>

Alternatively, we can use the chain rule and the formula for $\dfrac{dy}{dx}$ from Practice Problem 1.

$$\frac{dy}{dt} = \frac{dy}{dx}\frac{dx}{dt} = \frac{6x}{6y+1}\frac{dx}{dt}$$

</td></tr>
</table>

Review of Fundamental Concepts of Chapter 3

1. State the product rule and quotient rule.

2. State the chain rule. Give an example.

3. What is the relationship between the chain rule and the general power rule?

4. What does it mean for a function to be defined implicitly by an equation?

5. State the formula for $\dfrac{d}{dx}y^r$, where y is defined implicitly as a function of x.

6. Outline the procedure for solving a related rates problem.

▶ Chapter 3 Supplementary Exercises

Differentiate the following functions.

1. $(4x-1)(3x+1)^4$

2. $2(5-x)^3(6x-1)$

3. $x(x^5-1)^3$

4. $(2x+1)^{5/2}(4x-1)^{3/2}$

5. $5(\sqrt{x}-1)^4(\sqrt{x}-2)^2$

6. $\dfrac{\sqrt{x}}{\sqrt{x}+4}$

7. $3(x^2-1)^3(x^2+1)^5$

8. $\dfrac{1}{(x^2+5x+1)^6}$

9. $\dfrac{x^2-6x}{x-2}$

10. $\dfrac{2x}{2-3x}$

11. $\left(\dfrac{3-x^2}{x^3}\right)^2$

12. $\dfrac{x^3+x}{x^2-x}$

13. Let $f(x) = (3x+1)^4(3-x)^5$. Find all x such that $f'(x) = 0$.

14. Let $f(x) = (x^2+1)/(x^2+5)$. Find all x such that $f'(x) = 0$.

15. Find the equation of the line tangent to the graph of $y = (x^3-1)(x^2+1)^4$ at the point where $x = -1$.

16. Find the equation of the line tangent to the graph of $y = \dfrac{x-3}{\sqrt{4+x^2}}$ at the point where $x = 0$.

17. A botanical display is to be constructed as a rectangular region with a river as one side and a sidewalk 2 meters wide along the inside edges of the other three sides. (See Fig. 1.) The area for the plants must be 800 square meters. Find the outside dimensions of the region that minimizes the area of the sidewalk (and hence minimizes the amount of concrete needed for the sidewalk).

18. Repeat Exercise 17, with the sidewalk on the inside of all four sides. In this case, the 800-square-meter planted region has dimensions $x-4$ meters by $y-4$ meters.

19. A store estimates that its cost when selling x lamps per day is C dollars, where $C = 40x + 30$ (i.e., the marginal cost per lamp is $40). Suppose that daily sales are rising at the rate of three lamps per day. How fast are the costs rising? Explain your answer using the chain rule.

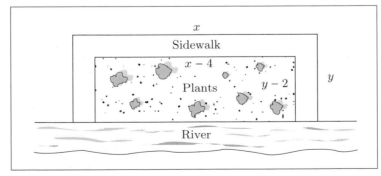

Figure 1. A botanical display.

20. A company pays y dollars in taxes when its annual profit is P dollars. Suppose that y is some (differentiable) function of P and P is some function of time t. Give a chain rule formula for the time rate of change of taxes $\dfrac{dy}{dt}$.

Exercises 21–26 refer to the graphs of the functions $f(x)$ and $g(x)$ in Fig. 2.

21. Let $h(x) = 2f(x) - 3g(x)$. Determine $h(1)$ and $h'(1)$.

22. Let $h(x) = f(x) \cdot g(x)$. Determine $h(1)$ and $h'(1)$.

23. Let $h(x) = \dfrac{f(x)}{g(x)}$. Determine $h(1)$ and $h'(1)$.

24. Let $h(x) = [f(x)]^2$. Determine $h(1)$ and $h'(1)$.

25. Let $h(x) = f(g(x))$. Determine $h(1)$ and $h'(1)$.

26. Let $h(x) = g(f(x))$. Determine $h(1)$ and $h'(1)$.

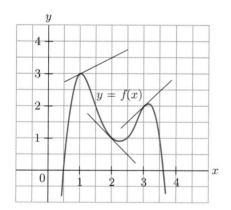

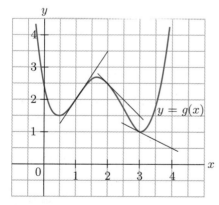

Figure 2.

In Exercises 27–29, find a formula for $\dfrac{d}{dx}f(g(x))$, where $f(x)$ is a function such that $f'(x) = 1/(x^2 + 1)$.

27. $g(x) = x^3$

28. $g(x) = \dfrac{1}{x}$

29. $g(x) = x^2 + 1$

In Exercises 30–32, find a formula for $\dfrac{d}{dx}f(g(x))$, where $f(x)$ is a function such that $f'(x) = x\sqrt{1 - x^2}$.

30. $g(x) = x^2$ **31.** $g(x) = \sqrt{x}$ **32.** $g(x) = x^{3/2}$

In Exercises 33–35, find $\dfrac{dy}{dx}$, where y is a function of u such that $\dfrac{dy}{du} = \dfrac{u}{u^2 + 1}$.

33. $u = x^{3/2}$ **34.** $u = x^2 + 1$ **35.** $u = \dfrac{5}{x}$

In Exercises 36–38, find $\dfrac{dy}{dx}$, where y is a function of u such that $\dfrac{dy}{du} = \dfrac{u}{\sqrt{1 + u^4}}$.

36. $u = x^2$ **37.** $u = \sqrt{x}$ **38.** $u = \dfrac{2}{x}$

39. The revenue R a company receives is a function of the weekly sales x. Also, the sales level x is a function of the weekly advertising expenditures A, and A in turn is a varying function of time.

(a) Write the derivative symbols for the following quantities: rate of change of revenue with respect to advertising expenditures, time rate of change of advertising expenditures, marginal revenue, and rate of change of sales with respect to advertising expenditures. Select your answers from the following:

$$\frac{dR}{dx}, \quad \frac{dR}{dt}, \quad \frac{dA}{dt}, \quad \frac{dA}{dR}, \quad \frac{dA}{dx}, \quad \frac{dx}{dA}, \quad \text{and} \quad \frac{dR}{dA}.$$

(b) Write a type of chain rule that expresses the time rate of change of revenue, $\dfrac{dR}{dt}$, in terms of three of the derivatives described in part (a).

40. The amount A of anesthetics that a certain hospital uses each week is a function of the number S of surgical operations performed each week. Also, S in turn is a function of the population P of the area served by the hospital, while P is a function of time t.

(a) Write the derivative symbols for the following quantities: population growth rate, rate of change of anesthetic usage with respect to the population size, rate of change of surgical operations with respect to the population size, and rate of change of anesthetic usage with respect to the number of surgical operations. Select your answers from the following:

$$\frac{dS}{dP}, \quad \frac{dS}{dt}, \quad \frac{dP}{dS}, \quad \frac{dP}{dt}, \quad \frac{dA}{dS}, \quad \frac{dA}{dP}, \quad \text{and} \quad \frac{dS}{dA}.$$

(b) Write a type of chain rule that expresses the time rate of change of anesthetic usage, $\dfrac{dA}{dt}$, in terms of three of the derivatives described in part (a).

41. The graph of $x^{2/3} + y^{2/3} = 8$ is the *astroid* in Fig. 3.

  (a) Find $\dfrac{dy}{dx}$ by implicit differentiation.

  (b) Find the slope of the tangent line at $(8, -8)$.

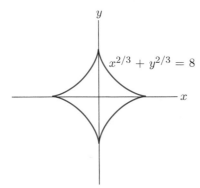

Figure 3. Astroid.

42. The graph of $x^3 + y^3 = 9xy$ is the folium of Descartes, shown in Fig. 4.

  (a) Find $\dfrac{dy}{dx}$ by implicit differentiation.

  (b) Find the slope of the curve at $(2, 4)$.

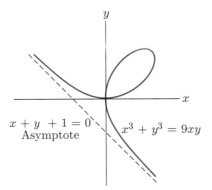

Figure 4. Folium of Descartes.

In Exercises 43–46, x and y are related by the given equation. Use implicit differentiation to calculate the value of $\dfrac{dy}{dx}$ for the given values of x and y.

43. $x^2 y^2 = 9$; $x = 1$, $y = 3$

44. $xy^4 = 48$; $x = 3$, $y = 2$

45. $x^2 - xy^3 = 20$; $x = 5$, $y = 1$

46. $xy^2 - x^3 = 10$; $x = 2$, $y = 3$

47. A factory's weekly production costs y and its weekly production quantity x are related by the equation $y^2 - 5x^3 = 4$, where y is in thousands of dollars and x is in thousands of units of output.

  (a) Use implicit differentiation to find a formula for $\dfrac{dy}{dx}$, the marginal cost of production.

  (b) Find the marginal cost of production when $x = 4$ and $y = 18$.

(c) Suppose that the factory begins to vary its weekly production level. Assuming that x and y are differentiable functions of time t, use the method of related rates to find a formula for $\dfrac{dy}{dt}$, the time rate of change of production costs.

(d) Compute $\dfrac{dy}{dt}$ when $x = 4$, $y = 18$, and the production level is rising at the rate of .3 thousand units per week (i.e., when $\dfrac{dx}{dt} = .3$).

48. A town library estimates that when the population is x thousand persons, approximately y thousand books will be checked out of the library during one year, where x and y are related by the equation $y^3 - 8000x^2 = 0$.

  (a) Use implicit differentiation to find a formula for $\dfrac{dy}{dx}$, the rate of change of library circulation with respect to population size.

  (b) Find the value of $\dfrac{dy}{dx}$ when $x = 27$ thousand persons and $y = 180$ thousand books per year.

  (c) Assume that x and y are both differentiable functions of time t, and use the method of related rates to find a formula for $\dfrac{dy}{dt}$, the time rate of change of library circulation.

  (d) Compute $\dfrac{dy}{dt}$ when $x = 27$, $y = 180$, and the population is rising at the rate of 1.8 thousand persons per year $\left(\text{i.e., } \dfrac{dx}{dt} = 1.8\right)$. Either use part (c), or use part (b) and the chain rule.

49. Suppose that the price p and quantity x of a certain commodity satisfy the demand equation $6p + 5x + xp = 50$, and suppose that p and x are functions of time, t. Determine the rate at which the quantity x is changing when $x = 4$, $p = 3$, and $\dfrac{dp}{dt} = -2$.

50. An offshore oil well is leaking oil onto the ocean surface, forming a circular oil slick about .005 meter thick. If the radius of the slick is r meters, then the volume of oil spilled is $V = .005\pi r^2$ cubic meters. Suppose that the oil is leaking at a constant rate of 20 cubic meters per hour, so that $\dfrac{dV}{dt} = 20$. Find the rate at which the radius of the oil slick is increasing, at a time when the radius is 50 meters. $\left(Hint\text{: Find a relation between } \dfrac{dV}{dt} \text{ and } \dfrac{dr}{dt}.\right)$

51. Animal physiologists have determined experimentally that the weight W (in kilograms) and the surface area S (in square meters) of a typical horse are related by the empirical equation $S = 0.1W^{2/3}$. How fast is the surface area of a horse increasing at a time when the horse weighs 350 kg and is gaining weight at the rate of 200 kg per year? (*Hint*: Use the chain rule.)

52. Suppose that a kitchen appliance company's monthly sales and advertising expenses are approximately related by the equation $xy - 6x + 20y = 0$, where x is thousands of dollars spent on advertising and y is thousands of dishwashers sold. Currently, the company is spending 10 thousand dollars on advertising and is selling 2 thousand dishwashers each month. If the company plans to increase monthly advertising expenditures at the rate of $1.5 thousand per month, how fast will sales rise? Use implicit differentiation to answer the question.

Chapter Project

The following table gives data on personal entertainment activities in each of the years 1990–2001. (The values for 1999–2001 are projections.)

Table 1	Relative Popularity of Various Forms of Entertainment		
Year	Annual Online Internet Access (hours)	Annual Expenditure on Home Video Games (dollars)	Time Spent Reading Daily Newspaper (hours)
1990	1	12	175
1991	1	18	169
1992	2	19	172
1993	2	19	170
1994	3	22	169
1995	7	24	165
1996	16	26	161
1997	22	29	158
1998	30	31	157
1999	33	33	155
2000	37	35	154
2001	39	37	153

(a) Plot the data using three scatter plots on a single coordinate system.

(b) Describe in words the behavior of the graphs.

(c) Use numerical differentiation to approximate, for each year, the rate of change of each activity.

(d) Use numerical differentiation to approximate, for each year, the rate of change of the rate of change.

(e) For each activity, use the table to determine the years for which the activity is becoming more popular. Do these results agree with those predicted by the approximate values of the derivatives?

(f) For each activity, determine the years for which the activity is becoming less popular. Do these results agree with those predicted by the approximate values of the derivatives?

CHAPTER

The Exponential and Natural Logarithm Functions

4

When an investment grows steadily at 15% per year, the rate of growth of the investment at any time is proportional to the value of the investment at that time. When a bacteria culture grows in a laboratory dish, the rate of growth of the culture at any moment is proportional to the total number of bacteria in the dish at that moment. These situations are examples of what is called *exponential growth*. A pile of radioactive uranium ^{235}U decays at a rate that at each moment is proportional to the amount of ^{235}U present. This decay of uranium (and of radioactive elements in general) is called *exponential decay*. Both exponential growth and exponential decay can be described and studied in terms of exponential functions and the natural logarithm function. The properties of these functions are investigated in this chapter. Subsequently, we shall explore a wide range of applications, in fields such as business, biology, archeology, public health, and psychology.

4.1 Exponential Functions

Throughout this section b will denote a positive number. The function

$$f(x) = b^x$$

is called an *exponential function*, because the variable x is in the exponent. The number b is called the *base* of the exponential function. In Section 0.5 we reviewed the definition of b^x for various values of b and x (although we used the letter r there instead of x). For instance, if $f(x)$ is the exponential function with base 2,

$$f(x) = 2^x,$$

then

$$f(0) = 2^0 = 1, \qquad f(1) = 2^1 = 2, \qquad f(4) = 2^4 = 2\cdot2\cdot2\cdot2 = 16,$$

and

$$f(-1) = 2^{-1} = \tfrac{1}{2}, \qquad f(\tfrac{1}{2}) = 2^{1/2} = \sqrt{2}, \qquad f(\tfrac{3}{5}) = (2^{1/5})^3 = (\sqrt[5]{2})^3.$$

Actually, in Section 0.5, we defined b^x only for rational (i.e., integer or fractional) values of x. For other values of x (such as $\sqrt{3}$ or π), it is possible to define b^x by first approximating x with rational numbers and then applying a limiting process. We shall omit the details and simply assume henceforth that b^x can be defined for all numbers x in such a way that the usual laws of exponents remain valid.

Let us state the laws of exponents for reference.

(i) $b^x \cdot b^y = b^{x+y}$

(ii) $b^{-x} = \dfrac{1}{b^x}$

(iii) $\dfrac{b^x}{b^y} = b^x \cdot b^{-y} = b^{x-y}$

(iv) $(b^y)^x = b^{xy}$

(v) $a^x b^x = (ab)^x$

(vi) $\dfrac{a^x}{b^x} = \left(\dfrac{a}{b}\right)^x$

Property (iv) may be used to change the appearance of an exponential function. For instance, the function $f(x) = 8^x$ may also be written as $f(x) = (2^3)^x = 2^{3x}$, and $g(x) = \left(\frac{1}{9}\right)^x$ may be written as $g(x) = (1/3^2)^x = (3^{-2})^x = 3^{-2x}$.

▶ **Example 1** Use properties of exponents to write the following functions in the form 2^{kx} for a suitable constant k.

(a) $4^{5x/2}$ (b) $(2^{4x} \cdot 2^{-x})^{1/2}$ (c) $8^{x/3} \cdot 16^{3x/4}$ (d) $\dfrac{10^x}{5^x}$

Solution (a) First express the base 4 as a power of 2, and then use Property (iv):

$$4^{5x/2} = (2^2)^{5x/2} = 2^{2(5x/2)} = 2^{5x}.$$

(b) First use Property (i) to simplify the quantity inside the parentheses, and then use Property (iv):

$$(2^{4x} \cdot 2^{-x})^{1/2} = (2^{4x-x})^{1/2} = (2^{3x})^{1/2} = 2^{(3/2)x}.$$

(c) First express the bases 8 and 16 as powers of 2, and then use (iv) and (i):

$$8^{x/3} \cdot 16^{3x/4} = (2^3)^{x/3} \cdot (2^4)^{3x/4} = 2^x \cdot 2^{3x} = 2^{4x}.$$

(d) Use (v) to change the numerator 10^x, and then cancel the common term 5^x:

$$\frac{10^x}{5^x} = \frac{(2 \cdot 5)^x}{5^x} = \frac{2^x \cdot 5^x}{5^x} = 2^x.$$

An alternative method is to use Property (vi):

$$\frac{10^x}{5^x} = \left(\frac{10}{5}\right)^x = 2^x. \qquad \blacklozenge$$

Let us now study the graph of the exponential function $y = b^x$ for various values of b. We begin with the special case $b = 2$.

We have tabulated the values of 2^x for $x = 0, \pm 1, \pm 2, \pm 3$ and plotted these values in Fig. 1. Other intermediate values of 2^x for $x = \pm .1, \pm .2, \pm .3, \ldots$, may be obtained from a table or from a graphing calculator. [See Fig. 2(a).] By passing a smooth curve through these points, we obtain the graph of $y = 2^x$, Fig. 2(b).

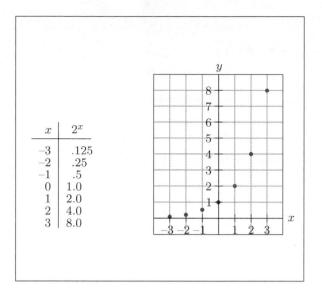

Figure 1. Values of 2^x.

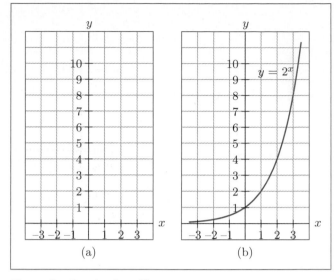

Figure 2. Graph of $y = 2^x$.

In the same manner, we have sketched the graph of $y = 3^x$ (Fig. 3). The graphs of $y = 2^x$ and $y = 3^x$ have the same basic shape. Also note that they both pass through the point $(0, 1)$ (because $2^0 = 1$, $3^0 = 1$).

In Fig. 4 we have sketched the graphs of several more exponential functions. Notice that the graph of $y = 5^x$ has a large slope at $x = 0$, since the graph at $x = 0$ is quite steep; however, the graph of $y = (1.1)^x$ is nearly horizontal at $x = 0$, and hence the slope is close to zero.

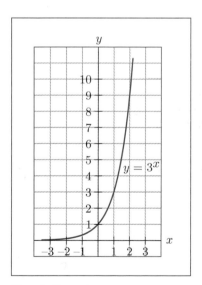

Figure 3.

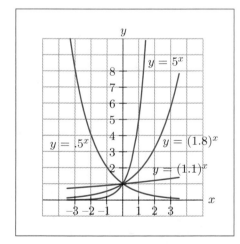

Figure 4.

There is an important property of the function 3^x that is readily apparent from its graph. Since the graph is always increasing, the function 3^x never assumes the same y-value twice. That is, the only way 3^r can equal 3^s is to have $r = s$. This fact is useful when solving certain equations involving exponentials.

▶ **Example 2**   Let $f(x) = 3^{5x}$. Determine all x for which $f(x) = 27$.

Solution   Since $27 = 3^3$, we must determine all x for which

$$3^{5x} = 3^3.$$

Equating exponents, we have

$$5x = 3$$
$$x = \tfrac{3}{5}.$$

◆

In general, for $b > 1$, the equation $b^r = b^s$ implies that $r = s$. This is because the graph of $y = b^x$ has the same basic shape as $y = 2^x$ and $y = 3^x$. Similarly, when $0 < b < 1$, the equation $b^r = b^s$ implies that $r = s$, because the graph of $y = b^x$ resembles the graph of $y = .5^x$ and is always decreasing.

There is no need at this point to become familiar with the graphs of the function b^x. We have shown a few graphs merely to make the reader more comfortable with the concept of an exponential function. The main purpose of this section has been to review properties of exponents in a context that is appropriate for our future work.

**Practice Problems
4.1**

1. Can a function such as $f(x) = 5^{3x}$ be written in the form $f(x) = b^x$? If so, what is b?

2. Solve the equation $7 \cdot 2^{6-3x} = 28$.

▶ Exercises 4.1

Write each function in Exercises 1–14 in the form 2^{kx} or 3^{kx}, for a suitable constant k.

1. 4^x, $(\sqrt{3})^x$, $\left(\tfrac{1}{9}\right)^x$

2. 27^x, $(\sqrt[3]{2})^x$, $\left(\tfrac{1}{8}\right)^x$

3. $8^{2x/3}$, $9^{3x/2}$, $16^{-3x/4}$

4. $9^{-x/2}$, $8^{4x/3}$, $27^{-2x/3}$

5. $\left(\tfrac{1}{4}\right)^{2x}$, $\left(\tfrac{1}{8}\right)^{-3x}$, $\left(\tfrac{1}{81}\right)^{x/2}$

6. $\left(\tfrac{1}{9}\right)^{2x}$, $\left(\tfrac{1}{27}\right)^{x/3}$, $\left(\tfrac{1}{16}\right)^{-x/2}$

7. $6^x \cdot 3^{-x}$, $\dfrac{15^x}{5^x}$, $\dfrac{12^x}{2^{2x}}$

8. $7^{-x} \cdot 14^x$, $\dfrac{2^x}{6^x}$, $\dfrac{3^{2x}}{18^x}$

9. $\dfrac{3^{4x}}{3^{2x}}$, $\dfrac{2^{5x+1}}{2 \cdot 2^{-x}}$, $\dfrac{9^{-x}}{27^{-x/3}}$

10. $\dfrac{2^x}{6^x}$, $\dfrac{3^{-5x}}{3^{-2x}}$, $\dfrac{16^x}{8^{-x}}$

11. $2^{3x} \cdot 2^{-5x/2}$, $3^{2x} \cdot \left(\tfrac{1}{3}\right)^{2x/3}$

12. $2^{5x/4} \cdot \left(\tfrac{1}{2}\right)^x$, $3^{-2x} \cdot 3^{5x/2}$

13. $(2^{-3x} \cdot 2^{-2x})^{2/5}$, $(9^{1/2} \cdot 9^4)^{x/9}$

14. $(3^{-x} \cdot 3^{x/5})^5$, $(16^{1/4} \cdot 16^{-3/4})^{3x}$

15. Find a number b such that the function $f(x) = 3^{-2x}$ can be written in the form b^x.

16. Find b so that $8^{-x/3} = b^x$ for all x.

Solve the following equations for x.

17. $5^{2x} = 5^2$

18. $10^{-x} = 10^2$

19. $(2.5)^{2x+1} = (2.5)^5$

20. $(3.2)^{x-3} = (3.2)^5$

21. $10^{1-x} = 100$

22. $2^{4-x} = 8$

23. $3(2.7)^{5x} = 8.1$

24. $4(2.7)^{2x-1} = 10.8$

25. $(2^{x+1} \cdot 2^{-3})^2 = 2$

26. $(3^{2x} \cdot 3^2)^4 = 3$

27. $2^{3x} = 4 \cdot 2^{5x}$

28. $3^{5x} \cdot 3^x - 3 = 0$

29. $(1 + x)2^{-x} - 5 \cdot 2^{-x} = 0$ *top here*

30. $(2 - 3x)5^x + 4 \cdot 5^x = 0$

The expressions in Exercises 31–36 may be factored as shown. Find the missing factors.

31. $2^{3+h} = 2^3(\quad)$

32. $5^{2+h} = 25(\quad)$

33. $2^{x+h} - 2^x = 2^x(\quad)$

34. $5^{x+h} + 5^x = 5^x(\quad)$

35. $3^{x/2} + 3^{-x/2} = 3^{-x/2}(\quad)$

36. $5^{7x/2} - 5^{x/2} = \sqrt{5^x}(\quad)$

Technology Exercises

37. Graph the function $f(x) = 2^x$ in the window $[-1, 2]$ by $[-1, 4]$, and estimate the slope of the graph at $x = 0$.

38. Graph the function $f(x) = 3^x$ in the window $[-1, 2]$ by $[-1, 8]$, and estimate the slope of the graph at $x = 0$.

39. By trial and error, find a number of the form $b = 2.\square$ (just one decimal place) with the property that the slope of the graph of b^x at $x = 0$ is as close to 1 as possible.

**Solutions to
Practice Problems
4.1**

1. If $5^{3x} = b^x$, then when $x = 1$, $5^{3(1)} = b^1$, which says that $b = 125$. This value of b certainly works, because

$$5^{3x} = (5^3)^x = 125^x.$$

2. Divide both sides of the equation by 7. We then obtain

$$2^{6-3x} = 4.$$

Now 4 can be written as 2^2. So we have

$$2^{6-3x} = 2^2.$$

Equate exponents to obtain

$$6 - 3x = 2$$
$$4 = 3x$$
$$x = \tfrac{4}{3}.$$

4.2 The Exponential Function e^x

Let us begin by examining the graphs of the exponential functions shown in Fig. 1. They all pass through $(0,1)$, but with different slopes there. Notice that the graph of 5^x is quite steep at $x = 0$, while the graph of $(1.1)^x$ is nearly horizontal at $x = 0$. It turns out that at $x = 0$, the graph of 2^x has a slope of approximately .693, while the graph of 3^x has a slope of approximately 1.1.

Evidently, there is a particular value of the base b, between 2 and 3, where the graph of b^x has slope *exactly* 1 at $x = 0$. We denote this special value of b by the letter e, and we call

$$f(x) = e^x$$

the exponential function. The number e is an important constant of nature that has been calculated to thousands of decimal places. To 10 significant digits, we have $e = 2.718281828$. For our purposes, it is usually sufficient to think of e as "approximately 2.7."

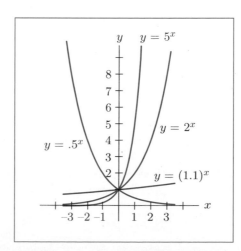

Figure 1. Several exponential functions.

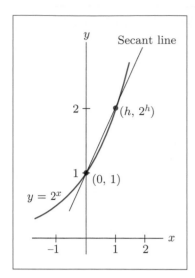

Figure 2. A secant line to the graph of $y = 2^x$.

Our goal in this section is to find a formula for the derivative of e^x. It turns out that the calculations for e^x and 2^x are very similar. Since many people are more comfortable working with 2^x rather than e^x, we shall first analyze the graph of 2^x. Then we shall draw the appropriate conclusions about the graph of e^x.

Before computing the slope of $y = 2^x$ at an arbitrary x, let us consider the special case $x = 0$. Denote the slope at $x = 0$ by m. We shall use the secant-line approximation of the derivative to approximate m. We proceed by constructing the secant line in Fig. 2. The slope of the secant line through $(0, 1)$ and $(h, 2^h)$ is $\dfrac{2^h - 1}{h}$. As h approaches zero, the slope of the secant line approaches the slope of $y = 2^x$ at $x = 0$. That is,

$$m = \lim_{h \to 0} \frac{2^h - 1}{h}. \tag{1}$$

We can estimate the value of m by taking h smaller and smaller. Table 1 shows the values of the expressions for $h = .1, .01, \ldots, .0000001$. From the table it is

Table 1	$Y_1 = (2^\wedge X - 1)/X$

X	Y1
.1	.71773
.01	.69556
.001	.69339
1E-4	.69317
1E-5	.69315
1E-6	.69315
1E-7	.69315

$Y_1 = .693147$

reasonable to conclude that $m \approx .693$. Since m equals the slope of $y = 2^x$ at $x = 0$, we have

$$m = \frac{d}{dx}(2^x)\Big|_{x=0} \approx .693. \tag{2}$$

Now that we have estimated the slope of $y = 2^x$ at $x = 0$, let us compute the slope for an arbitrary value of x. We construct a secant line through $(x, 2^x)$ and a nearby point $(x + h, 2^{x+h})$ on the graph. The slope of the secant line is

$$\frac{2^{x+h} - 2^x}{h}. \tag{3}$$

By a law of exponents, we have $2^{x+h} - 2^x = 2^x(2^h - 1)$ so that, by (1), we see that

$$\lim_{h \to 0} \frac{2^{x+h} - 2^x}{h} = \lim_{h \to 0} 2^x \frac{2^h - 1}{h} = 2^x \lim_{h \to 0} \frac{2^h - 1}{h} = m\, 2^x. \tag{4}$$

However, the slope of the secant (3) approaches the derivative of 2^x as h approaches zero. Consequently, we have

$$\frac{d}{dx}(2^x) = m\, 2^x, \quad \text{where } m = \frac{d}{dx}(2^x)\Big|_{x=0}. \tag{5}$$

▶ Example 1   Calculate (a) $\dfrac{d}{dx}(2^x)\Big|_{x=3}$ and (b) $\dfrac{d}{dx}(2^x)\Big|_{x=-1}$.

Solution   (a) $\dfrac{d}{dx}(2^x)\Big|_{x=3} = m \cdot 2^3 = 8m \approx 8(.693) = 5.544.$

       (b) $\dfrac{d}{dx}(2^x)\Big|_{x=-1} = m \cdot 2^{-1} = .5m \approx .5(.693) = .3465.$ ◆

The calculations just carried out for $y = 2^x$ can be carried out for $y = b^x$, where b is any positive number. Equation (5) will read exactly the same except that 2 will be replaced by b. Thus we have the following formula for the derivative of the function $f(x) = b^x$.

$$\frac{d}{dx}(b^x) = m\, b^x, \quad \text{where } m = \frac{d}{dx}(b^x)\Big|_{x=0} \tag{6}$$

Our calculations showed that if $b = 2$, then $m \approx .693$. If $b = 3$, then it turns out that $m \approx 1.1$. (See Exercise 1.) The derivative formula in (6) is simple when $m = 1$, that is, when the graph of b^x has slope 1 at $x = 0$. As we said earlier, this special value of b is denoted by the letter e. Thus the number e has the property that

$$\frac{d}{dx}(e^x)\Big|_{x=0} = 1 \tag{7}$$

and

$$\frac{d}{dx}(e^x) = 1 \cdot e^x = e^x. \tag{8}$$

The graphical interpretation of (7) is that the curve $y = e^x$ has slope 1 at $x = 0$. The graphical interpretation of (8) is that the slope of the curve $y = e^x$ at an arbitrary value of x is exactly equal to the value of the function e^x at that point. (See Fig. 3.)

The function e^x is the same type of function as 2^x and 3^x except that differentiating e^x is much easier. In fact, the next section shows that 2^x can be written as e^{kx} for a suitable constant k. The same is true for 3^x. For this reason, functions of the form e^{kx} are used in almost all applications that require an exponential type of function to describe a physical, economic, or biological phenomenon.

Incorporating Technology   Calculators usually have e^x as the secondary function of the LN key. Figs. 4(a) and 4(b) show three computations with e^x.

Parentheses are often necessary when evaluating e^x. In Fig. 5, **e^2/3** is calculated as $e^2/3$ instead of $e^{2/3}$.

Practice Problems 4.2

In the following problems use the number 20 as the (approximate) value of e^3.

1. Find the equation of the tangent line to the graph of $y = e^x$ at $x = 3$.

2. Solve the following equation for x:

$$4e^{6x} = 80.$$

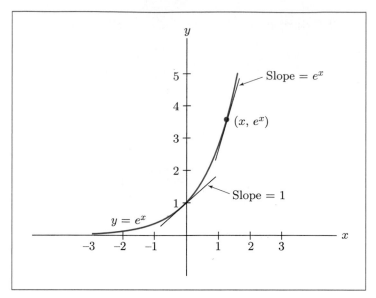

Figure 3. Fundamental properties of e^x.

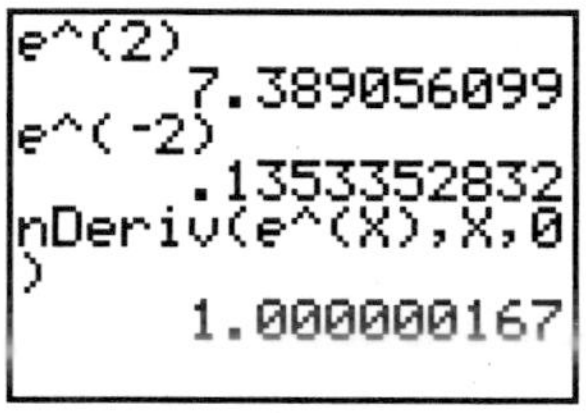

Figure 4(a) TI-83

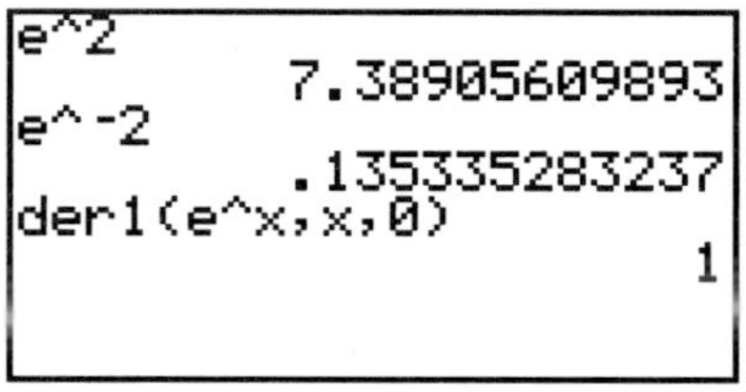

Figure 4(b) TI-86

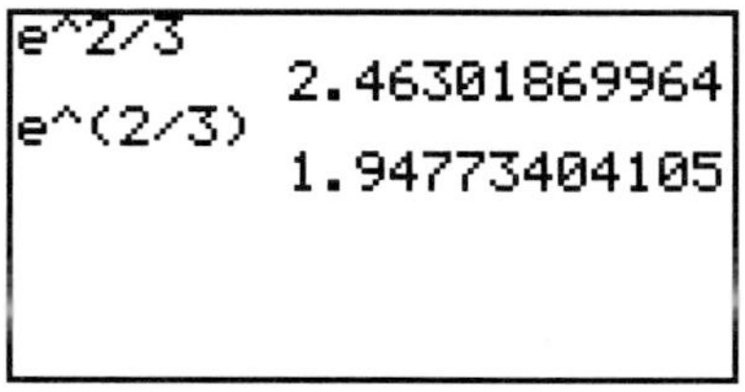

Figure 5. TI-86

▶ Exercises 4.2

1. Show that

$$\left.\frac{d}{dx}(3^x)\right|_{x=0} \approx 1.1$$

by calculating the slope $\dfrac{3^h - 1}{h}$ of the secant line passing through the points $(0,1)$ and $(h, 3^h)$. Take $h = .1$, .01, and .001.

2. Show that

$$\left.\frac{d}{dx}(2.7^x)\right|_{x=0} \approx .99$$

by calculating the slope $\dfrac{2.7^h - 1}{h}$ of the secant line passing through the points $(0,1)$ and $(h, 2.7^h)$. Take $h = .1$, .01, and .001.

3. Estimate the slope of e^x at $x = 0$ by calculating the slope $\dfrac{e^h - 1}{h}$ of the secant line passing through the points $(0,1)$ and (h, e^h). Take $h = .01$, .001, and .0001.

4. Use (8) and a familiar rule for differentiation to find

$$\frac{d}{dx}(5e^x).$$

5. Use (8) and a familiar rule for differentiation to find

$$\frac{d}{dx}(e^x)^{10}.$$

6. Use the fact that $e^{2+x} = e^2 \cdot e^x$ to find

$$\frac{d}{dx}(e^{2+x}).$$

[Remember that e^2 is just a constant—approximately $(2.7)^2$.]

7. (a) Use the fact that $e^{4x} = (e^x)^4$ to find $\dfrac{d}{dx}(e^{4x})$. Simplify the derivative as much as possible.

(b) Let k represent any constant. Generalize the calculations in part (a) to find a simple formula for $\dfrac{d}{dx}(e^{kx})$.

8. Find $\dfrac{d}{dx}(e^x + x^2)$.

Write each function in the form e^{kx} for a suitable constant k.

9. $(e^2)^x$, $\left(\dfrac{1}{e}\right)^x$ **10.** $(e^3)^{x/5}$, $\left(\dfrac{1}{e^2}\right)^x$

11. $\left(\dfrac{1}{e^3}\right)^{2x}$, $e^{1-x} \cdot e^{3x-1}$ **12.** $\left(\dfrac{e^5}{e^3}\right)^x$, $e^{4x+2} \cdot e^{x-2}$

13. $(e^{4x} \cdot e^{6x})^{3/5}$, $\dfrac{1}{e^{-2x}}$ **14.** $\sqrt{e^{-x} \cdot e^{7x}}$, $\dfrac{e^{-3x}}{e^{-4x}}$

Solve each equation for x.

15. $e^{5x} = e^{20}$ **16.** $e^{1-x} = e^2$

17. $e^{x^2 - 2x} = e^8$ **18.** $e^{-x} = 1$

Differentiate the following functions.

19. xe^x **20.** $\dfrac{e^x}{x}$

21. $\dfrac{e^x}{1 + e^x}$ **22.** $(1 + x^2)e^x$

23. $(1 + 5e^x)^4$ **24.** $(xe^x - 1)^{-3}$

25. The atmospheric pressure at an altitude of x kilometers is $f(x)$ g/cm^2 (grams per square centimeter), where $f(x) = 1035e^{-.12x}$. Give approximate answers to the following questions using the graphs of $f(x)$ and $f'(x)$ shown in Fig. 6.

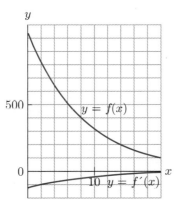

Figure 6.

(a) What is the pressure at an altitude of 2 kilometers?

(b) At what altitude is the pressure 200 g/cm^2?

(c) At an altitude of 8 kilometers, at what rate is the atmospheric pressure changing (with respect to change in altitude)?

(d) At what altitude is atmospheric pressure falling at the rate of 100 g/cm^2 per kilometer?

26. The national health expenditures (in billions of dollars) from 1960 to 1980 are given approximately by $f(t) = 27e^{.106t}$, with time in years measured from 1960. Give approximate answers to the following questions using the graphs of $f(t)$ and $f'(t)$ shown in Fig. 7.

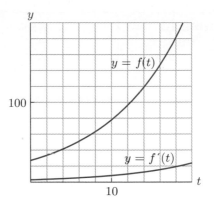

Figure 7.

(a) How much money was spent in 1978?

(b) How fast were expenditures rising in 1972?

(c) When did expenditures reach 120 billion dollars?

(d) When were expenditures rising at the rate of 20 billion dollars per year?

Technology Exercises

27. Find the equation of the tangent line to the graph of $y = e^x$ at $x = 0$. Then graph the function and the tangent line together to confirm that your answer is correct.

28. (a) Graph $y = e^x$.

(b) Zoom in on the region near $x = 0$ until the curve appears as a straight line and estimate the slope of the line. This number is an estimate of $\dfrac{d}{dx}e^x$ at $x = 0$. Compare your answer with the actual slope, 1.

(c) Repeat parts (a) and (b) for $y = 2^x$. Observe that the slope at $x = 0$ is not 1.

29. Set $Y_1 = e^x$ and use your calculator's derivative command to specify Y_2 as the derivative of Y_1. Graph the two functions simultaneously in the window $[-1, 3]$ by $[-3, 20]$ and observe that the graphs overlap.

30. Demonstrate that e^x grows faster than any power function. For instance, graph $\dfrac{x^n}{e^x}$ for $n = 3$ with the window $[0, 16]$ by $[0, 25]$ and observe that the function approaches zero as x gets large. Repeat for $n = 4$ and $n = 5$.

31. Calculate values of $\dfrac{10^x - 1}{x}$ for small values of x and use them to estimate $\dfrac{d}{dx}(10^x)\Big|_{x=0}$. What is the formula for $\dfrac{d}{dx}(10^x)$?

32. The graph of e^{x-2} may be obtained by translating the graph of e^x two units to the right. Find a constant k such that the graph of ke^x is the same as the graph of e^{x-2}. Verify your result by graphing both functions.

<table>
<tr><td>

Solutions to Practice Problems 4.2

</td><td>

1. When $x = 3$, $y = e^3 \approx 20$. So the point $(3, 20)$ is on the tangent line. Since $\frac{d}{dx}(e^x) = e^x$, the slope of the tangent line is e^3 or 20. Therefore, the equation of the tangent line in point-slope form is $y - 20 = 20(x - 3)$.

2. This problem is similar to Practice Problem 2 of Section 4.1. First divide both sides of the equation by 4.

$$e^{6x} = 20$$

The idea is to express 20 as a power of e and then equate exponents.

$$e^{6x} = e^3$$
$$6x = 3$$
$$x = \tfrac{1}{2}$$

</td></tr>
</table>

4.3 Differentiation of Exponential Functions

We have shown that $\frac{d}{dx}(e^x) = e^x$. Using this fact and the chain rule, we can differentiate functions of the form $e^{g(x)}$, where $g(x)$ is any differentiable function. This is because $e^{g(x)}$ is the composite of two functions. Indeed, if $f(x) = e^x$, then

$$e^{g(x)} = f(g(x)).$$

Thus, by the chain rule, we have

$$\begin{aligned}
\frac{d}{dx}(e^{g(x)}) &= f'(g(x))g'(x) \\
&= f(g(x))g'(x) \qquad [\text{since } f'(x) = f(x)] \\
&= e^{g(x)}g'(x).
\end{aligned}$$

So we have the following result:

Chain Rule for Exponential Functions Let $g(x)$ be any differentiable function. Then

$$\frac{d}{dx}(e^{g(x)}) = e^{g(x)}g'(x). \tag{1}$$

If we write $u = g(x)$, then (1) can be written in the form

$$\frac{d}{dx}(e^u) = e^u \frac{du}{dx}. \tag{1a}$$

▶ **Example 1** Differentiate e^{x^2+1}.

Solution Here $g(x) = x^2 + 1$, $g'(x) = 2x$, so

$$\frac{d}{dx}(e^{x^2+1}) = e^{x^2+1} \cdot 2x = 2xe^{x^2+1}. \qquad \blacklozenge$$

▶ **Example 2** Differentiate $e^{3x^2-(1/x)}$.

Solution $\frac{d}{dx}\left(e^{3x^2-(1/x)}\right) = e^{3x^2-(1/x)} \cdot \frac{d}{dx}\left(3x^2 - \frac{1}{x}\right) = e^{3x^2-(1/x)}\left(6x + \frac{1}{x^2}\right) \qquad \blacklozenge$

▶ **Example 3**   Differentiate e^{5x}.

Solution $\dfrac{d}{dx}(e^{5x}) = e^{5x} \cdot \dfrac{d}{dx}(5x) = e^{5x} \cdot 5 = 5e^{5x}$ ◆

Using a computation similar to that used in Example 3, we may differentiate e^{kx} for any constant k. (In Example 3 we have $k = 5$.) The result is the following useful formula.

$$\frac{d}{dx}(e^{kx}) = ke^{kx} \tag{2}$$

Many applications involve exponential functions of the form $y = Ce^{kx}$, where C and k are constants. In the next example we differentiate such functions.

▶ **Example 4**   Differentiate the following exponential functions.

(a) $3e^{5x}$

(b) $3e^{kx}$, where k is a constant

(c) Ce^{kx}, where C and k are constants

Solution   (a) $\dfrac{d}{dx}(3e^{5x}) = 3\dfrac{d}{dx}(e^{5x}) = 3 \cdot 5e^{5x} = 15e^{5x}$

(b) $\dfrac{d}{dx}(3e^{kx}) = 3\dfrac{d}{dx}(e^{kx})$

$\qquad\qquad = 3 \cdot ke^{kx} \qquad$ [by (2)]

$\qquad\qquad = 3ke^{kx}$

(c) $\dfrac{d}{dx}(Ce^{kx}) = C\dfrac{d}{dx}(e^{kx}) = Cke^{kx}$ ◆

The result of part (c) may be summarized in an extremely useful fashion as follows: Suppose that we let $y = Ce^{kx}$. By part (c) we have

$$y' = Cke^{kx} = k \cdot (Ce^{kx}) = ky.$$

In other words, the derivative of the function Ce^{kx} is k times the function itself. Let us record this fact.

Let C, k be any constants and let $y = Ce^{kx}$. Then y satisfies the equation

$$y' = ky.$$

The equation $y' = ky$ expresses a relationship between the function y and its derivative y'. Any equation expressing a relationship between a function y and one or more of its derivatives is called a *differential equation*.

Very often an applied problem will involve a function $y = f(x)$ that satisfies the differential equation $y' = ky$. It can be shown that y must then necessarily be an exponential function of the form Ce^{kx}. That is, we have the following result.

Suppose that $y = f(x)$ satisfies the differential equation

$$y' = ky. \tag{3}$$

Then y is an exponential function of the form

$$y = Ce^{kx}, \qquad C \text{ a constant.}$$

We leave the verification of this result to Exercise 45.

▶ **Example 5** Determine all functions $y = f(x)$ such that $y' = -.2y$.

Solution The equation $y' = -.2y$ has the form $y' = ky$ with $k = -.2$. Therefore, any solution of the equation has the form

$$y = Ce^{-.2x},$$

where C is a constant. ◆

▶ **Example 6** Determine all functions $y = f(x)$ such that $y' = y/2$ and $f(0) = 4$.

Solution The equation $y' = y/2$ has the form $y' = ky$ with $k = \frac{1}{2}$. Therefore,

$$f(x) = Ce^{(1/2)x}$$

for some constant C. We also require that $f(0) = 4$. That is,

$$4 = f(0) = Ce^{(1/2)0} = Ce^0 = C.$$

So $C = 4$ and

$$f(x) = 4e^{(1/2)x}. \qquad ◆$$

The Functions e^{kx}   Exponential functions of the form e^{kx} occur in many applications. Figure 1 shows the graphs of several functions of this type when k is a positive number. These curves $y = e^{kx}$, k positive, have several properties in common:

1. $(0, 1)$ is on the graph.
2. The graph lies strictly above the x-axis (e^{kx} is never zero).
3. The x-axis is an asymptote as x becomes large negatively.
4. The graph is always increasing and concave up.

When k is negative, the graph of $y = e^{kx}$ is decreasing. (See Fig. 2.) Note the following properties of the curves $y = e^{kx}$, k negative:

1. $(0, 1)$ is on the graph.
2. The graph lies strictly above the x-axis.
3. The x-axis is an asymptote as x becomes large positively.
4. The graph is always decreasing and concave up.

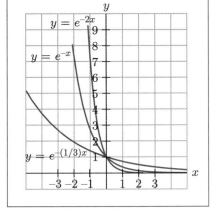

Figure 1.   **Figure 2.**

The Functions b^x If b is a positive number, then the function b^x may be written in the form e^{kx} for some k. For example, take $b = 2$. From Fig. 3 of the preceding section it is clear that there is some value of x such that $e^x = 2$. Call this value k, so that $e^k = 2$. Then

$$2^x = (e^k)^x = e^{kx}$$

for all x. In general, if b is any positive number, there is a value of x, say $x = k$, such that $e^k = b$. In this case, $b^x = (e^k)^x = e^{kx}$. Thus all the curves $y = b^x$ discussed in Section 4.1 can be written in the form $y = e^{kx}$. This is one reason why we have focused on exponential functions with base e instead of studying 2^x, 3^x, and so on.

Incorporating Technology Parentheses are often necessary when evaluating exponential functions on graphing calculators. With some calculators, **e^2x** is evaluated as $e^2 x$ instead of the intended e^{2x}. See Fig. 3.

```
3→x:e^2 x
           22.1671682968
3→x:e^(2x)
           403.428793493
3→x:(e^2)x
           22.1671682968
```

Figure 3. TI-86

Practice Problems 4.3

1. Differentiate $[e^{-3x}(1 + e^{6x})]^{12}$.

2. Determine all functions $y = f(x)$ such that $y' = -y/20$, $f(0) = 2$.

▶ Exercises 4.3

Differentiate the following functions.

1. $f(x) = 4e^{2x}$

2. $f(x) = e^{5x}$

3. $f(t) = 4 + e^{-t}$

4. $y = \dfrac{e^x + e^{-x}}{2}$

5. $y = e^{-2x} - x^2$

6. $f(t) = 2e^{1-3t}$

7. $y = (e^x + e^{-x})^3$

8. $f(x) = (e^{-x})^2$

9. $y = \frac{1}{3}e^{3+2x}$

10. $y = (e^{6x} + x^6)^3$

11. $g(t) = e^{1/t}$

12. $y = \frac{1}{10}e^{-x^2}$

13. $y = e^{x^2 - 5x + 4}$

14. $g(t) = e^{3/t}$

15. $y = 5(x^3 + e^{-3x})^4$

16. $f(t) = 1/e^{1-2t}$

[*Hint*: In Exercises 16–18, simplify $f(t)$ before differentiating.]

17. $f(t) = \dfrac{e^{3t} + e^{-3t}}{e^t}$

18. $f(t) = e^t(e^{2t} - e^{-2t})$

19. $(x + 1)e^{-x+2}$

20. $x\sqrt{4 + e^x}$

21. $f(x) = x^3 e^{2x}$

22. $f(x) = x^2 e^{-3x}$

23. $e^{4x}/(4 + x)$

24. $e^{-3x}/(1 - 4x)$

25. $\left(\dfrac{1}{x} + 3\right)e^{2x}$

26. $\dfrac{3x - 1}{e^{4x}}$

In Exercises 27–32, find the values of x at which the function has a possible relative maximum or minimum point. (Recall that e^x is positive for all x.) Use the second derivative to determine the nature of the function at these points.

27. $(1+x)e^{-3x}$

28. $(1-x)e^{2x}$

29. $\dfrac{3-4x}{e^{2x}}$

30. $\dfrac{4x-1}{e^{x/2}}$

31. $(5x-2)e^{1-2x}$

32. $(2x-5)e^{3x-1}$

33. A painting purchased in 1998 for \$100,000 is estimated to be worth $v(t) = 100{,}000e^{t/5}$ dollars after t years. At what rate will the painting be appreciating in 2003?

34. The value of a computer t years after purchase is $v(t) = 2000e^{-.35t}$ dollars. At what rate is the computer's value falling after 3 years?

35. The velocity of a parachutist during free fall is
$$f(t) = 60(1 - e^{-.17t})$$
meters per second. Answer the following questions by reading the graph in Fig. 4. (Recall that acceleration is the derivative of velocity.)

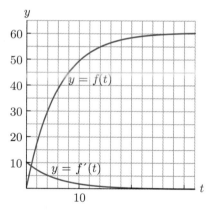

Figure 4.

(a) What is the velocity when $t = 8$ seconds?

(b) What is the acceleration when $t = 0$?

(c) When is the parachutist's velocity 30 m/sec?

(d) When is the acceleration 5 m/sec^2?

36. Suppose that the velocity of a parachutist is
$$v(t) = 65(1 - e^{-.16t})$$
meters per second. The graph of $v(t)$ is similar to that in Fig. 4. Calculate the parachutist's velocity and acceleration when $t = 9$ seconds.

37. The height of a certain plant after t weeks is
$$f(t) = \frac{1}{.05 + e^{-.4t}}$$
inches. The graph of $f(t)$ resembles the graph in Fig. 5. Calculate the rate of growth of the plant after 7 weeks.

38. The length of a certain weed after t weeks is
$$f(t) = \frac{6}{.2 + 5e^{-.5t}}$$
centimeters. Answer the following questions by reading the graph in Fig. 5.

(a) How fast is the weed growing after 10 weeks?

(b) When is the weed 10 centimeters long?

(c) When is the weed growing at the rate of 2 cm/week?

(d) What is the maximum rate of growth?

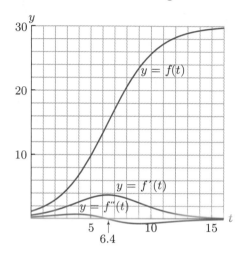

Figure 5.

39. Let a and b be positive numbers. A curve whose equation is $y = e^{-ae^{-bx}}$ is called a *Gompertz growth curve*. These curves are used in biology to describe certain types of population growth. Compute the derivative of $y = e^{-2e^{-.01x}}$.

40. Find $\dfrac{dy}{dx}$ if $y = e^{-(1/10)e^{-x/2}}$.

41. Determine all solutions of the differential equation $y' = -4y$.

42. Determine all solutions of the differential equation $y' = \frac{1}{3}y$.

43. Determine all functions $y = f(x)$ such that $y' = -.5y$ and $f(0) = 1$

44. Determine all functions $y = f(x)$ such that $y' = 3y$ and $f(0) = \frac{1}{2}$.

45. Verify the result (3). [*Hint:* Let $g(x) = f(x)e^{-kx}$. Show that $g'(x) = 0$.] You may assume that only a constant function has a zero derivative.

46. Let $f(x)$ be a function with the property that $f'(x) = 1/x$. Let $g(x) = f(e^x)$, and compute $g'(x)$.

47. As h approaches 0, what value is approached by the difference quotient $\dfrac{e^h - 1}{h}$? (*Hint:* $1 = e^0$.)

48. As h approaches 0, what value is approached by $\dfrac{e^{2h} - 1}{h}$? (*Hint:* $1 = e^0$.)

Technology Exercises

49. In a study, a cancerous tumor was found to have a volume of

$$f(t) = 1.825^3(1 - 1.6e^{-.4196t})^3$$

milliliters after t weeks, with $t > 1$.*

(a) Sketch the graphs of $f(t)$ and $f'(t)$ for $1 \le t \le 15$. What do you notice about the tumor's volume?

(b) How large is the tumor after 5 weeks?

(c) When will the tumor have a volume of 5 milliliters?

(d) How fast is the tumor growing after 5 weeks?

(e) When is the tumor growing at the fastest rate?

(f) What is the fastest rate of growth of the tumor?

50. Let $f(t)$ be the function from Exercise 37 that gives the height (inches) of a plant at time t (weeks).

(a) When is the plant 11 inches tall?

(b) When is the plant growing at the rate of 1 inch per week?

(c) What is the fastest rate of growth of the plant, and when does this occur?

Solutions to Practice Problems 4.3

1. We must use the general power rule. However, this is most easily done if we first use the laws of exponents to simplify the function inside the brackets.

$$e^{-3x}(1 + e^{6x}) = e^{-3x} + e^{-3x} \cdot e^{6x} = e^{-3x} + e^{3x}$$

Now

$$\frac{d}{dx}[e^{-3x} + e^{3x}]^{12} = 12 \cdot [e^{-3x} + e^{3x}]^{11} \cdot (-3e^{-3x} + 3e^{3x})$$

$$= 36 \cdot [e^{-3x} + e^{3x}]^{11} \cdot (-e^{-3x} + e^{3x}).$$

2. The differential equation $y' = -y/20$ is of the type $y' = ky$, where $k = -\frac{1}{20}$. Therefore, any solution has the form $f(x) = Ce^{-(1/20)x}$. Now $f(0) = Ce^{-(1/20) \cdot 0} = Ce^0 = C$, so that $C = 2$ when $f(0) = 2$. Therefore, the desired function is $f(x) = 2e^{-(1/20)x}$.

4.4 The Natural Logarithm Function

As a preparation for the definition of the natural logarithm, we shall make a geometric digression. In Fig. 1 we have plotted several pairs of points. Observe how they are related to the line $y = x$.

The points $(5, 7)$ and $(7, 5)$, for example, are the same distance from the line $y = x$. If we were to plot the point $(5, 7)$ with wet ink and then fold the page along the line $y = x$, the ink blot would produce a second blot at the point $(7, 5)$. If we think of the line $y = x$ as a mirror, then $(7, 5)$ is the mirror image of $(5, 7)$. We say that $(7, 5)$ is the *reflection* of $(5, 7)$ through the line $y = x$. Similarly, $(5, 7)$ is the reflection of $(7, 5)$ through the line $y = x$.

Now let us consider all points lying on the graph of the exponential function $y = e^x$ [see Fig. 2(a)]. If we reflect each such point through the line $y = x$, we obtain a new graph [see Fig. 2(b)]. For each positive x, there is exactly one value of y such that (x, y) is on the new graph. We call this value of y the *natural logarithm of x*, denoted $\ln x$. Thus the reflection of the graph of $y = e^x$ through the line $y = x$ is the graph of the natural logarithm function $y = \ln x$.

We may deduce some properties of the natural logarithm function from an inspection of its graph.

---

*Baker, Goddard, Clark, and Whimster, "Proportion of Necrosis in Transplanted Murine Adenocarcinomas and Its Relationship to Tumor Growth," *Growth, Development and Aging*, 54 (1990), 85–93.

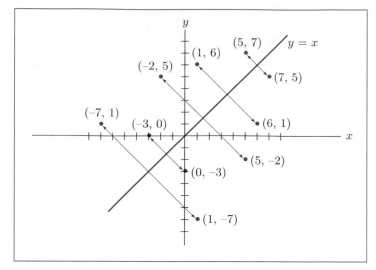

Figure 1. Reflections of points through the line y = x.

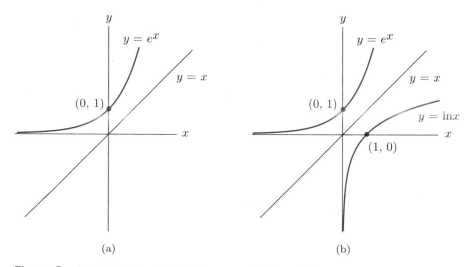

Figure 2. Obtaining the graph of ln x as a reflection of e^x.

1. The point $(1, 0)$ is on the graph of $y = \ln x$ [because $(0, 1)$ is on the graph of $y = e^x$]. In other words,

$$\ln 1 = 0. \tag{1}$$

2. $\ln x$ is defined only for positive values of x.

3. $\ln x$ is negative for x between 0 and 1.

4. $\ln x$ is positive for x greater than 1.

5. $\ln x$ is an increasing function and concave down.

Let us study the relationship between the natural logarithm and exponential functions more closely. From the way in which the graph of $\ln x$ was obtained we know that (a, b) is on the graph of $\ln x$ if and only if (b, a) is on the graph of

e^x. However, a typical point on the graph of $\ln x$ is of the form $(a, \ln a)$, $a > 0$. So for any positive value of a, the point $(\ln a, a)$ is on the graph of e^x. That is,

$$e^{\ln a} = a.$$

Since a was an arbitrary positive number, we have the following important relationship between the natural logarithm and exponential functions.

$$e^{\ln x} = x \quad \text{for } x > 0 \qquad (2)$$

Equation (2) can be put into verbal form.

For each positive number x, $\ln x$ is that exponent to which we must raise e in order to get x.

If b is any number, then e^b is positive and hence $\ln(e^b)$ makes sense. What is $\ln(e^b)$? Since (b, e^b) is on the graph of e^x, we know that (e^b, b) must be on the graph of $\ln x$. That is, $\ln(e^b) = b$. Thus we have shown that

$$\ln(e^x) = x \quad \text{for any } x. \qquad (3)$$

The identities (2) and (3) express the fact that the natural logarithm is the *inverse* of the exponential function (for $x > 0$). For instance, if we take a number x and compute e^x, then, by (3), we can undo the effect of the exponential by taking the natural logarithm; that is, the logarithm of e^x equals the original number x. Similarly, if we take a positive number x and compute $\ln x$, then, by (2), we can undo the effect of the logarithm by raising e to the $\ln x$ power; that is, $e^{\ln x}$ equals the original number x.

Scientific calculators have an "LN" key that will compute the natural logarithm of a number to as many as ten significant figures. For instance, pressing the LN key and entering the number 2 into the calculator, one obtains $\ln 2 = .6931471806$ (to 10 significant figures).

The relationships (2) and (3) between e^x and $\ln x$ may be used to solve equations, as the next examples show.

▶ **Example 1**   Solve the equation $5e^{x-3} = 4$ for x.

Solution   First divide each side by 5,

$$e^{x-3} = .8.$$

Taking the logarithm of each side and using (3), we have

$$\ln(e^{x-3}) = \ln .8$$
$$x - 3 = \ln .8$$
$$x = 3 + \ln .8.$$

[If desired, the numerical value of x can be obtained by using a scientific calculator, namely, $x = 3 - .22314 = 2.77686$ (to five decimal places).]   ◆

▶ **Example 2**   Solve the equation $2\ln x + 7 = 0$ for x.

Solution

$$2 \ln x = -7$$
$$\ln x = -3.5$$
$$e^{\ln x} = e^{-3.5}$$
$$x = e^{-3.5} \qquad \text{[by (2)]} \qquad \blacklozenge$$

Other Exponential and Logarithm Functions In our discussion of the exponential function, we mentioned that all exponential functions of the form b^x, where b is a fixed positive number, can be expressed in terms of *the* exponential function e^x. Now we can be quite explicit. For since $b = e^{\ln b}$, we see that

$$b^x = (e^{\ln b})^x = e^{(\ln b)x}.$$

Hence we have shown that

$$b^x = e^{kx}, \quad \text{where } k = \ln b.$$

The natural logarithm function is sometimes called the *logarithm to the base e*, for it is the inverse of the exponential function e^x. If we reflect the graph of the function $y = 2^x$ through the line $y = x$, we obtain the graph of a function called the *logarithm to the base* 2, denoted by $\log_2 x$. Similarly, if we reflect the graph of $y = 10^x$ through the line $y = x$, we obtain the graph of a function called the *logarithm to the base* 10, denoted by $\log_{10} x$. (See Fig. 3.)

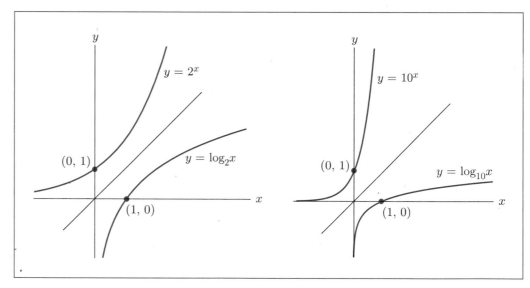

Figure 3. Graphs of $\log_2 x$ and $\log_{10} x$ as reflections of 2^x and 10^x.

Logarithms to the base 10 are sometimes called *common* logarithms. Common logarithms are usually introduced into algebra courses for the purpose of simplifying certain arithmetic calculations. However, with the advent of the modern computer and the widespread availability of graphing calculators, the need for common logarithms has diminished considerably. It can be shown that

$$\log_{10} x = \frac{1}{\ln 10} \cdot \ln x,$$

so that $\log_{10} x$ is simply a constant multiple of $\ln x$. However, we shall not need this fact.

The natural logarithm function is used in calculus because differentiation and integration formulas are simpler than for $\log_{10} x$ or $\log_2 x$, and so on. (Recall that we prefer the function e^x over the functions 10^x and 2^x for the same reason.) Also, $\ln x$ arises "naturally" in the process of solving certain differential equations that describe various growth processes.

Practice Problems 4.4

1. Find $\ln e$.

2. Solve $e^{-3x} = 2$ using the natural logarithm function.

▶ Exercises 4.4

1. Find $\ln(\sqrt{e})$.

2. Find $\ln(1/e)$.

3. If $e^x = 4.1$, write x in terms of the natural logarithm.

4. If $e^{-x} = 2.9$, write x in terms of the natural logarithm.

5. If $\ln x = -3.8$, write x using the exponential function.

6. If $\ln x = 2.3$, write x using the exponential function.

Simplify the following expressions.

7. $\ln e^{-3}$

8. $e^{\ln 2.88}$

9. $e^{e^{\ln 1}}$

10. $\ln(e^{-2\ln e})$

11. $\ln(\ln e)$

12. $e^{4\ln 1}$

Solve the following equations for x.

13. $e^{2x} = 5$

14. $e^{1-3x} = 4$

15. $\ln(4 - x) = \frac{1}{2}$

16. $\ln 3x = 2$

17. $\ln x^2 = 9$

18. $e^{x^2} = 25$

19. $6e^{-.00012x} = 3$

20. $4 - \ln x = 0$

21. $\ln 3x = \ln 5$

22. $\ln(x^2 - 5) = 0$

23. $\ln(\ln 3x) = 0$

24. $2\ln x = 7$

25. $2e^{x/3} - 9 = 0$

26. $4 - 3e^{x+6} = 0$

27. $5\ln 2x = 8$

28. $750e^{-.4x} = 375$

29. $(e^2)^x \cdot e^{\ln 1} = 4$

30. $e^{5x} \cdot e^{\ln 5} = 2$

31. $4e^x \cdot e^{-2x} = 6$

32. $(e^x)^2 \cdot e^{2-3x} = 4$

In Exercises 33–36, find the coordinates of each relative extreme point of the given function and determine if the point is a relative maximum point or a relative minimum point.

33. $f(x) = e^{-x} + 3x$

34. $f(x) = 5x - 2e^x$

35. $f(x) = \frac{1}{3}e^{2x} - x + \frac{1}{2}\ln\frac{3}{2}$

36. $f(x) = 5 - \frac{1}{2}x - e^{-3x}$

37. When a drug or vitamin is administered intramuscularly (into a muscle), the concentration in the blood at time t after injection can be approximated by a function of the form $f(t) = c(e^{-k_1 t} - e^{-k_2 t})$. The graph of $f(t) = 5(e^{-.01t} - e^{-.51t})$, for $t \geq 0$, has the general shape shown in Fig. 16 of Section 0.1. Find the value of t at which this function reaches its maximum value.

38. Under certain geographic conditions, the wind velocity v at a height x centimeters above the ground is given by $v = K\ln(x/x_0)$, where K is a positive constant (depending on the air density, average wind velocity, etc.), and x_0 is a roughness parameter (depending on the roughness of the vegetation on the ground).* Suppose that $x_0 = .7$ centimeter (a value that applies to lawn grass 3 centimeters high) and $K = 300$ centimeters per second.

(a) At what height above the ground is the wind velocity zero?

(b) At what height is the wind velocity 1200 centimeters per second?

39. Find k such that $2^x = e^{kx}$ for all x.

40. Find k such that $2^{-x/5} = e^{kx}$ for all x.

Technology Exercises

41. Graph the function $y = \ln(e^x)$ and use TRACE to convince yourself that it is the same as the function $y = x$. What do you observe about the graph of $y = e^{\ln x}$?

42. Graph $y = e^{2x}$ and $y = 5$ together and determine the x-coordinate of their point of intersection (to four decimal places). Express this number in terms of a logarithm.

43. Graph $y = \ln 5x$ and $y = 2$ together and determine the x-coordinate of their point of intersection (to four decimal places). Express this number in terms of a power of e.

---

*G. Cox, B. Collier, A. Johnson, and P. Miller, *Dynamic Ecology* (Englewood Cliffs, N.J.: Prentice-Hall, 1973), pp. 113–115.

1. Answer: 1. The number $\ln e$ is that exponent to which e must be raised in order to obtain e.

2. Take the logarithm of each side and use (3) to simplify the left side:

$$\ln e^{-3x} = \ln 2$$
$$-3x = \ln 2$$
$$x = -\frac{\ln 2}{3}.$$

4.5 The Derivative of ln x

Let us now compute the derivative of $\ln x$ for $x > 0$. Since $e^{\ln x} = x$, we have

$$\frac{d}{dx}(e^{\ln x}) = \frac{d}{dx}(x) = 1. \tag{1}$$

On the other hand, if we differentiate $e^{\ln x}$ by the chain rule, we find that

$$\frac{d}{dx}(e^{\ln x}) = e^{\ln x} \cdot \frac{d}{dx}(\ln x) = x \cdot \frac{d}{dx}(\ln x), \tag{2}$$

where the last equality used the fact that $e^{\ln x} = x$. By combining equations (1) and (2) we obtain

$$x \cdot \frac{d}{dx}(\ln x) = 1.$$

In other words,

$$\frac{d}{dx}(\ln x) = \frac{1}{x}, \qquad x > 0. \tag{3}$$

By combining this differentiation formula with the chain rule, product rule, and quotient rule, we can differentiate many functions involving $\ln x$.

▶ **Example 1** Differentiate.

(a) $(\ln x)^5$ \qquad\qquad (b) $x \ln x$ \qquad\qquad (c) $\ln(x^3 + 5x^2 + 8)$

Solution \quad (a) By the general power rule,

$$\frac{d}{dx}(\ln x)^5 = 5(\ln x)^4 \cdot \frac{d}{dx}(\ln x) = 5(\ln x)^4 \cdot \frac{1}{x} = \frac{5(\ln x)^4}{x}.$$

(b) By the product rule,

$$\frac{d}{dx}(x \ln x) = x \cdot \frac{d}{dx}(\ln x) + (\ln x) \cdot 1 = x \cdot \frac{1}{x} + \ln x = 1 + \ln x.$$

(c) By the chain rule,

$$\frac{d}{dx}\ln(x^3 + 5x^2 + 8) = \frac{1}{x^3 + 5x^2 + 8} \cdot \frac{d}{dx}(x^3 + 5x^2 + 8) = \frac{3x^2 + 10x}{x^3 + 5x^2 + 8}. \quad \blacklozenge$$

Let $g(x)$ be any differentiable function. For any value of x for which $g(x)$ is positive, the function $\ln(g(x))$ is defined. For such a value of x, the derivative is given by the chain rule as

$$\frac{d}{dx}[\ln g(x)] = \frac{1}{g(x)} \cdot \frac{d}{dx}g(x) = \frac{g'(x)}{g(x)} \tag{4}$$

If $u = g(x)$, equation (4) can be written in the form

$$\frac{d}{dx}[\ln u] = \frac{1}{u}\frac{du}{dx}. \tag{4a}$$

Example 1(c) illustrates a special case of this formula.

▶ **Example 2**   The function $f(x) = (\ln x)/x$ has a relative extreme point for some $x > 0$. Find the point and determine whether it is a relative maximum or a relative minimum point.

Solution   By the quotient rule,

$$f'(x) = \frac{x \cdot \dfrac{1}{x} - (\ln x) \cdot 1}{x^2} = \frac{1 - \ln x}{x^2}$$

$$f''(x) = \frac{x^2 \cdot \left(-\dfrac{1}{x}\right) - (1 - \ln x)(2x)}{x^4} = \frac{2\ln x - 3}{x^3}.$$

If we set $f'(x) = 0$, then

$$1 - \ln x = 0$$
$$\ln x = 1$$
$$e^{\ln x} = e^1 = e$$
$$x = e.$$

Therefore, the only possible relative extreme point is at $x = e$. When $x = e$, $f(e) = (\ln e)/e = 1/e$. Furthermore,

$$f''(e) = \frac{2\ln e - 3}{e^3} = -\frac{1}{e^3} < 0,$$

which implies that the graph of $f(x)$ is concave down at $x = e$. Therefore, $(e, 1/e)$ is a relative maximum point of the graph of $f(x)$.   ◆

The next example introduces a function that will be needed later when we study integration.

▶ **Example 3**   The function $\ln|x|$ is defined for all nonzero values of x. Its graph is sketched in Fig. 1. Compute the derivative of $\ln|x|$.

Solution   If x is positive, then $|x| = x$, so

$$\frac{d}{dx}\ln|x| = \frac{d}{dx}\ln x = \frac{1}{x}.$$

If x is negative, then $|x| = -x$; and, by the chain rule,

$$\frac{d}{dx}\ln|x| = \frac{d}{dx}\ln(-x) = \frac{1}{-x} \cdot \frac{d}{dx}(-x) = \frac{1}{-x} \cdot (-1) = \frac{1}{x}.$$ ◆

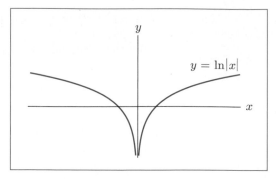

Figure 1. Graph of ln |x|.

Therefore, we have established the following useful fact.

$$\frac{d}{dx}\ln|x| = \frac{1}{x}, \qquad x \neq 0$$

Practice Problems 4.5

Differentiate.

1. $f(x) = \dfrac{1}{\ln(x^4 + 5)}$

2. $f(x) = \ln(\ln x)$

▶ Exercises 4.5

Differentiate the following functions.

1. $\ln 2x$

2. $\ln x^2$

3. $\ln(x + 5)$

4. $x^2 \ln x$

5. $\dfrac{1}{x}\ln(x + 1)$

6. $\sqrt{\ln x}$

7. $e^{\ln x + x}$

8. $\ln\left(\dfrac{x}{x - 3}\right)$

9. $4 + \ln\left(\dfrac{x}{2}\right)$

10. $\ln\sqrt{x}$

11. $(\ln x)^2 + \ln x$

12. $\ln(x^3 + 2x + 1)$

13. $\ln(kx)$, k constant

14. $\dfrac{x}{\ln x}$

15. $\dfrac{x}{(\ln x)^2}$

16. $(\ln x)e^{-x}$

17. $e^{2x}\ln x$

18. $(\ln x + 1)^3$

19. $\ln(e^{5x} + 1)$

20. $\ln(e^{e^x})$

Find.

21. $\dfrac{d}{dt}(t^2 \ln 4)$

22. $\dfrac{d^2}{dx^2}\ln(1 + x^2)$

23. $\dfrac{d^2}{dt^2}(\ln t)^3$ ~~so there~~

24. Find the slope of the graph of $y = \ln|x|$ at $x = 3$ and $x = -3$.

25. Write the equation of the tangent line to the graph of $y = \ln(x^2 + e)$ at $x = 0$.

26. The function $f(x) = (\ln x + 1)/x$ has a relative extreme point for $x > 0$. Find the coordinates of the point. Is it a relative maximum point?

27. The graph of $f(x) = (\ln x)/\sqrt{x}$ is shown in Fig. 2. Find the coordinates of the maximum point.

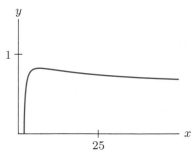

Figure 2.

28. The graph of $f(x) = x/(\ln x + x)$ is shown in Fig. 3. Find the coordinates of the minimum point.

29. Sketch the graph of $y = \ln x + \dfrac{1}{x} - \dfrac{1}{2}$.

30. Sketch the graph of $y = 1 + \ln(x^2 - 6x + 10)$.

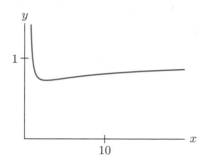

Figure 3.

31. If a cost function is $C(x) = (100 \ln x)/(40 - 3x)$, find the marginal cost when $x = 10$.

32. If a cost function is $C(x) = (1000 \ln x)/\sqrt{100 - 3x}$, find the marginal cost when $x = 25$.

33. Suppose that the demand equation for a certain commodity is $p = 45/(\ln x)$. Determine the marginal revenue function for this commodity, and compute the marginal revenue when $x = 20$.

34. Suppose that the total revenue function for a manufacturer is $R(x) = 300 \ln(x + 1)$, so that the sale of x units of a product brings in about $R(x)$ dollars. Suppose also that the total cost of producing x units is $C(x)$ dollars, where $C(x) = 2x$. Find the value of x at which the profit function $R(x) - C(x)$ will be maximized. Show that the profit function has a relative maximum and not a relative minimum point at this value of x.

35. Evaluate $\lim\limits_{h \to 0} \dfrac{\ln(7 + h) - \ln 7}{h}$.

36. Find the maximum area of a rectangle in the first quadrant with one corner at the origin, two sides on the coordinate axes, and one corner on the graph of $y = -\ln x$.

Technology Exercises

37. Graph the function $f(x) = \ln|x|$ in the window $[-5, 5]$ by $[-2, 2]$, and verify that the derivative formula gives the correct values of $f'(x)$ at $x = \pm 1$, ± 2, and ± 4.

38. (*Analysis of the effectiveness of an insect repellent*) Human hands covered with cotton fabrics that had been impregnated with the insect repellent DEPA were inserted for five minutes into a test chamber containing 200 female mosquitoes.[*] The function $f(x) = 26.48 - 14.09 \ln x$ gives the number of mosquito bites received when the concentration was x percent. [*Note:* The answers to parts (b)–(e) can be obtained either algebraically or from the graphs. You might consider trying both methods.]

(a) Graph $f(x)$ and $f'(x)$ for $0 \le x \le 6$.

(b) How many bites were received when the concentration was 3.25%?

(c) What concentration resulted in 15 bites?

(d) At what rate is the number of bites changing with respect to concentration of DEPA when $x = 2.75$?

(e) For what concentration does the rate of change of bites with respect to concentration equal -10 bites per percentage increase in concentration?

Solutions to Practice Problems 4.5	**1.** Here $f(x) = [\ln(x^4 + 5)]^{-1}$. By the chain rule,

$$f'(x) = (-1) \cdot [\ln(x^4 + 5)]^{-2} \cdot \frac{d}{dx} \ln(x^4 + 5)$$

$$= -[\ln(x^4 + 5)]^{-2} \cdot \frac{4x^3}{x^4 + 5}.$$

2. $f'(x) = \dfrac{d}{dx} \ln(\ln x) = \dfrac{1}{\ln x} \cdot \dfrac{d}{dx} \ln x = \dfrac{1}{\ln x} \cdot \dfrac{1}{x} = \dfrac{1}{x \ln x}.$

*Rao, K. M., Prakash, S., Kumar, S., Suryanarayana, M. V. S., Bhagwat, M. M., Gharia, M. M., and Bhavsar, R. B., "N-diethylphenylacetamide in Treated Fabrics, as a Repellent Against *Aedes aegypti* and *Culex quinquefasclatus* (Deptera: Culiciae)," *Journal of Medical Entomology*, 28 (January 1991), 1.

4.6   Properties of the Natural Logarithm Function

The natural logarithm function $\ln x$ possesses many of the familiar properties of logarithms to base 10 (or common logarithms) that are encountered in algebra.

Let x and y be positive numbers, b any number.

LI $\ln(xy) = \ln x + \ln y$.

LII $\ln\left(\dfrac{1}{x}\right) = -\ln x$.

LIII $\ln\left(\dfrac{x}{y}\right) = \ln x - \ln y$.

LIV $\ln(x^b) = b \ln x$.

Verification of LI   By equation (2) of Section 4.4 we have $e^{\ln(xy)} = xy$, $e^{\ln x} = x$, and $e^{\ln y} = y$. Therefore,

$$e^{\ln(xy)} = xy = e^{\ln x} \cdot e^{\ln y} = e^{\ln x + \ln y}.$$

By equating exponents, we get LI.

Verification of LII   Since $e^{\ln(1/x)} = 1/x$, we have

$$e^{\ln(1/x)} = \frac{1}{x} = \frac{1}{e^{\ln x}} = e^{-\ln x}.$$

By equating exponents, we get LII.

Verification of LIII   By LI and LII, we have

$$\ln\left(\frac{x}{y}\right) = \ln\left(x \cdot \frac{1}{y}\right) = \ln x + \ln\left(\frac{1}{y}\right) = \ln x - \ln y.$$

Verification of LIV   Since $e^{\ln(x^b)} = x^b$, we have

$$e^{\ln(x^b)} = x^b = \left(e^{\ln x}\right)^b = e^{b \ln x}.$$

Equating exponents, we get LIV.

These properties of the natural logarithm should be learned thoroughly. You will find them useful in many calculations involving $\ln x$ and the exponential function.

▶ **Example 1**   Write $\ln 5 + 2 \ln 3$ as a single logarithm.

Solution
$$\ln 5 + 2 \ln 3 = \ln 5 + \ln 3^2 \qquad \text{(LIV)}$$
$$= \ln 5 + \ln 9$$
$$= \ln(5 \cdot 9) \qquad \text{(LI)}$$
$$= \ln 45 \qquad\qquad ◆$$

▶ **Example 2**   Write $\frac{1}{2}\ln(4t) - \ln(t^2 + 1)$ as a single logarithm.

Solution

$$\frac{1}{2}\ln(4t) - \ln(t^2 + 1) = \ln\left[(4t)^{1/2}\right] - \ln(t^2 + 1) \qquad \text{(LIV)}$$

$$= \ln(2\sqrt{t}) - \ln(t^2 + 1)$$

$$= \ln\left(\frac{2\sqrt{t}}{t^2 + 1}\right) \qquad \text{(LIII)}$$

◆

▶ **Example 3**   Simplify $\ln x + \ln 3 + \ln y - \ln 5$.

Solution   Use (LI) twice and (LIII) once.

$$(\ln x + \ln 3) + \cancel{\ln y} - \ln 5 = \ln 3x + \cancel{\ln y} - \ln 5$$

$$= \ln 3xy - \ln 5$$

$$= \ln\left(\frac{3xy}{5}\right) \qquad \qquad \text{◆}$$

▶ **Example 4**   Differentiate $f(x) = \ln[x(x + 1)(x + 2)]$.

Solution   First rewrite $f(x)$, using (LI).

$$f(x) = \ln[x(x + 1)(x + 2)] = \ln x + \ln(x + 1) + \ln(x + 2).$$

Then $f'(x)$ is easily calculated:

$$f'(x) = \frac{1}{x} + \frac{1}{x + 1} + \frac{1}{x + 2}. \qquad \qquad \text{◆}$$

The natural logarithm function can be used to simplify the task of differentiating products. Suppose, for example, that we wish to differentiate the function

$$g(x) = x(x + 1)(x + 2).$$

As we showed in Example 4,

$$\frac{d}{dx}\ln g(x) = \frac{1}{x} + \frac{1}{x + 1} + \frac{1}{x + 2}.$$

However,

$$\frac{d}{dx}\ln g(x) = \frac{g'(x)}{g(x)}.$$

Therefore, equating the two expressions for $\dfrac{d}{dx}\ln g(x)$, we have

$$\frac{g'(x)}{g(x)} = \frac{1}{x} + \frac{1}{x + 1} + \frac{1}{x + 2}.$$

Finally, we solve for $g'(x)$:

$$g'(x) = g(x) \cdot \left(\frac{1}{x} + \frac{1}{x + 1} + \frac{1}{x + 2}\right)$$

$$= x(x + 1)(x + 2)\left(\frac{1}{x} + \frac{1}{x + 1} + \frac{1}{x + 2}\right).$$

In a similar way, we differentiate the product of any number of factors by first taking natural logarithms, then differentiating, and finally solving for the desired derivative. This procedure is called *logarithmic differentiation*.

▶ **Example 5**   Differentiate the function $g(x) = (x^2 + 1)(x^3 - 3)(2x + 5)$ using logarithmic differentiation.

Solution   Begin by taking the natural logarithm of both sides of the given equation:

$$\ln g(x) = \ln[(x^2 + 1)(x^3 - 3)(2x + 5)]$$
$$= \ln(x^2 + 1) + \ln(x^3 - 3) + \ln(2x + 5).$$

Now differentiate and solve for $g'(x)$:

$$\frac{d}{dx}(\ln g(x)) = \frac{g'(x)}{g(x)} = \frac{2x}{x^2 + 1} + \frac{3x^2}{x^3 - 3} + \frac{2}{2x + 5}$$

$$g'(x) = g(x)\left(\frac{2x}{x^2 + 1} + \frac{3x^2}{x^3 - 3} + \frac{2}{2x + 5}\right)$$

$$= (x^2 + 1)(x^3 - 3)(2x + 5)\left(\frac{2x}{x^2 + 1} + \frac{3x^2}{x^3 - 3} + \frac{2}{2x + 5}\right). \quad ◆$$

Let us now use logarithmic differentiation to finally establish the power rule:

$$\frac{d}{dx}(x^r) = rx^{r-1}.$$

Verification of the Power Rule   Let $f(x) = x^r$. Then

$$\ln f(x) = \ln x^r = r \ln x.$$

Differentiation of this equation yields

$$\frac{f'(x)}{f(x)} = r \cdot \frac{1}{x}$$

$$f'(x) = r \cdot \frac{1}{x} \cdot f(x) = r \cdot \frac{1}{x} \cdot x^r = rx^{r-1}.$$

Practice Problems 4.6

1. Differentiate $f(x) = \ln\left[\dfrac{e^x \sqrt{x}}{(x + 1)^6}\right]$.

2. Use logarithmic differentiation to differentiate

$$f(x) = (x + 1)^7 (x + 2)^8 (x + 3)^9.$$

▶ **Exercises 4.6**

Simplify the following expressions.

1. $\ln 5 + \ln x$

2. $\ln x^5 - \ln x^3$

3. $\frac{1}{2} \ln 9$

4. $3 \ln \frac{1}{2} + \ln 16$

5. $\ln 4 + \ln 6 - \ln 12$

6. $\ln 2 - \ln x + \ln 3$

7. $e^{2 \ln x}$

8. $\frac{3}{2} \ln 4 - 5 \ln 2$

9. $5 \ln x - \frac{1}{2} \ln y + 3 \ln z$

10. $e^{\ln x^2 + 3\ln y}$

11. $\ln x - \ln x^2 + \ln x^4$

12. $\frac{1}{2}\ln xy + \frac{3}{2}\ln\frac{x}{y}$

13. Which is larger, $2\ln 5$ or $3\ln 3$?

14. Which is larger, $\frac{1}{2}\ln 16$ or $\frac{1}{3}\ln 27$?

15. Which of the following is the same as $4\ln 2x$?

   (a) $\ln 8x$ (b) $8\ln x$

   (c) $\ln 8 + \ln x$ (d) $\ln 16x^4$

16. Which of the following is the same as $\ln(9x) - \ln(3x)$?

   (a) $\ln 6x$ (b) $\ln(9x)/\ln(3x)$

   (c) $6\cdot\ln(x)$ (d) $\ln 3$

17. Which of the following is the same as $\dfrac{\ln 8x^2}{\ln 2x}$?

   (a) $\ln 4x$ (b) $4x$

   (c) $\ln 8x^2 - \ln 2x$ (d) none of these

18. Which of the following is the same as $\ln 9x^2$?

   (a) $2\cdot\ln 9x$ (b) $3x\cdot\ln 3x$

   (c) $2\cdot\ln 3x$ (d) none of these

Differentiate.

19. $\ln[(x+5)(2x-1)(4-x)]$

20. $\ln[(x+1)(2x+1)(3x+1)]$

21. $\ln\left[(1+x)^2(2+x)^3(3+x)^4\right]$

22. $\ln\left[e^{2x}(x^3+1)(x^4+5x)\right]$

23. $\ln\left[\dfrac{e^{5x}(x+4)(3x-2)}{x+1}\right]$

24. $\ln\left[\dfrac{\sqrt{x}\,(x+1)^2(x+2)^3}{4x+1}\right]$

25. $\ln\left[\dfrac{(5x+1)(4x+1)(\ln x)}{\sqrt{2x+1}}\right]$

26. $\ln\left[\dfrac{x^5 e^{4x}\sqrt{3x+1}}{1-x^2}\right]$

27. $\ln(3x+1)\ln(5x+1)$

28. $(\ln 4x)(\ln 2x)$

Use logarithmic differentiation to differentiate the following functions.

29. $f(x) = (x+1)^4(4x-1)^2$

30. $f(x) = e^x(3x-4)^8$

31. $f(x) = \dfrac{(x+1)(2x+1)(3x+1)}{\sqrt{4x+1}}$

32. $f(x) = \dfrac{(x-2)^3(x-3)^4}{(x+4)^5}$

33. $f(x) = 2^x$ **34.** $f(x) = 10^x$

35. $f(x) = x^x$ **36.** $f(x) = x^{1/x}$

37. There are substantial empirical data to show that if x and y measure the sizes of two organs of a particular animal, then x and y are related by an *allometric equation* of the form

$$\ln y - k\ln x = \ln c,$$

where k and c are positive constants that depend only on the type of parts or organs that are measured, and are constant among animals belonging to the same species.* Solve this equation for y in terms of x, k, and c.

38. In the study of epidemics, one finds the equation

$$\ln(1-y) - \ln y = C - rt,$$

where y is the fraction of the population that has a specific disease at time t. Solve the equation for y in terms of t and the constants C and r.

39. Determine the values of h and k for which the graph of $y = he^{kx}$ passes through the points $(1,6)$ and $(4,48)$.

40. Find values of k and r for which the graph of $y = kx^r$ passes through the points $(2,3)$ and $(4,15)$.

Solutions to Practice Problems 4.6	**1.** Use the properties of the natural logarithm to express $f(x)$ as a sum of simple functions before differentiating.

$$f(x) = \ln\left[\frac{e^x\sqrt{x}}{(x+1)^6}\right]$$

$$f(x) = \ln e^x + \ln\sqrt{x} - \ln(x+1)^6$$
$$= x + \tfrac{1}{2}\ln x - 6\ln(x+1)$$

$$f'(x) = 1 + \frac{1}{2x} - \frac{6}{x+1}$$

---

*E. Batschelet, *Introduction to Mathematics for Life Scientists* (New York: Springer-Verlag, 1971), pp. 305–307.

Solutions to Practice Problems 4.6 (Continued)	**2.** $f(x) = (x+1)^7(x+2)^8(x+3)^9$

$$\ln f(x) = 7\ln(x+1) + 8\ln(x+2) + 9\ln(x+3)$$

Now we differentiate both sides of the equation.

$$\frac{f'(x)}{f(x)} = \frac{7}{x+1} + \frac{8}{x+2} + \frac{9}{x+3}$$

$$f'(x) = f(x)\left(\frac{7}{x+1} + \frac{8}{x+2} + \frac{9}{x+3}\right)$$

$$= (x+1)^7(x+2)^8(x+3)^9\left(\frac{7}{x+1} + \frac{8}{x+2} + \frac{9}{x+3}\right)$$

Review of Fundamental Concepts of Chapter 4

1. State as many laws of exponents as you can recall.
2. What is e?
3. Write the differential equation satisfied by $y = Ce^{kt}$.
4. State the properties that graphs of the form $y = e^{kx}$ have in common when k is positive. When k is negative.
5. What are the coordinates of the reflection of the point (a, b) in the line $y = x$?
6. What is a logarithm?
7. What is the x-intercept of the natural logarithm function?
8. State the main features of the graph of $y = \ln x$.
9. State the two key equations giving the relationships between e^x and $\ln x$. (*Hint*: The right side of each equation is just x.)

10. What is the difference between a natural logarithm and a common logarithm?
11. Give the formula that converts a function of the form b^x to an exponential function with base e.
12. State the differentiation formula for each of the following functions:
    (a) $f(x) = e^{kx}$ (b) $f(x) = e^{g(x)}$
    (c) $f(x) = \ln g(x)$
13. State the four algebraic properties of the natural logarithm function.
14. Give an example of the use of logarithmic differentiation.

◗ Chapter 4 Supplementary Exercises

Calculate the following.

1. $27^{4/3}$
2. $4^{1.5}$
3. 5^{-2}
4. $16^{-.25}$
5. $(2^{5/7})^{14/5}$
6. $8^{1/2} \cdot 2^{1/2}$
7. $\dfrac{9^{5/2}}{9^{3/2}}$
8. $4^{.2} \cdot 4^{.3}$

Simplify the following.

9. $\left(e^{x^2}\right)^3$
10. $e^{5x} \cdot e^{2x}$
11. $\dfrac{e^{3x}}{e^x}$
12. $2^x \cdot 3^x$
13. $(e^{8x} + 7e^{-2x})e^{3x}$
14. $\dfrac{e^{5x/2} - e^{3x}}{\sqrt{e^x}}$

Solve the following equations for x.

15. $e^{-3x} = e^{-12}$
16. $e^{x^2 - x} = e^2$
17. $(e^x \cdot e^2)^3 = e^{-9}$
18. $e^{-5x} \cdot e^4 = e$

Differentiate the following functions.

19. $10e^{7x}$
20. $e^{\sqrt{x}}$
21. xe^{x^2}
22. $\dfrac{e^x + 1}{x - 1}$
23. e^{e^x}
24. $(\sqrt{x} + 1)e^{-2x}$
25. $\dfrac{x^2 - x + 5}{e^{3x} + 3}$
26. x^e
27. Determine all solutions of the differential equation $y' = -y$.

28. Determine all functions $y = f(x)$ such that $y' = -1.5y$ and $f(0) = 2000$.

29. Determine all solutions of the differential equation $y' = 1.5y$ and $f(0) = 2$.

30. Determine all solutions of the differential equation $y' = \frac{1}{3}y$.

Graph the following functions.

31. $e^{-x} + x$

32. $e^x - x$

33. $e^{-(1/2)x^2}$

34. $100(x-2)e^{-x}$ for $x \geq 2$

Simplify the following expressions.

35. $\dfrac{\ln x^2}{\ln x^3}$

36. $e^{2\ln 2}$

37. $e^{-5\ln 1}$

38. $\left[e^{\ln x}\right]^2$

39. $e^{(\ln 5)/2}$

40. $e^{\ln(x^2)}$

Solve the following equations for t.

41. $3e^{2t} = 15$

42. $3e^{t/2} - 12 = 0$

43. $2\ln t = 5$

44. $2e^{-.3t} = 1$

45. $t^{\ln t} = e$

46. $\ln(\ln 3t) = 0$

Differentiate the following functions.

47. $\ln(5x - 7)$

48. $\ln(9x)$

49. $(\ln x)^2$

50. $(x \ln x)^3$

51. $\ln(x^6 + 3x^4 + 1)$

52. $\dfrac{x}{\ln x}$

53. $\ln\left(\dfrac{xe^x}{\sqrt{1+x}}\right)$

54. $\ln\left[e^{6x}(x^2 + 3)^5(x^3 + 1)^{-4}\right]$

55. $\ln(\ln\sqrt{x})$

56. $\dfrac{1}{\ln x}$

57. $x\ln x - x$

58. $e^{2\ln(x+1)}$

59. $e^x \ln x$

60. $\ln(x^2 + e^x)$

Use logarithmic differentiation to differentiate the following functions.

61. $f(x) = (x^2 + 5)^6(x^3 + 7)^8(x^4 + 9)^{10}$

62. $f(x) = x^{1+x}$

63. $f(x) = 10^x$

64. $f(x) = \sqrt{x^2 + 5}\, e^{x^2}$

Graph the following functions.

65. $y = x - \ln x$

66. $y = \ln(x^2 + 1)$

67. $y = (\ln x)^2$

68. $y = (\ln x)^3$

Chapter Project

The following table gives the value of the Standard & Poor's 500 Stock Index (S&P Average) at the end of each year 1980–1997.

Table 1	End-of-Year Price of Standard & Poor's Stock Average
End of year	**S&P Stock Index (normalized so that 1941–43 = 100)**
1980	118.8
1981	128.0
1982	119.7
1983	160.4
1984	160.5
1985	186.8
1986	236.3
1987	286.8
1988	265.8
1989	322.8
1990	334.6
1991	376.2
1992	415.7
1993	451.4
1994	460.3
1995	541.6
1996	670.8
1997	872.7

Let t measure time, with $t = 0$ corresponding to 1980. Let $S(t)$ denote the value of the S&P Average at time t.

(a) Determine an exponential model approximating $S(t)$.

    (i) Assume that $S(t) = Ae^{kt}$ for constants A and k. First show that

$$\frac{\Delta S}{S} = \frac{Ae^{k(t+\Delta t)} - Ae^{kt}}{Ae^{kt}} = e^{k\Delta t} - 1.$$

    (ii) For each pair of consecutive data points in the table, determine the value of $\frac{\Delta S}{S}$. Take the average c of all these values.

    (iii) Solve the equation $c = e^{k\Delta t} - 1$ for k.

    (iv) The approximate midpoint of the data in the table corresponds to 1988 (when $t = 8$). Set $t_0 = 8$ and $S_0 = 265.8$ and solve the equation for A.

(b) Plot the given table of data and your exponential model on the same coordinate system. How well does your model fit the data?

Chapter Project   (Continued)

(c) According to your model, what values for the S&P Average are predicted for Dec. 31, 1998 and Dec. 31, 1999?

(d) Compare the projections of part (c) with the actual values of the S&P average on Dec. 31, 1998 and Dec. 31, 1999. (You will need to look up values of the S&P average. You may use the Internet to locate these values.)

(e) Critique your model. Does it give an accurate picture of the function $S(t)$ for the current year?

CHAPTER

Applications of the Exponential and Natural Logarithm Functions

5

E arlier, we introduced the exponential function e^x and the natural logarithm function $\ln x$ and studied their most important properties. From the way we introduced these functions, it is by no means clear that they have any substantial connection with the physical world. However, as this chapter will demonstrate, the exponential and natural logarithm functions are involved in the study of many physical problems, often in a very curious and unexpected way.

Here the most significant fact that we require is that the exponential function is uniquely characterized by its differential equation. In other words, we will constantly make use of the following fact, stated previously.

The function* $y = Ce^{kt}$ satisfies the differential equation

$$y' = ky.$$

Conversely, if $y = f(t)$ satisfies the differential equation, then $y = Ce^{kt}$ for some constant C.

If $f(t) = Ce^{kt}$, then, by setting $t = 0$, we have

$$f(0) = Ce^0 = C.$$

Therefore, C is the value of $f(t)$ at $t = 0$.

5.1 Exponential Growth and Decay

In biology, chemistry, and economics it is often necessary to study the behavior of a quantity that is increasing as time passes. If, at every instant, the rate of increase of the quantity is proportional to the quantity at that instant, then we say that the quantity is *growing exponentially* or is *exhibiting exponential growth*.

---

*Note that we use the independent variable t instead of x throughout this chapter. The reason is that, in most applications, the variable of our exponential function is time.

A simple example of exponential growth is exhibited by the growth of bacteria in a culture. Under ideal laboratory conditions a bacteria culture grows at a rate proportional to the number of bacteria present. It does so because the growth of the culture is accounted for by the division of the bacteria. The more bacteria there are at a given instant, the greater the possibilities for division and hence the greater the rate of growth.

Let us study the growth of a bacteria culture as a typical example of exponential growth. Suppose that $P(t)$ denotes the number of bacteria in a certain culture at time t. The rate of growth of the culture at time t is $P'(t)$. We assume that this rate of growth is proportional to the size of the culture at time t, so that

$$P'(t) = kP(t), \tag{1}$$

where k is a positive constant of proportionality. If we let $y = P(t)$, then (1) can be written as

$$y' = ky.$$

Therefore, from our discussion at the beginning of this chapter we see that

$$y = P(t) = P_0 e^{kt}, \tag{2}$$

where P_0 is the number of bacteria in the culture at time $t = 0$. The number k is called the *growth constant*.

▶ Example 1

Suppose that a certain bacteria culture grows at a rate proportional to its size. At time $t = 0$, approximately 20,000 bacteria are present. In 5 hours there are 400,000 bacteria. Determine a function that expresses the size of the culture as a function of time, measured in hours.

Solution

Let $P(t)$ be the number of bacteria present at time t. By assumption, $P(t)$ satisfies a differential equation of the form $y' = ky$, so $P(t)$ has the form

$$P(t) = P_0 e^{kt},$$

where the constants P_0 and k must be determined. The value of P_0 and k can be obtained from the data that give the population size at two different times. We are told that

$$P(0) = 20{,}000, \qquad P(5) = 400{,}000. \tag{3}$$

The first condition immediately implies that $P_0 = 20{,}000$, so

$$P(t) = 20{,}000 e^{kt}.$$

Using the second condition in (3), we have

$$20{,}000 e^{k(5)} = P(5) = 400{,}000$$
$$e^{5k} = 20$$
$$5k = \ln 20$$
$$k = \frac{\ln 20}{5} \approx .60. \tag{4}$$

So we may take

$$P(t) = 20{,}000 e^{.6t}.$$

This function is a mathematical model of the growth of the bacteria culture. (See Fig. 1.)

◆

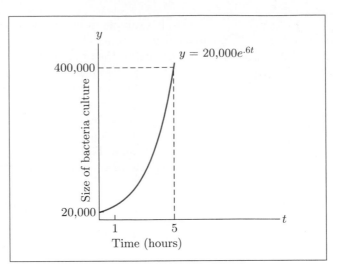

Figure 1. A model for bacteria growth.

▸ **Example 2** Suppose that a colony of fruit flies is growing according to the exponential law $P(t) = P_0 e^{kt}$, and suppose that the size of the colony doubles in 9 days. Determine the growth constant k.

Solution We do not know the initial size of the population at $t = 0$. However, we are told that $P(9) = 2P(0)$; that is,

$$P_0 e^{k(9)} = 2P_0$$
$$e^{9k} = 2$$
$$9k = \ln 2$$
$$k = \frac{\ln 2}{9} \approx .077.$$ ◆

Notice that the initial size P_0 of the population was not given in Example 2. We were able to determine the growth constant because we were told the amount of time required for the colony to double in size. Thus the growth constant does not depend on the initial size of the population. This property is characteristic of exponential growth.

▸ **Example 3** Suppose that the initial size of the colony in Example 2 was 100.

(a) How large will the colony be after 41 days?

(b) How fast will the colony be growing at that time?

(c) At what time will the colony contain 800 fruit flies?

Solution (a) From Example 2 we have $P(t) = P_0 e^{.077t}$. Since $P(0) = 100$, we conclude that

$$P(t) = 100 e^{.077t}.$$

Therefore, after 41 days the size of the colony is

$$P(41) = 100 e^{.077(41)} = 100 e^{3.157} = 2350.$$

(b) Since the function satisfies the differential equation $y' = .077y$,

$$P'(t) = .077 P(t).$$

In particular, when $t = 41$,

$$P'(41) = .077P(41) = .077 \cdot 2350 \approx 181.$$

Therefore, after 41 days the colony is growing at the rate of about 181 fruit flies per day.

(c) Use the formula for $P(t)$ and set $P(t) = 800$.

$$100e^{.077t} = 800$$
$$e^{.077t} = 8$$
$$.077t = \ln 8$$
$$t = \frac{\ln 8}{.077} \approx 27 \text{ days.}$$ ◆

Exponential Decay An example of negative exponential growth, or *exponential decay*, is given by the disintegration of a radioactive element such as uranium 235. It is known that, at any instant, the rate at which a radioactive substance is decaying is proportional to the amount of the substance that has not yet disintegrated. If $P(t)$ is the quantity present at time t, then $P'(t)$ is the rate of decay. Of course, $P'(t)$ must be negative, since $P(t)$ is decreasing. Thus we may write $P'(t) = kP(t)$ for some negative constant k. To emphasize the fact that the constant is negative, k is often replaced by $-\lambda$, where λ is a positive constant.[*] Then $P(t)$ satisfies the differential equation

$$P'(t) = -\lambda P(t). \tag{5}$$

The general solution of (5) has the form

$$P(t) = P_0 e^{-\lambda t}$$

for some positive number P_0. We call such a function an *exponential decay function*. The constant λ is called the *decay constant*.

▶ **Example 4** The decay constant for strontium 90 is $\lambda = .0244$, where time is measured in years. How long will it take for a quantity P_0 of strontium 90 to decay to one-half its original mass?

Solution We have

$$P(t) = P_0 e^{-.0244t}.$$

Next, set $P(t)$ equal to $\frac{1}{2}P_0$ and solve for t:

$$P_0 e^{-.0244t} = \tfrac{1}{2}P_0$$
$$e^{-.0244t} = \tfrac{1}{2} = .5$$
$$-.0244t = \ln .5$$
$$t = \frac{\ln .5}{-.0244} \approx 28 \text{ years.}$$ ◆

The *half-life* of a radioactive element is the length of time required for a given quantity of that element to decay to one-half its original mass. Thus strontium 90 has a half-life of about 28 years. This means that it takes 28 years for it to decay to half its original mass and another 28 years for it to decay to $\frac{1}{4}$ its original mass, another 28 years to decay to $\frac{1}{8}$, and so forth. (See Fig. 2.) Notice from Example 4 that the half-life does not depend on the initial amount P_0.

_____

[*]λ is the Greek lowercase letter lambda.

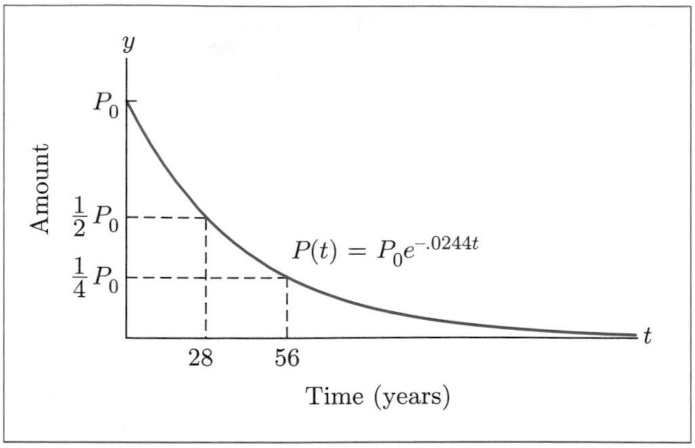

Figure 2. Half-life of radioactive strontium 90.

▶ **Example 5** Radioactive carbon 14 has a half-life of about 5730 years. Find its decay constant.

Solution If P_0 denotes the initial amount of carbon 14, then the amount after t years will be

$$P(t) = P_0 e^{-\lambda t}.$$

After 5730 years, $P(t)$ will equal $\frac{1}{2}P_0$. That is,

$$P_0 e^{-\lambda(5730)} = P(5730) = \tfrac{1}{2}P_0 = .5P_0.$$

Solving for λ gives

$$e^{-5730\lambda} = .5$$
$$-5730\lambda = \ln .5$$
$$\lambda = \frac{\ln .5}{-5730} \approx .00012. \qquad \blacklozenge$$

One of the problems connected with aboveground nuclear explosions is the radioactive debris ("fallout") that falls on plants and grass, thereby contaminating the food supply of animals. Strontium 90 is one of the most dangerous components of radioactive debris because it has a relatively long half-life and because it is chemically similar to calcium and is absorbed into the bone structure of animals (including humans) who eat contaminated food. Iodine 131 is also produced by nuclear explosions, but it presents less of a hazard because it has a half-life of 8 days.

▶ **Example 6** If dairy cows eat hay containing too much iodine 131, their milk will be unfit to drink. Suppose that some hay contains 10 times the maximum allowable level of iodine 131. How many days should the hay be stored before it is fed to dairy cows?

Solution Let P_0 be the amount of iodine 131 present in the hay. Then the amount at time t is $P(t) = P_0 e^{-\lambda t}$ (t in days). The half-life of iodine 131 is 8 days, so

$$P_0 e^{-\lambda(8)} = .5P_0$$
$$e^{-8\lambda} = .5$$
$$-8\lambda = \ln .5$$
$$\lambda = \frac{\ln .5}{-8} \approx .087$$

and

$$P(t) = P_0 e^{-.087t}.$$

Now that we have the formula for $P(t)$, we want to find t such that $P(t) = \frac{1}{10}P_0$. We have

$$P_0 e^{-.087t} = .1 P_0,$$

so

$$e^{-.087t} = .1$$
$$-.087t = \ln .1$$
$$t = \frac{\ln .1}{-.087} \approx 26 \text{ days.}$$

◆

Radiocarbon Dating Knowledge about radioactive decay is valuable to archaeologists and anthropologists who want to estimate the age of objects belonging to ancient civilizations. Several different substances are useful for radioactive-dating techniques; the most common is radiocarbon, ^{14}C. Carbon 14 is produced in the upper atmosphere when cosmic rays react with atmospheric nitrogen. Because the ^{14}C eventually decays, the concentration of ^{14}C cannot rise above certain levels. An equilibrium is reached where ^{14}C is produced at the same rate as it decays. Scientists usually assume that the total amount of ^{14}C in the biosphere has remained constant over the past 50,000 years. Consequently, it is assumed that the *ratio* of ^{14}C to ordinary nonradioactive carbon 12, ^{12}C, has been constant during this same period. (The ratio is about one part ^{14}C to 10^{12} parts of ^{12}C.) Both ^{14}C and ^{12}C are in the atmosphere as constituents of carbon dioxide. All living vegetation and most forms of animal life contain ^{14}C and ^{12}C in the same proportion as the atmosphere. The reason is that plants absorb carbon dioxide through photosynthesis. The ^{14}C and ^{12}C in plants are distributed through the food chain to almost all animal life.

When an organism dies, it stops replacing its carbon, and therefore the amount of ^{14}C begins to decrease through radioactive decay. (The ^{12}C in the dead organism remains constant.) The ratio of ^{14}C to ^{12}C can be later measured in order to determine when the organism died. (See Fig. 3.)

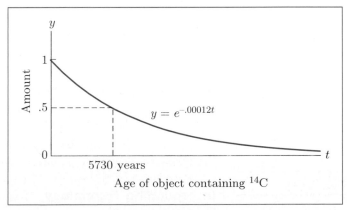

Figure 3. ^{14}C -^{12}C ratio compared to the ratio in living plants.

▶ Example 7   A parchment fragment was discovered that had about 80% of the ^{14}C level found today in living matter. Estimate the age of the parchment.

Solution   We assume that the original ^{14}C level in the parchment was the same as the level in living organisms today. Consequently, about eight-tenths of the original ^{14}C remains. From Example 5 we obtain the formula for the amount of ^{14}C present t years after the parchment was made from an animal skin:

$$P(t) = P_0 e^{-.00012t},$$

where P_0 = initial amount. We want to find t such that $P(t) = .8P_0$.

$$P_0 e^{-.00012t} = .8P_0$$
$$e^{-.00012t} = .8$$
$$-.00012t = \ln .8$$
$$t = \frac{\ln .8}{-.00012} \approx 1860 \text{ years old.} \qquad \blacklozenge$$

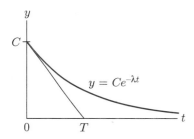

Figure 4. Exponential decay of sales.

A Sales Decay Curve   Marketing studies[*] have demonstrated that if advertising and other promotions of a particular product are stopped and if other market conditions remain fairly constant, then, at any time t, the sales of that product will be declining at a rate proportional to the amount of current sales at time t. (See Fig. 4.) If S_0 is the number of sales in the last month during which advertising occurred and if $S(t)$ is the monthly level of sales t months after the cessation of promotional effort, then a good mathematical model for $S(t)$ is

$$S(t) = S_0 e^{-\lambda t},$$

where λ is a positive number called the *sales decay constant*. The value of λ depends on many factors, such as the type of product, the number of years of prior advertising, the number of competing products, and other characteristics of the market.

Figure 5. The time constant T in exponential decay: $T = 1/\lambda$.

The Time Constant   Consider an exponential decay function $y = Ce^{-\lambda t}$. In Fig. 5, we have drawn the tangent line to the decay curve when $t = 0$. The slope there is the initial rate of decay. If the decay process were to continue at this rate, the decay curve would follow the tangent line and y would be zero at some time T. This time is called the *time constant* of the decay curve. It can be shown (see Exercise 30) that $T = 1/\lambda$ for the curve $y = Ce^{-\lambda t}$. Thus $\lambda = 1/T$, and the decay curve can be written in the form

$$y = Ce^{-t/T}.$$

If one has experimental data that tend to lie along an exponential decay curve, then the numerical constants for the curve may be obtained from Fig. 5. First, sketch the curve and estimate the y-intercept, C. Then sketch an approximate tangent line and from this estimate the time constant, T. This procedure is sometimes used in biology and medicine.

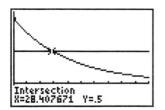

Figure 6. $Y_1 = .5$
$Y_2 = e^{\wedge}(-.0244X)$

Incorporating Technology   Exponential equations also can be solved graphically with a graphing calculator. Figure 6 shows the solution to the equation $e^{-.0244x} = .5$ from Example 4.

---

[*]M. Vidale and H. Wolfe, "An Operations-Research Study of Sales Response to Advertising," *Operations Research*, 5 (1957), 370–381. Reprinted in F. Bass et al., *Mathematical Models and Methods in Marketing* (Homewood, Ill.: Richard D. Irwin, Inc., 1961).

<table>
<tr><td>

**Practice Problems
5.1**

</td><td>

1. (a) Solve the differential equation $P'(t) = -.6P(t)$, $P(0) = 50$.

(b) Solve the differential equation $P'(t) = kP(t)$, $P(0) = 4000$, where k is some constant.

(c) Find the value of k in part (b) for which $P(2) = 100P(0)$.

2. Under ideal conditions a colony of *Escherichia coli* bacteria can grow by a factor of 100 every 2 hours. If initially 4000 bacteria are present, how long will it take before there are 1,000,000 bacteria?

</td></tr>
</table>

▶ Exercises 5.1

1. Let $P(t)$ be the population (in millions) of a certain city t years after 1990, and suppose that $P(t)$ satisfies the differential equation

$$P'(t) = .02P(t), \qquad P(0) = 3.$$

(a) Find the formula for $P(t)$.

(b) What was the initial population, that is, the population in 1990?

(c) What is the growth constant?

(d) What was the population in 1998?

(e) Use the differential equation to determine how fast the population is growing when it reaches 4 million people.

(f) How large is the population when it is growing at the rate of 70,000 people per year?

2. Approximately 10,000 bacteria are placed in a culture. Let $P(t)$ be the number of bacteria present in the culture after t hours, and suppose that $P(t)$ satisfies the differential equation

$$P'(t) = .55P(t).$$

(a) Find the formula for $P(t)$.

(b) What is $P(0)$?

(c) How many bacteria are there after 5 hours?

(d) What is the growth constant?

(e) Use the differential equation to determine how fast the bacteria culture is growing when it reaches 100,000.

(f) What is the size of the bacteria culture when it is growing at the rate of 34,000 bacteria per hour?

3. Suppose that after t hours there are $P(t)$ cells present in a culture, where $P(t) = 5000e^{.2t}$.

(a) How many cells were present initially?

(b) Give a differential equation satisfied by $P(t)$.

(c) When will the population double?

(d) When will 20,000 cells be present?

4. The size of a certain insect population is given by $P(t) = 300e^{.01t}$, where t is measured in days.

(a) How many insects were present initially?

(b) Give a differential equation satisfied by $P(t)$.

(c) At what time will the population double?

(d) At what time will the population equal 1200?

5. Determine the growth constant of a population that is growing at a rate proportional to its size, where the population doubles in size every 40 days.

6. Determine the growth constant of a population that is growing at a rate proportional to its size, where the population triples in size every 10 years.

7. Suppose a population is growing exponentially with growth constant .05. In how many years will the current population triple?

8. Suppose a population is growing exponentially with growth constant .04. In how many years will the current population double?

9. The rate of growth of a certain cell culture is proportional to its size. In 10 hours a population of 1 million cells grew to 9 million. How large will the cell culture be after 15 hours?

10. The world's population was 5.51 billion on January 1, 1993 and 5.88 billion on January 1, 1998. Assume that at any time the population grows at a rate proportional to the population at that time. In what year will the world's population reach 7 billion?

11. At the beginning of 1990, 20.2 million people lived in the metropolitan area of Mexico City, and the population was growing exponentially. The 1995 population was 23 million. (Part of the growth is due to immigration.) If this trend continues, how large will the population be in the year 2010?

12. The population (in millions) of a state t years after 1970 is given by the graph of the exponential function $y = P(t)$ with growth constant .025 in Fig. 7. [In parts (c) and (d) use the differential equation satisfied by $P(t)$.]

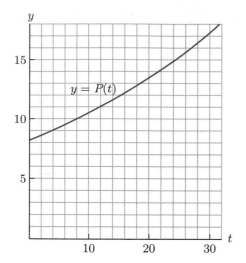

Figure 7.

(a) What was the population in 1974?

(b) When was the population 10 million?

(c) How fast was the population growing in 1974?

(d) When is the population growing at the rate of 275,000 people per year?

13. A sample of 8 grams of radioactive material is placed in a vault. Let $P(t)$ be the amount remaining after t years, and suppose that $P(t)$ satisfies the differential equation $P'(t) = -.021P(t)$.

(a) Find the formula for $P(t)$.

(b) What is $P(0)$?

(c) What is the decay constant?

(d) How much of the material will remain after 10 years?

(e) Use the differential equation to determine how fast the sample is disintegrating when just one gram remains.

(f) What amount of radioactive material remains when it is disintegrating at the rate of .105 grams per year?

(g) The radioactive material has a half-life of 33 years. How much will remain after 33 years? 66 years? 99 years?

14. Radium-226 is used in cancer radiotherapy, as a neutron source for some research purposes, and as a constituent of luminescent paints. Let $P(t)$ be the number of grams of radium-226 in a sample remaining after t years, and suppose that $P(t)$ satisfies the differential equation

$$P'(t) = -.00043P(t), \qquad P(0) = 12.$$

(a) Find the formula for $P(t)$.

(b) What was the initial amount?

(c) What is the decay constant?

(d) Approximately how much of the radium will remain after 943 years?

(e) How fast is the sample disintegrating when just one gram remains? Use the differential equation.

(f) What is the weight of the sample when it is disintegrating at the rate of .004 grams per year?

(g) The radioactive material has a half-life of about 1612 years. How much will remain after 1612 years? 3224 years? 4836 years?

15. Suppose a person is given an injection of 300 milligrams of penicillin at time $t = 0$, and let $f(t)$ be the amount (in milligrams) of penicillin present in the person's bloodstream t hours after the injection. Then the amount of penicillin decays exponentially and a typical formula is $f(t) = 300e^{-.6t}$.

(a) Give the differential equation satisfied by $f(t)$.

(b) How much will remain at time $t = 5$ hours?

(c) What is the biological half-life of the penicillin (that is, the time required for half of a given amount to decompose) in this case?

16. Ten grams of a radioactive substance with decay constant .04 is stored in a vault. Assume that time is measured in days and let $P(t)$ be the amount remaining at time t.

(a) Give the formula for $P(t)$.

(b) Give the differential equation satisfied by $P(t)$.

(c) How much will remain after 5 days?

(d) What is the half-life of this radioactive substance?

17. The decay constant for the radioactive element cesium-137 is .023 when time is measured in years. Find the half-life of cesium-137.

18. Radioactive cobalt-60 has a half-life of 5.3 years. Find the decay constant of cobalt-60.

19. A sample of radioactive material disintegrates from 5 grams to 2 grams in 100 days. After how many days will just 1 gram remain?

20. Ten grams of a radioactive material disintegrates to 3 grams in 5 years. What is the half-life of the radioactive material?

21. In an animal hospital, 8 units of a sulfate were injected into a dog. After 50 minutes only 4 units remained in the dog. Let $f(t)$ be the amount of sulfate present after t minutes. At any time, the rate of change of $f(t)$ is proportional to the value of $f(t)$. Find the formula for $f(t)$.

22. Five grams of a certain radioactive material decays to 3 grams in 1 year. After how many years will just 1 gram remain?

23. A sample of radioactive material decays over time (measured in hours) with decay constant .2. The graph of the exponential function $y = P(t)$ in Fig. 8 gives the number of grams remaining after t hours. [*Hint*: In parts (c) and (d) use the differential equation satisfied by $P(t)$.]

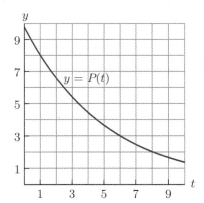

Figure 8.

(a) How much was remaining after 1 hour?

(b) Approximate the half-life of the material.

(c) How fast was the sample decaying after 6 hours?

(d) When was the population decaying at the rate of .4 grams per hour?

24. A sample of radioactive material has decay constant .25, where time is measured in hours. How fast will the sample be disintegrating when the sample size is 8 grams? For what sample size will the sample size be decreasing at the rate of 2 grams per day?

25. In 1947, a cave with beautiful prehistoric wall paintings was discovered in Lascaux, France. Some charcoal found in the cave contained 20% of the ^{14}C expected in living trees. How old are the Lascaux cave paintings? (Recall that the decay constant for ^{14}C is .00012.)

26. According to legend, in the fifth century King Arthur and his knights sat at a huge round table. A round table alleged to have belonged to King Arthur was found at Winchester Castle in England. In 1976 carbon dating revealed the amount of radiocarbon in the table to be 91% of the radiocarbon present in living wood. Could the table possibly have belonged to King Arthur? Why? (Recall that the decay constant for ^{14}C is .00012.)

27. A 4500-year-old wooden chest was found in the tomb of the twenty-fifth century B.C. Chaldean king Meskalumdug of Ur. What percentage of the original ^{14}C would you expect to find in the wooden chest?

28. In 1938, sandals woven from strands of tree bark were found in Fort Rock Creek Cave in Oregon. The bark contained 34% of the level of ^{14}C found in living bark. Approximately how old were the sandals? (*Note*: This discovery by University of Oregon anthropologist

Luther Cressman forced scientists to double their estimate of how long ago people came to the Pacific Northwest.)

29. Many scientists believe there have been four ice ages in the past 1 million years. Before the technique of carbon dating was known, geologists erroneously believed that the retreat of the Fourth Ice Age began about 25,000 years ago. In 1950, logs from ancient spruce trees were found under glacial debris near Two Creeks, Wisconsin. Geologists determined that these trees had been crushed by the advance of ice during the Fourth Ice Age. Wood from the spruce trees contained 27% of the level of ^{14}C found in living trees. Approximately how long ago did the Fourth Ice Age actually occur?

30. Let T be the time constant of the curve $y = Ce^{-\lambda t}$ as defined in Fig. 5. Show that $T = 1/\lambda$. (*Hint*: Express the slope of the tangent line in Fig. 5 in terms of C and T. Then set this slope equal to the slope of the curve $y = Ce^{-\lambda t}$ at $t = 0$.)

31. The amount (in grams) of a certain radioactive material present after t years is given by the function $P(t)$. Match each of the following answers with its corresponding question.

Answers:

   a. Solve $P(t) = .5P(0)$ for t.

   b. Solve $P(t) = .5$ for t.

   c. $P(.5)$

   d. $P'(.5)$

   e. $P(0)$

   f. Solve $P'(t) = -.5$ for t.

   g. $y' = ky$

   h. $P_0 e^{kt}$, $k < 0$

Questions:

   A. Give a differential equation satisfied by $P(t)$.

   B. How fast will the radioactive material be disintegrating in $\frac{1}{2}$ year?

   C. Give the general form of the function $P(t)$.

   D. Find the half-life of the radioactive material.

   E. How many grams of the material will remain after $\frac{1}{2}$ year?

   F. When will the radioactive material be disintegrating at the rate of $\frac{1}{2}$ gram per year?

   G. When will there be $\frac{1}{2}$ gram remaining?

   H. How much radioactive material was present initially?

Technology Exercises

32. Figure 9 shows the point of intersection of a horizontal line and the decay curve for one gram of plutonium 237. Make up a question that can be answered by Fig. 9.

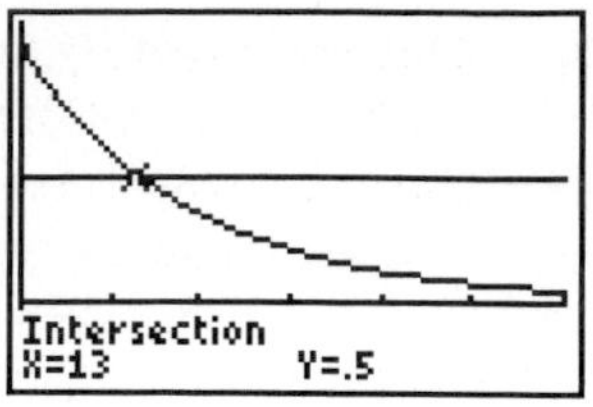

[Figure 9 graph]

Figure 9.

initially has 100 individuals. Make up a question that can be answered by Fig. 10.

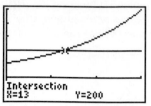

Figure 10.

33. Figure 10 shows the point of intersection of a horizontal line and the growth curve for a population that

---

Solutions to Practice Problems 5.1

1. **(a)** Answer: $P(t) = 50e^{-.6t}$. Differential equations of the type $y' = ky$ have as their solution $P(t) = Ce^{kt}$, where C is $P(0)$.

   (b) Answer: $P(t) = 4000e^{kt}$. This problem is like the previous one except that the constant is not specified. Additional information is needed if one wants to determine a specific value for k.

   (c) Answer: $P(t) = 4000e^{2.3t}$. From the solution to part (b) we know that $P(t) = 4000e^{kt}$. We are given that $P(2) = 100P(0) = 100(4000) = 400,000$. So

   $$P(2) = 4000e^{k(2)} = 400,000$$
   $$e^{2k} = 100$$
   $$2k = \ln 100$$
   $$k = \frac{\ln 100}{2} \approx 2.3.$$

2. Let $P(t)$ be the number of bacteria present after t hours. We must first find an expression for $P(t)$ and then determine the value of t for which $P(t) = 1,000,000$. From the discussion at the beginning of the section we know that $P'(t) = k \cdot P(t)$. Also, we are given that $P(2)$ (the population after 2 hours) is $100P(0)$ (100 times the initial population). From Problem 1(c) we have an expression for $P(t)$:

   $$P(t) = 4000e^{2.3t}.$$

   Now we must solve $P(t) = 1,000,000$ for t.

   $$4000e^{2.3t} = 1,000,000$$
   $$e^{2.3t} = 250$$
   $$2.3t = \ln 250$$
   $$t = \frac{\ln 250}{2.3} \approx 2.4.$$

   Therefore, after 2.4 hours there will be 1,000,000 bacteria.

5.2 Compound Interest

In Section 0.5, we introduced the notion of compound interest. Recall the fundamental formula developed there. If a principal amount P is compounded m times per year at an annual rate of interest r for t years, then the compound amount A, the balance at the end of time t, is given by the formula

$$A = P\left(1 + \frac{r}{m}\right)^{mt}.$$

For instance, suppose $1000 is invested at 6% interest for one year. In the formula this corresponds to $P = \$1000$, $m = 1$, $r = .06$, and $t = 1$ year. If interest is compounded annually, the amount in the account at the end of the year is

$$A = P(1 + r)^1 = 1000(1 + .06) = \$1060.00.$$

If interest is compounded quarterly, the amount in the account at the end of the year is

$$A = P\left(1 + \frac{r}{4}\right)^{4t} = 1000\left(1 + \frac{.06}{4}\right)^4 = \$1061.36.$$

Table 1 contains these results along with those for monthly and daily compounding for one year.

Table 1	Effect of Increased Compounding Periods			
Frequency of compounding	Annually	Quarterly	Monthly	Daily
m	1	4	12	360
Balance after one year ($)	1060.00	1061.36	1061.68	1061.83

Would the account balance in Table 1 be much more if the interest were compounded every hour? Every minute? It turns out that no matter how frequently interest is compounded, the balance will not exceed 1061.84 (to the nearest cent). To understand why, we connect the concept of compound interest to the exponential function e^{rt}. Start from the formula for the compound amount and write it in the form

$$A = P\left(1 + \frac{r}{m}\right)^{mt} = P\left(1 + \frac{r}{m}\right)^{(m/r)\cdot rt}.$$

If we set $h = r/m$, then $1/h = m/r$, and

$$A = P(1 + h)^{(1/h)\cdot rt}.$$

As the frequency of compounding is increased, m gets large and $h = r/m$ approaches 0. To determine what happens to the compound amount, we must therefore examine the limit

$$\lim_{h\to 0} P(1 + h)^{(1/h)\cdot rt}.$$

The following remarkable fact is proved in the appendix at the end of this section:

$$\lim_{h\to 0}(1 + h)^{1/h} = e.$$

Using this fact together with two limit theorems, we have

$$\lim_{h\to 0} P(1 + h)^{(1/h)rt} = P\left[\lim_{h\to 0}(1 + h)^{1/h}\right]^{rt} = Pe^{rt}.$$

These calculations show that the compound amount calculated from formula

$$P\left(1 + \frac{r}{m}\right)^{mt}$$

gets closer to Pe^{rt} as the number m of interest periods per year is increased. When the formula

$$A = Pe^{rt} \tag{1}$$

is used to calculate the compound amount, we say that the interest is *compounded continuously.*

When $1000 is invested for one year at 6% interest compounded continuously, $P = 1000$, $r = .06$, and $t = 1$. The compound amount is

$$1000e^{.06(1)} \approx \$1061.84.$$

This is only one cent more than the result of the daily compounding shown in Table 1. Consequently, frequent compounding (such as every hour or every second) will produce at most one cent more.

In many computations, it is simpler to use the formula for interest compounded continuously than the formula for ordinary compound interest. In these instances, it is commonplace to use interest compounded continuously as an approximation to ordinary compound interest.

When interest is compounded continuously, the compound amount $A(t)$ is an exponential function of the number of years t that interest is earned, $A(t) = Pe^{rt}$. Hence $A(t)$ satisfies the differential equation

$$\frac{dA}{dt} = rA.$$

The rate of growth of the compound amount is proportional to the amount of money present. Since the growth comes from the interest, we conclude that under continuous compounding, interest is earned continuously at a rate of growth proportional to the amount of money present.

The formula $A = Pe^{rt}$ contains four variables. (Remember that the letter e represents a specific constant, $e = 2.718\ldots$.) In a typical problem, we are given values for three of these variables and must solve for the remaining variable.

▶ Example 1   One thousand dollars is invested at 5% interest compounded continuously.

   (a) Give the formula for $A(t)$, the compound amount after t years.
   (b) How much will be in the account after 6 years?
   (c) After 6 years, at what rate will $A(t)$ be growing?
   (d) How long is required for the initial investment to double?

Solution   (a) $P = 1000$ and $r = .05$. By formula (1), $A(t) = 1000e^{.05t}$.
   (b) $A(6) = 1000e^{.05(6)} = 1000e^{.3} = \$1349.86$.
   (c) Rate of growth is different from interest rate. Interest rate is fixed at 5% and does not change with time. On the other hand, the rate of growth $A'(t)$ is always changing. Since $A(t) = 1000e^{.05t}$, $A'(t) = 1000 \cdot .05e^{.05t} = 50e^{.05t}$.

$$A'(6) = 50e^{.05(6)} = 50e^{.3} = \$67.49.$$

After 6 years the investment is growing at the rate of $67.49 per year.
There is an easier way to answer part (c). Since $A(t)$ satisfies the differential equation $A'(t) = rA(t)$,

$$A'(6) = .05A(6) = .05 \cdot 1349.86 = \$67.49$$

(d) We must find t such that $A(t) = \$2000$. So we set $1000e^{.05t} = 2000$ and solve for t.

$$1000e^{.05t} = 2000$$

$$e^{.05t} = 2$$

$$\ln e^{.05t} = \ln 2$$

$$.05t = \ln 2$$

$$t = \frac{\ln 2}{.05} \approx 13.86 \text{ years} \qquad \blacklozenge$$

Remark The calculations in Example 1(d) would be essentially unchanged after the first step if the initial amount of the investment were changed from $1000 to any arbitrary amount P. When this investment doubles, the compound amount will be $2P$. So one sets $2P = Pe^{.05t}$ and solves for t as we did previously, to conclude that, at 5% interest compounded continuously, any amount doubles in about 13.86 years.

▶ **Example 2**   Pablo Picasso's "The Dream" was purchased in 1941 for a war-distressed price of $7000. The painting was sold in 1997 for $48.4 million, the second highest price ever paid for a Picasso painting at auction. What rate of interest compounded continuously did this investment earn?

Solution   Let Pe^{rt} be the value (in millions) of the painting t years after 1941. Since the initial value is .007 million, $P = .007$. Since the value after 56 years is 48.4 million dollars, $.007e^{r(56)} = 48.4$. Divide both sides of the equation by .007, take the logarithms of both sides, and then solve for r.

$$e^{r(56)} = \frac{48.4}{.007} = 6914.29$$

$$r(56) = \ln 6914.29$$

$$r = \frac{\ln 6914.29}{56} \approx .158$$

Therefore, as an investment, the painting earned an interest rate of 15.8%.   ◆

If P dollars are invested today, the formula $A = Pe^{rt}$ gives the value of this investment after t years (assuming continuously compounded interest). We say that P is the *present value* of the amount A to be received in t years. If we solve for P in terms of A, we obtain

$$P = Ae^{-rt}. \tag{2}$$

The concept of the present value of money is an important theoretical tool in business and economics. Problems involving depreciation of equipment, for example, may be analyzed by calculus techniques when the present value of money is computed from (2) using continuously compounded interest.

▶ **Example 3** Find the present value of $5000 to be received in 2 years if money can be invested at 12% compounded continuously.

Solution Use (2) with $A = 5000$, $r = .12$, and $t = 2$.

$$P = 5000e^{-(.12)(2)} = 5000e^{-.24}$$

$$\approx \$3933.14 \qquad \blacklozenge$$

APPENDIX A Limit Formula for e

For $h \neq 0$, we have

$$\ln(1+h)^{1/h} = (1/h)\ln(1+h).$$

Taking the exponential of both sides, we find that

$$(1+h)^{1/h} = e^{(1/h)\ln(1+h)}.$$

Since the exponential function is continuous,

$$\lim_{h \to 0}(1+h)^{1/h} = e^{\left[\lim_{h \to 0}(1/h)\ln(1+h)\right]}. \qquad (3)$$

To examine the limit inside the exponential function, we note that $\ln 1 = 0$, and hence

$$\lim_{h \to 0}\left(\frac{1}{h}\right)\ln(1+h) = \lim_{h \to 0}\frac{\ln(1+h) - \ln 1}{h}.$$

The limit on the right is a difference quotient of the type used to compute a derivative. In fact,

$$\lim_{h \to 0}\frac{\ln(1+h) - \ln 1}{h} = \left.\frac{d}{dx}\ln x\right|_{x=1} = \left.\frac{1}{x}\right|_{x=1} = 1.$$

Thus the limit inside the exponential function in (3) is 1. That is,

$$\lim_{h \to 0}(1+h)^{1/h} = e^{[1]} = e.$$

Practice Problems **5.2**	**1.** One thousand dollars is to be invested in a bank for 4 years. Would 8% interest compounded semiannually be better than $7\frac{3}{4}\%$ interest compounded continuously?
	2. A building was bought for $150,000 and sold 10 years later for $400,000. What interest rate (compounded continuously) was earned on the investment?

▶ Exercises 5.2

1. Let $A(t) = 5000e^{.04t}$ be the balance in a savings account after t years.

   (a) How much money was originally deposited?

   (b) What is the interest rate?

   (c) How much money will be in the account after 10 years?

   (d) What differential equation is satisfied by $y = A(t)$?

   (e) Use the results of (c) and (d) to determine how fast the balance is growing after 10 years.

   (f) How large will the balance be when it is growing at the rate of $280 per year?

2. Let $A(t)$ be the balance in a savings account after t years, and suppose that $A(t)$ satisfies the differential equation

   $$A'(t) = .045A(t), \qquad A(0) = 3000.$$

   (a) How much money was originally deposited into the account?

   (b) What interest rate is being earned?

   (c) Find the formula for $A(t)$.

   (d) What is the balance after 5 years?

   (e) Use part (d) and the differential equation to determine how fast the balance is growing after 5 years.

   (f) How large will the balance be when it is growing at the rate of $270 per year?

3. Four thousand dollars is deposited into a savings account at 3.5% interest compounded continuously.

   (a) What is the formula for $A(t)$, the balance after t years?

   (b) What differential equation is satisfied by $A(t)$, the balance after t years?

   (c) How much money will be in the account after 2 years?

   (d) When will the balance reach $5000?

   (e) How fast is the balance growing when it reaches $5000?

4. Ten thousand dollars is deposited into a savings account at 4.6% interest compounded continuously.

   (a) What differential equation is satisfied by $A(t)$, the balance after t years?

   (b) What is the formula for $A(t)$, the balance after t years?

   (c) How much money will be in the account after 3 years?

   (d) When will the balance triple?

   (e) How fast is the balance growing when it triples?

5. An investment earns 4.2% interest compounded continuously. How fast is the investment growing when its value is $9000?

6. An investment earns 5.1% interest compounded continuously and is currently growing at the rate of $765 per year. What is the current value of the investment?

7. One thousand dollars is deposited in a savings account at 6% interest compounded continuously. How many years are required for the balance in the account to reach $2500?

8. Ten thousand dollars is invested at 6.5% interest compounded continuously. When will the investment be worth $41,787?

9. One hundred shares of a technology stock were purchased on January 2, 1990 for $1200 and sold on January 2, 1998 for $12,500. What rate of interest compounded continuously did this investment earn?

10. Pablo Picasso's "Angel Fernandez de Soto" was acquired in 1946 for a postwar splurge of $22,220. The painting was sold in 1995 for $29.1 million. What rate of interest compounded continuously did this investment earn?

11. How many years are required for an investment to double in value if it is appreciating at the rate of 4% compounded continuously?

12. What interest rate (compounded continuously) is earned by an investment that doubles in 10 years?

13. If an investment triples in 15 years, what interest rate (compounded continuously) does the investment earn?

14. If real estate in a certain city appreciates at the rate of 15% compounded continuously, when will a building purchased in 1998 triple in value?

15. In a certain town, property values tripled from 1980 to 1995. If this trend continues, when will property values be at five times their 1980 level? (Use an exponential model for the property value at time t.)

16. A lot purchased in 1980 for $5000 was appraised at $60,000 in 1998. If the lot continues to appreciate at the same rate, when will it be worth $100,000?

17. A farm purchased in 1985 for $1,000,000 was valued at $3,000,000 in 1995. If the farm continues to appreciate at the same rate (with continuous compounding), when will it be worth $10,000,000?

18. A parcel of land bought in 1990 for $10,000 was worth $16,000 in 1995. If the land continues to appreciate at this rate, in what year will it be worth $45,000?

19. Find the present value of $1000 payable at the end of 3 years, if money may be invested at 8% with interest compounded continuously.

20. Find the present value of $2000 to be received in 10 years, if money may be invested at 8% with interest compounded continuously.

21. How much money must you invest now at 4.5% interest compounded continuously in order to have $10,000 at the end of 5 years?

22. Suppose that the present value of $1000 to be received in 5 years is $559.90. What rate of interest, compounded continuously, was used to compute this present value?

23. Investment A is currently worth $70,200 and is growing at the rate of 13% per year compounded continuously. Investment B is currently worth $60,000 and is growing at the rate of 14% per year compounded continuously. After how many years will the two investments have the same value?

24. Ten thousand dollars is deposited into a money market fund paying 8% interest compounded continuously. How much interest will be earned during the second year of the investment?

25. A small amount of money is deposited in a savings account with interest compounded continuously. Let $A(t)$ be the balance in the account after t years. Match each of the following answers with its corresponding question.

Answers:
   a. Pe^{rt}
   b. $A(3)$
   c. $A(0)$
   d. $A'(3)$
   e. Solve $A'(t) = 3$ for t.
   f. Solve $A(t) = 3$ for t.
   g. $y' = ry$
   h. Solve $A(t) = 3A(0)$ for t.

Questions:
   A. How fast will the balance be growing in 3 years?
   B. Give the general form of the function $A(t)$.
   C. How long will it take for the initial deposit to triple?
   D. Find the balance after 3 years.
   E. When will the balance be 3 dollars?
   F. When will the balance be growing at the rate of 3 dollars per year?
   G. What was the principal amount?
   H. Give a differential equation satisfied by $A(t)$.

26. The curve in Fig. 1 shows the growth of money in a savings account with interest compounded continuously.

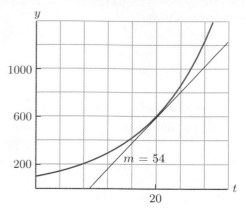

Figure 1. Growth of money in a savings account.

(a) What is the balance after 20 years?

(b) At what rate is the money growing after 20 years?

(c) Use the answers to (a) and (b) to determine the interest rate.

27. The function $A(t)$ in Fig. 2(a) gives the balance in a savings account after t years with interest compounded continuously. Fig. 2(b) shows the derivative of $A(t)$.

(a) What is the balance after 20 years?

(b) How fast is the balance increasing after 20 years?

(c) Use the answers to (a) and (b) to determine the interest rate.

(d) When is the balance $300?

(e) When is the balance increasing at the rate of $12 per year?

(f) Why do the graphs of $A(t)$ and $A'(t)$ look the same?

28. When $1000 is invested at r% interest (compounded continuously) for 10 years, the balance is $f(r)$ dollars, where f is the function shown in Fig. 3.

(a) What will the balance be at 7% interest?

(b) For what interest rate will the balance be $3000?

(c) If the interest rate is 9%, what is the growth rate of the balance with respect to a unit increase in interest?

Technology Exercises

29. Convince yourself that $\lim_{h \to 0}(1 + h)^{1/h} = e$ by graphing $f(x) = (1 + x)^{1/x}$ near $x = 0$ and examining values of $f(x)$ for x near 0. For instance, use a window such as $[-.5, .5]$ *by* $[-.5, 4]$.

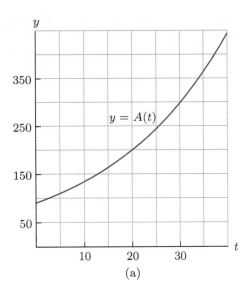

Figure 2.

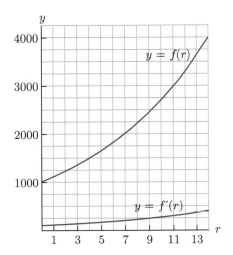

Figure 3. Effect of interest rate on balance.

30. Convince yourself that daily compounding is nearly the same as continuous compounding by graphing $100(1 + (.05/360))^{360x}$ together with $100e^{.05x}$ in the window $[0, 64]$ by $[250, 2500]$. The two graphs should appear the same on the screen. Approximately how far apart are they when $x = 32$? When $x = 64$?

31. Suppose that an investment of \$2000 yields payments of \$1200 in 3 years, \$800 in 4 years, and \$500 in 5 years. Thereafter, the investment is worthless. What constant rate of return r would the investment need to produce in order to yield the payments specified? The number r is called the *internal rate of return* on the investment. We can consider the investment as consisting of three parts, each part yielding one payment. The sum of the present values of the three parts must total \$2000. This yields the equation

$$2000 = 1200e^{-3r} + 800e^{-4r} + 500e^{-5r}.$$

Solve this equation to find the value of r.

1. Let us compute the balance after 4 years for each type of interest.

   8% compounded semiannually: Use the formula at the beginning of this section. Here $P = 1000$, $r = .08$, $m = 2$ (semiannually means there are two interest periods per year), and $t = 4$. Therefore,

   $$A = 1000 \left(1 + \frac{.08}{2}\right)^{2 \cdot 4} = 1000(1.04)^8 = \$1368.57.$$

   $7\frac{3}{4}\%$ *compounded continuously:* Use the formula $A = Pe^{rt}$, where $P = 1000$, $r = .0775$, and $t = 4$. Then

   $$A = 1000e^{(.0775) \cdot 4} = 1000e^{.31} = \$1363.43.$$

   Therefore, 8% compounded semiannually is better.

2. If the \$150,000 had been compounded continuously for 10 years at interest rate r, the balance would be $150{,}000e^{r \cdot 10}$. The question asks: For what value of r will the balance be 400,000? We need just to solve an equation for r.

   $$150{,}000e^{r \cdot 10} = 400{,}000$$
   $$e^{r \cdot 10} \approx 2.67$$
   $$r \cdot 10 = \ln 2.67$$
   $$r = \frac{\ln 2.67}{10} \approx .098$$

   Therefore, the investment earned 9.8% interest per year.

5.3 Applications of the Natural Logarithm Function to Economics

In this section we consider two applications of the natural logarithm to the field of economics. Our first application is concerned with relative rates of change and the second with elasticity of demand.

Relative Rates of Change   The *logarithmic derivative* of a function $f(t)$ is defined by the equation

$$\frac{d}{dt} \ln f(t) = \frac{f'(t)}{f(t)}. \tag{1}$$

The quantity on either side of equation (1) is often called the *relative rate of change of $f(t)$ per unit change of t.* Indeed, this quantity compares the rate of change of $f(t)$ [namely, $f'(t)$] with $f(t)$ itself. The *percentage rate of change* is the relative rate of change of $f(t)$ expressed as a percentage.

A simple example will illustrate these concepts. Suppose that $f(t)$ denotes the average price per pound of sirloin steak at time t and $g(t)$ denotes the average price of a new car (of a given make and model) at time t, where $f(t)$ and $g(t)$ are given in dollars and time is measured in years. Then the ordinary derivatives $f'(t)$ and $g'(t)$ may be interpreted as the rate of change of the price of a pound of sirloin steak and of a new car, respectively, where both are measured in dollars per year. Suppose that, at a given time t_0, we have $f(t_0) = \$5.25$ and $g(t_0) = \$12,000$. Moreover, suppose that $f'(t_0) = \$.75$ and $g'(t_0) = \$1500$. Then at time t_0 the price per pound of steak is increasing at a rate of \$.75 per year, while the price of a new car is increasing at a rate of \$1500 per year. Which price is increasing more quickly? It is not meaningful to say that the car price is increasing faster simply because $1500 is larger than $.75. We must take into account the vast difference between the actual cost of a car and the cost of steak. The usual basis of comparison of price increases is the percentage rate of increase. In other words, at $t = t_0$, the price of sirloin steak is increasing at the percentage rate

$$\frac{f'(t_0)}{f(t_0)} = \frac{.75}{5.25} \approx .143 = 14.3\%$$

per year, but at the same time the price of a new car is increasing at the percentage rate

$$\frac{g'(t_0)}{g(t_0)} = \frac{1500}{12,000} \approx .125 = 12.5\%$$

per year. Thus the price of sirloin steak is increasing at a faster percentage rate than the price of a new car.

Economists often use relative rates of change (or percentage rates of change) when discussing the growth of various economic quantities, such as national income or national debt, because such rates of change can be meaningfully compared.

▶ **Example 1** Suppose that a certain school of economists modeled the Gross Domestic Product of the United States at time t (measured in years from January 1, 1990) by the formula

$$f(t) = 3.4 + .04t + .13e^{-t},$$

where the Gross Domestic Product is measured in trillions of dollars. What was the predicted percentage rate of growth (or decline) of the economy at $t = 0$ and $t = 1$?

Solution Since

$$f'(t) = .04 - .13e^{-t},$$

we see that

$$\frac{f'(0)}{f(0)} = \frac{.04 - .13}{3.4 + .13} = -\frac{.09}{3.53} \approx -2.6\%.$$

$$\frac{f'(1)}{f(1)} = \frac{.04 - .13e^{-1}}{3.4 + .04 + .13e^{-1}} = -\frac{.00782}{3.4878} \approx -.2\%.$$

So on January 1, 1990, the economy is predicted to contract at a relative rate of 2.6% per year; on January 1, 1991, the economy is predicted to be still contracting but only at a relative rate of .2% per year. ◆

▶ **Example 2** Suppose that the value in dollars of a certain business investment at time t may be approximated empirically by the function $f(t) = 750{,}000e^{.6\sqrt{t}}$. Use a logarithmic derivative to describe how fast the value of the investment is increasing when $t = 5$ years.

Solution   We have

$$\frac{f'(t)}{f(t)} = \frac{d}{dt}\ln f(t) = \frac{d}{dt}\left(\ln 750{,}000 + \ln e^{.6\sqrt{t}}\right)$$

$$= \frac{d}{dt}(\ln 750{,}000 + .6\sqrt{t})$$

$$= (.6)\left(\frac{1}{2}\right)t^{-1/2} = \frac{.3}{\sqrt{t}}.$$

When $t = 5$,

$$\frac{f'(5)}{f(5)} = \frac{.3}{\sqrt{5}} \approx .134 = 13.4\%.$$

Thus, when $t = 5$ years, the value of the investment is increasing at the relative rate of 13.4% per year.   ◆

In certain mathematical models, it is assumed that for a limited period of time, the percentage rate of change of a particular function is constant. The following example shows that such a function must be an exponential function.

▶ Example 3   Suppose that the function $f(t)$ has a constant relative rate of change k. Show that $f(t) = Ce^{kt}$ for some constant C.

Solution   We are given that

$$\frac{d}{dt}\ln f(t) = k.$$

That is,

$$\frac{f'(t)}{f(t)} = k.$$

Hence $f'(t) = kf(t)$. But this is just the differential equation satisfied by the exponential function. Therefore, we must have $f(t) = Ce^{kt}$ for some constant C.   ◆

Elasticity of Demand   In Section 2.7 we considered demand equations for monopolists and for entire industries. Recall that a demand equation expresses, for each quantity q to be produced, the market price that will generate a demand of exactly q. For instance, the demand equation

$$p = 150 - .01x \tag{2}$$

says that in order to sell x units, the price must be set at $150 - .01x$ dollars. To be specific, in order to sell 6000 units, the price must be set at $150 - .01(6000) = \$90$ per unit.

Equation (2) may be solved for x in terms of p to yield

$$x = 100(150 - p). \tag{3}$$

This last equation gives quantity in terms of price. If we let the letter q represent quantity, equation (3) becomes

$$q = 100(150 - p). \tag{3'}$$

This equation is of the form $q = f(p)$, where in this case $f(p)$ is the function $f(p) = 100(150 - p)$. In what follows it will be convenient to always write our demand functions so that the quantity q is expressed as a function $f(p)$ of the price p.

Usually, raising the price of a commodity lowers demand. Therefore, the typical demand function $q = f(p)$ is decreasing and has a negative slope everywhere.

A demand function $q = f(p)$ relates the quantity demanded to the price. Therefore, the derivative $f'(p)$ compares the change in quantity demanded with the change in price. By way of contrast, the concept of elasticity is designed to compare the *relative* rate of change of the quantity demanded with the *relative* rate of change of price.

Let us be more explicit. Consider a particular demand function $q = f(p)$ and a particular price p. Then at this price, the ratio of the relative rates of change of the quantity demanded and the price is given by

$$\frac{[\text{relative rate of change of quantity}]}{[\text{relative rate of change of price}]} = \frac{\dfrac{d}{dp}\ln f(p)}{\dfrac{d}{dp}\ln p}$$

$$= \frac{f'(p)/f(p)}{1/p}$$

$$= \frac{pf'(p)}{f(p)}.$$

Since $f'(p)$ is always negative for a typical demand function, the quantity $pf'(p)/f(p)$ will be negative for all values of p. For convenience, economists prefer to work with positive numbers and therefore the *elasticity of demand* is taken to be this quantity multiplied by -1.

> The elasticity of demand $E(p)$ at price p for the demand function $q = f(p)$ is defined to be
>
> $$E(p) = \frac{-pf'(p)}{f(p)}.$$

▶ **Example 4** Suppose that the demand function for a certain metal is $q = 100 - 2p$, where p is the price per pound and q is the quantity demanded (in millions of pounds).

(a) What quantity can be sold at \$30 per pound?
(b) Determine the function $E(p)$.
(c) Determine and interpret the elasticity of demand at $p = 30$.
(d) Determine and interpret the elasticity of demand at $p = 20$.

Solution (a) In this case, $q = f(p)$, where $f(p) = 100 - 2p$. When $p = 30$, we have $q = f(30) = 100 - 2(30) = 40$. Therefore, 40 million pounds of the metal can be sold. We also say that the *demand* is 40 million pounds.

(b) $E(p) = \dfrac{-pf'(p)}{f(p)} = \dfrac{-p(-2)}{100 - 2p} = \dfrac{2p}{100 - 2p}.$

(c) The elasticity of demand at price $p = 30$ is $E(30)$.

$$E(30) = \frac{2(30)}{100 - 2(30)} = \frac{60}{40} = \frac{3}{2}$$

When the price is set at \$30 per pound, a small increase in price will result in a relative rate of decrease in quantity demanded of about $\frac{3}{2}$ times the relative rate of increase in price. For example, if the price is increased from \$30 by 1%, then the quantity demanded will decrease by about 1.5%.

(d) When $p = 20$, we have

$$E(20) = \frac{2(20)}{100 - 2(20)} = \frac{40}{60} = \frac{2}{3}.$$

When the price is set at \$20 per pound, a small increase in price will result in a relative rate of decrease in quantity demanded of only $\frac{2}{3}$ of the relative rate of increase of price. For example, if the price is increased from \$20 by 1%, the quantity demanded will decrease by $\frac{2}{3}$ of 1%. ◆

Economists say that demand is *elastic* at price p_0 if $E(p_0) > 1$ and *inelastic* at price p_0 if $E(p_0) < 1$. In Example 4, the demand for the metal is elastic at \$30 per pound and inelastic at \$20 per pound.

The significance of the concept of elasticity may perhaps best be appreciated by studying how revenue, $R(p)$, responds to changes in price. Recall that

$$[\text{revenue}] = [\text{quantity}] \cdot [\text{price per unit}],$$

that is,

$$R(p) = f(p) \cdot p.$$

If we differentiate $R(p)$ using the product rule, we find that

$$R'(p) = \frac{d}{dp}[f(p) \cdot p] = f(p) \cdot 1 + p \cdot f'(p)$$

$$= f(p)\left[1 + \frac{pf'(p)}{f(p)}\right]$$

$$= f(p)[1 - E(p)]. \tag{4}$$

Now suppose that demand is elastic at some price p_0. Then $E(p_0) > 1$ and $1 - E(p_0)$ is negative. Since $f(p)$ is always positive, we see from (4) that $R'(p_0)$ is negative. Therefore, by the first derivative rule, $R(p)$ is decreasing at p_0. So an increase in price will result in a decrease in revenue, and a decrease in price will result in an increase in revenue. On the other hand, if demand is inelastic at p_0, then $1 - E(p_0)$ will be positive and hence $R'(p_0)$ will be positive. In this case an increase in price will result in an increase in revenue, and a decrease in price will result in a decrease in revenue. We may summarize this as follows:

The change in revenue is in the opposite direction of the change in price when demand is elastic and in the same direction when demand is inelastic.

Practice Problems 5.3

The current toll for the use of a certain toll road is \$2.50. A study conducted by the state highway department determined that with a toll of p dollars, q cars will use the road each day, where $q = 60{,}000e^{-.5p}$.

1. Compute the elasticity of demand at $p = 2.5$.

2. Is demand elastic or inelastic at $p = 2.5$?

3. If the state increases the toll slightly, will the revenue increase or decrease?

▶ Exercises 5.3

Determine the percentage rate of change of the functions at the points indicated.

1. $f(t) = t^2$ at $t = 10$ and $t = 50$

2. $f(t) = t^{10}$ at $t = 10$ and $t = 50$

3. $f(x) = e^{.3x}$ at $x = 10$ and $x = 20$

4. $f(x) = e^{-.05x}$ at $x = 1$ and $x = 10$

5. $f(t) = e^{.3t^2}$ at $t = 1$ and $t = 5$

6. $G(s) = e^{-.05s^2}$ at $s = 1$ and $s = 10$

7. $f(p) = 1/(p+2)$ at $p = 2$ and $p = 8$

8. $g(p) = 5/(2p+3)$ at $p = 1$ and $p = 11$

9. Suppose that the annual sales S (in dollars) of a company may be approximated empirically by the formula

$$S = 50{,}000\sqrt{e^{\sqrt{t}}},$$

where t is the number of years beyond some fixed reference date. Use a logarithmic derivative to determine the percentage rate of growth of sales at $t = 4$.

10. Suppose that the price of wheat per bushel at time t (in months) is approximated by

$$f(t) = 4 + .001t + .01e^{-t}.$$

What is the percentage rate of change of $f(t)$ at $t = 0$? $t = 1$? $t = 2$?

11. Suppose that an investment grows at a continuous 12% rate per year. In how many years will the value of the investment double?

12. Suppose that the value of a piece of property is growing at a continuous $r\%$ rate per year and that the value doubles in 3 years. Find r.

For each demand function, find $E(p)$ and determine if demand is elastic or inelastic (or neither) at the indicated price.

13. $q = 700 - 5p$, $p = 80$

14. $q = 600e^{-.2p}$, $p = 10$

15. $q = 400(116 - p^2)$, $p = 6$

16. $q = (77/p^2) + 3$, $p = 1$

17. $q = p^2e^{-(p+3)}$, $p = 4$

18. $q = 700/(p+5)$, $p = 15$

19. Currently, 1800 people ride a certain commuter train each day and pay $4 for a ticket. The number of people q willing to ride the train at price p is $q = 600(5 - \sqrt{p})$. The railroad would like to increase its revenue.

 (a) Is demand elastic or inelastic at $p = 4$?

 (b) Should the price of a ticket be raised or lowered?

20. A company can sell $q = 9000/(p+60) - 50$ radios at a price of p dollars per radio. The current price is $30.

 (a) Is demand elastic or inelastic at $p = 30$?

 (b) If the price is lowered slightly, will revenue increase or decrease?

21. A movie theater has a seating capacity of 3000 people. The number of people attending a show at price p dollars per ticket is $q = (18{,}000/p) - 1500$. Currently the price is $6 per ticket.

 (a) Is demand elastic or inelastic at $p = 6$?

 (b) If the price is lowered, will revenue increase or decrease?

22. A subway charges 65 cents per person and has 10,000 riders each day. The demand function for the subway is $q = 2000\sqrt{90 - p}$.

 (a) Is demand elastic or inelastic at $p = 65$?

 (b) Should the price of a ride be raised or lowered in order to increase the amount of money taken in by the subway?

23. A country that is the major supplier of a certain commodity wishes to improve its balance of trade position by lowering the price of the commodity. The demand function is $q = 1000/p^2$.

 (a) Compute $E(p)$.

 (b) Will the country succeed in raising its revenue?

24. Show that any demand function of the form $q = a/p^m$ has constant elasticity m.

A cost function $C(x)$ gives the total cost of producing x units of a product. The *elasticity of cost at quantity x* is defined to be

$$E_c(x) = \frac{\dfrac{d}{dx}\ln C(x)}{\dfrac{d}{dx}\ln x}.$$

25. Show that $E_c(x) = x \cdot C'(x)/C(x)$.

26. Show that E_c is equal to the marginal cost divided by the average cost.

27. Let $C(x) = (1/10)x^2 + 5x + 300$. Show that $E_c(50) < 1$. (Hence when producing 50 units, a small relative increase in production results in an even smaller relative increase in total cost. Also, the average cost of producing 50 units is greater than the marginal cost at $x = 50$.)

28. Let $C(x) = 1000e^{.02x}$. Determine and simplify the formula for $E_c(x)$. Show that $E_c(60) > 1$ and interpret this result.

Technology Exercises

29. Consider the demand function $q = 60{,}000e^{-.5p}$ from the practice problems.

(a) Determine the value of p for which the value of $E(p)$ is 1. For what values of p is demand inelastic?

(b) Graph the revenue function in the window $[0, 4]$ by $[-5000, 50000]$ and determine where its maximum value occurs. For what values of p is the revenue an increasing function?

Solutions to Practice Problems 5.3

1. The demand function is $f(p) = 60{,}000e^{-.5p}$.

$$f'(p) = -30{,}000e^{-.5p}$$

$$E(p) = \frac{-pf'(p)}{f(p)} = \frac{-p(-30{,}000)e^{-.5p}}{60{,}000e^{-.5p}} = \frac{p}{2}$$

$$E(2.5) = \frac{2.5}{2} = 1.25$$

2. The demand is elastic, because $E(2.5) > 1$.

3. Since demand is elastic at \$2.50, a slight change in price causes revenue to change in the *opposite* direction. Hence revenue will decrease.

5.4 Further Exponential Models

A skydiver, on jumping out of an airplane, falls at an increasing rate. However, the wind rushing past the skydiver's body creates an upward force that begins to counterbalance the downward force of gravity. This air friction finally becomes so great that the skydiver's velocity reaches a limiting speed called the *terminal velocity*. If we let $v(t)$ be the downward velocity of the skydiver after t seconds of free fall, then a good mathematical model for $v(t)$ is given by

$$v(t) = M(1 - e^{-kt}), \tag{1}$$

where M is the terminal velocity and k is some positive constant (Fig. 1). When t is close to zero, e^{-kt} is close to one, and the velocity is small. As t increases, e^{-kt} becomes small and so $v(t)$ approaches M.

▶ Example 1   Show that the velocity given in (1) satisfies the equation

$$v'(t) = k[M - v(t)], \qquad v(0) = 0. \tag{2}$$

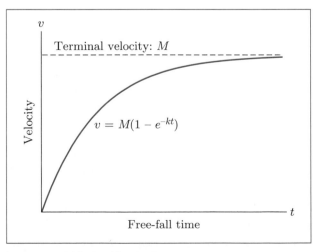

Figure 1. Velocity of a skydiver.

Solution   From (1) we have $v(t) = M - Me^{-kt}$. Then

$$v'(t) = Mke^{-kt}.$$

However,

$$k[M - v(t)] = k[M - (M - Me^{-kt})] = kMe^{-kt},$$

so that the differential equation $v'(t) = k[M - v(t)]$ holds. Also,

$$v(0) = M - Me^0 = M - M = 0. \qquad \blacklozenge$$

The differential equation (2) says that the rate of change in v is proportional to the difference between the terminal velocity M and the actual velocity v. It is not difficult to show that the only solution of (2) is given by the formula in (1).

The two equations (1) and (2) arise as mathematical models in a variety of situations. Some of these applications are described next.

The Learning Curve   Psychologists have found that in many learning situations a person's rate of learning is rapid at first and then slows down. Finally, as the task is mastered, the person's level of performance reaches a level above which it is almost impossible to rise. For example, within reasonable limits, each person seems to have a certain maximum capacity for memorizing a list of nonsense syllables. Suppose that a subject can memorize M syllables in a row if given sufficient time, say an hour, to study the list but cannot memorize $M + 1$ syllables in a row even if allowed several hours of study. By giving the subject different lists of syllables and varying lengths of time to study the lists, the psychologist can determine an empirical relationship between the number of nonsense syllables memorized accurately and the number of minutes of study time. It turns out that a good model for this situation is

$$f(t) = M(1 - e^{-kt})$$

for some appropriate positive constant k. (See Fig. 2.)

The *slope* of this learning curve at time t is approximately the number of additional syllables that can be memorized if the subject is given one more minute

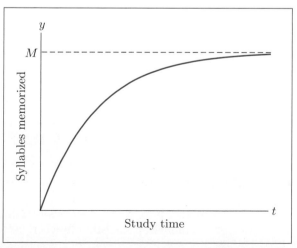

Figure 2. Learning curve, $f(t) = M(1 - e^{-kt})$.

of study time. Thus the slope is a measure of the *rate of learning*. The differential equation satisfied by the function $y = f(t)$ is

$$y' = k(M - y), \qquad f(0) = 0.$$

This equation says that if the subject is given a list of M nonsense syllables, then the rate of memorization is proportional to the number of syllables remaining to be memorized.

Diffusion of Information by Mass Media   Sociologists have found that the differential equation (2) provides a good model for the way information is spread (or "diffused") through a population when the information is being propagated constantly by mass media, such as television or magazines.* Given a fixed population P, let $f(t)$ be the number of people who have already heard a certain piece of information by time t. Then $P - f(t)$ is the number who have not yet heard the information by time t. Also, $f'(t)$ is the rate of increase of the number of people who have heard the news (the "rate of diffusion" of the information). If the information is being publicized often by some mass media, then it is likely that the number of *newly informed* people per unit time is proportional to the number of people who have not yet heard the news. Therefore,

$$f'(t) = k[P - f(t)]. \tag{3}$$

Assume that $f(0) = 0$ (i.e., there was a time $t = 0$ when nobody had heard the news). Then the remark following Example 1 shows that

$$f(t) = P(1 - e^{-kt}). \tag{4}$$

(See Fig. 3.)

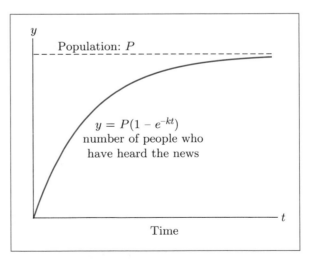

Figure 3. Diffusion of information by mass media.

▶ Example 2   Suppose that the news of resignation of a public official is broadcast frequently by radio and television stations. Also suppose that one-half of the residents of a city have heard the news within 4 hours of its initial release. Use the exponential model (4) to estimate when 90% of the residents will have heard the news.

---

*J. Coleman, *Introduction to Mathematical Sociology* (New York: The Free Press, 1964), p. 43.

Solution   We must find the value of k in (4). If P is the number of residents, then the number who will have heard the news in the first four hours is given by (4) with $t = 4$. By assumption, this number is half the population. So

$$\tfrac{1}{2}P = P(1 - e^{-k4})$$
$$.5 = 1 - e^{-4k}$$
$$e^{-4k} = 1 - .5 = .5.$$

Solving for k, we find that $k \approx .17$. So the model for this particular situation is

$$f(t) = P(1 - e^{-.17t}).$$

Now we want to find t such that $f(t) = .90P$. We solve for t:

$$.90P = P(1 - e^{-.17t})$$
$$.90 = 1 - e^{-.17t}$$
$$e^{-.17t} = 1 - .90 = .10$$
$$-.17t = \ln .10$$
$$t = \frac{\ln .10}{-.17} \approx 14.$$

Therefore, 90% of the residents will hear the news within 14 hours of its initial release.   ◆

Intravenous Infusion of Glucose   The human body both manufactures and uses glucose ("blood sugar"). Usually, there is a balance in these two processes, so that the bloodstream has a certain "equilibrium level" of glucose. Suppose that a patient is given a single intravenous injection of glucose and let $A(t)$ be the amount of glucose (in milligrams) above the equilibrium level. Then the body will start using up the excess glucose at a rate proportional to the amount of excess glucose; that is,

$$A'(t) = -\lambda A(t), \tag{5}$$

where λ is a positive constant called the *velocity constant of elimination*. This constant depends on how fast an individual patient's metabolic processes eliminate the excess glucose from the blood. Equation (5) describes a simple exponential decay process.

Now suppose that, instead of a single shot, the patient receives a continuous intravenous infusion of glucose. A bottle of glucose solution is suspended above the patient, and a small tube carries the glucose down to a needle that runs into a vein. In this case, there are two influences on the amount of excess glucose in the blood: the glucose being added steadily from the bottle and the glucose being removed from the body by metabolic processes. Let r be the rate of infusion of glucose (often from 10 to 100 milligrams per minute). If the body did not remove any glucose, the excess glucose would increase at a constant rate of r milligrams per minute; that is,

$$A'(t) = r. \tag{6}$$

Taking into account the two influences on $A'(t)$ described by (5) and (6), we can write

$$A'(t) = r - \lambda A(t). \tag{7}$$

Define M to be r/λ and note that initially there is no excess glucose; then

$$A'(t) = \lambda(M - A(t)), \qquad A(0) = 0.$$

As stated in Example 1, a solution of this differential equation is given by

$$A(t) = M(1 - e^{-\lambda t}) = \frac{r}{\lambda}(1 - e^{-\lambda t}). \qquad (8)$$

Note that M is the limiting value of the glucose level. Reasoning as in Example 1, we conclude that the amount of excess glucose rises until it reaches a stable level. (See Fig. 4).

The Logistic Growth Curve The model for simple exponential growth discussed in Section 5.1 is adequate for describing the growth of many types of populations, but obviously a population cannot increase exponentially forever. The simple exponential growth model becomes inapplicable when the environment begins to inhibit the growth of the population. The logistic growth curve is an important exponential model that takes into account some of the effects of the environment on a population (Fig. 5). For small values of t, the curve has the same basic shape as an exponential growth curve. Then when the population begins to suffer from overcrowding or lack of food, the growth rate (the slope of the population curve) begins to slow down. Eventually, the growth rate tapers off to zero as the population reaches the maximum size that the environment will support. This latter part of the curve resembles the growth curves studied earlier in this section.

The equation for logistic growth has the general form

$$y = \frac{M}{1 + Be^{-Mkt}}, \qquad (9)$$

where B, M, and k are positive constants. We can show that y satisfies the differential equation

$$y' = ky(M - y). \qquad (10)$$

The factor y reflects the fact that the growth rate (y') depends in part on the size y of the population. The factor $M - y$ reflects the fact that the growth rate also depends on how close y is to the maximum level M.

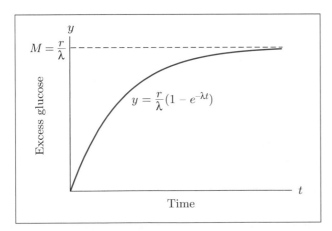

Figure 4. Continuous infusion of glucose.

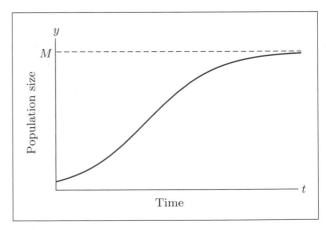

Figure 5. Logistic growth.

The logistic curve is often used to fit experimental data that lie along an "S-shaped" curve. Examples are given by the growth of a fish population in a lake and the growth of a fruit fly population in a laboratory container. Also, certain enzyme reactions in animals follow a logistic law. One of the earliest applications of the logistic curve occurred in about 1840 when the Belgian sociologist P. Verhulst fit a logistic curve to six U.S. census figures, 1790 to 1840, and predicted the U.S. population for 1940. His prediction missed by less than 1 million persons (an error of about 1%).

▶ **Example 3**   Suppose that a lake is stocked with 100 fish. After 3 months there are 250 fish. A study of the ecology of the lake predicts that the lake can support 1000 fish. Find a formula for the number $P(t)$ of fish in the lake t months after it has been stocked.

Solution   The limiting population M is 1000. Therefore, we have

$$P(t) = \frac{1000}{1 + Be^{-1000kt}}.$$

At $t = 0$ there are 100 fish, so that

$$100 = P(0) = \frac{1000}{1 + Be^0} = \frac{1000}{1 + B}.$$

Thus $1 + B = 10$, or $B = 9$. Finally, since $P(3) = 250$, we have

$$250 = \frac{1000}{1 + 9e^{-3000k}}$$

$$1 + 9e^{-3000k} = 4$$

$$e^{-3000k} = \tfrac{1}{3}$$

$$-3000k = \ln \tfrac{1}{3}$$

$$k \approx .00037.$$

Therefore,

$$P(t) = \frac{1000}{1 + 9e^{-.37t}}. \qquad\qquad ◆$$

Several theoretical justifications can be given for using (9) and (10) in situations where the environment prevents a population from exceeding a certain size. A discussion of this topic may be found in *Mathematical Models and Applications* by D. Maki and M. Thompson (Englewood Cliffs, N.J.: Prentice-Hall, Inc., 1973), pp. 312–317.

An Epidemic Model   It will be instructive to actually "build" a mathematical model. Our example concerns the spread of a highly contagious disease. We begin by making several simplifying assumptions:

1. The population is a fixed number P and each member of the population is susceptible to the disease.

2. The duration of the disease is long, so that no cures occur during the time period under study.

3. All infected individuals are contagious and circulate freely among the population.

4. During each time period (such as 1 day or 1 week) each infected person makes c contacts, and each contact with an uninfected person results in transmission of the disease.

Consider a short period of time from t to $t + h$. Each infected person makes $c \cdot h$ contacts. How many of these contacts are with uninfected persons? If $f(t)$ is the number of infected persons at time t, then $P - f(t)$ is the number of uninfected persons, and $[P - f(t)]/P$ is the fraction of the population that is uninfected. Thus, of the $c \cdot h$ contacts made,

$$\left[\frac{P - f(t)}{P} \right] \cdot c \cdot h$$

will be with uninfected persons. This is the number of new infections produced by one infected person during the time period of length h. The total number of *new* infections during this period is

$$f(t) \left[\frac{P - f(t)}{P} \right] ch.$$

But this number must equal $f(t + h) - f(t)$, where $f(t + h)$ is the total number of infected persons at time $t + h$. So

$$f(t + h) - f(t) = f(t) \left[\frac{P - f(t)}{P} \right] ch.$$

Dividing by h, the length of the time period, we obtain the average number of new infections per unit time (during the small time period):

$$\frac{f(t + h) - f(t)}{h} = \frac{c}{P} f(t)[P - f(t)].$$

If we let h approach zero and let y stand for $f(t)$, the left-hand side approaches the rate of change in the number of infected persons and we derive the following equation:

$$\frac{dy}{dt} = \frac{c}{P} y(P - y). \tag{11}$$

This is the same type of equation as that used in (10) for logistic growth, although the two situations leading to this model appear to be quite dissimilar.

Comparing (11) with (10), we see that the number of infected individuals at time t is described by a logistic curve with $M = P$ and $k = c/P$. Therefore, by (9), we can write

$$f(t) = \frac{P}{1 + Be^{-ct}}.$$

B and c can be determined from the characteristics of the epidemic. (See Example 4.)

The logistic curve has an inflection point at that value of t for which $f(t) = P/2$. The position of this inflection point has great significance for applications of the logistic curve. From inspecting a graph of the logistic curve, we see that the inflection point is the point at which the curve has greatest slope. In other words, the inflection point corresponds to the instant of fastest growth of the logistic curve. This means, for example, that in the foregoing epidemic model the disease is spreading with the greatest rapidity precisely when half the population is infected. Any attempt at disease control (through immunization, for example) must strive to reduce the incidence of the disease to as low a point as possible, but in any case at least below the inflection point at $P/2$, at which point the epidemic is spreading fastest.

▶ **Example 4**   The Public Health Service monitors the spread of an epidemic of a particularly long-lasting strain of flu in a city of 500,000 people. At the beginning of the first week of monitoring, 200 cases have been reported; during the first week 300 new cases are reported. Estimate the number of infected individuals after 6 weeks.

Solution   Here $P = 500{,}000$. If $f(t)$ denotes the number of cases at the end of t weeks, then

$$f(t) = \frac{P}{1 + Be^{-ct}} = \frac{500{,}000}{1 + Be^{-ct}}.$$

Moreover, $f(0) = 200$, so that

$$200 = \frac{500{,}000}{1 + Be^0} = \frac{500{,}000}{1 + B},$$

and $B = 2499$. Consequently, since $f(1) = 300 + 200 = 500$, we have

$$500 = f(1) = \frac{500{,}000}{1 + 2499e^{-c}},$$

so that $e^{-c} \approx .4$ and $c \approx .92$. Finally,

$$f(t) = \frac{500{,}000}{1 + 2499e^{-.92t}}$$

and

$$f(6) = \frac{500{,}000}{1 + 2499e^{-.92(6)}} \approx 45{,}000.$$

After 6 weeks, about 45,000 individuals are infected.   ◆

This epidemic model is used by sociologists (who still call it an epidemic model) to describe the spread of a rumor. In economics the model is used to describe the diffusion of knowledge about a product. An "infected person" represents an individual who possesses knowledge of the product. In both cases, it is assumed that the members of the population are themselves primarily responsible for the spread of the rumor or knowledge of the product. This situation is in contrast to the model described earlier where information was spread through a population by external sources, such as radio and television.

There are several limitations to this epidemic model. Each of the four simplifying assumptions made at the outset is unrealistic in varying degrees. More complicated models can be constructed that rectify one or more of these defects, but they require more advanced mathematical tools.

The Exponential Function in Lung Physiology   Let us conclude this section by deriving a useful model for the pressure in a person's lungs when the air is allowed to escape passively from the lungs with no use of the person's muscles.

Let V be the volume of air in the lungs and let P be the relative pressure in the lungs when compared with the pressure in the mouth. The *total compliance* (of the respiratory system) is defined to be the derivative $\dfrac{dV}{dP}$. For normal values of V and P we may assume that the total compliance is a positive constant, say C. That is,

$$\frac{dV}{dP} = C. \tag{12}$$

We shall assume that the airflow during the passive respiration is smooth and not turbulent. Then Poiseuille's law of fluid flow says that the rate of change of volume as a function of time (i.e., the rate of airflow) satisfies

$$\frac{dV}{dt} = -\frac{P}{R}, \tag{13}$$

where R is a (positive) constant called the airway resistance. Under these conditions, we may derive a formula for P as a function of time. Since the volume is a function of the pressure, and the pressure is in turn a function of time, we may use the chain rule to write

$$\frac{dV}{dt} = \frac{dV}{dP} \cdot \frac{dP}{dt}.$$

From (12) and (13),

$$-\frac{P}{R} = C \cdot \frac{dP}{dt},$$

so

$$\frac{dP}{dt} = -\frac{1}{RC} \cdot P.$$

From this differential equation we conclude that P must be an exponential function of t. In fact,

$$P = P_0 e^{kt},$$

where P_0 is the initial pressure at time $t = 0$ and $k = -1/RC$. This relation between k and the product RC is useful to lung specialists, because they can experimentally compute k and the compliance C, and then use the formula $k = -1/RC$ to determine the airway resistance R.

Practice Problem 5.4

1. A sociological study[*] was made to examine the process by which doctors decide to adopt a new drug. The doctors were divided into two groups. The doctors in group A had little interaction with other doctors and so received most of their information through mass media. The doctors in group B had extensive interaction with other doctors and so received most of their information through word of mouth. For each group, let $f(t)$ be the number who have learned about the new drug after t months. Examine the appropriate differential equations to explain why the two graphs were of the types shown in Fig. 6.

---

[*]James S. Coleman, Eliku Katz, and Herbert Menzel, "The Diffusion of an Innovation Among Physicians," *Sociometry*, 20 (1957), 253–270.

Practice Problem 5.4 (Continued)

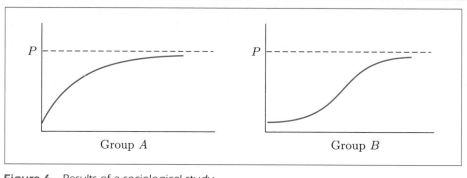

Figure 6. Results of a sociological study.

▶ Exercises 5.4

1. Consider the function $f(x) = 5(1 - e^{-2x})$, $x \geq 0$.
   (a) Show that $f(x)$ is increasing and concave down for all $x \geq 0$.
   (b) Explain why $f(x)$ approaches 5 as x gets large.
   (c) Sketch the graph of $f(x)$, $x \geq 0$.

2. Consider the function $g(x) = 10 - 10e^{-.1x}$, $x \geq 0$.
   (a) Show that $g(x)$ is increasing and concave down for $x \geq 0$.
   (b) Explain why $g(x)$ approaches 10 as x gets large.
   (c) Sketch the graph of $g(x)$, $x \geq 0$.

3. Suppose that $y = 2(1 - e^{-x})$. Compute y' and show that $y' = 2 - y$.

4. Suppose that $y = 5(1 - e^{-2x})$. Compute y' and show that $y' = 10 - 2y$.

5. Suppose that $f(x) = 3(1 - e^{-10x})$. Show that $y = f(x)$ satisfies the differential equation

$$y' = 10(3 - y), \qquad f(0) = 0.$$

6. (*Ebbinghaus Model for Forgetting*) Suppose that a student learns a certain amount of material for some class. Let $f(t)$ denote the percentage of the material that the student can recall t weeks later. The psychologist Ebbinghaus found that this percent retention can be modeled by a function of the form

$$f(t) = (100 - a)e^{-\lambda t} + a,$$

where λ and a are positive constants and $0 < a < 100$. Sketch the graph of the function $f(t) = 85e^{-.5t} + 15$, $t \geq 0$.

7. When a grand jury indicted the mayor of a certain town for accepting bribes, the newspaper, radio, and television immediately began to publicize the news. Within an hour, one-quarter of the citizens heard about the indictment. Estimate when three-quarters of the town heard the news.

8. Examine formula (8) for the amount $A(t)$ of excess glucose in the bloodstream of a patient at time t. Describe what would happen if the rate r of infusion of glucose were doubled.

9. Describe an experiment that a doctor could perform in order to determine the velocity constant of elimination of glucose for a particular patient.

10. Physiologists usually describe the continuous intravenous infusion of glucose in terms of the excess *concentration* of glucose, $C(t) = A(t)/V$, where V is the total volume of blood in the patient. In this case, the rate of increase in the concentration of glucose due to the continuous injection is r/V. Find a differential equation that gives a model for the rate of change of the excess concentration of glucose.

11. A news item is spread by word of mouth to a potential audience of 10,000 people. After t days,

$$f(t) = \frac{10{,}000}{1 + 50e^{-.4t}}$$

people will have heard the news. The graph of this function is shown in Fig. 7.
   (a) Approximately how many people will have heard the news after 7 days?
   (b) At approximately what rate will the news be spreading after 14 days?
   (c) Approximately when will 7000 people have heard the news?
   (d) Approximately when will the news be spreading at the rate of 600 people per day?
   (e) When will the news be spreading at the greatest rate?
   (f) Use (9) and (10) to determine the differential equation satisfied by $f(t)$.
   (g) At what rate will the news be spreading when half of the potential audience has heard the news?

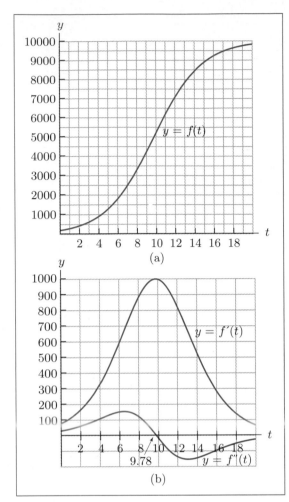

Figure 7.

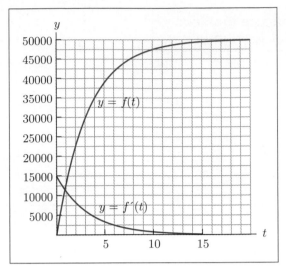

Figure 8.

Technology Exercises

13. After a drug is taken orally, the amount of the drug in the bloodstream after t hours is $f(t) = 122(e^{-.2t} - e^{-t})$ units.

(a) Graph $f(t)$, $f'(t)$, and $f''(t)$ in the window $[0, 12]$ *by* $[-20, 75]$.

(b) How many units of the drug are in the bloodstream after 7 hours?

(c) At what rate is the level of drug in the bloodstream increasing after 1 hour?

(d) When, while the level is decreasing, is the level of the drug in the bloodstream 20 units?

(e) What is the greatest level of drug in the bloodstream and when is this level reached?

(f) When is the level of drug in the bloodstream decreasing the fastest?

14. A model incorporating growth restrictions for the number of bacteria in a culture after t days is given by $f(t) = 5000(20 + te^{-.04t})$.

(a) Graph $f'(t)$ and $f''(t)$ in the window $[0, 100]$ *by* $[-700, 300]$.

(b) How fast is the culture changing after 100 days?

(c) Approximately when is the culture growing at the rate of 76.6 bacteria per day?

(d) When is the size of the culture greatest?

(e) When is the size of the culture decreasing the fastest?

12. A news item is broadcast by mass media to a potential audience of 50,000 people. After t days,

$$f(t) = 50{,}000(1 - e^{-.3t})$$

people will have heard the news. The graph of this function is shown in Fig. 8.

(a) How many people will have heard the news after 10 days?

(b) At what rate is the news spreading initially?

(c) When will 22,500 people have heard the news?

(d) Approximately when will the news be spreading at the rate of 2500 people per day?

(e) Use equations (3) and (4) on page 300 to determine the differential equation satisfied by $f(t)$.

(f) At what rate will the news be spreading when half of the potential audience has heard the news?

<table>
<tr><td>**Solution to Practice Problem 5.4**</td><td>1. The difference between transmission of information via mass media and via word of mouth is that in the second case the rate of transmission depends not only on the number of people who have not yet received the information, but also on the number of people who know the information and therefore are capable of spreading it. Therefore, for group A, $f'(t) = k[P - f(t)]$, and for group B, $f'(t) = kf(t)[P - f(t)]$. Note that the spread of information by word of mouth follows the same pattern as the spread of an epidemic.</td></tr>
</table>

Review of Fundamental Concepts of Chapter 5

1. What differential equation is key to solving exponential growth and decay problems? State a result about the solution to this differential equation.

2. What is a growth constant? Decay constant?

3. What is meant by the half-life of a radioactive element?

4. Explain how carbon dating works.

5. State the formula for each of the following quantities:
   (a) the compound amount of P dollars in t years at interest rate r compounded continuously

   (b) the present value of A dollars in n years at interest rate r compounded continuously

6. What is the difference between a relative rate of change and a percentage rate of change?

7. Define the elasticity of demand, $E(p)$, for a demand function. How is $E(p)$ used?

8. Describe an application of the differential equation $y' = k(M - y)$.

9. Describe an application of the differential equation $y' = ky(M - y)$.

▶ Chapter 5 Supplementary Exercises

1. The atmospheric pressure $P(x)$ (measured in inches of mercury) at height x miles above sea level satisfies the differential equation $P'(x) = -.2P(x)$. Find the formula for $P(x)$ if the atmospheric pressure at sea level is 29.92.

2. The herring gull population in North America has been doubling every 13 years since 1900. Give a differential equation satisfied by $P(t)$, the population t years after 1900.

3. Find the present value of $10,000 payable at the end of 5 years if money can be invested at 12% with interest compounded continuously.

4. One thousand dollars is deposited in a savings account at 10% interest compounded continuously. How many years are required for the balance in the account to reach $3000?

5. The half-life of the radioactive element tritium is 12 years. Find its decay constant.

6. A piece of charcoal found at Stonehenge contained 63% of the level of ^{14}C found in living trees. Approximately how old is the charcoal?

7. From January 1, 1990 to January 1, 1997, the population of Texas grew from 17 million to 19.3 million.
   (a) Give the formula for the population t years after 1990.
   (b) If this growth continues, how large will the population be in 2000?

(c) In what year will the population reach 25 million?

8. A stock portfolio increased in value from $100,000 to $117,000 in 2 years. What rate of interest, compounded continuously, did this investment earn?

9. An investor initially invests $10,000 in a speculative venture. Suppose that the investment earns 20% interest compounded continuously for 5 years and then 6% interest compounded continuously for 5 years thereafter.
   (a) How much does the $10,000 grow to after 10 years?
   (b) Suppose that the investor has the alternative of an investment paying 14% interest compounded continuously. Which investment is superior over a 10-year period, and by how much?

10. Two different bacteria colonies are growing near a pool of stagnant water. Suppose that the first colony initially has 1000 bacteria and doubles every 21 minutes. The second colony has 710,000 bacteria and doubles every 33 minutes. How much time will elapse before the first colony becomes as large as the second?

11. The population of a city t years after 1990 satisfies the differential equation $y' = .02y$. What is the growth constant? How fast will the population be growing when the population reaches 3 million people? At what level of population will the population be growing at the rate of 100,000 people per year?

12. A colony of bacteria is growing exponentially with growth constant .4, with time measured in hours. Determine the size of the colony when the colony is growing at the rate of 200,000 bacteria per hour. Determine the rate at which the colony will be growing when its size is 1,000,000.

13. The population of a certain country is growing exponentially. The total population (in millions) in t years is given by the function $P(t)$. Match each of the following answers with its corresponding question.

Answers:

  a. Solve $P(t) = 2$ for t.

  b. $P(2)$

  c. $P'(2)$

  d. Solve $P'(t) = 2$ for t.

  e. $y' = ky$

  f. Solve $P(t) = 2P(0)$ for t.

  g. $P_0 e^{kt}$, $k > 0$

  h. $P(0)$

Questions:

  A. How fast will the population be growing in 2 years?

  B. Give the general form of the function $P(t)$.

  C. How long will it take for the current population to double?

  D. What will be the size of the population in 2 years?

  E. What is the initial size of the population?

  F. When will the size of the population be 2 million?

  G. When will the population be growing at the rate of 2 million people per year?

  H. Give a differential equation satisfied by $P(t)$.

14. Suppose you have 80 grams of a certain radioactive material and the amount remaining after t years is given by the function $f(t)$ shown in Fig. 1.

(a) How much will remain after 5 years?

(b) When will 10 grams remain?

(c) What is the half-life of this radioactive material?

(d) At what rate will the radioactive material be disintegrating after 1 year?

(e) After how many years will the radioactive material be disintegrating at the rate of about 5 grams per year?

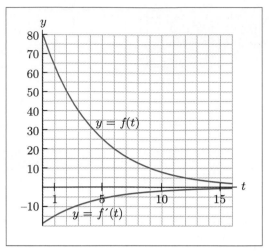

Figure 1.

15. A few years after money is deposited into the bank, the compound amount is \$1000 and it is growing at the rate of \$60 per year. What interest rate (compounded continuously) is the money earning?

16. The current balance in a savings account is \$1230 and the interest rate is 4.5%. At what rate is the compound amount currently growing?

17. Find the percentage rate of change of the function $f(t) = 50e^{.2t^2}$ at $t = 10$.

18. Find $E(p)$ for the demand function $q = 4000 - 40p^2$, and determine if demand is elastic or inelastic at $p = 5$.

19. Suppose that for a certain demand function, $E(8) = 1.5$. If the price is increased to \$8.16, estimate the percentage decrease in the quantity demanded. Will the revenue increase or decrease?

20. Find the percentage rate of change of the function
$$f(p) = \frac{1}{3p + 1} \text{ at } p = 1.$$

21. A company can sell $q = 1000p^2 e^{-.02(p+5)}$ calculators at a price of p dollars per calculator. The current price is \$200. If the price is decreased, will the revenue increase or decrease?

22. Consider a demand function of the form $q = ae^{-bp}$, where a and b are positive numbers. Find $E(p)$ and show that the elasticity equals 1 when $p = 1/b$.

23. Refer to Practice Problem 5.4. Out of 100 doctors in group A, none knew about the drug at time $t = 0$, but 66 of them were familiar with the drug after 13 months. Find the formula for $f(t)$.

24. The growth of the yellow nutsedge weed is described by a formula $f(t)$ of type (9) in Section 5.4. A typical weed has length 8 centimeters after 9 days, length 48 centimeters after 25 days, and reaches length 55 centimeters at maturity. Find the formula for $f(t)$.

25. When a rod of molten steel with a temperature of 1800°F is placed in a large vat of water at temperature 60°F, the temperature of the rod after t seconds is

$$f(t) = 60(1 + 29e^{-.15t})$$

degrees Fahrenheit. The graph of this function is shown in Fig. 2.

(a) What is the temperature of the rod after 11 seconds?

(b) At what rate is the temperature of the rod changing after 6 seconds?

(c) Approximately when is the temperature of the rod 200 degrees?

(d) Approximately when is the rod cooling at the rate of 200 degrees per second?

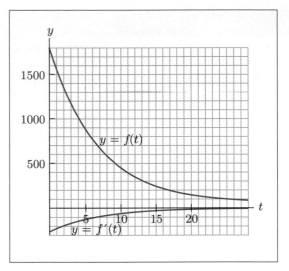

Figure 2.

Chapter Project

The following table of data summarizes the experiments of C. F. Gause, on the growth of colonies of yeast cells, which was cited at the beginning of Chapter 1.

Table 1	Growth of Yeast Cell Colonies		
Age of culture (hours)	Volume of *saccharomyces* yeast cells	Volume of mixed yeast cells	Volume of *schizosaccharomyces* yeast cells
6	0.37	0.5	
16	8.87	6.83	1.00
24	10.66	8.66	
29	12.50	9.07	1.70
40	13.27		
48	12.87	10.23	2.73
53	12.70	10.5	
72			4.87
93			5.67
117			5.80
141			5.83

See C. F. Gause, "Experimental Studies on the Struggle of Existence," *J. Exp. Biology,* 9 (1932), 389–402.

(a) Graph scatter plots corresponding to each type of yeast colony.

(b) For each type of yeast colony, make a table of its rate of growth for the various culture ages.

(c) Does the table of growth rates support a logistic growth model for this data? Explain your answer.

(d) Fit a logistic curve for each type of yeast colony as follows: Estimate the maximum colony size P by examining the data. Next, rewrite the logistic equation in the form

$$\frac{y}{P} = \frac{1}{1 + Ae^{-kt}}$$

$$\frac{P}{y} = 1 + Ae^{-kt}.$$

Substitute two data points for (t, y) to get two equations for A and k. Solve the resulting system of equations to obtain the values of A and k.

(e) How should you choose the data points in part (d) so that the resulting logistic curve is most likely to provide a good fit to the data?

The Definite Integral 6

There are two fundamental problems of calculus. The first is to find the slope of a curve at a point, and the second is to find the area of a region under a curve. These problems are quite simple when the curve is a straight line, as in Fig. 1. Both the slope of the line and the area of the shaded trapezoid can be calculated by geometric principles. When the graph consists of several line segments, as in Fig. 2, the slope of each line segment can be computed separately, and the area of the region can be found by adding the areas of the regions under each line segment.

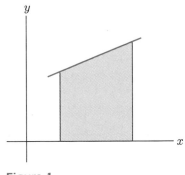

Figure 1.

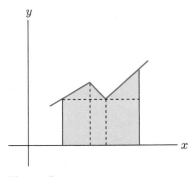

Figure 2.

Calculus is needed when the curves are not straight lines. We have seen that the slope problem is resolved with the derivative of a function. In this chapter, we describe how the area problem is connected with the notion of the "integral" of a function. Both the slope problem and the area problem were studied by the ancient Greeks and solved in special cases. But it was not until the development of calculus in the seventeenth century that the intimate connection between the two problems was discovered. In this chapter we will discuss this connection as stated in the fundamental theorem of calculus.

6.1 Antidifferentiation

We have developed several techniques for calculating the derivative $F'(x)$ of a function $F(x)$. In many applications, however, it is necessary to proceed in reverse. We are given the derivative $F'(x)$ and must determine the function $F(x)$. The process of determining $F(x)$ from $F'(x)$ is called *antidifferentiation*. The next example gives a typical application involving antidifferentiation.

▶ **Example 1** During the early 1970s, the annual worldwide rate of oil consumption grew exponentially with a growth constant of about .07. At the beginning of 1970, the rate was about 16.1 billion barrels of oil per year. Let $R(t)$ denote the rate of oil consumption at time t, where t is the number of years since the beginning of 1970. Then a reasonable model for $R(t)$ is given by

$$R(t) = 16.1e^{.07t}. \tag{1}$$

Use this formula for $R(t)$ to determine the total amount of oil that would have been consumed from 1970 to 1980 had this rate of consumption continued throughout the decade.

Solution Let $T(t)$ be the total amount of oil consumed from time 0 (1970) until time t. We wish to calculate $T(10)$, the amount of oil consumed from 1970 to 1980. We do this by first determining a formula for $T(t)$. Since $T(t)$ is the total oil consumed, the derivative $T'(t)$ is the *rate* of oil consumption, namely, $R(t)$. Thus, although we do not yet have a formula for $T(t)$, we do know that

$$T'(t) = 16.1e^{.07t}.$$

Thus the problem of determining a formula for $T(t)$ has been reduced to a problem of antidifferentiation: Find a function whose derivative is $R(t)$. We shall solve this particular problem after developing some techniques for solving antidifferentiation problems in general.   ◆

Suppose that $f(x)$ is a given function and $F(x)$ is a function having $f(x)$ as its derivative—that is, $F'(x) = f(x)$. We call $F(x)$ an *antiderivative* of $f(x)$.

▶ **Example 2** Find an antiderivative of $f(x) = x^2$.

Solution The derivative of x^3 is $3x^2$, which is almost the same as x^2 except for a factor of 3. To make this factor 1 instead of 3, take $F(x) = \frac{1}{3}x^3$. Then

$$F'(x) = \underbrace{\frac{d}{dx}\left(\frac{1}{3}x^3\right) = \frac{1}{3}\left(\frac{d}{dx}x^3\right)}_{\text{constant multiple rule}} = \frac{1}{3} \cdot 3x^2 = x^2.$$

So $F(x)$ is an antiderivative of x^2. Another antiderivative is $\frac{1}{3}x^3 + 5$, because the derivative of a constant function is zero:

$$\frac{d}{dx}\left(\frac{1}{3}x^3 + 5\right) = \frac{d}{dx}\left(\frac{1}{3}x^3\right) + \frac{d}{dx}(5) = x^2 + 0 = x^2.$$

In fact, if C is any constant, the function $F(x) = \frac{1}{3}x^3 + C$ is also an antiderivative of x^2, since

$$\frac{d}{dx}\left(\frac{1}{3}x^3 + C\right) = \frac{1}{3} \cdot 3x^2 + 0 = x^2.$$

(The derivative of a constant function is zero.)   ◆

▶ Example 3   Find an antiderivative of $f(x) = e^{-2x}$.

Solution   Recall that the derivative of e^{rx} is just a constant times e^{rx}. For an antiderivative of e^{-2x}, try a function of the form ke^{-2x}, where k is some constant to be determined. Then

$$\frac{d}{dx} ke^{-2x} = k \cdot (-2e^{-2x}) = -2ke^{-2x}.$$

Choose k to make $-2k = 1$; that is, choose $k = -\frac{1}{2}$. Then

$$\frac{d}{dx} \left(-\frac{1}{2} e^{-2x} \right) = \left(-\frac{1}{2} \right)(-2e^{-2x}) = 1 \cdot e^{-2x} = e^{-2x}.$$

Thus $-\frac{1}{2} e^{-2x}$ is an antiderivative of e^{-2x}. Also, for any constant C, the function $-\frac{1}{2} e^{-2x} + C$ is an antiderivative of e^{-2x}, because

$$\frac{d}{dx} \left(-\frac{1}{2} e^{-2x} + C \right) = e^{-2x} + 0 = e^{-2x}. \qquad \blacklozenge$$

Examples 2 and 3 illustrate the fact that if $F(x)$ is an antiderivative of $f(x)$, then so is $F(x) + C$, where C is any constant. (The derivative of the constant function C is zero.) The next theorem says that *all* antiderivatives of $f(x)$ can be produced in this way.

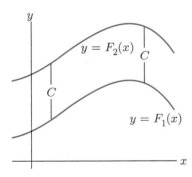

Figure 1. Two antiderivatives of the same function.

> **Theorem I**   If $F_1(x)$ and $F_2(x)$ are two antiderivatives of the same function $f(x)$, then $F_1(x)$ and $F_2(x)$ differ by a constant. In other words, there is a constant C such that
>
> $$F_2(x) = F_1(x) + C.$$

Geometrically, the graph of any antiderivative $F_2(x)$ is obtained by shifting the graph of $F_1(x)$ vertically. See Fig. 1.

Our verification of Theorem I is based on the following fact, which is important in its own right.

> **Theorem II**   If $F'(x) = 0$ for all x, then $F(x) = C$ for some constant C.

It is easy to see why Theorem II is reasonable. (A formal proof of the theorem requires an important theoretical result called the mean value theorem.) If $F'(x) = 0$ for all x, then the curve $y = F(x)$ has slope equal to zero at every point. Thus the tangent line to $y = F(x)$ at any point is horizontal, which implies that the graph of $y = F(x)$ is a horizontal line. (Try to draw the graph of a function with a horizontal tangent everywhere. There is no choice but to keep your pencil moving on a constant, horizontal line!) If the horizontal line is $y = C$, then $F(x) = C$ for all x.

Verification of Theorem I If $F_1(x)$ and $F_2(x)$ are two antiderivatives of $f(x)$, then the function $F(x) = F_2(x) - F_1(x)$ has the derivative

$$F'(x) = F_2'(x) - F_1'(x) = f(x) - f(x) = 0.$$

So, by Theorem II, we know that $F(x) = C$ for some constant C. In other words, $F_2(x) - F_1(x) = C$, so that

$$F_2(x) = F_1(x) + C,$$

which is Theorem I.

Using Theorem I, we can find *all* antiderivatives of a given function once we know one antiderivative. For instance, since one antiderivative of x^2 is $\frac{1}{3}x^3$ (Example 2), all antiderivatives of x^2 have the form $\frac{1}{3}x^3 + C$, where C is a constant.

Suppose that $f(x)$ is a function whose antiderivatives are $F(x) + C$. The standard way to express this fact is to write

$$\int f(x)\,dx = F(x) + C.$$

The symbol $\int$ is called an *integral sign*. The entire notation $\int f(x)\,dx$ is called an *indefinite integral* and stands for antidifferentiation of the function $f(x)$. We always record the variable of interest by prefacing it by the letter d. For example, if the variable of interest is t rather than x, then we write $\int f(t)\,dt$ for the antiderivative.

▶ **Example 4** Determine

(a) $\displaystyle\int x^r\,dx$, r a constant $\neq -1$ \qquad (b) $\displaystyle\int e^{kx}\,dx$, k a constant $\neq 0$

Solution   (a) By the constant-multiple and power rules,

$$\frac{d}{dx}\left(\frac{1}{r+1}x^{r+1}\right) = \frac{1}{r+1}\cdot\frac{d}{dx}x^{r+1} = \frac{1}{r+1}\cdot(r+1)x^r = x^r.$$

Thus $x^{r+1}/(r+1)$ is an antiderivative of x^r. Letting C represent any constant, we have

$$\int x^r\,dx = \frac{1}{r+1}x^{r+1} + C, \qquad r \neq -1. \tag{2}$$

(b) An antiderivative of e^{kx} is e^{kx}/k, since

$$\frac{d}{dx}\left(\frac{1}{k}e^{kx}\right) = \frac{1}{k}\cdot\frac{d}{dx}e^{kx} = \frac{1}{k}(ke^{kx}) = e^{kx}.$$

Hence

$$\int e^{kx}\,dx = \frac{1}{k}e^{kx} + C, \qquad k \neq 0. \tag{3}$$

◆

Formula (2) does not give an antiderivative of x^{-1} because $1/(r+1)$ is undefined for $r = -1$. However, we know that for $x \neq 0$, the derivative of $\ln|x|$ is $1/x$. Hence $\ln|x|$ is an antiderivative of $1/x$, and we have

$$\int \frac{1}{x}\, dx = \ln|x| + C, \qquad x \neq 0. \tag{4}$$

Formulas (2), (3), and (4) are each obtained by "reversing" a familiar differentiation rule. In a similar fashion, one may use the sum rule and constant-multiple rule for derivatives to obtain corresponding rules for antiderivatives:

$$\int [f(x) + g(x)]\, dx = \int f(x)\, dx + \int g(x)\, dx \tag{5}$$

$$\int k f(x)\, dx = k \int f(x)\, dx, \quad k \text{ a constant.} \tag{6}$$

In words, (5) says that a sum of functions may be antidifferentiated term by term, and (6) says that a constant multiple may be moved through the integral sign.

▶ **Example 5** Compute:

$$\int \left(x^{-3} + 7e^{5x} + \frac{4}{x} \right) dx.$$

Solution Using the preceding rules, we have

$$\int \left(x^{-3} + 7e^{5x} + \frac{4}{x} \right) dx = \int x^{-3}\, dx + \int 7e^{5x}\, dx + \int \frac{4}{x}\, dx$$

$$= \int x^{-3}\, dx + 7 \int e^{5x}\, dx + 4 \int \frac{1}{x}\, dx$$

$$= \frac{1}{-2} x^{-2} + 7 \left(\frac{1}{5} e^{5x} \right) + 4 \ln|x| + C$$

$$= -\frac{1}{2} x^{-2} + \frac{7}{5} e^{5x} + 4 \ln|x| + C. \qquad \blacklozenge$$

After some practice, most of the intermediate steps shown in the solution of Example 5 can be omitted.

A function has infinitely many different antiderivatives, corresponding to various choices of the constant C. In applications, it is often necessary to satisfy an additional condition, which then determines a specific value of C.

▶ **Example 6** Find the function $f(x)$ for which $f'(x) = x^2 - 2$ and $f(1) = \frac{4}{3}$.

Solution The unknown function $f(x)$ is an antiderivative of $x^2 - 2$. One antiderivative of $x^2 - 2$ is $\frac{1}{3} x^3 - 2x$. Therefore, by Theorem I,

$$f(x) = \tfrac{1}{3} x^3 - 2x + C, \quad C \text{ a constant.}$$

Figure 2 shows the graphs of $f(x)$ for several choices of C. We want the function whose graph passes through $(1, \frac{4}{3})$. To find the value of C that makes $f(1) = \frac{4}{3}$, we set

$$\tfrac{4}{3} = f(1) = \tfrac{1}{3}(1)^3 - 2(1) + C = -\tfrac{5}{3} + C$$

and find $C = \frac{4}{3} + \frac{5}{3} = 3$. Therefore, $f(x) = \frac{1}{3}x^3 - 2x + 3$. ◆

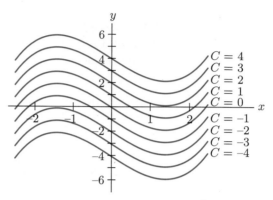

Figure 2. Several antiderivatives of $x^2 - 2$.

Having introduced the basics of antidifferentiation, let us now solve the oil-consumption problem.

Solution of Example 1 (Continued) The rate of oil consumption at time t is $R(t) = 16.1e^{.07t}$ billion barrels per year. Moreover, we observed that the total consumption $T(t)$, from time 0 to time t, is an antiderivative of $R(t)$. Using (3) and (6), we have

$$T(t) = \int 16.1e^{.07t}\, dt = \frac{16.1}{.07}e^{.07t} + C = 230e^{.07t} + C,$$

where C is a constant. However, in our particular example, $T(0) = 0$, since $T(0)$ is the amount of oil used from time 0 to time 0. Therefore, the constant C must satisfy

$$0 = T(0) = 230e^{.07(0)} + C = 230 + C$$
$$C = -230.$$

Therefore,

$$T(t) = 230e^{.07t} - 230 = 230(e^{.07t} - 1).$$

The total amount of oil that would have been consumed from 1970 to 1980 is

$$T(10) = 230(e^{.07(10)} - 1) \approx 233 \text{ billion barrels.} \qquad ◆$$

Antidifferentiation can be used to solve a variety of applied problems, of which the next two examples are typical.

▶ **Example 7** A rocket is fired vertically into the air. Its velocity at t seconds after lift-off is $v(t) = 6t + .5$ meter per second. Before launch, the top of the rocket is 8 meters above the launch pad. Find the height of the rocket (measured from the top of the rocket to the launch pad) at time t.

Solution   If $s(t)$ denotes the height of the rocket at time t, then $s'(t)$ is the rate at which the height is changing. That is, $s'(t) = v(t)$, and therefore $s(t)$ is an antiderivative of $v(t)$. Thus

$$s(t) = \int v(t)\, dt = \int (6t + .5)\, dt = 3t^2 + .5t + C,$$

where C is a constant. When $t = 0$, the rocket's height is 8 meters. That is, $s(0) = 8$ and

$$8 = s(0) = 3(0)^2 + .5(0) + C = C.$$

Thus $C = 8$ and

$$s(t) = 3t^2 + .5t + 8.$$ ◆

▶ **Example 8**   A company's marginal cost function is $.015x^2 - 2x + 80$ dollars, where x denotes the number of units produced in one day. The company has fixed costs of \$1000 per day.

(a) Find the cost of producing x units per day.

(b) Suppose the current production level is $x = 30$. Determine the amount costs will rise if the production level is raised to $x = 60$ units.

Solution   (a) Let $C(x)$ be the cost of producing x units in one day. The derivative $C'(x)$ is the marginal cost. In other words, $C(x)$ is an antiderivative of the marginal cost function. Thus

$$C(x) = \int (.015x^2 - 2x + 80)\, dx = .005x^3 - x^2 + 80x + C.$$

The \$1000 fixed costs are the costs incurred when producing 0 units. That is, $C(0) = 1000$. So

$$1000 = C(0) = .005(0)^3 - (0)^2 + 80(0) + C.$$

Therefore, $C = 1000$, and

$$C(x) = .005x^3 - x^2 + 80x + 1000.$$

(b) The cost when $x = 30$ is $C(30)$, and the cost when $x = 60$ is $C(60)$. So the *increase* in cost when production is raised from $x = 30$ to $x = 60$ is $C(60) - C(30)$. We compute

$$C(60) = .005(60)^3 - (60)^2 + 80(60) + 1000 = 3280$$
$$C(30) = .005(30)^3 - (30)^2 + 80(30) + 1000 = 2635.$$

Thus the increase in cost is $3280 - 2635 = \$645$.   ◆

Incorporating Technology   The graph of an antiderivative of a function can be obtained without actually antidifferentiating the function. See Appendices A–D for details. Figure 3 shows the graph of the solution to Example 6.

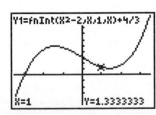

Figure 3.

1. Determine each of the following.

(a) $\int t^{7/2}\, dt$

(b) $\int \left(\dfrac{x^3}{3} + \dfrac{3}{x^3} + \dfrac{3}{x} \right) dx$

2. Find the value of k that makes the following antidifferentiation formula true.

$$\int (1 - 2x)^3\, dx = k(1 - 2x)^4 + C.$$

▶ Exercises 6.1

Find all antiderivatives of each of the following functions.

1. $f(x) = x$

2. $f(x) = 9x^8$

3. $f(x) = e^{3x}$

4. $f(x) = e^{-3x}$

5. $f(x) = 3$

6. $f(x) = -4x$

In Exercises 7–22, find the value of k that makes the antidifferentiation formula true. (*Note:* You can check your answer without looking in the answer section. How?)

7. $\int x^{-5}\, dx = kx^{-4} + C$

8. $\int x^{1/3}\, dx = kx^{4/3} + C$

9. $\int \sqrt{x}\, dx = kx^{3/2} + C$

10. $\int \dfrac{6}{x^3}\, dx = \dfrac{k}{x^2} + C$

11. $\int \dfrac{10}{t^6}\, dt = kt^{-5} + C$

12. $\int \dfrac{3}{\sqrt{t}}\, dt = k\sqrt{t} + C$

13. $\int 5e^{-2t}\, dt = ke^{-2t} + C$

14. $\int 3e^{t/10}\, dt = ke^{t/10} + C$

15. $\int 2e^{4x-1}\, dx = ke^{4x-1} + C$

16. $\int \dfrac{4}{e^{3x+1}}\, dx = \dfrac{k}{e^{3x+1}} + C$

17. $\int (5x - 7)^{-2}\, dx = k(5x - 7)^{-1} + C$

18. $\int \sqrt{x + 1}\, dx = k(x + 1)^{3/2} + C$

19. $\int (4 - x)^{-1}\, dx = k \ln |4 - x| + C$

20. $\int \dfrac{7}{(8 - x)^4}\, dx = \dfrac{k}{(8 - x)^3} + C$

21. $\int (3x + 2)^4\, dx = k(3x + 2)^5 + C$

22. $\int (2x - 1)^3\, dx = k(2x - 1)^4 + C$

Determine the following.

23. $\int (x^2 - x - 1)\, dx$

24. $\int (x^3 + 6x^2 - x)\, dx$

25. $\int \left(\dfrac{2}{\sqrt{x}} - 3\sqrt{x} \right) dx$

26. $\int \left[\dfrac{\sqrt{t}}{4} - 4(t - 3)^{-2} \right] dt$

27. $\int \left(4 - 5e^{-5t} + \dfrac{e^{2t}}{3} \right) dt$

28. $\int (e^2 + 3t^2 - 2e^{3t})\, dt$

Find all functions $f(t)$ with the following property.

29. $f'(t) = t^{3/2}$

30. $f'(t) = \dfrac{4}{6 + t}$

31. $f'(t) = 0$

32. $f'(t) = t^2 - 5t - 7$

Find all functions $f(x)$ with the following properties.

33. $f'(x) = x$, $f(0) = 3$

34. $f'(x) = 8x^{1/3}$, $f(1) = 4$

35. $f'(x) = \sqrt{x} + 1$, $f(4) = 0$

36. $f'(x) = x^2 + \sqrt{x}$, $f(1) = 3$

37. Figure 4 shows the graphs of several functions $f(x)$ for which $f'(x) = \dfrac{2}{x}$. Find the expression for the function $f(x)$ whose graph passes through $(1, 2)$.

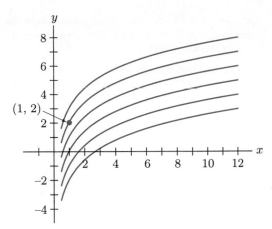

Figure 4.

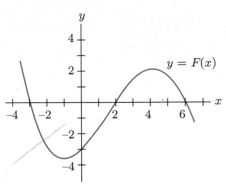

Figure 6.

38. Figure 5 shows the graphs of several functions $f(x)$ for which $f'(x) = \frac{1}{3}$. Find the expression for the function $f(x)$ whose graph passes through $(6,3)$.

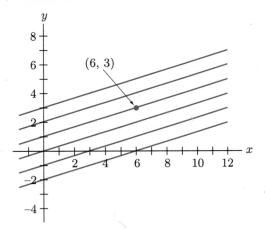

Figure 5.

39. Which of the following is $\int \ln x \, dx$?

(a) $\dfrac{1}{x} + C$

(b) $x \cdot \ln x - x + C$

(c) $\dfrac{1}{2} \cdot (\ln x)^2 + C$

40. Which of the following is $\int x\sqrt{x+1} \, dx$?

(a) $\frac{2}{5}(x+1)^{5/2} - \frac{2}{3}(x+1)^{3/2} + C$

(b) $\frac{1}{2}x^2 \cdot \frac{2}{3}(x+1)^{3/2} + C$

41. Figure 6 contains the graph of a function $F(x)$. On the same coordinate system, draw the graph of the function $G(x)$ having the properties $G(0) = 0$ and $G'(x) = F'(x)$ for each x.

42. Figure 7 contains an antiderivative of the function $f(x)$. Draw the graph of another antiderivative of $f(x)$.

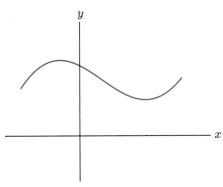

Figure 7.

43. The function $g(x)$ in Fig. 8 was obtained by shifting the graph of $f(x)$ up three units. If $f'(5) = \frac{1}{4}$, what is $g'(5)$?

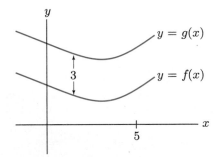

Figure 8.

44. The function $g(x)$ in Fig. 9 was obtained by shifting the graph of $f(x)$ up two units. What is the derivative of $h(x) = g(x) - f(x)$?

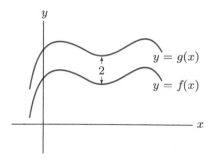

Figure 9.

45. A ball is thrown upward from a height of 256 feet above the ground, with an initial velocity of 96 feet per second. From physics it is known that the velocity at time t is $96 - 32t$ feet per second.

(a) Find $s(t)$, the function giving the height of the ball at time t.

(b) How long will the ball take to reach the ground?

(c) How high will the ball go?

46. A rock is dropped from the top of a 400-foot cliff. Its velocity at time t seconds is $v(t) = -32t$ feet per second.

(a) Find $s(t)$, the height of the rock above the ground at time t.

(b) How long will the rock take to reach the ground?

(c) What will be its velocity when it hits the ground?

47. Let $P(t)$ be the total output of a factory assembly line after t hours of work. Suppose that the rate of production at time t is $60 + 2t - \frac{1}{4}t^2$ units per hour. Find the formula for $P(t)$. [*Hint:* The rate of production is $P'(t)$ and $P(0) = 0$.]

48. After t hours of operation a coal mine is producing coal at the rate of $40 + 2t - \frac{1}{5}t^2$ tons of coal per hour. Find a formula for the total output of the coal mine after t hours of operation.

49. A package of frozen strawberries is taken from a freezer at $-5°\text{C}$ into a room at $20°\text{C}$. At time t the average temperature of the strawberries is increasing at the rate of $10e^{-.4t}$ degrees Celsius per hour. Find the temperature of the strawberries at time t.

50. A flu epidemic hits a town. Let $P(t)$ be the number of persons sick with the flu at time t, where time is measured in days from the beginning of the epidemic

and $P(0) = 100$. Suppose that after t days the flu is spreading at the rate of $120t - 3t^2$ people per day. Find the formula for $P(t)$.

51. A small tie shop finds that at a sales level of x ties per day its marginal profit is $MP(x)$ dollars per tie, where $MP(x) = 1.30 + .06x - .0018x^2$. Also, the shop will lose $95 per day at a sales level of $x = 0$. Find the profit from operating the shop at a sales level of x ties per day.

52. A soap manufacturer estimates that its marginal cost of producing soap powder is $.2x + 1$ hundred dollars per ton at a production level of x tons per day. Fixed costs are $200 per day. Find the cost of producing x tons of soap powder per day.

53. The United States has been consuming iron ore at the rate of $R(t)$ million metric tons per year at time t, where $t = 0$ corresponds to 1980 and $R(t) = 94e^{.016t}$. Find a formula for the total U.S. consumption of iron ore from 1980 until time t.

54. Since 1987, the rate of production of natural gas in the United States has been approximately $R(t)$ quadrillion British thermal units per year at time t, with $t = 0$ corresponding to 1987 and $R(t) = 17.04e^{.016t}$. Find a formula for the total U.S. production of natural gas from 1987 until time t.

55. Suppose the drilling of an oil well has a fixed cost of $10,000 and a marginal cost of $C'(x) = 1000 + 50x$ dollars per foot, where x is the depth in feet. Find the expression for $C(x)$, the total cost of drilling x feet. [*Note:* $C(0) = 10,000$.]

Technology Exercises

In Exercises 56–59, find an antiderivative of $f(x)$—call it $F(x)$—and then compare the graphs of $F(x)$ and $f(x)$ in the given window to check that the expression for $F(x)$ is reasonable. [That is, determine whether the two graphs are consistent. When $F(x)$ has a relative extreme point, $f(x)$ should be zero; when $F(x)$ is increasing, $f(x)$ should be positive, and so on.]

56. $f(x) = 2x - e^{-.02x}$, $[-10, 10]$ *by* $[-20, 100]$

57. $f(x) = e^{2x} + e^{-x} + \frac{1}{2}x^2$, $[-2.4, 1.7]$ *by* $[-10, 10]$

58. $f(x) = 8x^2 - 3e^{-x} + x^3$, $[-3, 2]$ *by* $[-4, 7]$

59. $f(x) = \frac{1}{10}(x^3 - 9x^2 + 5x) + 3$, $[-4, 11]$ *by* $[-14, 7]$

1. (a) $\int t^{7/2}\,dt = \frac{1}{\frac{9}{2}}t^{9/2} + C = \frac{2}{9}t^{9/2} + C$

   (b) $\int \left(\frac{x^3}{3} + \frac{3}{x^3} + \frac{3}{x} \right) dx = \int \left(\frac{1}{3}\cdot x^3 + 3x^{-3} + 3\cdot\frac{1}{x} \right) dx$

   $$= \frac{1}{3}\left(\frac{1}{4}x^4 \right) + 3\left(-\frac{1}{2}x^{-2} \right) + 3\ln|x| + C$$

   $$= \frac{1}{12}x^4 - \frac{3}{2}x^{-2} + 3\ln|x| + C$$

2. Since we are told that the antiderivative has the general form $k(1-2x)^4$, all we have to do is determine the value of k. Differentiating, we obtain

   $$4k(1-2x)^3(-2) \quad \text{or} \quad -8k(1-2x)^3,$$

   which is supposed to equal $(1-2x)^3$. Therefore, $-8k = 1$, so $k = -\frac{1}{8}$.

6.2 Areas and Riemann Sums

This section and the next reveal the important connection between antiderivatives and areas of regions under curves. Although the full story will have to wait until the next section, we can give a hint now by describing how "area" is related to a problem solved in Section 6.1.

Example 8 in Section 6.1 concerned a company's marginal cost function, $f(x) = .015x^2 - 2x + 80$. A calculation with an antiderivative of $f(x)$ showed that if the production level is increased from $x = 30$ to $x = 60$ units per day, the change in total costs is \$645. As we'll see, this change in cost exactly equals the area of the region under the graph of the marginal cost curve in Fig. 1 from $x = 30$ to $x = 60$. First, however, we need to learn how to find areas of such regions.

Area Under a Graph If $f(x)$ is a continuous nonnegative function on the interval $a \le x \le b$, we refer to the area of the region shown in Fig. 2 as the *area under the graph of $f(x)$ from a to b.*

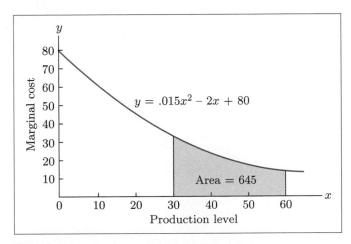

Figure 1. Area under a marginal cost curve.

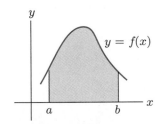

Figure 2. Area under a graph.

The computation of the area in Fig. 2 is not a trivial matter when the top boundary of the region is curved. However, we can *estimate* the area to any desired degree of accuracy. The basic idea is to construct rectangles whose total area is approximately the same as the area to be computed. The area of each rectangle, of course, is easy to compute.

Figure 3 shows three rectangular approximations to the area under a graph. When the rectangles are thin, the mismatch between the rectangles and the region under the graph is quite small. In general, a rectangular approximation can be made as close as desired to the exact area simply by making the width of the rectangles sufficiently small.

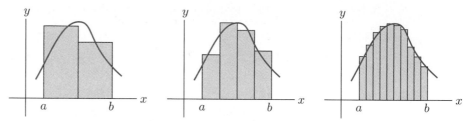

Figure 3. Approximating a region with rectangles.

Given a continuous nonnegative function $f(x)$ on the interval $a \leq x \leq b$, divide the x-axis interval into n equal subintervals, where n represents some positive integer. Such a subdivision is called a *partition* of the interval from a to b. Since the entire interval is of width $b - a$, the width of each of the n subintervals is $(b - a)/n$. For brevity, denote this width by Δx. That is,

$$\Delta x = \frac{b - a}{n} \quad \text{(width of one subinterval)}.$$

In each subinterval, select a point. (Any point in the subinterval will do.) Let x_1 be the point selected from the first subinterval, x_2 the point from the second subinterval, and so on. These points are used to form rectangles that approximate the region under the graph of $f(x)$. Construct the first rectangle with height $f(x_1)$ and the first subinterval as base, as in Fig. 4. The top of the rectangle touches the graph directly above x_1. Notice that

$$[\text{area of first rectangle}] = [\text{height}][\text{width}] = f(x_1)\,\Delta x.$$

The second rectangle rests on the second subinterval and has height $f(x_2)$. Thus

$$[\text{area of second rectangle}] = [\text{height}][\text{width}] = f(x_2)\,\Delta x.$$

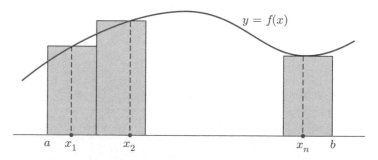

Figure 4. Rectangles with heights $f(x_1), \ldots, f(x_n)$.

Continuing in this way, we construct n rectangles with a combined area of

$$f(x_1)\,\Delta x + f(x_2)\,\Delta x + \cdots + f(x_n)\,\Delta x. \tag{1}$$

A sum as in (1) is called a *Riemann sum*. It provides an approximation to the area under the graph of $f(x)$ when $f(x)$ is nonnegative and continuous. In fact, as the number of subintervals increases indefinitely, the Riemann sums (1) approach a limiting value, the area under the graph.*

The value in (1) is easier to calculate when written as

$$[f(x_1) + f(x_2) + \cdots + f(x_n)]\,\Delta x.$$

This calculation requires only one multiplication.

▶ **Example 1**   Estimate the area under the graph of the marginal cost function $f(x) = .015x^2 - 2x + 80$ from $x = 30$ to $x = 60$. Use partitions of 5, 20, and 100 subintervals. Use the midpoints of the subintervals as $x_1, x_2, \ldots, x_n$ to construct the rectangles. See Fig. 5.

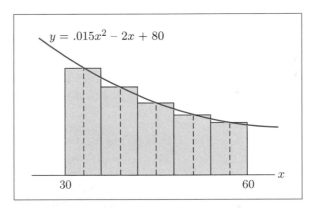

Figure 5. Estimating the area under a marginal cost curve.

Solution   The partition of $30 \le x \le 60$ with $n = 5$ is shown in Fig. 6. The length of each subinterval is

$$\Delta x = \frac{60 - 30}{5} = 6.$$

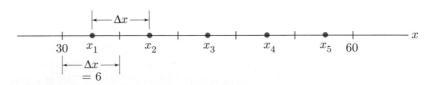

Figure 6. A partition of the interval $30 \le x \le 60$.

---

*Riemann sums are named after the nineteenth-century German mathematician G. B. Riemann (pronounced "Reemahn"), who used them extensively in his work on calculus. The concept of a Riemann sum has several uses: to approximate areas under curves, to construct mathematical models in applied problems, and to give a formal definition of area. In the next section, Riemann sums are used to define the definite integral of a function.

Observe that the first midpoint is $\Delta x/2$ units from the left endpoint, and the midpoints themselves are Δx units apart. The first midpoint is $x_1 = 30 + \Delta x/2 = 30 + 3 = 33$. Subsequent midpoints are found by successively adding $\Delta x = 6$.

$$\text{midpoints: } 33, 39, 45, 51, 57$$

The corresponding estimate for the area under the graph of $f(x)$ is

$$f(33)\,\Delta x + f(39)\,\Delta x + f(45)\,\Delta x + f(51)\,\Delta x + f(57)\,\Delta x$$
$$= [f(33) + f(39) + f(45) + f(51) + f(57)]\,\Delta x$$
$$= [30.335 + 24.815 + 20.375 + 17.015 + 14.735] \cdot 6$$
$$= 107.275 \cdot 6 = 643.65.$$

A similar calculation with 20 subintervals produces an area estimate of 644.916. With 100 subintervals the estimate is 644.997.   ◆

The approximations in Example 1 seem to confirm the claim made at the beginning of the section that the area under the marginal cost curve equals the change in total cost, \$645. The verification of this fact will be given in the next section.

Although the midpoints of subintervals are often selected as the $x_1, x_2, \ldots, x_n$ in a Riemann sum, left endpoints and right endpoints may also be convenient.

▶ Example 2   Use a Riemann sum with $n = 4$ to estimate the area under the graph of $f(x) = x^2$ from 1 to 3. Select the right endpoints of the subintervals as x_1, x_2, x_3, x_4.

Solution   Here $\Delta x = (3 - 1)/4 = .5$. The right endpoint of the first subinterval is $1 + \Delta x = 1.5$. Subsequent right endpoints are obtained by successively adding .5, as follows:

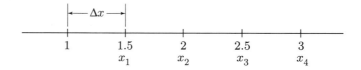

The corresponding Riemann sum is

$$f(x_1)\,\Delta x + f(x_2)\,\Delta x + f(x_3)\,\Delta x + f(x_4)\,\Delta x$$
$$= [f(x_1) + f(x_2) + f(x_3) + f(x_4)]\Delta x$$
$$= [(1.5)^2 + (2)^2 + (2.5)^2 + (3)^2](.5)$$
$$= [2.25 + 4 + 6.25 + 9](.5)$$
$$= 21.5 \cdot (.5) = 10.75.$$

The rectangles used for this Riemann sum are shown in Fig. 7. The right endpoints here give an area estimate that is obviously greater than the exact area. Midpoints would work better. But if the rectangles are sufficiently narrow, even a Riemann sum using right endpoints will be close to the exact area.   ◆

▶ Example 3   To estimate the area of a 100-foot-wide waterfront lot, a surveyor measured the distance from the street to the waterline at 20-foot intervals, starting 10 feet from one corner of the lot. Use the data to construct a Riemann sum approximation to the area of the lot. See Fig. 8.

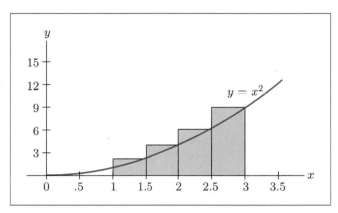

Figure 7. A Riemann sum using right endpoints.

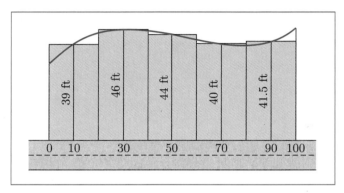

Figure 8. Survey of a waterfront property.

Solution   Treat the street as the x-axis, and consider the waterline along the property as the graph of a function $f(x)$ over the interval from 0 to 100. The five "vertical" distances give $f(x_1), \ldots, f(x_5)$, where $x_1 = 10, \ldots, x_5 = 90$. Since there are five points $x_1, \ldots, x_5$ spread across the interval $0 \leq x \leq 100$, we partition the interval into five subintervals, with $\Delta x = \frac{100}{5} = 20$. Fortunately, each subinterval contains one x_i. (In fact, each x_i is the midpoint of a subinterval.) Thus the area of the lot is approximated by the Riemann sum

$$
\begin{aligned}
f(x_1)\,\Delta x + \cdots + f(x_5)\,\Delta x &= [f(x_1) + \cdots + f(x_5)]\,\Delta x \\
&= [39 + 46 + 44 + 40 + 41.5] \cdot 20 \\
&= 210.5 \cdot 20 = 4210 \text{ square feet.}
\end{aligned}
$$

For a better estimate of the area, the surveyor will have to make more measurements from the street to the waterline. Note that we are able to approximate the area without ever knowing an analytic expression for the function $f(x)$.   ◆

The final example shows how Riemann sums arise in an application. Since the variable is time, t, we write Δt in place of Δx.

▶ **Example 4**   The velocity of a rocket at time t is $v(t)$ feet per second. Construct a Riemann sum that estimates how far the rocket travels in the first 10 seconds. (*Note*: Don't calculate the Riemann sum. Just set it up.) What happens when the number of subintervals in the partition increases without bound?

Solution   Partition the interval $0 \leq t \leq 10$ into n subintervals of width Δt, and select points $t_1, t_2, \ldots, t_n$ from these subintervals. Although the rocket's velocity is not constant, it does not change much during a small subinterval of time. So we may use $v(t_1)$ as an approximation of the rocket's velocity during the first subinterval. Over this time subinterval of length $\Delta t = (10 - 0)/n$,

$$\text{[distance traveled]} \approx \text{[velocity]} \cdot \text{[time]}$$
$$= v(t_1) \, \Delta t. \tag{2}$$

The distance traveled during the second time subinterval is approximately $v(t_2)\Delta t$, and so on. Thus an estimate for the total distance traveled is

$$v(t_1) \, \Delta t + v(t_2) \, \Delta t + \cdots + v(t_n) \, \Delta t. \tag{3}$$

The sum in (3) is a Riemann sum for the velocity function on the interval $0 \le t \le 10$. As n increases, such a Riemann sum approaches the area under the graph of the velocity function. However, from our derivation of (3), it seems reasonable that as n increases, the sums become a better and better estimate of the total distance traveled. We conclude that

$$\begin{bmatrix} \text{total distance rocket travels} \\ \text{during the first 10 seconds} \end{bmatrix} = \begin{bmatrix} \text{area under graph of} \\ v(t) \text{ over } 0 \le t \le 10 \end{bmatrix}.$$

See Fig. 9. ◆

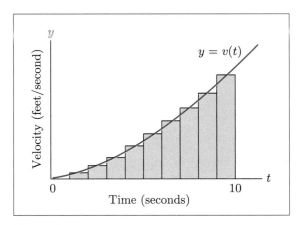

Figure 9. Area under a velocity curve.

It is helpful to look at Example 4 from a slightly different point of view. The height of the rocket is an increasing quantity and $v(t)$ is the *rate* of change of height at time t. The area under the graph of $v(t)$ from time $a = 0$ to time $b = 10$ is the *amount* of increase in the height during the first 10 seconds.

This connection between the rate of change of a function and the amount of increase of the function generalizes to a variety of situations.

If a quantity is increasing, then the area under the rate of change function from a to b is the amount of increase in the quantity from a to b.

Table 1 shows four instances of this principle. The result is justified in the next section. The first example in the table was discussed in Example 1.

Table 1	Interpretation of Areas		
Function		*a* to *b*	**Area Under the Graph from *a* to *b***
Marginal cost at production level x		30 to 60	Additional cost when production is increased from 30 to 60 units
Rate of sulfur emissions from a power plant t years after 1990		1 to 3	Amount of sulfur released from 1991 to 1993
Birth rate t years after 1980 (in babies per year)		0 to 10	Number of babies born from 1980 to 1990
Rate of gas consumption t years after 1985		3 to 6	Amount of gas used from 1988 to 1991

```
sum(seq(Y₁,X,33,
57,6))*6
                643.65
sum(seq(Y₂,X,1.5
,3,.5))*.5
                 10.75
```

Figure 10.
$Y_1 = .015X^2 - 2X + 80$
$Y_2 = X^2$

Incorporating Technology When the points selected for a Riemann sum are all midpoints, all left endpoints, or all right endpoints, then $[f(x_1) + f(x_2) + \cdots + f(x_n)]$ is the sum of the sequence of values $f(x_1), f(x_2), \ldots, f(x_n)$ where successive numbers $x_1, x_2, \ldots, x_n$ each differ by the value Δx. If $\mathbf{Y_1}$ represents the function $f(x)$, then this sum of a sequence can be evaluated on a graphing calculator as $\mathbf{sum(seq(Y_1,X,x_1,x_n,\Delta x))}$. Therefore, the entire Riemann sum can be evaluated as

$$\mathbf{sum(seq(Y_1,X,x_1,x_n,\Delta x)) * \Delta x}.$$

Figure 10 shows the computations for Examples 1 and 2. See Appendices A–D.

Practice Problems 6.2

1. Determine Δx and the midpoints of the subintervals formed by partitioning the interval $-2 \le x \le 2$ into five subintervals.

2. The graph in Fig. 11 gives the rate at which new jobs were created (in millions of jobs per year) in the United States, where $t = 0$ corresponds to 1983. For instance, on January 1, 1986, jobs were being created at the rate of 2.4 million new jobs per year. Interpret the area of the shaded region.

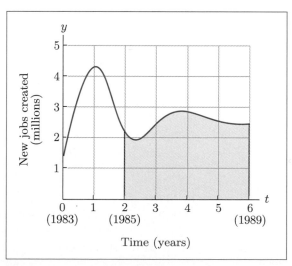

Figure 11. Rate of job creation.

▶ Exercises 6.2

Determine Δx and the midpoints of the subintervals formed by partitioning the given interval into n subintervals. (*Hint*: Decimals are sometimes easier to use than fractions.)

1. $0 \le x \le 2$; $n = 4$ **2.** $0 \le x \le 3$; $n = 6$

3. $1 \le x \le 4$; $n = 5$ **4.** $3 \le x \le 5$; $n = 5$

In Exercises 5–10, use a Riemann sum to approximate the area under the graph of $f(x)$ on the given interval, with selected points as specified.

5. $f(x) = x^2$; $1 \le x \le 3$, $n = 4$, midpoints of subintervals

6. $f(x) = x^2$; $-2 \le x \le 2$, $n = 4$, midpoints of subintervals

7. $f(x) = x^3$; $1 \le x \le 3$, $n = 5$, left endpoints

8. $f(x) = x^3$; $0 \le x \le 1$, $n = 5$, right endpoints

9. $f(x) = e^{-x}$; $2 \le x \le 3$, $n = 5$, right endpoints

10. $f(x) = \ln x$; $2 \le x \le 4$, $n = 5$, left endpoints

In Exercises 11–14, use a Riemann sum to approximate the area under the graph of $f(x)$ in Fig. 12 on the given interval, with selected points as specified. Draw the approximating rectangles.

11. $0 \le x \le 8$, $n = 4$, midpoints of subintervals

12. $3 \le x \le 7$, $n = 4$, left endpoints

13. $4 \le x \le 9$, $n = 5$, right endpoints

14. $1 \le x \le 7$, $n = 3$, midpoints of subintervals

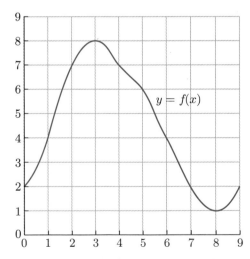

Figure 12. Graph for Exercises 11–14.

15. Use a Riemann sum with $n = 4$ and left endpoints to estimate the area under the graph of $f(x) = 4 - x$ on the interval $1 \le x \le 4$. Then repeat with $n = 4$ and midpoints. Compare the answers with the exact answer, 4.5, which can be computed from the formula for the area of a triangle.

16. Use a Riemann sum with $n = 4$ and right endpoints to estimate the area under the graph of $f(x) = 2x - 4$ on the interval $2 \le x \le 3$. Then repeat with $n = 4$ and midpoints. Compare the answers with the exact answer, 1, which can be computed from the formula for the area of a triangle.

17. The graph of the function $f(x) = \sqrt{1 - x^2}$ on the interval $-1 \le x \le 1$ is a semicircle. The area under the graph is $\frac{1}{2}\pi(1)^2 = \pi/2 = 1.57080$, to five decimal places. Use a Riemann sum with $n = 5$ and midpoints to estimate the area under the graph. See Fig. 13. Carry out the calculations to five decimal places and compute the error (the difference between the estimate and 1.57080).

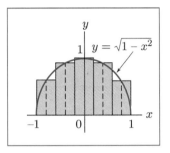

Figure 13.

18. Use a Riemann sum with $n = 5$ and midpoints to estimate the area under the graph of $f(x) = \sqrt{1 - x^2}$ on the interval $0 \le x \le 1$. The graph is a quarter circle, and the area under the graph is .78540, to five decimal places. See Fig. 14. Carry out the calculations to five decimal places and compute the error. If you double the estimate from this exercise, will it be more accurate than the estimate computed in Exercise 17?

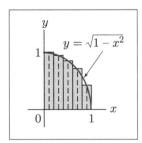

Figure 14.

19. Lung physiologists measure the velocity of air passing through a patient's throat by having the patient breathe into a pneumotachograph. This machine produces a graph that plots airflow rate as a function of time. The graph in Fig. 15 shows the flow rate while a patient is breathing out. The area under the graph gives the total volume of air during exhalation. Estimate this volume with a Riemann sum. Use $n = 5$ and midpoints.

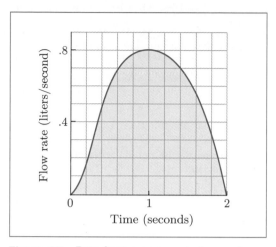

Figure 15. Data from a pneumotachograph.

20. Estimate the area (in square feet) of the piece of land shown in Fig. 16.

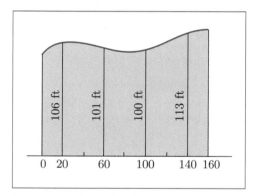

Figure 16. Area of residential property.

21. The velocity of a car (in feet per second) is recorded from the speedometer every 10 seconds, beginning 5 seconds after the car starts to move. See Table 2. Use a Riemann sum to estimate the distance the car travels during the first 60 seconds. (*Note:* Each velocity is

given at the *middle* of a ten-second interval. The first interval extends from 0 to 10, and so on.)

Table 2	A Car's Velocity					
Time	5	15	25	35	45	55
Velocity	20	44	32	39	65	80

22. Table 3 shows the velocity (in feet per second) at the end of each second for a person starting a morning jog. Make three Riemann sum estimates of the total distance jogged during the time from $t = 2$ to $t = 8$.

(a) $n = 6$, left endpoints

(b) $n = 6$, right endpoints

(c) $n = 3$, midpoints

Table 3	A Jogger's Velocity								
Time	0	1	2	3	4	5	6	7	8
Velocity	0	2	3	5	5	6	6	7	8

23. Complete the three missing entries in Table 4.

24. Interpret the area of the shaded region in Fig. 17.

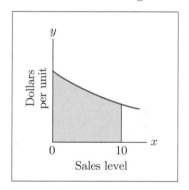

Figure 17. Marginal revenue.

Table 4	Interpretation of Areas	
Function	**a to b**	**Area Under the Graph from a to b**
Rate of growth of a population t years after 1900 (in millions of people per year)	10 to 50	
	5 to 7	Number of cigarettes smoked from 1990 to 1992
Marginal profit at production level x		Additional profit created by increasing production from 20 to 50 units

25. Interpret the area of the shaded region in Fig. 18.

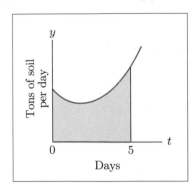

Figure 18. Rate of soil erosion.

26. Interpret the area of the shaded region in Fig. 19.

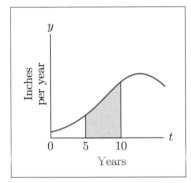

Figure 19. A child's growth rate.

In Exercises 27 and 28, let A be the area under the graph of $y = f(x)$ from $x = a$ to $x = b$.

27. Suppose that the function $f(x)$ is always increasing on the interval $a \leq x \leq b$. Explain why [approximation given by the Riemann sum with left endpoints] $< A <$ [approximation given by the Riemann sum with right endpoints].

28. Suppose that the function $f(x)$ is always decreasing on the interval $a \leq x \leq b$. Explain why [approximation given by the Riemann sum with left endpoints] $> A >$ [approximation given by the Riemann sum with right endpoints].

Technology Exercises

Evaluate a Riemann sum to approximate the area under the graph of $f(x)$ on the given interval, with points selected as specified.

29. $f(x) = x\sqrt{1 + x^2}$; $1 \leq x \leq 3$, $n = 20$, midpoints of subintervals

30. $f(x) = \sqrt{1 - x^2}$; $-1 \leq x \leq 1$, $n = 20$, midpoints of subintervals

31. $f(x) = e^{-x^2}$; $-2 \leq x \leq 2$, $n = 100$, left endpoints

32. $f(x) = x \ln x$; $2 \leq x \leq 6$, $n = 100$, right endpoints

33. $f(x) = (\ln x)^2$; $2 \leq x \leq 3$, $n = 200$, right endpoints

34. $f(x) = (1 + x)^2 e^{.4x}$; $2 \leq x \leq 4$, $n = 200$, left endpoints

---

Solutions to Practice Problems 6.2

1. Since $n = 5$, $\Delta x = \frac{2 - (-2)}{5} = \frac{4}{5} = .8$. The first midpoint is $x_1 = -2 + .8/2 = -1.6$. Subsequent midpoints are found by successively adding .8, to obtain $x_2 = -.8$, $x_3 = 0$, $x_4 = .8$, and $x_5 = 1.6$.

2. The area under the curve from $t = 2$ to $t = 6$ is the number of new jobs created from 1985 to 1989.

6.3 Definite Integrals and the Fundamental Theorem

In Section 6.2 we saw that the area under the graph of a continuous nonnegative function $f(x)$ from a to b is the limiting value of Riemann sums of the form

$$f(x_1) \Delta x + f(x_2) \Delta x + \cdots + f(x_n) \Delta x$$

as n increases without bound or, equivalently, as Δx approaches zero. (Recall that $x_1, x_2, \ldots, x_n$ are selected points from a partition of $a \leq x \leq b$ and Δx is the width of each of the n subintervals.) It can be shown that even if $f(x)$ has negative values, the Riemann sums still approach a limiting value as $\Delta x \to 0$. This number is called the *definite integral of $f(x)$ from a to b* and is denoted by

$$\int_a^b f(x)\, dx.$$

That is,

$$\int_a^b f(x)\,dx = \lim_{\Delta x \to 0}\left[f(x_1)\,\Delta x + f(x_2)\,\Delta x + \cdots + f(x_n)\,\Delta x\right]. \qquad (1)$$

If $f(x)$ is a nonnegative function, we know from Section 6.2 that the Riemann sum on the right side of (1) approaches the area under the graph of $f(x)$ from a to b. Thus, *the definite integral of a nonnegative function $f(x)$ equals the area under the graph of $f(x)$.*

▶ Example 1   Calculate $\displaystyle\int_1^4 \left(\tfrac{1}{3}x + \tfrac{2}{3}\right)dx$.

Solution   Figure 1 shows the graph of the function $f(x) = \tfrac{1}{3}x + \tfrac{2}{3}$. Since $f(x)$ is nonnegative for $1 \le x \le 4$, the definite integral of $f(x)$ equals the area of the shaded region in Fig. 1. The region consists of a rectangle and a triangle. By geometry,

$$[\text{area of rectangle}] = [\text{width}]\cdot[\text{height}] = 3\cdot 1 = 3,$$
$$[\text{area of triangle}] = \tfrac{1}{2}[\text{width}]\cdot[\text{height}] = \tfrac{1}{2}\cdot 3\cdot 1 = \tfrac{3}{2}.$$

Thus the area under the graph is $4\tfrac{1}{2}$, and hence

$$\int_1^4 \left(\tfrac{1}{3}x + \tfrac{2}{3}\right)dx = 4\tfrac{1}{2}. \qquad \blacklozenge$$

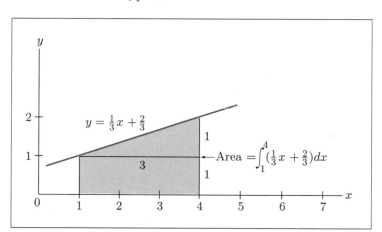

Figure 1. A definite integral as the area under a curve.

In case $f(x)$ is negative at some points in the interval, we may also give a geometric interpretation of the definite integral. Consider the function $f(x)$ shown in Fig. 2. It shows a rectangular approximation of the region between the graph and the x-axis from a to b. Consider a typical rectangle located above or below the selected point x_i. If $f(x_i)$ is nonnegative, the area of the rectangle equals $f(x_i)\,\Delta x$. In case $f(x_i)$ is negative, the area of the rectangle equals $(-f(x_i))\,\Delta x$. So the expression $f(x_i)\,\Delta x$ equals either the area of the corresponding rectangle or the negative of the area, according to whether $f(x_i)$ is nonnegative or negative, respectively. In particular, the Riemann sum

$$f(x_1)\,\Delta x + f(x_2)\,\Delta x + \cdots + f(x_n)\,\Delta x$$

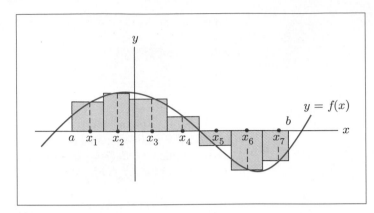

Figure 2.

is equal to the area of the rectangles above the x-axis minus the area of the rectangles below the x-axis. Now take the limit as Δx approaches 0. On the one hand, the Riemann sum approaches the definite integral. On the other hand, the rectangular approximations approach the area bounded by the graph that is above the x-axis minus the area bounded by the graph that is below the x-axis. This gives us the following geometric interpretation of the definite integral.

Suppose that $f(x)$ is continuous on the interval $a \le x \le b$. Then

$$\int_a^b f(x)\, dx$$

is equal to the area above the x-axis bounded by the graph of $y = f(x)$ from $x = a$ to $x = b$ minus the area below the x-axis. Referring to Fig. 3, we have

$$\int_a^b f(x)\, dx = [\text{area of } B \text{ and } D] - [\text{area of } A \text{ and } C]$$

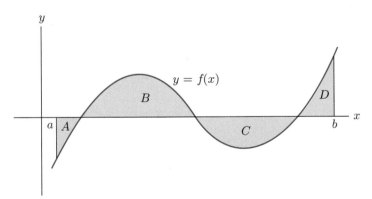

Figure 3. Regions above and below the x-axis.

▶ **Example 2**   Calculate $\displaystyle\int_0^5 (2x - 4)\, dx$.

Solution   Figure 4 shows the graph of the function $f(x) = 2x - 4$ on the interval $0 \le x \le 5$. The area of the triangle above the x-axis is 9 and the area of the triangle below

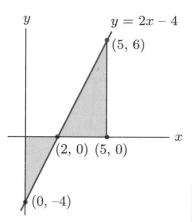

y

$y = 2x - 4$

(5, 6)

(2, 0) (5, 0)

x

(0, –4)

Figure 4.

the x-axis is 4. Therefore, from geometry, we find that

$$\int_0^5 (2x - 4)\, dx = 9 - 4 = 5.$$ ◆

The values of the definite integrals in Examples 1 and 2 follow from simple geometric formulas. For integrals of more complex functions, analogous area formulas are not available. Of course, Riemann sums may always be used to estimate the value of a definite integral to any desired degree of accuracy. However, for most of the integrals in this text, the following fundamental theorem of calculus will rescue us from such calculations.

Fundamental Theorem of Calculus Suppose that $f(x)$ is continuous on the interval $a \leq x \leq b$, and let $F(x)$ be an antiderivative of $f(x)$. Then

$$\int_a^b f(x)\, dx = F(b) - F(a).$$ (2)

This theorem connects the two key concepts of calculus—the integral and the derivative. An explanation of why the theorem is true is given later in this section. First we show how to use the theorem to evaluate definite integrals.

▶ **Example 3** Use the fundamental theorem of calculus to evaluate the following definite integrals.

(a) $\displaystyle\int_1^4 \left(\tfrac{1}{3}x + \tfrac{2}{3}\right) dx$ (b) $\displaystyle\int_0^5 (2x - 4)\, dx$

Solution (a) An antiderivative of $\tfrac{1}{3}x + \tfrac{2}{3}$ is $F(x) = \tfrac{1}{6}x^2 + \tfrac{2}{3}x$. Therefore, by the fundamental theorem,

$$\int_1^4 \left(\tfrac{1}{3}x + \tfrac{2}{3}\right) dx = F(4) - F(1)$$

$$= \left[\tfrac{1}{6}(4)^2 + \tfrac{2}{3}(4)\right] - \left[\tfrac{1}{6}(1)^2 + \tfrac{2}{3}(1)\right]$$

$$= \left[\tfrac{16}{6} + \tfrac{8}{3}\right] - \left[\tfrac{1}{6} + \tfrac{2}{3}\right] = 4\tfrac{1}{2}.$$

This result is the same as in Example 1.

(b) An antiderivative of $2x - 4$ is $F(x) = x^2 - 4x$. Therefore,

$$\int_0^5 (2x - 4)\, dx = F(5) - F(0) = \left[5^2 - 4(5)\right] - \left[0^2 - 4(0)\right] = 5.$$

This result is the same as in Example 2.   ◆

▶ **Example 4** Evaluate $\displaystyle\int_2^5 3x^2\, dx$.

Solution An antiderivative of $f(x) = 3x^2$ is $F(x) = x^3 + C$, where C is any constant. Then

$$\int_2^5 3x^2\, dx = F(5) - F(2) = \left[5^3 + C\right] - \left[2^3 + C\right] = 5^3 + C - 2^3 - C = 117.$$

Notice how the C in $F(2)$ is subtracted from the C in $F(5)$. Thus the value of the definite integral does not depend on the choice of the constant C. For convenience, we may omit C when evaluating a definite integral.   ◆

The quantity $F(b) - F(a)$ is called the *net change of $F(x)$ from $x = a$ to $x = b$*. It is abbreviated by the symbol $F(x)\big|_a^b$. For instance, the net change of $F(x) = \frac{1}{3}e^{3x}$ from $x = 0$ to $x = 2$ is written as $\frac{1}{3}e^{3x}\big|_0^2$ and is evaluated as $F(2) - F(0)$.

▶ **Example 5**   Evaluate $\displaystyle\int_0^2 e^{3x}\, dx$.

Solution   An antiderivative of e^{3x} is $\frac{1}{3}e^{3x}$. Therefore,

$$\int_0^2 e^{3x}\, dx = \frac{1}{3}e^{3x}\bigg|_0^2 = \frac{1}{3}e^{3(2)} - \frac{1}{3}e^{3(0)} = \frac{1}{3}e^6 - \frac{1}{3}.$$ ◆

▶ **Example 6**   Compute the area under the curve $y = x^2 - 4x + 5$ from $x = -1$ to $x = 3$.

Solution   The graph of $f(x) = x^2 - 4x + 5$ is shown in Fig. 5. Since $f(x)$ is nonnegative for $-1 \le x \le 3$, the area under the curve is given by the definite integral

$$\int_{-1}^3 (x^2 - 4x + 5)\, dx = \left(\frac{x^3}{3} - 2x^2 + 5x\right)\bigg|_{-1}^3$$

$$= \left[\frac{(3)^3}{3} - 2(3)^2 + 5(3)\right] - \left[\frac{(-1)^3}{3} - 2(-1)^2 + 5(-1)\right]$$

$$= \left[9 - 18 + 15\right] - \left[-\tfrac{1}{3} - 2 - 5\right]$$

$$= \left[6\right] - \left[-\tfrac{22}{3}\right] = \tfrac{18}{3} + \tfrac{22}{3} = \tfrac{40}{3}.$$ ◆

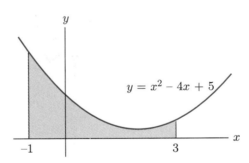

Figure 5.

Study this solution carefully. The calculations show how to use parentheses to avoid errors in arithmetic. It is particularly important to include the outside brackets around the value of the antiderivative at -1.

Areas in Applications   At the end of Section 6.2, we described how the area under a graph can represent the amount of change (now called the "net change") in a quantity. This interpretation of area follows immediately from the fundamental theorem of calculus, which we may rewrite as

$$\int_a^b F'(x)\, dx = F(b) - F(a) \qquad (3)$$

because the $f(x)$ in (2) is the derivative of $F(x)$. If the derivative is nonnegative, then the integral in (3) is the area under the graph of the "rate of change" function $F'(x)$. Equation (3) says that this area equals the net change in $F(x)$ over the interval from a to b.

▶ Example 7   A rocket is fired vertically into the air. Its velocity at t seconds after lift-off is $v(t) = 6t + .5$ meters per second.

    (a) Describe a region whose area represents the distance the rocket travels from time $t = 40$ to $t = 100$ seconds.

    (b) Compute the distance in (a).

Solution   (a) Let $s(t)$ be the position of the rocket at time t, measured from some reference point. (Example 7 in Section 6.1 measured distances from the launch pad, for example.) Then the required distance is $s(100) - s(40)$, the net change in position over the interval $40 \leq t \leq 100$. This distance is represented by the area under the velocity curve from 40 to 100. See Fig. 6.

    (b) Now that we know the connection between antiderivatives and area, we can easily compute the area in (a) corresponding to the distance traveled by the rocket:

$$s(100) - s(40) = \int_{40}^{100} s'(t)\, dt = \int_{40}^{100} v(t)\, dt$$

$$= \int_{40}^{100} (6t + .5)\, dt = \left. \left(3t^2 + .5t \right) \right|_{40}^{100}$$

$$= \left[3(100)^2 + .5(100) \right] - \left[3(40)^2 + .5(40) \right]$$

$$= 30{,}050 - 4820 = 25{,}230 \text{ meters.} \qquad \blacklozenge$$

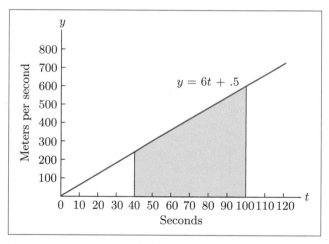

Figure 6. Area under velocity curve is distance traveled.

▶ Example 8   During the early 1970s, the annual worldwide rate of oil consumption was $R(t) = 16.1e^{.07t}$ billion barrels of oil per year, where t is the number of years since the beginning of 1970.

    (a) Determine the amount of oil consumed from 1972 to 1974.

    (b) Represent the answer to part (a) as an area.

Solution   (a) We are interested in $T(4) - T(2)$, where $T(t)$ is the total consumption of oil since 1970. This difference is the net change in $T(t)$ over the time period from $t = 2$ (1972) to $t = 4$ (1974). Now, $T(t)$ is an antiderivative of the rate

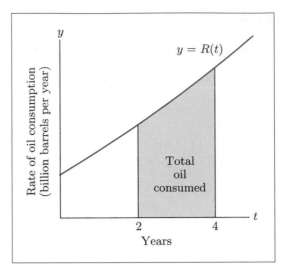

Figure 7. Total oil consumed.

function $R(t)$. Hence

$$T(4) - T(2) = \int_2^4 R(t)\,dt = \int_2^4 16.1e^{.07t}\,dt = \frac{16.1}{.07}e^{.07t}\Big|_2^4$$

$$= 230e^{.07(4)} - 230e^{.07(2)} \approx 39.76 \text{ billion barrels of oil.}$$

(b) The net change in $T(t)$ is the area of the region under the graph of the rate function $T'(t) = R(t)$ from $t = 2$ to $t = 4$. See Fig. 7. ◆

Verification of the Fundamental Theorem of Calculus   We shall give two explanations of why the fundamental theorem of calculus is true. Both convey important ideas about definite integrals and the fundamental theorem.

The first explanation of the fundamental theorem is based directly on the Riemann sum definition of the integral. As we observed earlier, the fundamental theorem may be written in the form

$$\int_a^b F'(x)\,dx = F(b) - F(a), \tag{4}$$

where $F(x)$ is any function with a continuous derivative on $a \leq x \leq b$. There are three key ideas in an explanation of why (4) is true.

I. If the interval $a \leq x \leq b$ is partitioned into n subintervals of width $\Delta x = (b - a)/n$, then the net change in $F(x)$ over $a \leq x \leq b$ is the sum of the net changes in $F(x)$ over each subinterval.

For example, partition $a \leq x \leq b$ into three subintervals and denote the left endpoints by x_1, x_2, x_3, as follows:

Then

$$[\text{change in } F(x) \text{ over 1st interval}] = F(x_2) - F(a),$$

$$[\text{change in } F(x) \text{ over 2nd interval}] = F(x_3) - F(x_2),$$

$$[\text{change in } F(x) \text{ over 3rd interval}] = F(b) - F(x_3).$$

When these changes are summed, the intermediate terms cancel:

$$F(b) - F(x_3) + F(x_3) - F(x_2) + F(x_2) - F(a) = F(b) - F(a).$$

II. If Δx is small, then the change in $F(x)$ over the ith subinterval is approximately $F'(x_i) \, \Delta x$.

This is the approximation of the change in a function discussed in Section 1.8. Here x_i is the left endpoint of the ith subinterval, as follows:

$$\longleftarrow \Delta x \longrightarrow$$

$$x_i \qquad x_i + \Delta x$$

III. The sum of the approximations in (II) is a Riemann sum for the definite integral $\int_a^b F'(x) \, dx$.

That is,

$$F(b) - F(a)$$

$$= \begin{bmatrix} \text{change over} \\ \text{1st subinterval} \end{bmatrix} + \begin{bmatrix} \text{change over} \\ \text{2nd subinterval} \end{bmatrix} + \cdots + \begin{bmatrix} \text{change over} \\ n\text{th subinterval} \end{bmatrix}$$

$$\approx F'(x_1) \, \Delta x + F'(x_2) \, \Delta x + \cdots + F'(x_n) \, \Delta x.$$

As $\Delta x \to 0$, these approximations improve. Thus, in the limit

$$F(b) - F(a) = \int_a^b F'(x) \, dx.$$

To summarize, the derivative of $F(x)$ determines the approximate change of $F(x)$ on small subintervals. The definite integral sums these approximate changes and in the limit gives the exact change of $F(x)$ over the entire interval $a \le x \le b$.

If you look back at the solution of Example 4 of Section 6.2, you will see essentially the same argument. The distance a rocket travels is the sum of the distances it travels over small intervals of time. On each subinterval, this distance is approximately the velocity (the derivative) times Δt.

An Area Function as an Antiderivative   The second explanation of the fundamental theorem of calculus applies only when $f(x)$ is nonnegative. We begin with the following theorem, which describes the basic relationship between area and antiderivatives. In fact, this theorem is sometimes referred to as an alternative version of the fundamental theorem of calculus.

Theorem III   Let $f(x)$ be a continuous nonnegative function for $a \leq x \leq b$. Let $A(x)$ be the area of the region under the graph of the function from a to the number x. (See Fig. 8.) Then $A(x)$ is an antiderivative of $f(x)$.

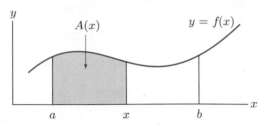

Figure 8.

The fundamental theorem of calculus for a nonnegative function follows easily from Theorem III. Let $F(x)$ be an antiderivative of $f(x)$. Since the "area function" $A(x)$ is also an antiderivative of $f(x)$, by Theorem III, we have

$$A(x) = F(x) + C$$

for some constant C. Notice that $A(a)$ is 0 and $A(b)$ equals the area of the region under the graph of $f(x)$ for $a \leq x \leq b$. Therefore,

$$\int_a^b f(x)\,dx = A(b) = A(b) - A(a)$$
$$= [F(b) + C] - [F(a) + C]$$
$$= F(b) - F(a).$$

It is not difficult to explain why Theorem III is reasonable, although we shall not give a formal proof of the theorem. If h is a small positive number, then $A(x + h) - A(x)$ is the area of the shaded region in Fig. 9. This shaded region is

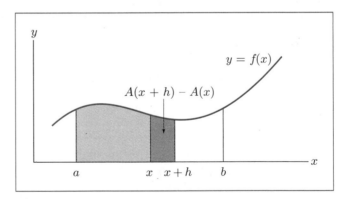

Figure 9.

approximately a rectangle of width h, height $f(x)$, and area $h \cdot f(x)$. Thus

$$A(x + h) - A(x) \approx h \cdot f(x),$$

where the approximation becomes better as h approaches zero. Dividing by h, we have

$$\frac{A(x + h) - A(x)}{h} \approx f(x).$$

Since the approximation improves as h approaches zero, the quotient must approach $f(x)$. However, the limit definition of the derivative tells us that the quotient approaches $A'(x)$ as h approaches zero. Therefore, we have $A'(x) = f(x)$. Since x represented any number between a and b, this shows that $A(x)$ is an antiderivative of $f(x)$.

Incorporating Technology   In Fig. 10(a) the region from Example 6 is shaded and its area evaluated, and in Fig. 10(b) the definite integral is evaluated. In both cases the values obtained are good approximations, but are not exact. See Appendices A–D.

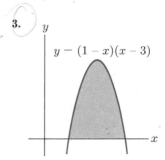

(a)   (b)

Figure 10.

Practice Problems 6.3

1. Find the area under the curve $y = e^{x/2}$ from $x = -3$ to $x = 2$.

2. Let $MR(x)$ be a company's marginal revenue at production level x. Give an economic interpretation of the number $\int_{75}^{80} MR(x)\, dx$.

▶ Exercises 6.3

In Exercises 1–3, set up the definite integral that gives the area of the shaded region.

1.

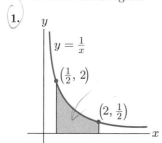

$y = \frac{1}{x}$

$\left(\frac{1}{2}, 2\right)$

$\left(2, \frac{1}{2}\right)$

2.

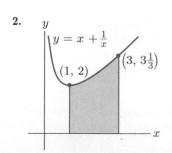

$y = x + \frac{1}{x}$

$\left(3, 3\frac{1}{3}\right)$

$(1, 2)$

3.

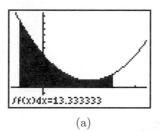

$y - (1 - x)(x - 3)$

In Exercises 4–6, draw the region whose area is given by the definite integral.

4. $\int_{2}^{4} x^2\, dx$

5. $\int_{0}^{4} (8 - 2x)\, dx$

6. $\int_{0}^{4} \sqrt{x}\, dx$

Calculate the following definite integrals.

7. $\int_{-1}^{1} x\, dx$

8. $\int_{4}^{5} e^{2x}\, dx$

9. $\int_1^2 5\,dx$

10. $\int_{-1}^{-1/2} \frac{1}{x^2}\,dx$

11. $\int_1^2 8x^3\,dx$

12. $\int_0^1 e^{x/3}\,dx$

13. $\int_0^1 4e^{-3x}\,dx$

14. $\int_1^3 \frac{5}{x}\,dx$

15. $\int_1^4 3\sqrt{x}\,dx$

16. $\int_1^8 2x^{1/3}\,dx$

17. $\int_0^5 e^{-2t}\,dt$

18. $\int_0^1 \frac{5}{e^{3t}}\,dt$

19. $\int_3^6 x^{-1}\,dx$

20. $\int_1^3 (5t-1)^3\,dt$

21. $\int_{-1}^1 \frac{4}{(t+2)^3}\,dt$

22. $\int_{-3}^0 \sqrt{25+3t}\,dt$

23. $\int_2^3 (5-2t)^4\,dt$

24. $\int_4^9 \frac{3}{t-2}\,dt$

25. $\int_0^3 (x^3 + x - 7)\,dx$

26. $\int_{-5}^5 (e^{x/10} - x^2 - 1)\,dx$

27. $\int_2^4 \left(x^2 + \frac{2}{x^2} - \frac{1}{x+5}\right)dx$

28. $\int_1^2 (4x^3 + 3x^{-4} - 5)\,dx$

Find the area under each of the given curves.

29. $y = 4x$; $x = 2$ to $x = 3$

30. $y = 3x^2$; $x = -1$ to $x = 1$

31. $y = e^{x/2}$; $x = 0$ to $x = 1$

32. $y = \sqrt{x}$; $x = 0$ to $x = 4$

33. $y = (x-3)^4$; $x = 1$ to $x = 4$

34. $y = e^{3x}$; $x = -\frac{1}{3}$ to $x = 0$

35. Let $f(x)$ be the function pictured in Fig. 11. Determine whether $\int_0^7 f(x)\,dx$ is positive, negative, or zero.

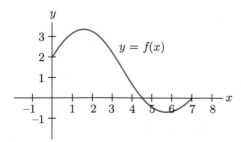

Figure 11.

36. Let $g(x)$ be the function pictured in Fig. 12. Determine whether $\int_0^7 g(x)\,dx$ is positive, negative, or zero.

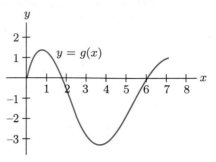

Figure 12.

37. The worldwide rate of cigarette consumption (in trillions of cigarettes per year) since 1960 is given approximately by the function $c(t) = .1t + 2.4$, where $t = 0$ corresponds to 1960. Determine the number of cigarettes sold from 1980 to 1998.

38. Suppose $p(t)$ is the rate (in tons per year) at which pollutants are discharged into a lake, where t is the number of years since 1990. Interpret $\int_5^7 p(t)\,dt$.

39. A helicopter is rising straight up in the air. Its velocity at time t is $2t + 1$ feet per second.
   (a) How high does the helicopter rise during the first 5 seconds?
   (b) Represent the answer to part (a) as an area.

40. After t hours of operation, an assembly line is producing power lawn mowers at the rate of $21 - \frac{4}{5}t$ mowers per hour.
   (a) How many mowers are produced during the time from $t = 2$ to $t = 5$ hours?
   (b) Represent the answer to part (a) as an area.

41. Suppose that the marginal cost function of a handbag manufacturer is $\frac{3}{32}x^2 - x + 200$ dollars per unit at production level x (where x is measured in units of 100 handbags).
   (a) Find the total cost of producing 6 additional units if 2 units are currently being produced.
   (b) Describe the answer to part (a) as an area. (Give a written description rather than a sketch.)

42. Suppose that the marginal profit function for a company is $100 + 50x - 3x^2$ at production level x.
   (a) Find the extra profit earned from the sale of 3 additional units if 5 units are currently being produced.
   (b) Describe the answer to part (a) as an area. (Do not make a sketch.)

43. Let $MP(x)$ be a company's marginal profit at production level x. Give an economic interpretation of the number $\int_{44}^{48} MP(x)\,dx$.

44. Let $MC(x)$ be a company's marginal cost at production level x. Give an economic interpretation of the number $\int_0^{100} MC(x)\,dx$. (*Note*: At any production level, the total cost equals the fixed cost plus the total variable cost.)

45. Some food is placed in a freezer. After t hours the temperature of the food is dropping at the rate of $r(t)$ degrees Fahrenheit per hour, where $r(t) = 12 + 4/(t+3)^2$.

(a) Compute the area under the graph of $y = r(t)$ over the interval $0 \le t \le 2$.

(b) What does the area in part (a) represent?

46. Suppose that the velocity of a car at time t is $40 + 8/(t+1)^2$ kilometers per hour.

(a) Compute the area under the velocity curve from $t = 1$ to $t = 9$.

(b) What does the area in part (a) represent?

47. Deforestation is one of the major problems facing sub-Saharan Africa. Although the clearing of land for farming has been the major cause, the steadily increasing demand for fuelwood has become a significant factor. Figure 13 summarizes projections of the World Bank. The rate of fuelwood consumption (in millions of cubic meters per year) in the Sudan t years after 1980 is given approximately by the function $c(t) = 76.2e^{.03t}$. Determine the amount of fuelwood that will be consumed from 1980 to 2000.

48. (a) Compute $\int_1^b \frac{1}{t}\,dt$, where $b > 1$.

(b) Explain how the logarithm of a number greater than 1 may be interpreted as the area of a region under a curve. (What is the curve?)

49. For each positive number x, let $A(x)$ be the area of the region under the curve $y = x^2 + 1$ from 0 to x. Find $A'(3)$.

50. For each number $x > 2$, let $A(x)$ be the area of the region under the curve $y = x^3$ from 2 to x. Find $A'(6)$.

Technology Exercises

In Exercises 51–54 first use the fundamental theorem of calculus to compute the definite integral and then verify your answer numerically with a graphing utility. (For a TI calculator, use **fnInt** as discussed in Appendices A–D.)

51. $\int_2^3 \frac{1}{x}\,dx$

52. $\int_1^3 10x^4\,dx$

53. $\int_0^4 5\sqrt{x}\,dx$

54. $\int_0^3 6e^{2x}\,dx$

In Exercises 55–58 sketch the region corresponding to the definite integral, and compute its area numerically to four decimal places.

55. $\int_1^4 \ln x\,dx$

56. $\int_{-1}^1 e^{-x^2}\,dx$

57. $\int_0^{.5} \frac{1}{1+x^2}\,dx$

58. $\int_0^3 2^x\,dx$

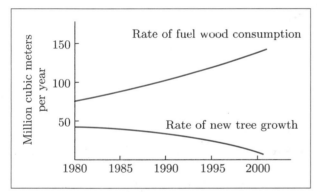

Rate of fuel wood consumption

Rate of new tree growth

Million cubic meters per year

150

100

50

1980 1985 1990 1995 2000

Figure 13. Data from "Sudan and Options in the Energy Sector," World Bank/UNDP, July 1983.

Solutions to Practice Problems 6.3

1. The desired area is

$$\int_{-3}^2 e^{x/2}\,dx = 2e^{x/2}\Big|_{-3}^2 = 2e - 2e^{-3/2} \approx 4.99.$$

2. The number $\int_{75}^{80} MR(x)\,dx$ is the net change in revenue received when the production level is raised from $x = 75$ to $x = 80$ units.

6.4 Areas in the xy-Plane

In this section we show how to use the definite integral to compute the area of a region that lies between the graphs of two or more functions. Three simple but important properties of the integral will be used repeatedly.

Let $f(x)$ and $g(x)$ be functions and a, b, and k be any constants. Then

$$\int_a^b f(x)\,dx + \int_a^b g(x)\,dx = \int_a^b [f(x) + g(x)]\,dx \tag{1}$$

$$\int_a^b f(x)\,dx - \int_a^b g(x)\,dx = \int_a^b [f(x) - g(x)]\,dx \tag{2}$$

$$\int_a^b kf(x)\,dx = k \int_a^b f(x)\,dx. \tag{3}$$

To verify (1), let $F(x)$ and $G(x)$ be antiderivatives of $f(x)$ and $g(x)$, respectively. Then $F(x) + G(x)$ is an antiderivative of $f(x) + g(x)$. By the fundamental theorem of calculus,

$$\int_a^b [f(x) + g(x)]\,dx = [F(x) + G(x)]\Big|_a^b$$

$$= [F(b) + G(b)] - [F(a) + G(a)]$$

$$= [F(b) - F(a)] + [G(b) - G(a)]$$

$$= \int_a^b f(x)\,dx + \int_a^b g(x)\,dx.$$

The verifications of (2) and (3) are similar and use the facts that $F(x) - G(x)$ is an antiderivative of $f(x) - g(x)$ and $kF(x)$ is an antiderivative of $kf(x)$.

Let us now consider regions that are bounded both above and below by graphs of functions. Referring to Fig. 1, we would like to find a simple expression for the area of the shaded region under the graph of $y = f(x)$ and above the graph

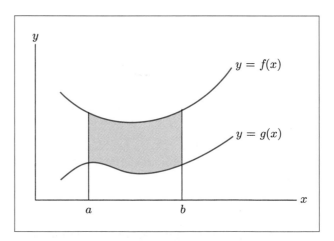

Figure 1.

of $y = g(x)$ from $x = a$ to $x = b$. It is the region under the graph of $f(x)$ with the region under the graph of $g(x)$ taken away. Therefore,

[area of shaded region] = [area under $f(x)$] − [area under $g(x)$]

$$= \int_a^b f(x)\, dx - \int_a^b g(x)\, dx$$

$$= \int_a^b [f(x) - g(x)]\, dx \qquad \text{[by property (2)]}.$$

> **Area Between Two Curves** If $y = f(x)$ lies above $y = g(x)$ from $x = a$ to $x = b$, the area of the region between $f(x)$ and $g(x)$ from $x = a$ to $x = b$ is
>
> $$\int_a^b [f(x) - g(x)]\, dx.$$

▶ **Example 1** Find the area of the region between $y = 2x^2 - 4x + 6$ and $y = -x^2 + 2x + 1$ from $x = 1$ to $x = 2$.

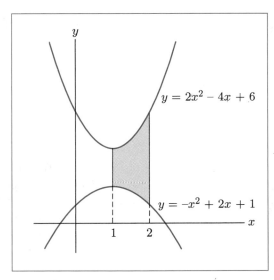

Figure 2.

Solution Upon sketching the two graphs (Fig. 2), we see that $f(x) = 2x^2 - 4x + 6$ lies above $g(x) = -x^2 + 2x + 1$ for $1 \le x \le 2$. Therefore, our formula gives the area of the shaded region as

$$\int_1^2 [(2x^2 - 4x + 6) - (-x^2 + 2x + 1)]\, dx = \int_1^2 (3x^2 - 6x + 5)\, dx$$

$$= (x^3 - 3x^2 + 5x)\Big|_1^2 = 6 - 3 = 3. \quad \blacklozenge$$

▶ **Example 2** Find the area of the region between $y = x^2$ and $y = (x - 2)^2 = x^2 - 4x + 4$ from $x = 0$ to $x = 3$.

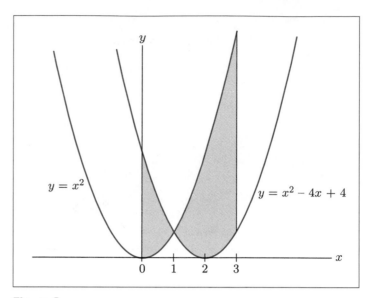

Figure 3.

Solution   Upon sketching the graphs (Fig. 3), we see that the two graphs cross; by setting $x^2 = x^2 - 4x + 4$, we find that they cross when $x = 1$. Thus one graph does not always lie above the other from $x = 0$ to $x = 3$, so that we cannot directly apply our rule for finding the area between two curves. However, the difficulty is easily surmounted if we break the region into two parts, namely, the area from $x = 0$ to $x = 1$ and the area from $x = 1$ to $x = 3$. For from $x = 0$ to $x = 1$, $y = x^2 - 4x + 4$ is on top; and from $x = 1$ to $x = 3$, $y = x^2$ is on top. Consequently,

$$[\text{area from } x = 0 \text{ to } x = 1] = \int_0^1 [(x^2 - 4x + 4) - (x^2)]\,dx$$

$$= \int_0^1 (-4x + 4)\,dx$$

$$= (-2x^2 + 4x)\Big|_0^1 = 2 - 0 = 2.$$

$$[\text{area from } x = 1 \text{ to } x = 3] = \int_1^3 [(x^2) - (x^2 - 4x + 4)]\,dx$$

$$= \int_1^3 (4x - 4)\,dx$$

$$= (2x^2 - 4x)\Big|_1^3 = 6 - (-2) = 8.$$

Thus the total area is $2 + 8 = 10$.   ◆

In our derivation of the formula for the area between two curves, we examined functions that are nonnegative. However, the statement of the rule does not contain this stipulation, and rightfully so. Consider the case where $f(x)$ and $g(x)$ are not always positive. Let us determine the area of the shaded region in Fig. 4(a). Select some constant c such that the graphs of the functions $f(x) + c$ and $g(x) + c$ lie completely above the x-axis [Fig. 4(b)]. The region between them will have the same area as the original region. Using the rule as applied to

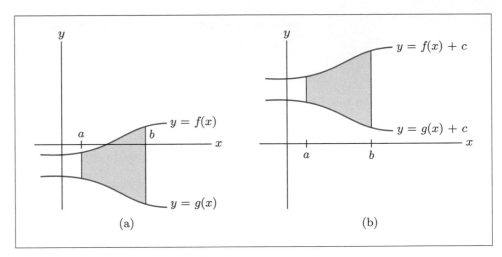

Figure 4.

nonnegative functions, we have

$$[\text{area of the region}] = \int_a^b \left[(f(x) + c) - (g(x) + c) \right] dx$$

$$= \int_a^b \left[f(x) - g(x) \right] dx.$$

Therefore, we see that our rule is valid for any functions $f(x)$ and $g(x)$ as long as the graph of $f(x)$ lies above the graph of $g(x)$ for all x from $x = a$ to $x = b$.

▶ **Example 3** Set up the integral that gives the area between the curves $y = x^2 - 2x$ and $y = -e^x$ from $x = -1$ to $x = 2$.

Solution Since $y = x^2 - 2x$ lies above $y = -e^x$ (Fig. 5), the rule for finding the area

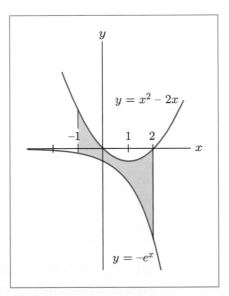

Figure 5.

between two curves can be applied directly. The area between the curves is

$$\int_{-1}^{2} (x^2 - 2x + e^x)\,dx. \qquad \blacklozenge$$

Sometimes we are asked to find the area between two curves without being given the values of a and b. In these cases there is a region that is completely enclosed by the two curves. As the next examples illustrate, we must first find the points of intersection of the two curves in order to obtain the values of a and b. In such problems careful curve sketching is especially important.

▶ Example 4   Set up the integral that gives the area bounded by the curves $y = x^2 + 2x + 3$ and $y = 2x + 4$.

Solution   The two curves are sketched in Fig. 6, and the region bounded by them is shaded. In order to find the points of intersection, we set $x^2 + 2x + 3 = 2x + 4$ and solve for x. We obtain $x^2 = 1$, or $x = -1$ and $x = +1$. When $x = -1$, $2x + 4 = 2(-1) + 4 = 2$. When $x = 1$, $2x + 4 = 2(1) + 4 = 6$. Thus the curves intersect at the points $(1, 6)$ and $(-1, 2)$.

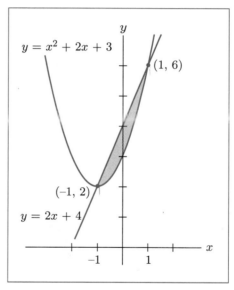

Figure 6.

Since $y = 2x + 4$ lies above $y = x^2 + 2x + 3$ from $x = -1$ to $x = 1$, the area between the curves is given by

$$\int_{-1}^{1} [(2x + 4) - (x^2 + 2x + 3)]\,dx = \int_{-1}^{1} (1 - x^2)\,dx. \qquad \blacklozenge$$

▶ Example 5   Set up the integral that gives the area bounded by the two curves $y = 2x^2$ and $y = x^3 - 3x$.

Solution   First we make a rough sketch of the two curves, as in Fig. 7. The curves intersect where $x^3 - 3x = 2x^2$, or $x^3 - 2x^2 - 3x = 0$. Note that

$$x^3 - 2x^2 - 3x = x(x^2 - 2x - 3) = x(x - 3)(x + 1).$$

So the solutions to $x^3 - 2x^2 - 3x = 0$ are $x = 0, 3, -1$, and the curves intersect at $(-1, 2)$, $(0, 0)$, and $(3, 18)$. From $x = -1$ to $x = 0$, the curve $y = x^3 - 3x$ lies

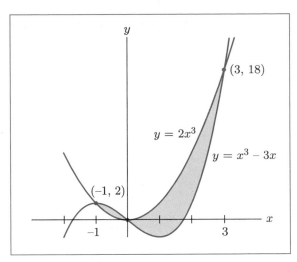

Figure 7.

above $y = 2x^2$. But from $x = 0$ to $x = 3$, the reverse is true. Thus the area between the curves is given by

$$\int_{-1}^{0} (x^3 - 3x - 2x^2)\, dx + \int_{0}^{3} (2\dot{x}^2 - x^3 + 3x)\, dx. \qquad \blacklozenge$$

▶ Example 6   Beginning in 1974, with the advent of dramatically higher oil prices, the exponential rate of growth of world oil consumption slowed down from a growth constant of 7% to a growth constant of 4% per year. A fairly good model for the annual rate of oil consumption since 1974 is given by

$$R_1(t) = 21.3e^{.04(t-4)}, \qquad t \geq 4,$$

where $t = 0$ corresponds to 1970. Determine the total amount of oil saved between 1976 and 1980 by not consuming oil at the rate predicted by the model of Example 1, Section 6.1, namely,

$$R(t) = 16.1e^{.07t}, \qquad t \geq 0.$$

Solution   If oil consumption had continued to grow as it did prior to 1974, then the total oil consumption between 1976 and 1980 would have been

$$\int_{6}^{10} R(t)\, dt. \qquad (4)$$

However, taking into account the slower increase in the rate of oil consumption since 1974, we find that the total oil consumed between 1976 and 1980 was approximately

$$\int_{6}^{10} R_1(t)\, dt. \qquad (5)$$

The integrals in (4) and (5) may be interpreted as the areas under the curves $y = R(t)$ and $y = R_1(t)$, respectively, from $t = 6$ to $t = 10$. (See Fig. 8.) By superimposing the two curves we see that the area between them from $t = 6$ to

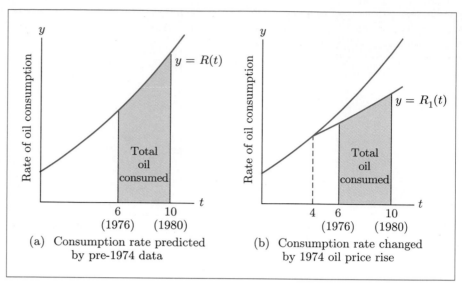

Figure 8.

$t = 10$ represents the total oil that was saved by consuming oil at the rate given by $R_1(t)$ instead of $R(t)$. (See Fig. 9.) The area between the two curves equals

$$\int_6^{10} [R(t) - R_1(t)]\, dt = \int_6^{10} \left[16.1e^{.07t} - 21.3e^{.04(t-4)}\right] dt$$

$$= \left(\frac{16.1}{.07}e^{.07t} - \frac{21.3}{.04}e^{.04(t-4)}\right)\Bigg|_6^{10}$$

$$\approx 13.02.$$

Thus about 13 billion barrels of oil were saved between 1976 and 1980. ◆

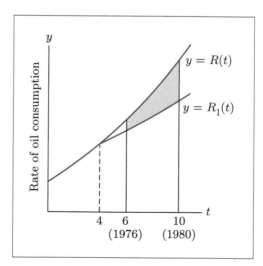

Figure 9.

1. Find the area between the curves $y = x + 3$ and $y = \frac{1}{2}x^2 + x - 7$ from $x = -2$ to $x = 1$.

2. A company plans to increase its production from 10 to 15 units per day. The present marginal cost function is $MC_1(x) = x^2 - 20x + 108$. By redesigning the production process and purchasing new equipment, the company can change the marginal cost function to $MC_2(x) = \frac{1}{2}x^2 - 12x + 75$. Determine the area between the graphs of the two marginal cost curves from $x = 10$ to $x = 15$. Interpret this area in economic terms.

▶ Exercises 6.4

1. Write down a definite integral or sum of definite integrals that gives the area of the shaded portion of Fig. 10.

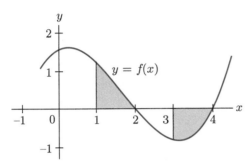

Figure 10.

2. Write down a definite integral or sum of definite integrals that gives the area of the shaded portion of Fig. 11.

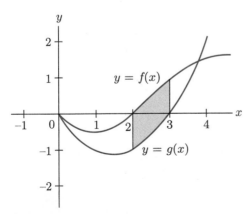

Figure 11.

3. Shade the portion of Fig. 12 whose area is given by the integral

$$\int_0^2 [f(x) - g(x)]\, dx + \int_2^4 [h(x) - g(x)]\, dx.$$

4. Shade the portion of Fig. 13 whose area is given by the integral

$$\int_0^1 [f(x) - g(x)] + \int_1^2 [g(x) - f(x)]\, dx.$$

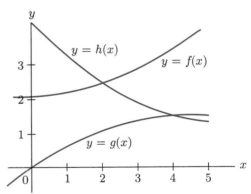

Figure 12.

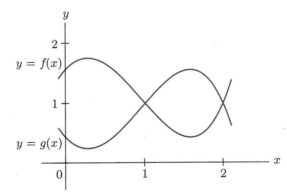

Figure 13.

Find the area of the region between the curves.

5. $y = 2x^2$ and $y = 8$ (a horizontal line) from $x = -2$ to $x = 2$

6. $y = 13 - 3x^2$ and $y = 1$ from $x = -2$ to $x = 2$

7. $y = x^2 - 6x + 12$ and $y = 1$ from $x = 0$ to $x = 4$

8. $y = x(2 - x)$ and $y = 4$ from $x = 0$ to $x = 2$

9. $y = 3x^2$ and $y = -3x^2$ from $x = -1$ to $x = 2$

10. $y = e^{2x}$ and $y = -e^{2x}$ from $x = -1$ to $x = 1$

Find the area of the region bounded by the curves.

11. $y = x^2 + x$ and $y = 3 - x$

12. $y = 3x - x^2$ and $y = 4 - 2x$

13. $y = -x^2 + 6x - 5$ and $y = 2x - 5$

14. $y = 2x^2 + x - 7$ and $y = x + 1$

15. $y = x^2$ and $y = 18 - x^2$

16. $y = 4x^2 - 24x + 20$ and $y = 2 - 2x^2$

17. $y = x^2 - 6x - 7$ and $y = -2x^2 + 6x + 8$

18. $y = 2x^2 - 8x + 8$ and $y = -x^2 + 7x + 8$

19. Find the area of the region between $y = x^2 - 3x$ and the x-axis

   (a) from $x = 0$ to $x = 3$,

   (b) from $x = 0$ to $x = 4$,

   (c) from $x = -2$ to $x = 3$.

20. Find the area of the region between $y = x^2$ and $y = 1/x^2$

   (a) from $x = 1$ to $x = 4$,

   (b) from $x = \frac{1}{2}$ to $x = 4$.

21. Find the area of the region bounded by $y = 1/x^2$, $y = x$, and $y = 8x$, for $x \geq 0$. (See Fig. 14.)

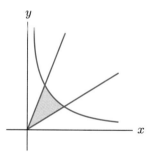

Figure 14.

22. Find the area of the region bounded by $y = 1/x$, $y = 4x$, and $y = x/2$, for $x \geq 0$. (The region resembles the shaded region in Fig. 14.)

23. Find the area of the shaded region in Fig. 15(a) bounded by $y = 12/x$, $y = \frac{3}{2}\sqrt{x}$, and $y = x/3$.

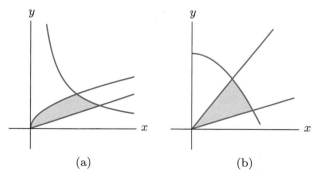

(a)   (b)

Figure 15.

24. Find the area of the shaded region in Fig. 15(b) bounded by $y = 12 - x^2$, $y = 4x$, and $y = x$.

25. Refer to Exercise 47 of Section 6.3. The rate of new tree growth (in millions of cubic meters per year) in the Sudan t years after 1980 is given approximately by the function $g(t) = 50 - 6.03e^{.09t}$. Set up the definite integral giving the amount of depletion of the forests due to the excess of fuel wood consumption over new growth from 1980 to 2000.

26. Refer to the oil-consumption data in Example 1, Section 6.1. Suppose that in 1970 the growth constant for the annual rate of oil consumption had been held to .04. What effect would this action have had on oil consumption from 1970 to 1974?

27. The marginal profit for a certain company is $MP_1(x) = -x^2 + 14x - 24$. The company expects the daily production level to rise from $x = 6$ to $x = 8$ units. The management is considering a plan that would have the effect of changing the marginal profit to $M_2(x) = -x^2 + 12x - 20$. Should the company adopt the plan? Determine the area between the graphs of the two marginal profit functions from $x = 6$ to $x = 8$. Interpret this area in economic terms.

28. Two rockets are fired simultaneously straight up into the air. Their velocities (in meters per second) are $v_1(t)$ and $v_2(t)$, respectively, and $v_1(t) \geq v_2(t)$ for $t \geq 0$. Let A denote the area of the region between the graphs of $y = v_1(t)$ and $y = v_2(t)$ for $0 \leq t \leq 10$. What physical interpretation may be given to the value of A?

29. Cars A and B start at the same place and travel in the same direction, with velocities after t hours given by the functions $v_A(t)$ and $v_B(t)$ in Fig. 16.

   (a) What does the area between the two curves from $t = 0$ to $t = 1$ represent?

   (b) At what time will the distance between the cars be greatest?

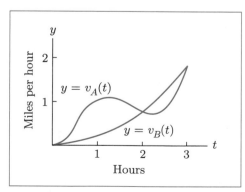

Figure 16.

Technology Exercises

In Exercises 30–33, use a graphing utility to find the intersection points of the curves, and then use the utility to find the area of the region bounded by the curves.

30. $y = e^x$, $y = 4x + 1$

31. $y = 5 - (x - 2)^2$, $y = e^x$

32. $y = \sqrt{x + 1}$, $y = (x - 1)^2$

33. $y = 1/x$, $y = 3 - x$

Solutions to Practice Problems 6.4

1. First graph the two curves, as shown in Fig. 17. The curve $y = x + 3$ lies on top. So the area between the curves is

$$\int_{-2}^{1} \left[(x + 3) - \left(\tfrac{1}{2}x^2 + x - 7 \right) \right] dx = \int_{-2}^{1} \left(-\tfrac{1}{2}x^2 + 10 \right) dx$$

$$= \left(-\tfrac{1}{6}x^3 + 10x \right) \Big|_{-2}^{1} = 28.5.$$

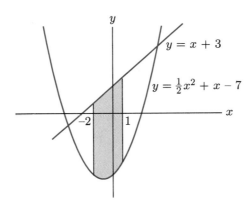

Figure 17.

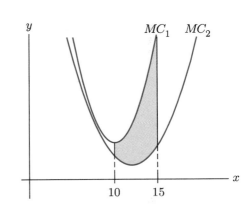

Figure 18.

2. Graphing the two marginal cost functions yields the results shown in Fig. 18. So the area between the curves equals

$$\int_{10}^{15} \left[MC_1(x) - MC_2(x) \right] dx$$

$$= \int_{10}^{15} \left[(x^2 - 20x + 108) - (\tfrac{1}{2}x^2 - 12x + 75) \right] dx$$

$$= \int_{10}^{15} \left[\tfrac{1}{2}x^2 - 8x + 33 \right] dx$$

$$= \left(\tfrac{1}{6}x^3 - 4x^2 + 33x \right) \Big|_{10}^{15}$$

$$= 60\tfrac{5}{6}.$$

This amount represents the cost savings on the increased production (from 10 to 15) provided the new production process is used.

6.5   Applications of the Definite Integral

The applications in this section have two features in common. First, each example contains a quantity that is computed by evaluating a definite integral. Second, the formula for the definite integral is derived by looking at Riemann sums.

In each application, we show that a certain quantity, call it Q, may be approximated by dividing an interval into equal subintervals and forming an appropriate sum. We then observe that this sum is a Riemann sum for some function $f(x)$. [Sometimes $f(x)$ is not given beforehand.] The Riemann sums approach Q as the number of subintervals becomes large. Since the Riemann sums also approach a definite integral, we conclude that the value of Q is given by the definite integral.

Once we have expressed Q as a definite integral, we may calculate its value using the fundamental theorem of calculus, which reduces the calculation to antidifferentiation. It is not necessary to calculate the value of any Riemann sum. Rather, we use the Riemann sums only as a device to express Q as a definite integral. The key step is to recognize a given sum as a Riemann sum and to determine the corresponding definite integral. Example 1 illustrates this step.

▶ **Example 1**   Suppose that the interval $1 \leq x \leq 2$ is divided into 50 subintervals, each of length Δx. Let $x_1, x_2, \ldots, x_{50}$ denote points selected from these subintervals. Find an approximate value for the sum

$$(8x_1^7 + 6x_1)\,\Delta x + (8x_2^7 + 6x_2)\,\Delta x + \cdots + (8x_{50}^7 + 6x_{50})\,\Delta x.$$

Solution   The sum is clearly a Riemann sum for the function $f(x) = 8x^7 + 6x$ on the interval $1 \leq x \leq 2$. Therefore, an approximation to the sum is given by the integral

$$\int_1^2 (8x^7 + 6x)\,dx.$$

We may evaluate this integral using the fundamental theorem of calculus:

$$\int_1^2 (8x^7 + 6x)\,dx = (x^8 + 3x^2)\Big|_1^2$$

$$= [2^8 + 3(2^2)] - [1^8 + 3(1^2)]$$

$$= 268 - 4 = 264.$$

Therefore, the sum is approximately 264.   ◆

The Average Value of a Function   Let $f(x)$ be a continuous function on the interval $a \leq x \leq b$. The definite integral may be used to define the *average value* of $f(x)$ on this interval. To calculate the average of a collection of numbers $y_1, y_2, \ldots, y_n$, we add the numbers and divide by n to obtain

$$\frac{y_1 + y_2 + \cdots + y_n}{n}.$$

To determine the average value of $f(x)$, we proceed similarly. Choose n values of x, say $x_1, x_2, \ldots, x_n$, and calculate the corresponding function values $f(x_1), f(x_2), \ldots, f(x_n)$. The average of these values is

$$\frac{f(x_1) + f(x_2) + \cdots + f(x_n)}{n}. \tag{1}$$

Our goal now is to obtain a reasonable definition of the average of all the values of $f(x)$ on the interval $a \le x \le b$. If the points $x_1, x_2, \ldots, x_n$ are spread "evenly" throughout the interval, then the average (1) should be a good approximation to our intuitive concept of the average value of $f(x)$. In fact, as n becomes large, the average (1) should approximate the average value of $f(x)$ to any arbitrary degree of accuracy. To guarantee that the points $x_1, x_2, \ldots, x_n$ are "evenly" spread out from a to b, let us divide the interval from $x = a$ to $x = b$ into n subintervals of equal length $\Delta x = (b - a)/n$. Then choose x_1 from the first subinterval, x_2 from the second, and so forth. The average (1) that corresponds to these points may be arranged in the form of a Riemann sum as follows:

$$\frac{f(x_1) + f(x_2) + \cdots + f(x_n)}{n}$$

$$= f(x_1) \cdot \frac{1}{n} + f(x_2) \cdot \frac{1}{n} + \cdots + f(x_n) \cdot \frac{1}{n}$$

$$= \frac{1}{b - a} \left[f(x_1) \cdot \frac{b - a}{n} + f(x_2) \cdot \frac{b - a}{n} + \cdots + f(x_n) \cdot \frac{b - a}{n} \right]$$

$$= \frac{1}{b - a} [f(x_1) \Delta x + f(x_2) \Delta x + \cdots + f(x_n) \Delta x].$$

The sum inside the brackets is a Riemann sum for the definite integral of $f(x)$. Thus we see that for a large number of points x_i, the average in (1) approaches the quantity

$$\frac{1}{b - a} \int_a^b f(x)\, dx.$$

This argument motivates the following definition.

The *average value* of a continuous function $f(x)$ over the interval $a \le x \le b$ is defined as the quantity

$$\frac{1}{b - a} \int_a^b f(x)\, dx \tag{2}$$

▶ **Example 2**   Compute the average value of $f(x) = \sqrt{x}$ over the interval $0 \le x \le 9$.

Solution   Using (2) with $a = 0$ and $b = 9$, the average value of $f(x) = \sqrt{x}$ over the interval $0 \le x \le 9$ is equal to

$$\frac{1}{9 - 0} \int_0^9 \sqrt{x}\, dx.$$

Since $\sqrt{x} = x^{1/2}$, an antiderivative of $\sqrt{x}$ is $\frac{2}{3} x^{3/2}$. Therefore,

$$\frac{1}{9} \int_0^9 \sqrt{x}\, dx = \frac{1}{9} \left(\frac{2}{3} x^{3/2} \right) \Big|_0^9 = \frac{1}{9} \left(\frac{2}{3} \cdot 9^{3/2} - 0 \right) = \frac{1}{9} \left(\frac{2}{3} \cdot 27 \right) = 2,$$

so that the average value of $\sqrt{x}$ over the interval $0 \le x \le 9$ is 2. The area of the shaded region is the same as the area of the rectangle pictured in Fig. 1.   ◆

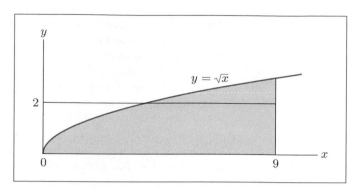

Figure 1. Average value of a function.

▶ Example 3   On January 1, 1998 the world population was 5.9 billion and growing at the rate of 1.34% annually. Assuming this growth rate continues, the population t years from then will be given by the exponential growth law

$$P(t) = 5.9e^{.0134t}.$$

Determine the average world population during the next 30 years. (This average is important in long-range planning for agricultural production and the allocation of goods and services.)

Solution   The average value of the population $P(t)$ from $t = 0$ to $t = 30$ is

$$\frac{1}{30-0}\int_0^{30} P(t)\,dt = \frac{1}{30}\int_0^{30} 5.9e^{.0134t}\,dt$$

$$= \frac{1}{30}\left(\frac{5.9}{.0134}e^{.0134t}\right)\Bigg|_0^{30} = \frac{5.9}{.402}(e^{.402}-1)$$

$$\approx 7.26 \text{ billion.}\qquad\blacklozenge$$

Consumers' Surplus   Using a demand curve from economics, we can derive a formula showing the amount that consumers benefit from an open system that has no price discrimination. Figure 2(a) is a *demand curve* for a commodity. It is determined by complex economic factors and gives a relationship between the quantity sold and the unit price of a commodity. Specifically, it says that, in order to sell x units, the price must be set at $f(x)$ dollars per unit. Since,

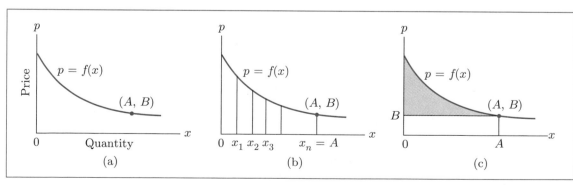

Figure 2. Consumers' surplus.

for most commodities, selling larger quantities requires a lowering of the price, demand curves are usually decreasing. Interactions between supply and demand determine the amount of a quantity available. Let A designate the amount of the commodity currently available and $B = f(A)$ the current selling price.

Divide the interval from 0 to A into n subintervals, each of length $\Delta x = (A-0)/n$, and take x_i to be the right-hand endpoint of the ith interval. Consider the first subinterval, from 0 to x_1. [See Fig. 2(b).] Suppose that only x_1 units had been available. Then the price per unit could have been set at $f(x_1)$ dollars and these x_1 units sold. Of course, at this price we could not have sold any more units. However, those people who paid $f(x_1)$ dollars had a great demand for the commodity. It was extremely valuable to them, and there was no advantage in substituting another commodity at that price. They were actually paying what the commodity was worth to them. In theory, then, the first x_1 units of the commodity could be sold to these people at $f(x_1)$ dollars per unit, yielding

$$\text{(price per unit)} \cdot \text{(number of units)} = f(x_1) \cdot (x_1) = f(x_1) \cdot \Delta x \text{ dollars.}$$

After selling the first x_1 units, suppose that more units become available, so that now a total of x_2 units have been produced. Setting the price at $f(x_2)$, the remaining $x_2 - x_1 = \Delta x$ units can be sold, yielding $f(x_2) \cdot \Delta x$ dollars. Here, again, the second group of buyers would have paid as much for the commodity as it was worth to them. Continuing this process of price discrimination, the amount of money paid by consumers would be

$$f(x_1) \Delta x + f(x_2) \Delta x + \cdots + f(x_n) \Delta x.$$

Taking n large, we see that this Riemann sum approaches $\int_0^A f(x)\, dx$. Since $f(x)$ is positive, this integral equals the area under the graph of $f(x)$ from $x = 0$ to $x = A$.

Of course, in an open system, everyone pays the same price, B, so the total amount paid is [price per unit] $\cdot$ [number of units] $= BA$. Since BA is the area of the rectangle under the graph of the line $p = B$ from $x = 0$ to $x = A$, the amount of money saved by the consumers is the area of the shaded region in Fig. 2(c). That is, the area between the curves $p = f(x)$ and $p = B$ gives a numerical value to one benefit of a modern efficient economy.

The *consumers' surplus* for a commodity having demand curve $p = f(x)$ is

$$\int_0^A [f(x) - B]\, dx,$$

where the quantity demanded is A and the price is $B = f(A)$.

▶ **Example 4** Find the consumers' surplus for the demand curve $p = 50 - .06x^2$ at the sales level $x = 20$.

Solution Since 20 units are sold, the price must be

$$B = 50 - .06(20)^2 = 50 - 24 = 26.$$

Therefore, the consumers' surplus is

$$\int_0^{20} [(50 - .06x^2) - 26]\,dx = \int_0^{20} (24 - .06x^2)\,dx$$

$$= (24x - .02x^3)\Big|_0^{20}$$

$$= 24(20) - .02(20)^3$$

$$= 480 - 160 = 320.$$

That is, the consumers' surplus is $320.   ◆

Future Value of an Income Stream   The next example shows how the definite integral can be used to approximate the sum of a large number of terms.

▶ **Example 5**   Suppose that money is deposited daily into a savings account at an annual rate of $1000. The account pays 6% interest compounded continuously. Approximate the amount of money in the account at the end of 5 years.

Solution   Divide the time interval from 0 to 5 years into daily subintervals. Each subinterval is then of duration $\Delta t = \frac{1}{365}$ years. Let $t_1, t_2, \ldots, t_n$ be points chosen from these subintervals. Since we deposit money at an annual rate of $1000, the amount deposited during one of the subintervals is $1000\Delta t$ dollars. If this amount is deposited at time t_i, the $1000\Delta t$ dollars will earn interest for the remaining $5 - t_i$ years. The total amount resulting from this one deposit at time t_i is then

$$1000\,\Delta t\,e^{.06(5-t_i)}.$$

Add the effects of the deposits at times $t_1, t_2, \ldots, t_n$ to arrive at the total balance in the account:

$$A = 1000e^{.06(5-t_1)}\Delta t + 1000e^{.06(5-t_2)}\Delta t + \cdots + 1000e^{.06(5-t_n)}\Delta t.$$

This is a Riemann sum for the function $f(t) = 1000e^{.06(5-t)}$ on the interval $0 \le t \le 5$. Since Δt is very small when compared with the interval, the total amount in the account, A, is approximately

$$\int_0^5 1000e^{.06(5-t)}\,dt = \frac{1000}{-.06}\,e^{.06(5-t)}\Big|_0^5 = \frac{1000}{-.06}(1 - e^{.3}) \approx 5831.$$

That is, the approximate balance in the account at the end of 5 years is $5831.◆

Note   The antiderivative was computed by observing that since the derivative of $e^{.06(5-t)}$ is $e^{.06(5-t)}(-.06)$, we must divide the integrand by $-.06$ to obtain an antiderivative.

In Example 5, money was deposited daily into the account. If the money had been deposited several times a day, the definite integral would have given an even better approximation to the balance. Actually, the more frequently the money is deposited, the better the approximation. Economists consider a hypothetical situation where money is deposited steadily throughout the year. This flow of money is called a *continuous income stream*, and the balance in the account is given exactly by the definite integral.

The *future value of a continuous income stream* of K dollars per year for N years at interest rate r compounded continuously is

$$\int_0^N Ke^{r(N-t)} \, dt.$$

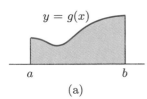

$y = g(x)$

a b

(a)

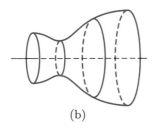

(b)

Figure 3.

Solids of Revolution When the region of Fig. 3(a) is revolved about the x-axis, it sweeps out a solid [Fig. 3(b)]. Riemann sums can be used to derive a formula for the volume of this *solid of revolution*. Let us break the x-axis between a and b into a large number n of equal subintervals, each of length $\Delta x = (b-a)/n$. Using each subinterval as a base, we can divide the region into strips (see Fig. 4).

Let x_i be a point in the ith subinterval. Then the volume swept out by revolving the ith strip is approximately the same as the volume of the cylinder swept out by revolving the rectangle of height $g(x_i)$ and base Δx around the x-axis (Fig. 5). The volume of the cylinder is

$$[\text{area of circular side}] \cdot [\text{width}] = \pi[g(x_i)]^2 \cdot \Delta x.$$

The total volume swept out by all the strips is approximated by the total volume swept out by the rectangles, which is

$$[\text{volume}] \approx \pi[g(x_1)]^2 \Delta x + \pi[g(x_2)]^2 \Delta x + \cdots + \pi[g(x_n)]^2 \Delta x.$$

As n gets larger and larger, this approximation becomes arbitrarily close to the true volume. The expression on the right is a Riemann sum for the definite integral of $f(x) = \pi[g(x)]^2$. Therefore, the volume of the solid equals the value of the definite integral.

The volume of the *solid of revolution* obtained from revolving the region below the graph of $y = g(x)$ from $x = a$ to $x = b$ about the x-axis is

$$\int_a^b \pi[g(x)]^2 \, dx.$$

▶ **Example 6** Find the volume of the solid of revolution obtained by revolving the region of Fig. 6 about the x-axis.

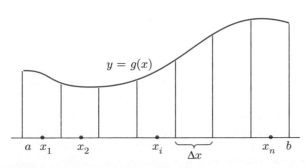

$y = g(x)$

a x_1 x_2 x_i x_n b

Δx

Figure 4.

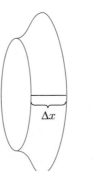

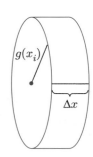

$(x_i, g(x_i))$

Δx Δx $g(x_i)$ Δx

Figure 5.

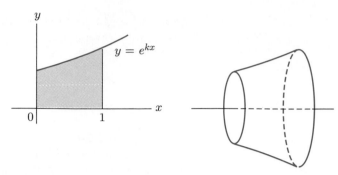

Figure 6.

Solution   Here $g(x) = e^{kx}$, and

$$[\text{volume}] = \int_0^1 \pi(e^{kx})^2 \, dx = \int_0^1 \pi e^{2kx} \, dx = \frac{\pi}{2k} e^{2kx} \Big|_0^1 = \frac{\pi}{2k}(e^{2k} - 1). \quad \blacklozenge$$

▶ Example 7   Find the volume of a right circular cone of radius r and height h.

Solution   The cone [Fig. 7(a)] is the solid of revolution swept out when the shaded region in Fig. 7(b) is revolved about the x-axis. Using the formula developed previously, the volume of the cone is

$$\int_0^h \pi \left(\frac{r}{h}x\right)^2 \, dx = \frac{\pi r^2}{h^2} \int_0^h x^2 \, dx = \frac{\pi r^2}{h^2} \frac{x^3}{3} \Big|_0^h = \frac{\pi r^2 h}{3}. \quad \blacklozenge$$

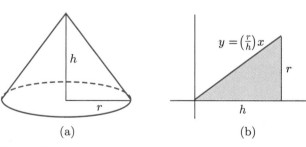

(a)   (b)

Figure 7.

Practice Problems 6.5

1. A rock dropped from a bridge has a velocity of $-32t$ feet per second after t seconds. Find the average velocity of the rock during the first three seconds.

2. An investment yields \$300 per year compounded continuously for 10 years at 10% interest. What is the (future) value of this income stream at the end of 10 years?

▶ Exercises 6.5

Determine the average value of $f(x)$ over the interval from $x = a$ to $x = b$, where

1. $f(x) = x^2$; $a = 0$, $b = 3$

2. $f(x) = e^{x/3}$; $a = 0$, $b = 3$

3. $f(x) = x^3$; $a = -1$, $b = 1$

4. $f(x) = 5$; $a = 1$, $b = 10$

5. $f(x) = 1/x^2$; $a = \frac{1}{4}$, $b = \frac{1}{2}$

6. $f(x) = 2x - 6$; $a = 2$, $b = 4$

7. During a certain 12-hour period the temperature at time t (measured in hours from the start of the period) was $47 + 4t - \frac{1}{3}t^2$ degrees. What was the average temperature during that period?

8. Assuming that a country's population is now 3 million and is growing exponentially with growth constant .02, what will be the average population during the next 50 years?

9. One hundred grams of radioactive radium having a half-life of 1690 years is placed in a concrete vault. What will be the average amount of radium in the vault during the next 1000 years?

10. One hundred dollars are deposited in the bank at 5% interest compounded continuously. What will be the average value of the money in the account during the next 20 years?

Find the consumers' surplus for each of the following demand curves at the given sales level x.

11. $p = 3 - \dfrac{x}{10}$; $x = 20$

12. $p = \dfrac{x^2}{200} - x + 50$; $x = 20$

13. $p = \dfrac{500}{x + 10} - 3$; $x = 40$

14. $p = \sqrt{16 - .02x}$; $x = 350$

Figure 8 shows a supply curve for a commodity. It gives the relationship between the selling price of the commodity and the quantity that producers will manufacture. At a higher selling price, a greater quantity will be produced. Therefore, the curve is increasing. If (A, B) is a point on the curve, then, in order to stimulate the production of A units of the commodity, the price per unit must be B dollars. Of course, some producers will be willing to produce the commodity even with a lower selling price. Since everyone receives the same price in an open efficient economy, most producers are receiving more than their minimal required price. The excess is called the *producers' surplus*. Using an argument analogous to that of the *consumers' surplus*, one can show that the total producers' surplus when the price is B is the area of the shaded region in Fig. 8. Find the producers' surplus for each of the following supply curves at the given sales level x.

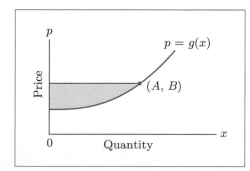

Figure 8. Producers' surplus.

15. $p = .01x + 3$; $x = 200$

16. $p = \dfrac{x^2}{9} + 1$; $x = 3$

17. $p = \dfrac{x}{2} + 7$; $x = 10$

18. $p = 1 + \frac{1}{2}\sqrt{x}$; $x = 36$

For a particular commodity, the quantity produced and the unit price are given by the coordinates of the point where the supply and demand curves intersect. For each pair of supply and demand curves, determine the point of intersection (A, B) and the consumers' and producers' surplus. (See Fig. 9.)

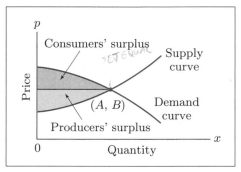

Figure 9.

19. Demand curve: $p = 12 - (x/50)$; supply curve: $p = (x/20) + 5$.

20. Demand curve: $p = \sqrt{25 - .1x}$; supply curve: $p = \sqrt{.1x + 9} - 2$.

21. Suppose that money is deposited daily into a savings account at an annual rate of $1000. If the account pays 5% interest compounded continuously, estimate the balance in the account at the end of 3 years.

22. Suppose that money is deposited daily into a savings account at an annual rate of $2000. If the account pays 6% interest compounded continuously, approximately how much will be in the account at the end of 2 years?

23. Suppose that money is deposited steadily into a savings account at the rate of $16,000 per year. Determine the balance at the end of 4 years if the account pays 8% interest compounded continuously.

24. Suppose that money is deposited steadily into a savings account at the rate of $14,000 per year. Determine the balance at the end of 6 years if the account pays 4.5% interest compounded continuously.

25. An investment pays 10% interest compounded continuously. If money is invested steadily at the rate of $5000 per year, how much time is required until the value of the investment reaches $140,000?

26. A savings account pays 4.25% interest compounded continuously. At what rate per year must money be deposited steadily into the account in order to accumulate a balance of $100,000 after 10 years?

27. Suppose money is to be deposited daily for 5 years into a savings account at an annual rate of $1000 and the account pays 4% interest compounded continuously. Let the interval from 0 to 5 be divided into daily subintervals, with each subinterval of duration $\Delta t = \frac{1}{365}$ years. Let $t_1, \ldots, t_n$ be points chosen from the subintervals.

(a) Show that the present value of a daily deposit at time t_i is $1000 \Delta t \, e^{-.04t_i}$.

(b) Find the Riemann sum corresponding to the sum of the present values of all the deposits.

(c) What is the function and interval corresponding to the Riemann sum in (b)?

(d) Give the definite integral that approximates the Riemann sum in (b).

(e) Evaluate the definite integral in (d). This number is the *present value of a continuous income stream.*

28. Use the result of Exercise 27 to calculate the present value of a continuous income stream of $5000 per year for 10 years at an interest rate of 5% compounded continuously.

Find the volume of the solid of revolution generated by revolving about the x-axis the region under each of the following curves.

29. $y = \sqrt{r^2 - x^2}$ from $x = -r$ to $x = r$ (generates a sphere of radius r)

30. $y = kx$ from $x = 0$ to $x = h$ (generates a cone)

31. $y = x^2$ from $x = 1$ to $x = 2$

32. $y = \dfrac{1}{x}$ from $x = 1$ to $x = 100$

33. $y = \sqrt{x}$ from $x = 0$ to $x = 4$ (The solid generated is called a *paraboloid.*)

34. $y = 2x - x^2$ from $x = 0$ to $x = 2$

35. $y = e^{-x}$ from $x = 0$ to $x = r$

36. $y = 2x + 1$ from $x = 0$ to $x = 1$ (The solid generated is called a *truncated cone.*)

For the Riemann sums in Exercises 37–40, determine n, b, and $f(x)$.

37. $\left[(8.25)^3 + (8.75)^3 + (9.25)^3 + (9.75)^3\right](.5); \ a = 8$

38. $\left[\dfrac{3}{1} + \dfrac{3}{1.5} + \dfrac{3}{2} + \dfrac{3}{2.5} + \dfrac{3}{3} + \dfrac{3}{3.5}\right](.5); \ a = 1$

39. $\left[(5 + e^5) + (6 + e^6) + (7 + e^7)\right](1); \ a = 4$

40. $\left[3(.3)^2 + 3(.9)^2 + 3(1.5)^2 + 3(2.1)^2 + 3(2.7)^2\right](.6); \ a = 0$

41. Suppose that the interval $0 \le x \le 3$ is divided into 100 subintervals of width $\Delta x = .03$. Let $x_1, x_2, \ldots, x_{100}$ be points in these subintervals. Suppose that in a particular application, one needs to estimate the sum

$$(3 - x_1)^2 \Delta x + (3 - x_2)^2 \Delta x + \cdots + (3 - x_{100})^2 \Delta x.$$

Show that this sum is close to 9.

42. Suppose that the interval $0 \le x \le 1$ is divided into 100 subintervals of width $\Delta x = .01$. Show that the following sum is close to 5/4.

$$\left[2(.01) + (.01)^3\right] \Delta x + \left[2(.02) + (.02)^3\right] \Delta x$$
$$+ \cdots + \left[2(1.0) + (1.0)^3\right] \Delta x$$

Technology Exercises

The following exercises ask for an unknown quantity, call it x. After setting up the appropriate formula involving a definite integral, use the fundamental theorem to evaluate the definite integral as an expression in x. Because the resulting equation will be too complicated to solve algebraically, you must use a graphing utility to obtain the solution. *Note*: If the quantity x is an interest rate paid by a savings account, it will most likely be between 0 and .10.

43. A single deposit of $1000 is to be made into a savings account and the interest (compounded continuously) is allowed to accumulate for 3 years. Therefore, the amount at the end of t years is $1000e^{rt}$.

(a) Find an expression (involving r) that gives the average value of the money in the account during the three-year time period $0 \le t \le 3$.

(b) Find the interest rate r at which the average amount in the account during the three-year period is $1070.60.

44. A single deposit of $100 is made into a savings account paying 4% interest compounded continuously. How long must the money be held in the account so that the average amount of money during that time period will be $122.96?

45. Money is deposited steadily at the rate of $1000 per year into a savings account.

(a) Find the expression (involving r) that gives the (future) balance in the account at the end of 6 years.

(b) Find the interest rate that will result in a balance of $6997.18 after 6 years.

46. Money is deposited steadily at the rate of $3000 per year into a savings account. After 10 years the balance is $36,887. What interest rate, with interest compounded continuously, did the money earn?

1. By definition, the average value of the function $v(t) = -32t$ for $t = 0$ to $t = 3$ is

$$\frac{1}{3-0}\int_0^3 -32t\,dt = \frac{1}{3}(-16t^2)\Big|_0^3 = \frac{1}{3}(-16\cdot 3^2)$$

$$= -48 \text{ feet per second.}$$

Note: There is another way to approach this problem:

$$[\text{average velocity}] = \frac{[\text{distance traveled}]}{[\text{time elapsed}]}.$$

As we discussed in Section 6.2, distance traveled equals the area under the velocity curve. Therefore,

$$[\text{average velocity}] = \frac{\int_0^3 -32t\,dt}{3}.$$

2. According to the formula developed in the text, the future value of the income stream after 10 years is equal to

$$\int_0^{10} 300e^{.1(10-t)}\,dt = -\frac{300}{.1}e^{.1(10-t)}\Big|_0^{10}$$

$$= -3000e^0 - (-3000e^1)$$

$$= 3000e - 3000$$

$$\approx \$5154.85.$$

6.6   Techniques of Integration

Integration by Substitution   Many integrals that appear complex have a simple form that becomes evident after making an appropriate change of variable. Integration by substitution is a technique for making such changes of variables.

Let $f(x)$ and $g(x)$ be two given functions and let $F(x)$ be an antiderivative for $f(x)$. The chain rule asserts that

$$\frac{d}{dx}[F(g(x))] = F'(g(x))g'(x)$$

$$= f(g(x))g'(x) \qquad [\text{since } F'(x) = f(x).]$$

Turning this formula into an integration formula, we have

$$\int f(g(x))g'(x)\,dx = F(g(x)) + C, \tag{1}$$

where C is any constant.

▶ **Example 1**   Determine $\displaystyle\int (x^2+1)^3\cdot 2x\,dx$.

Solution   If we set $f(x) = x^3$, $g(x) = x^2 + 1$, then $f(g(x)) = (x^2+1)^3$ and $g'(x) = 2x$. Therefore, we can apply formula (1). An antiderivative $F(x)$ of $f(x)$ is given by

$$F(x) = \frac{1}{4}x^4,$$

so that, by formula (1), we have

$$\int (x^2 + 1)^3 \cdot 2x \, dx = F(g(x)) + C = \frac{1}{4}(x^2 + 1)^4 + C.$$ ◆

Formula (1) can be elevated from the status of a sometimes-useful formula to a technique of integration by the introduction of a simple mnemonic device. Suppose that we are faced with integrating a function of the form $f(g(x))g'(x)$. Of course, we know the answer from formula (1). However, let us proceed somewhat differently. Replace the expression $g(x)$ by a new variable u, and replace $g'(x) \, dx$ by du. Such a substitution has the advantage that it reduces the generally complex expression $f(g(x))$ to the simpler form $f(u)$. In terms of u, the integration problem may be written

$$\int f(g(x))g'(x) \, dx = \int f(u) \, du.$$

However, the integral on the right is easy to evaluate, since

$$\int f(u) \, du = F(u) + C.$$

Since $u = g(x)$, we then obtain

$$\int f(g(x))g'(x) \, dx = F(u) + C = F(g(x)) + C,$$

which is the correct answer by (1). Remember, however, that replacing $g'(x) \, dx$ by du only has status as a correct mathematical statement because doing so leads to the correct answers. We do not, in this book, seek to explain in any deeper way what this replacement means.

Let us rework Example 1 using this method.

Second Solution of Example 1   Set $u = x^2 + 1$. Then $du = \dfrac{d}{dx}(x^2 + 1) \, dx = 2x \, dx$, and

$$\int (x^2 + 1)^3 \cdot 2x \, dx = \int u^3 \, du$$

$$= \frac{1}{4}u^4 + C$$

$$= \frac{1}{4}(x^2 + 1)^4 + C \quad \text{(since } u = x^2 + 1\text{).}$$ ◆

▶ **Example 2** Evaluate $\displaystyle\int 2xe^{x^2} \, dx$.

Solution Let $u = x^2$, so that $du = \dfrac{d}{dx}(x^2) \, dx = 2x \, dx$. Therefore,

$$\int 2xe^{x^2} \, dx = \int e^{x^2} \cdot 2x \, dx$$

$$= \int e^u \, du$$

$$= e^u + C$$

$$= e^{x^2} + C.$$ ◆

From Examples 1 and 2 we can deduce the following method for integration of functions of the form $f'(g(x))g'(x)$.

> **Integration by Substitution**
>
> *1.* Define a new variable $u = g(x)$, where $g(x)$ is chosen in such a way that, when written in terms of u, the integrand is simpler than when written in terms of x.
>
> *2.* Transform the integral with respect to x into an integral with respect to u by replacing $g(x)$ everywhere by u and $g'(x)\,dx$ by du.
>
> *3.* Integrate the resulting function of u.
>
> *4.* Rewrite the answer in terms of x by replacing u by $g(x)$.

Let us try a few more examples.

▶ **Example 3** Evaluate $\displaystyle\int 3x^2\sqrt{x^3 + 1}\,dx$.

Solution The first problem facing us is to find an appropriate substitution that will simplify the integral. An immediate possibility is offered by setting $u = x^3 + 1$. Then $\sqrt{x^3 + 1}$ will become $\sqrt{u}$, a significant simplification. If $u = x^3 + 1$, then $du = \dfrac{d}{dx}(x^3 + 1)\,dx = 3x^2\,dx$, so that

$$\int 3x^2\sqrt{x^3 + 1}\,dx = \int \sqrt{u}\,du$$

$$= \frac{2}{3}u^{3/2} + C$$

$$= \frac{2}{3}(x^3 + 1)^{3/2} + C. \qquad \blacklozenge$$

▶ **Example 4** Find $\displaystyle\int \frac{(\ln x)^2}{x}\,dx$.

Solution Let $u = \ln x$. Then $du = (1/x)\,dx$ and

$$\int \frac{(\ln x)^2}{x}\,dx = \int (\ln x)^2 \cdot \frac{1}{x}\,dx$$

$$= \int u^2\,du$$

$$= \frac{u^3}{3} + C$$

$$= \frac{(\ln x)^3}{3} + C \quad (\text{since } u = \ln x). \qquad \blacklozenge$$

Knowing the correct substitution to make is a skill that develops through practice. Basically, we look for an occurence of function composition, $f(g(x))$, where $f(x)$ is a function that we know how to integrate and where $g'(x)$ also appears in the integrand. Sometimes $g'(x)$ does not appear exactly, but can be obtained by multiplying by a constant. Such a shortcoming is easily remedied, as is illustrated in Examples 5 and 6.

▶ **Example 5** Find $\int x^2 e^{x^3}\, dx$.

Solution Let $u = x^3$; then $du = 3x^2\, dx$. The integrand involves $x^2\, dx$, not $3x^2\, dx$. To introduce the missing factor "3", we write

$$\int x^2 e^{x^3}\, dx = \int \frac{1}{3}\cdot 3x^2 e^{x^3}\, dx = \frac{1}{3}\int e^{x^3} 3x^2\, dx.$$

(Recall from Section 6.1 that constant multiples may be moved through the integral sign.) Substituting, we obtain

$$\int x^2 e^{x^3}\, dx = \frac{1}{3}\int e^u\, du = \frac{1}{3}e^u + C$$

$$= \frac{1}{3}e^{x^3} + C \quad \text{(since } u = x^3\text{).}$$

Another way to handle the missing factor "3" is to write

$$u = x^3, \quad du = 3x^2\, dx, \quad \text{and} \quad \frac{1}{3}\, du = x^2\, dx.$$

Then substitution yields

$$\int x^2 e^{x^3}\, dx = \int e^{x^3}\cdot x^2\, dx = \int e^u\cdot\frac{1}{3}\, du = \frac{1}{3}\int e^u\, du$$

$$= \frac{1}{3}e^u + C = \frac{1}{3}e^{x^3} + C.$$ ◆

▶ **Example 6** Find $\int \dfrac{2 - x}{\sqrt{2x^2 - 8x + 1}}\, dx$.

Solution Let $u = 2x^2 - 8x + 1$; then $du = (4x - 8)\, dx$. Observe that $4x - 8 = -4(2 - x)$. So we multiply the integrand by -4 and compensate by placing a factor of $-\frac{1}{4}$ in front of the integral.

$$\int \frac{1}{\sqrt{2x^2 - 8x + 1}}\cdot(2 - x)\, dx = -\frac{1}{4}\int \frac{1}{\sqrt{2x^2 - 8x + 1}}\cdot(-4)(2 - x)\, dx$$

$$= -\frac{1}{4}\int \frac{1}{\sqrt{u}}\, du = -\frac{1}{4}\int u^{-1/2}\, du$$

$$= -\frac{1}{4}\cdot 2u^{1/2} + C = -\frac{1}{2}u^{1/2} + C$$

$$= -\frac{1}{2}(2x^2 - 8x + 1)^{1/2} + C$$ ◆

▶ **Example 7** Find $\int \dfrac{x}{x^2 + 1}\, dx$.

Solution If $u = x^2 + 1$, then $du = 2x\, dx$, and

$$\int \frac{x}{x^2 + 1}\, dx = \int \frac{1}{2}\cdot\frac{1}{x^2 + 1}\cdot 2x\, dx$$

$$= \frac{1}{2}\int \frac{1}{x^2 + 1}\cdot 2x\, dx = \frac{1}{2}\int \frac{1}{u}\, du$$

$$= \frac{1}{2}\ln|u| + C = \frac{1}{2}\ln|x^2 + 1| + C.$$ ◆

Integration by Parts Integration by parts is a technique of integration that exchanges a given integral for another (less complicated) integral. Let $f(x)$ and $g(x)$ be given functions and let $G(x)$ be an antiderivative of $g(x)$. The product rule asserts that

$$\frac{d}{dx}[f(x)G(x)] = f(x)G'(x) + f'(x)G(x)$$
$$= f(x)g(x) + f'(x)G(x) \quad \text{[since } G'(x) = g(x)\text{]}.$$

Therefore,

$$f(x)G(x) = \int f(x)g(x)\, dx + \int f'(x)G(x)\, dx.$$

This last formula can be rewritten in the following more useful form.

$$\int f(x)g(x)\, dx = f(x)G(x) - \int f'(x)G(x)\, dx. \tag{2}$$

▶ **Example 8** Evaluate $\int xe^x\, dx$.

Solution Set $f(x) = x$, $g(x) = e^x$. Then $f'(x) = 1$, $G(x) = e^x$, and equation (2) yields

$$\int xe^x\, dx = xe^x - \int 1 \cdot e^x\, dx = xe^x - e^x + C. \qquad \blacklozenge$$

The following principles underlie Example 1 and also illustrate general features of situations to which integration by parts may be applied:

1. The integrand is the product of two functions $f(x) = x$ and $g(x) = e^x$.
2. It is easy to compute $f'(x)$ and $G(x)$. That is, we can differentiate $f(x)$ and integrate $g(x)$.
3. The integral $\int f'(x)G(x)\, dx$ can be calculated.

Let us consider another example to see how these three principles work.

▶ **Example 9** Evaluate $\int x(x+5)^8\, dx$.

Solution Our calculations can be set up as follows:

$$f(x) = x, \qquad g(x) = (x+5)^8,$$

$$f'(x) = 1, \qquad G(x) = \frac{1}{9}(x+5)^9.$$

Then

$$\int x(x+5)^8\, dx = x \cdot \frac{1}{9}(x+5)^9 - \int 1 \cdot \frac{1}{9}(x+5)^9\, dx$$

$$= \frac{1}{9}x(x+5)^9 - \frac{1}{9}\int (x+5)^9\, dx$$

$$= \frac{1}{9}x(x+5)^9 - \frac{1}{9} \cdot \frac{1}{10}(x+5)^{10} + C$$

$$= \frac{1}{9}x(x+5)^9 - \frac{1}{90}(x+5)^{10} + C. \qquad \blacklozenge$$

We were led to try integration by parts because our integrand is the product of two functions. We choose $f(x) = x$ [and not $(x + 5)^8$] because $f'(x) = 1$, so that the factor x in the integrand is made to disappear, thereby simplifying the integral.

▶ **Example 10** Evaluate $\int x^2 \ln x \, dx$.

Solution Set

$$f(x) = \ln x, \qquad g(x) = x^2,$$

$$f'(x) = \frac{1}{x}, \qquad G(x) = \frac{x^3}{3}.$$

Then

$$\int x^2 \ln x \, dx = \frac{x^3}{3} \ln x - \int \frac{1}{x} \cdot \frac{x^3}{3} \, dx$$

$$= \frac{x^3}{3} \ln x - \frac{1}{3} \int x^2 \, dx$$

$$= \frac{x^3}{3} \ln x - \frac{1}{9} x^3 + C. \qquad \blacklozenge$$

▶ **Example 11** Evaluate $\int \ln x \, dx$.

Solution Since $\ln x = 1 \cdot \ln x$, we may view $\ln x$ as a product $f(x)g(x)$, where $f(x) = \ln x$, $g(x) = 1$. Then

$$f'(x) = \frac{1}{x}, \qquad G(x) = x.$$

Finally,

$$\int \ln x \, dx = x \ln x - \int \frac{1}{x} \cdot x \, dx$$

$$= x \ln x - \int 1 \, dx$$

$$= x \ln x - x + C. \qquad \blacklozenge$$

Present Value of an Income Stream Let us illustrate how an integration technique such as integration by parts can arise in a fairly simple business situation. Recall from our discussion of compound interest that the present value of A dollars to be received t years from now is given by

$$P = Ae^{-rt},$$

where r is a specified annual rate of interest, with interest compounded continuously. Now suppose that a sum of money is to be received in a series of frequent payments over a period of time instead of in one lump sum at the end of the period. Such a series of payments is often viewed as a "continuous stream of income." If $K(t)$ is the annual rate of income at time t, and if the income is to be received over the next N years, then the *present value P of the stream of income* at interest rate r is defined by the integral

$$P = \int_0^N K(t)e^{-rt}\,dt. \tag{3}$$

A Riemann sum argument can be used to show that it takes a fund of P dollars available today in order to create a continuous stream of income of $K(t)$ dollars (annual rate) at time t. (See Exercise 27 in Section 6.5.)

The concept of the present value of a continuous stream of income is an important tool in management decision processes involving the selection or replacement of equipment. Even when $K(t)$ is a simple function, the evaluation of the integral in (3) usually requires special techniques such as integration by parts, as we see in the following example.

▶ **Example 12** A printing company estimates that the rate of revenue generated by one of its printing presses at time t will be $80 - 2t$ thousand dollars per year. Find the present value of this continuous stream of income over the next 4 years at a 10% interest rate.

Solution We use (3) with $K(t) = 80 - 2t$, $N = 4$, and $r = .1$. Our first task is to find an antiderivative of $K(t)e^{-rt} = (80 - 2t)e^{-.1t}$. We integrate by parts, with $f(t) = 80 - 2t$, $g(t) = e^{-.1t}$, $f'(t) = -2$, and $G(t) = -10e^{-.1t}$.

$$\int (80 - 2t)e^{-.1t}\,dt = (80 - 2t)(-10e^{-.1t}) - \int 20e^{-.1t}\,dt$$

$$= -800e^{-.1t} + 20te^{-.1t} + 200e^{-.1t} + C$$

$$= 20te^{-.1t} - 600e^{-.1t} + C.$$

Then

$$P = \int_0^4 (80 - 2t)e^{-.1t}\,dt = \left(20te^{-.1t} - 600e^{-.1t}\right)\Big|_0^4$$

$$= (80e^{-.4} - 600e^{-.4}) - (0 - 600e^0)$$

$$= 600 - 520e^{-.4}$$

$$\approx 251.$$

Thus the present value of the machine's earnings is approximately \$251,000. ◆

Practice Problems 6.6

Calculate the following integrals.

1. $\displaystyle\int \frac{e^{5x} + x^4}{(e^{5x} + x^5)^2}\,dx$ [*Hint:* Let $u = e^{5x} + x^5$.]

2. $\displaystyle\int \frac{2x - 1}{(x + 4)^{1/3}}\,dx$ [*Hint:* Let $f(x) = 2x - 1$, $g(x) = (x + 4)^{-1/3}$.]

3. $\displaystyle\int \ln\sqrt{x}\,dx$

▶ Exercises 6.6

Calculate each of the following indefinite integrals.

1. $\displaystyle\int 2x(x^2 + 4)^5 \, dx$

2. $\displaystyle\int 3x^2(x^3 + 1)^2 \, dx$

3. $\displaystyle\int (x^2 - 5x)^3(2x - 5) \, dx$

4. $\displaystyle\int 2x\sqrt{x^2 + 3} \, dx$

5. $\displaystyle\int 5e^{5x-3} \, dx$

6. $\displaystyle\int 2xe^{-x^2} \, dx$

7. $\displaystyle\int \frac{3x^2}{x^3 - 1} \, dx$

8. $\displaystyle\int \frac{2x + 1}{(x^2 + x + 3)^6} \, dx$

Calculate the following integrals, making the indicated substitutions.

9. $\displaystyle\int \frac{x^2}{\sqrt{x^3 - 1}} \, dx; \; u = x^3 - 1$

10. $\displaystyle\int \frac{1}{\sqrt{2x + 1}} \, dx; \; u = 2x + 1$

11. $\displaystyle\int \frac{e^{1/x}}{x^2} \, dx; \; u = \frac{1}{x}$

12. $\displaystyle\int \frac{e^{3x}}{e^{3x} + 1} \, dx; \; u = e^{3x} + 1$

13. $\displaystyle\int \frac{x^2 + 1}{x^3 + 3x + 2} \, dx; \; u = x^3 + 3x + 2$

14. $\displaystyle\int \frac{\ln x}{x} \, dx; \; u = \ln x$

Determine the following integrals by making an appropriate substitution.

15. $\displaystyle\int (x^5 - 2x + 1)^{10}(5x^4 - 2) \, dx$

16. $\displaystyle\int (x + 1)e^{x^2 + 2x + 4} \, dx$

17. $\displaystyle\int \frac{3}{2x - 4} \, dx$

18. $\displaystyle\int \frac{e^{\sqrt{x}}}{\sqrt{x}} \, dx$

19. $\displaystyle\int \frac{x}{e^{x^2}} \, dx$

20. $\displaystyle\int \frac{3x - x^3}{x^4 - 6x^2 + 5} \, dx$

Use integration by parts to determine the following integrals.

21. $\displaystyle\int xe^{5x} \, dx$

22. $\displaystyle\int xe^{-x/2} \, dx$

23. $\displaystyle\int x(2x + 1)^4 \, dx$

24. $\displaystyle\int (x + 1)e^x \, dx$

25. $\displaystyle\int x\sqrt{x + 1} \, dx$

26. $\displaystyle\int x(x + 5)^{-3} \, dx$

27. $\displaystyle\int \frac{3x}{e^x} \, dx$

28. $\displaystyle\int x^3 \ln x \, dx$

29. $\displaystyle\int \ln 3x \, dx$ [Let $f(x) = \ln 3x$, $g(x) = 1$.]

30. $\displaystyle\int \frac{x}{\sqrt{3 + 2x}} \, dx$ [Let $f(x) = x$, $g(x) = (3 + 2x)^{-1/2}$.]

Determine the following indefinite integrals.

31. $\displaystyle\int xe^{2x} \, dx$

32. $\displaystyle\int x\sqrt{x^2 - 1} \, dx$

33. $\displaystyle\int xe^{x^2} \, dx$

34. $\displaystyle\int x\sqrt{x - 1} \, dx$

35. $\displaystyle\int \frac{2x - 1}{\sqrt{3x - 3}} \, dx$

36. $\displaystyle\int \frac{2x - 1}{3x^2 - 3x + 1} \, dx$

37. $\displaystyle\int (4x + 3)(6x^2 + 9x)^{-7} \, dx$

38. $\displaystyle\int (4x + 3)(6x + 9)^{-7} \, dx$

39. $\displaystyle\int \frac{\ln x}{\sqrt{x}} \, dx$

40. $\displaystyle\int \frac{x + 4}{e^{4x}} \, dx$

41. $\displaystyle\int \frac{1}{x \ln 5x} \, dx$

42. $\displaystyle\int x \ln 5x \, dx$

43. $\displaystyle\int \ln x^4 \, dx$

44. $\displaystyle\int \frac{(\ln x)^4}{x} \, dx$

45. $\displaystyle\int \frac{\ln \sqrt{x}}{x} \, dx$

46. $\displaystyle\int \sqrt{x} \ln \sqrt{x} \, dx$

47. Figure 1 shows graphs of several functions $f(x)$ whose slope at each x is $x/\sqrt{x^2 + 9}$. Find the expression for the function $f(x)$ whose graph passes through $(4, 8)$.

48. Figure 2 shows graphs of several functions $f(x)$ whose slope at each x is $(2\sqrt{x} + 1)/\sqrt{x}$. Find the expression for the function $f(x)$ whose graph passes through $(4, 15)$.

49. Suppose that an investment produces a continuous stream of income at the rate of $300t + 500$ dollars per year at time t. Find the present value of the investment income over the next 5 years, using a 10% interest rate.

50. Recompute the present value of the stream of income in Exercise 49 over 5 years using a 5% interest rate.

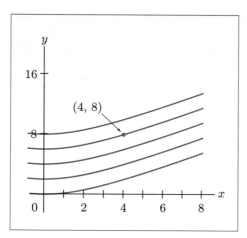

Figure 1.

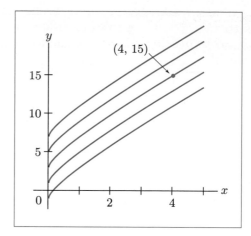

Figure 2.

Solutions to Practice Problems 6.6

1. Let $u = e^{5x} + x^5$ and $du = (5e^{5x} + 5x^4)\,dx$, and note that $5e^{5x} + 5x^4 = 5(e^{5x} + x^4)$. Then

$$\int \frac{e^{5x} + x^4}{(e^{5x} + x^5)^2}\,dx = \int \frac{1}{5} \cdot \frac{5e^{5x} + 5x^4}{(e^{5x} + x^5)^2}\,dx$$

$$= \frac{1}{5}\int \frac{1}{(e^{5x} + x^5)^2} \cdot (5e^{5x} + 5x^4)\,dx$$

$$= \frac{1}{5}\int \frac{1}{u^2}\,du$$

$$= \frac{1}{5}(-u^{-1}) + C$$

$$= -\frac{1}{5}(e^{5x} + x^5)^{-1} + C.$$

2. Use integration by parts with $f(x) = 2x - 1$, $g(x) = (x+4)^{-1/3}$, $f'(x) = 2$, and $G(x) = \frac{3}{2}(x+4)^{2/3}$.

$$\int \frac{2x - 1}{(x+4)^{1/3}}\,dx = \int (2x - 1)(x+4)^{-1/3}\,dx$$

$$= (2x - 1) \cdot \frac{3}{2}(x+4)^{2/3} - \int 2 \cdot \frac{3}{2}(x+4)^{2/3}\,dx$$

$$= \frac{3}{2}(2x - 1)(x+4)^{2/3} - 3\int (x+4)^{2/3}\,dx$$

$$= \frac{3}{2}(2x - 1)(x+4)^{2/3} - \frac{9}{5}(x+4)^{5/3} + C.$$

Solutions to Practice Problems 6.6 (Continued)

3. This problem is similar to Example 11, which asks for $\int \ln x \, dx$, and can be approached in the same way by letting $f(x) = \ln \sqrt{x}$ and $g(x) = 1$. Another approach is to use a property of logarithms to simplify the integrand.

$$\int \ln \sqrt{x} \, dx = \int \ln(x^{1/2}) \, dx = \int \frac{1}{2} \ln x \, dx = \frac{1}{2} \int \ln x \, dx.$$

Since we know $\int \ln x \, dx$ from Example 11,

$$\int \ln \sqrt{x} \, dx = \frac{1}{2} \int \ln x \, dx = \frac{1}{2}(x \ln x - x) + C.$$

6.7 Improper Integrals

In applications of calculus, especially to statistics, it is often necessary to consider the area of a region that extends infinitely far to the right or left along the x-axis. We have drawn several such regions in Fig. 1. The areas of such "infinite" regions may be computed using *improper integrals*.

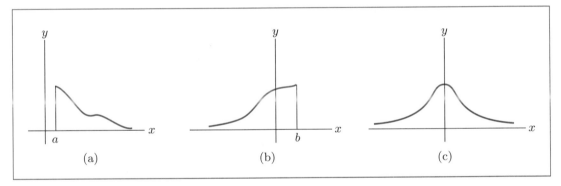

Figure 1.

In order to motivate the idea of an improper integral, let us attempt to calculate the area under the curve $y = 3/x^2$ to the right of $x = 1$ (Fig. 2.)

First, we shall compute the area under the graph of this function from $x = 1$ to $x = b$, where b is some number greater than 1. [See Fig. 3(a).] Then we shall examine how the area increases as we let b get larger, as in Figs. 3(b) and 3(c).

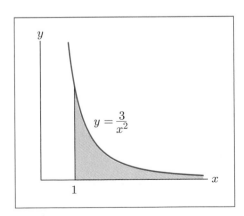

Figure 2.

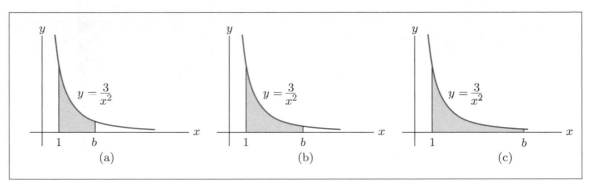

Figure 3.

The area from 1 to b is given by

$$\int_1^b \frac{3}{x^2}\, dx = -\frac{3}{x}\Big|_1^b = \left(-\frac{3}{b}\right) - \left(-\frac{3}{1}\right) = 3 - \frac{3}{b}.$$

When b is large, $3/b$ is small and the integral nearly equals 3. That is, the area under the curve from 1 to b nearly equals 3. (See Table 1.) In fact, the area gets arbitrarily close to 3 as b gets larger. Thus it is reasonable to say that the region under the curve $y = 3/x^2$ for $x \geq 1$ has area 3.

Table 1	Value of an "Infinitely Long" Area as a Limit
b	$Area = \int_1^b \frac{3}{x^2}\, dx = 3 - \frac{3}{b}$
10	2.7000
100	2.9700
1,000	2.9970
10,000	2.9997

Recall from Chapter 1 that we write $b \to \infty$ as shorthand for "b gets arbitrarily large, without bound." Then, to express the fact that the value of $\int_1^b \frac{3}{x^2}\, dx$ approaches 3 as $b \to \infty$, we write

$$\int_1^\infty \frac{3}{x^2}\, dx = \lim_{b \to \infty} \int_1^b \frac{3}{x^2}\, dx = 3.$$

We call $\int_1^\infty \frac{3}{x^2}\, dx$ an *improper* integral because the upper limit of the integral is ∞ (infinity) rather than a finite number.

Definition Let a be fixed and suppose that $f(x)$ is a nonnegative function for $x \geq a$. If $\lim_{b \to \infty} \int_a^b f(x)\, dx = L$, we define

$$\int_a^\infty f(x)\, dx = \lim_{b \to \infty} \int_a^b f(x)\, dx = L.$$

We say that the improper integral $\int_a^\infty f(x)\,dx$ is *convergent* and that the region under the curve $y = f(x)$ for $x \geq a$ has area L. (See Fig. 4.)

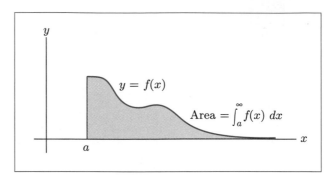

Figure 4. Area defined by an improper integral.

It is possible to consider improper integrals in which $f(x)$ is both positive and negative. However, we shall consider only nonnegative functions, since this is the case occurring in most applications.

▶ **Example 1** Find the area under the curve $y = e^{-x}$ for $x \geq 0$ (Fig. 5).

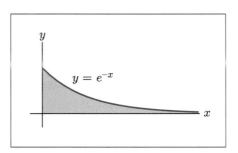

Figure 5.

Solution We must calculate the improper integral

$$\int_0^\infty e^{-x}\,dx.$$

We take $b > 0$ and compute

$$\int_0^b e^{-x}\,dx = -e^{-x}\Big|_0^b = (-e^{-b}) - (-e^0) = 1 - e^{-b} = 1 - \frac{1}{e^b}.$$

We now consider the limit as $b \to \infty$ and note that $1/e^b$ approaches zero. Thus

$$\int_0^\infty e^{-x}\,dx = \lim_{b \to \infty} \int_0^b e^{-x}\,dx = \lim_{b \to \infty} \left(1 - \frac{1}{e^b} \right) = 1.$$

Therefore, the region in Fig. 5 has area 1. ◆

▶ **Example 2** Evaluate the improper integral $\int_7^\infty \dfrac{1}{(x-5)^2}\,dx.$

Solution
$$\int_7^b \frac{1}{(x-5)^2}\,dx = -\frac{1}{x-5}\Big|_7^b = -\frac{1}{b-5} - \left(-\frac{1}{7-5}\right) = \frac{1}{2} - \frac{1}{b-5}.$$

As $b \to \infty$, the fraction $1/(b-5)$ approaches zero, so

$$\int_7^\infty \frac{1}{(x-5)^2}\,dx = \lim_{b\to\infty} \int_7^b \frac{1}{(x-5)^2}\,dx = \lim_{b\to\infty} \left(\frac{1}{2} - \frac{1}{b-5}\right) = \frac{1}{2}. \quad \blacklozenge$$

Not every improper integral is convergent. If the value $\int_a^b f(x)\,dx$ does not have a limit as $b \to \infty$, we cannot assign any numerical value to $\int_a^\infty f(x)\,dx$, and we say that the improper integral $\int_a^\infty f(x)\,dx$ is *divergent*.

▶ Example 3   Show that $\displaystyle\int_1^\infty \frac{1}{\sqrt{x}}\,dx$ is divergent.

Solution   For $b > 1$ we have

$$\int_1^b \frac{1}{\sqrt{x}}\,dx = 2\sqrt{x}\,\Big|_1^b = 2\sqrt{b} - 2. \tag{1}$$

As $b \to \infty$, the quantity $2\sqrt{b} - 2$ increases without bound. That is, $2\sqrt{b} - 2$ can be made larger than any specific number. Therefore $\displaystyle\int_a^b \frac{1}{\sqrt{x}}\,dx$ has no limit as $b \to \infty$, so $\displaystyle\int_1^\infty \frac{1}{\sqrt{x}}\,dx$ is divergent. $\quad \blacklozenge$

In some cases it is necessary to consider improper integrals of the form

$$\int_{-\infty}^b f(x)\,dx.$$

Let b be fixed and examine the value of $\int_a^b f(x)\,dx$ as $a \to -\infty$, that is, as a moves arbitrarily far to the left on the number line. If $\lim\limits_{a\to-\infty} \int_a^b f(x)\,dx = L$, we say that the improper integral $\int_{-\infty}^b f(x)\,dx$ is *convergent* and we write

$$\int_{-\infty}^b f(x)\,dx = L.$$

Otherwise, the improper integral is divergent. An integral of the form $\int_{-\infty}^b f(x)\,dx$ may be used to compute the area of a region such as that shown in Fig. 1(b).

▶ Example 4   Determine if $\int_{-\infty}^0 e^{5x}\,dx$ is convergent. If convergent, find its value.

Solution
$$\int_{-\infty}^0 e^{5x}\,dx = \lim_{a\to-\infty} \int_a^0 e^{5x}\,dx = \lim_{a\to-\infty} \frac{1}{5}e^{5x}\Big|_a^0 = \lim_{a\to-\infty} \left(\frac{1}{5} - \frac{1}{5}e^{5a}\right).$$

As $a \to -\infty$, e^{5a} approaches 0 so that $\frac{1}{5} - \frac{1}{5}e^{5a}$ approaches $\frac{1}{5}$. Thus the improper integral converges and has value $\frac{1}{5}$. $\quad \blacklozenge$

Areas of regions that extend infinitely far to the left *and* right, such as the region in Fig. 1(c), are calculated using improper integrals of the form

$$\int_{-\infty}^\infty f(x)\,dx.$$

We define such an integral to have the value

$$\int_{-\infty}^{0} f(x)\,dx + \int_{0}^{\infty} f(x)\,dx,$$

provided that both of the latter improper integrals are convergent.

An important area that arises in probability theory is the area under the so-called standard normal curve, whose equation is

$$y = \frac{1}{\sqrt{2\pi}}e^{-x^2/2}.$$

(See Fig. 6.) It is of fundamental importance for probability theory that this area is 1. In terms of an improper integral, this fact may be written

$$\int_{-\infty}^{\infty} \frac{1}{\sqrt{2\pi}}e^{-x^2/2}\,dx = 1.$$

The proof of this result is beyond the scope of this book.

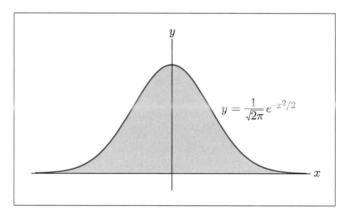

Figure 6. The standard normal curve.

Incorporating Technology Although graphing calculators cannot tell you whether an improper integral converges or not, you can use the calculator to obtain a reliable indication of the behavior of the inetgral. Just look at values of $\int_{a}^{b} f(x)\,dx$ as b increases. Figures. 7(a) and 7(b), which were created by setting $\mathbf{Y_1 = fnInt(e^{\wedge}(-X), X, 0, X)}$, give convincing evidence that the value of the improper integral in Example 1 is 1. The final $\mathbf{X}$ in $\mathbf{Y_1 = fnInt(e^{\wedge}(-X), X, 0, X)}$ is the upper limit of integration; that is, b.

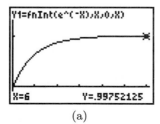
(a)

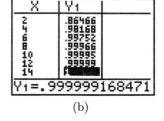
(b)

Figure 7.

1. Does $1 - 2(1 - 3b)^{-4}$ approach a limit as $b \to \infty$?

2. Evaluate $\displaystyle\int_1^\infty \frac{x^2}{x^3 + 8}\, dx$.

3. Evaluate $\displaystyle\int_{-\infty}^{-2} \frac{1}{x^4}\, dx$.

▶ Exercises 6.7

In Exercises 1–12, determine if the given expression approaches a limit as $b \to \infty$, and find that number when it exists.

1. $\dfrac{5}{b}$

2. b^2

3. $-3e^{2b}$

4. $\dfrac{1}{b} + \dfrac{1}{3}$

5. $\dfrac{1}{4} - \dfrac{1}{b^2}$

6. $\dfrac{1}{2}\sqrt{b}$

7. $2 - (b+1)^{-1/2}$

8. $5 - (b-1)^{-1}$

9. $5(b^2 + 3)^{-1}$

10. $4(1 - b^{-3/4})$

11. $e^{-b/2} + 5$

12. $2 - e^{-3b}$

13. Find the area under the graph of $y = 1/x^2$ for $x \geq 2$.

14. Find the area under the graph of $y = (x+1)^{-2}$ for $x \geq 0$.

15. Find the area under the graph of $y = e^{-x/2}$ for $x \geq 0$.

16. Find the area under the graph of $y = 4e^{-4x}$ for $x \geq 0$.

17. Find the area under the graph of $y = (x+1)^{-3/2}$ for $x \geq 3$.

18. Find the area under the graph of $y = (2x + 6)^{-4/3}$ for $x \geq 1$. See Fig. 8.

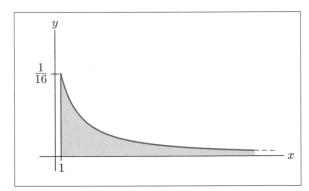

Figure 8.

19. Show that the region under the graph of $y = (14x + 18)^{-4/5}$ for $x \geq 1$ cannot be assigned any finite number as its area. See Fig. 9.

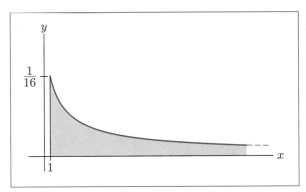

Figure 9.

20. Show that the region under the graph of $y = (x-1)^{-1/3}$ for $x \geq 2$ cannot be assigned any finite number as its area.

Evaluate the following improper integrals whenever they are convergent.

21. $\displaystyle\int_1^\infty \frac{1}{x^3}\, dx$

22. $\displaystyle\int_1^\infty \frac{2}{x^{3/2}}\, dx$

23. $\displaystyle\int_0^\infty \frac{1}{(2x+3)^2}\, dx$

24. $\displaystyle\int_0^\infty e^{-3x}\, dx$

25. $\displaystyle\int_0^\infty e^{2x}\, dx$

26. $\displaystyle\int_0^\infty (x^2 + 1)\, dx$

27. $\displaystyle\int_2^\infty \frac{1}{(x-1)^{5/2}}\, dx$

28. $\displaystyle\int_2^\infty e^{2-x}\, dx$

29. $\displaystyle\int_0^\infty .01e^{-.01x}\, dx$

30. $\displaystyle\int_0^\infty \frac{4}{(2x+1)^3}\, dx$

31. $\displaystyle\int_0^\infty 6e^{1-3x}\, dx$

32. $\displaystyle\int_1^\infty e^{-.2x}\, dx$

33. $\displaystyle\int_3^\infty \frac{x^2}{\sqrt{x^3 - 1}}\, dx$

34. $\displaystyle\int_2^\infty \frac{1}{x \ln x}\, dx$

35. $\displaystyle\int_0^\infty xe^{-x^2}\, dx$

36. $\displaystyle\int_0^\infty \frac{x}{x^2 + 1}\, dx$

37. $\displaystyle\int_0^\infty 2x(x^2 + 1)^{-3/2}\, dx$

38. $\displaystyle\int_1^\infty (5x + 1)^{-4}\, dx$

39. $\displaystyle\int_{-\infty}^0 e^{4x}\, dx$

40. $\displaystyle\int_{-\infty}^0 \frac{8}{(x-5)^2}\, dx$

41. $\displaystyle\int_{-\infty}^{0} \frac{6}{(1-3x)^2}\, dx$

42. $\displaystyle\int_{-\infty}^{0} \frac{1}{\sqrt{4-x}}\, dx$

43. $\displaystyle\int_{0}^{\infty} \frac{e^{-x}}{(e^{-x}+2)^2}\, dx$

44. $\displaystyle\int_{-\infty}^{\infty} \frac{e^{-x}}{(e^{-x}+2)^2}\, dx$

45. If $k > 0$, show that $\displaystyle\int_{0}^{\infty} ke^{-kx}\, dx = 1$.

46. If $k > 0$, show that $\displaystyle\int_{1}^{\infty} \frac{k}{x^{k+1}}\, dx = 1$.

47. If $k > 0$, show that $\displaystyle\int_{e}^{\infty} \frac{k}{x(\ln x)^{k+1}}\, dx = 1$.

The *capital value* of an asset such as a machine is sometimes defined as the present value of all future net earnings. The actual lifetime of the asset may not be known, and since some assets may last indefinitely, the capital value of the asset may be written in the form

$$[\text{capital value}] = \int_{0}^{\infty} K(t)e^{-rt}\, dt,$$

where r is the annual rate of interest, compounded continuously, and $K(t)$ is the income stream of the asset.

48. Find the capital value of an asset that generates income at the rate of $5000 per year, assuming an interest rate of 10%.

49. Construct a formula for the capital value of a rental property that will generate a fixed income at the rate of K dollars per year indefinitely, assuming an annual interest rate r.

50. Suppose that a large farm with a known reservoir of gas beneath the ground sells the gas rights to a company for a guaranteed payment at the rate of $10{,}000e^{.04t}$ dollars per year. Find the present value of this perpetual stream of income, assuming an interest rate of 12%, compounded continuously.

Solutions to Practice Problems 6.7

1. The expression $1 - 2(1-3b)^{-4}$ may also be written in the form

$$1 - \frac{2}{(1-3b)^4}.$$

When b is large, $(1-3b)^4$ is very large, so $2/(1-3b)^4$ is very small. Thus $1 - 2(1-3b)^{-4}$ approaches 1 as $b \to \infty$.

2. The first step is to find an antiderivative of $x^2/(x^3 + 8)$. Using the substitution $u = x^3 + 8$, $du = 3x^2\, dx$, we obtain

$$\int \frac{x^2}{x^3 + 8}\, dx = \frac{1}{3} \int \frac{1}{u}\, du = \frac{1}{3} \ln|u| + C = \frac{1}{3} \ln|x^3 + 8| + C.$$

Now,

$$\int_{1}^{b} \frac{x^2}{x^3 + 8}\, dx = \frac{1}{3} \ln|x^3 + 8|\Big|_{1}^{b} = \frac{1}{3} \ln(b^3 + 8) - \frac{1}{3} \ln 9.$$

Finally, we examine what happens as $b \to \infty$. Certainly, $b^3 + 8$ gets arbitrarily large, so $\ln(b^3 + 8)$ must also get arbitrarily large. Hence

$$\int_{1}^{b} \frac{x^2}{x^3 + 8}\, dx$$

has no limit as $b \to \infty$, so the improper integral

$$\int_{1}^{\infty} \frac{x^2}{x^3 + 8}\, dx$$

is divergent.

**Solutions to
Practice Problems
6.7 (Continued)**

3.

$$\int_a^{-2} \frac{1}{x^4}\,dx = \int_a^{-2} x^{-4}\,dx = \left.\frac{x^{-3}}{-3}\right|_a^{-2} = \left.\frac{1}{-3x^3}\right|_a^{-2}$$

$$= \frac{1}{-3(-2)^3} - \left(\frac{1}{-3\cdot a^3}\right)$$

$$= \frac{1}{24} + \frac{1}{3a^3}$$

$$\int_{-\infty}^{-2} \frac{1}{x^4}\,dx = \lim_{a\to-\infty} \int_a^{-2} \frac{1}{x^4}\,dx = \lim_{a\to-\infty}\left(\frac{1}{24} + \frac{1}{3a^3}\right) = \frac{1}{24}.$$

6.8 Applications of Calculus to Probability

Consider a cell population that is growing vigorously. Suppose that when a cell is T days old it divides and forms two new "daughter" cells. If the population is sufficiently large, it will contain cells of many different ages between 0 and T. It turns out that the proportion of cells of various ages remains constant. That is, if a and b are any two numbers between 0 and T, with $a < b$, the proportion of cells whose ages lie between a and b is essentially constant from one moment to the next, even though individual cells are aging and new cells are being formed all the time. In fact, biologists have found that under the ideal circumstances described, the proportion of cells whose ages are between a and b is given by the area under the graph of the function $f(x) = 2ke^{-kx}$ from $x = a$ to $x = b$, where $k = (\ln 2)/T$.* (See Fig. 1.)

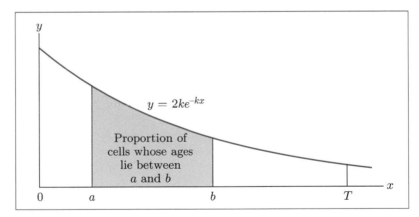

Figure 1. Age distribution of cells.

Now consider an experiment where we select a cell at random from the population and observe its age, X. Then the probability or likelihood[†] that X lies between a and b is given by the area under the graph of $f(x) = 2ke^{-kx}$ from a to b, as in Fig. 1. Let us denote this probability by $\Pr(a \le X \le b)$. Using the

---

[*] See J. R. Cook and T. W. James, "Age Distribution of Cells in Logarithmically Growing Cell Populations," in *Synchrony in Cell Division and Growth*, Erik Zeuthen, ed. (New York: John Wiley & Sons, 1964), pp. 485–495.

[†] For our purposes, it is sufficient to think of probability in the following intuitive terms. Suppose that an experiment with observed outcome X is repeated very often. Then the probability $\Pr(a \le X \le b)$ is given (approximately) as the fraction of repetitions in which X was between a and b.

fact that the area under the graph of $f(x)$ is given by a definite integral, we have

$$\Pr(a \leq X \leq b) = \int_a^b f(x)\,dx = \int_a^b 2ke^{-kx}\,dx. \tag{1}$$

The function $f(x)$ that determines the probability in (1) for each a and b is called the *probability density function* of X (or of the experiment whose outcome is X).

The general situation we wish to describe in this section concerns an experiment whose outcome is a number X in a certain interval, say between A and B. For the cell population above, $A = 0$ and $B = T$. Another typical experiment might consist of choosing a real number X at random between $A = 5$ and $B = 6$. Or, one could select a random telephone call at some telephone switchboard and observe its duration, X. If we have no way of knowing how long a call might last, then X might be any nonnegative number. In this case it is convenient to say that X lies between 0 and ∞ and to take $A = 0$ and $B = \infty$. A similar situation arises in reliability studies where one measures the lifetime X of a transistor selected at random from a manufacturer's production line. Again, the possible values of X lie between 0 and ∞.

When we are dealing with experiments such as those described above, many questions can be reduced to calculating the probability that the outcome X lies between two specified numbers, say a and b. This probability, $\Pr(a \leq X \leq b)$, is a measure of the likelihood that an outcome of the experiment will lie between a and b. If the experiment is repeated many times, then the proportion of times X has a value between a and b will be close to $\Pr(a \leq X \leq b)$. In experiments of practical interest, it is often possible to find a density function $f(x)$ such that

$$\Pr(a \leq X \leq b) = \int_a^b f(x)\,dx, \tag{2}$$

for all a and b in the range of possible values of X.

Any function $f(x)$ with the following two properties is said to be a (*probability*) *density function*:

    (I) $f(x) \geq 0$ for $A \leq x \leq B$.

    (II) $\displaystyle\int_A^B f(x) = 1$.

Equation (2) relates a density function to the outcome X of a specific experiment. Graphically, properties I and II mean that for x between A and B, the graph of $f(x)$ must lie above or on the x-axis and the area under the graph must equal 1. Property II simply says that there is probability 1 (certainty) that X has a value between A and B. Of course, if $B = \infty$, then the integral in property II is an improper integral.

▶ **Example 1**   Consider the cell population described earlier. Let $f(x) = 2ke^{-kx}$, where $k = (\ln 2)/3$. Show that $f(x)$ is indeed a probability density function on the interval from $x = 0$ to $x = 3$.

Solution   Clearly $f(x) \geq 0$, since the exponential function is never negative. Thus property

I is satisfied. For property II, we check that

$$\int_0^3 f(x)\,dx = \int_0^3 2ke^{-kx}\,dx = -2e^{-kx}\Big|_0^3 = -2e^{-k3} + 2e^0$$

$$= 2 - 2e^{-[(\ln 2)/3]3} = 2 - 2e^{-\ln 2}$$

$$= 2 - 2(e^{\ln 2})^{-1} = 2 - 2(2)^{-1} = 2 - 1 = 1. \qquad \blacklozenge$$

The simplest probability density function is that which assumes a constant value for $A \le x \le B$. In order for property II to hold, this constant value must be $1/(B - A)$; that is,

$$f(x) = \frac{1}{B - A}, \qquad A \le x \le B.$$

For in this case the area under the graph of $f(x)$ in Fig. 2 is 1. Such a probability density function is said to be *uniform*. Experiments having uniform density functions generalize those experiments with a finite number of equally likely outcomes.

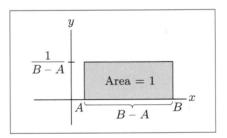

Figure 2. A uniform density function.

▶ **Example 2**   Suppose that a subway train leaves the station every 15 minutes. A person who arrives at a random time during the day must wait between 0 and 15 minutes for the next train. What are the chances the person must wait at least 10 minutes?

Solution   Let X be the number of minutes the person must wait. The statement that the person arrives at a "random" time is usually interpreted to mean that X has a uniform probability density function. Since the possible values of X lie between 0 and 15, we take $f(x) = \frac{1}{15}$. Then the probability that the person waits at least 10 mintues is given by

$$\Pr(10 \le X \le 15) = \int_{10}^{15} \frac{1}{15}\,dx = \frac{1}{15}x\Big|_{10}^{15} = \frac{15}{15} - \frac{10}{15} = \frac{1}{3}.$$

(See Fig. 3.) $\blacklozenge$

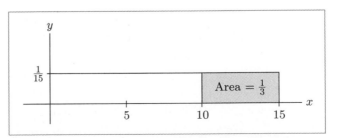

Figure 3.

As we mentioned earlier, for some experiments the possible outcomes are the numbers between 0 and ∞. In such a case the probability density function $f(x)$ is defined for all $x \geq 0$, and property II is written as

$$\int_0^\infty f(x)\,dx = 1.$$

The most important function of this type has the form $f(x) = \lambda e^{-\lambda x}$, where λ is a positive constant. Just as in Example 1 of Section 6.7, one easily verifies that

$$\int_0^\infty \lambda e^{-\lambda x}\,dx = 1.$$

(See Fig. 4.) If the outcome X of an experiment has such a probability density function, the experiment is said to be *exponential* (or *exponentially distributed*). It can be shown that the constant λ may be interpreted as

$$\lambda = \frac{1}{a}, \quad \text{where } a = \text{average value of } X.$$

Typical uses of exponential probability density functions are given in the next two examples.

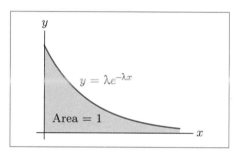

Figure 4. Exponential density function.

▶ Example 3   Experiment has shown that the lifetime of a light bulb is exponentially distributed. Let X be the lifetime of a light bulb selected at random from the production line of a light bulb manufacturer. For simplicity, let us measure the lifetime in years rather than hours and suppose that the average light bulb produced by the manufacturer burns out in $\frac{1}{4}$ year of continuous use.

(a) What proportion of the light bulbs will burn out within $\frac{1}{2}$ year?

(b) What proportion will continue to burn for at least 1 year?

Solution   The average value of X is $\frac{1}{4}$, so we let $\lambda = 4$ and $f(x) = 4e^{-4x}$.

(a) The proportion of light bulbs where X is less than or equal to $\frac{1}{2}$ is

$$\Pr\left(0 \leq X \leq \tfrac{1}{2}\right) = \int_0^{1/2} 4e^{-4x}\,dx = -e^{-4x}\Big|_0^{1/2} = -e^{-2} + 1$$

$$= 1 - .13534 = .86466.$$

(b) The proportion of light bulbs that do not burn out for at least 1 year is

$$\Pr(1 \leq X < \infty) = \int_1^\infty 4e^{-4x}\,dx.$$

To evaluate this improper integral, we compute

$$\int_1^b 4e^{-4x}\, dx = -e^{-4x}\Big|_1^b = -e^{-4b} + e^{-4} \to e^{-4}$$

as $b \to \infty$. Hence $\Pr(1 \le X < \infty) = e^{-4} = .01832$. ◆

▶ Example 4   A company makes a survey of the duration of telephone calls made by its employees. It finds that the lengths of calls are exponentially distributed, with the average call lasting 5 minutes. What is the probability that a randomly chosen call will last between 5 and 10 minutes?

Solution   Let X be the length of the call. Since the average value of X is 5, we take $\lambda = \frac{1}{5} = .2$ and $f(x) = .2e^{-.2x}$. The desired probability is

$$\Pr(5 \le X \le 10) = \int_5^{10} .2e^{-.2x}\, dx = -e^{-.2x}\Big|_5^{10}$$

$$= -e^{-2} + e^{-1}$$

$$= -.13534 + .36788 = .23254.$$ ◆

Practice Problems 6.8

1. An experimenter determines that the probability density function for a certain experiment with outcomes between 0 and 1 is given by $f(x) = x$. Why might you doubt his conclusion?

2. An experiment with outcomes between 0 and 1 has probability density function $f(x) = 4x^3$. What is the probability of an outcome larger than $\frac{1}{2}$?

▶ **Exercises 6.8**

1. An experiment has the probability density function $f(x) = 6(x - x^2)$ and outcomes lying between 0 and 1. Determine the probability that an outcome
   (a) lies between $\frac{1}{4}$ and $\frac{1}{2}$,
   (b) lies between 0 and $\frac{1}{3}$,
   (c) is at least $\frac{1}{4}$,
   (d) is at most $\frac{3}{4}$.

2. Suppose that the outcome X of an experiment lies between 0 and 4, and the probability density function for X is $f(x) = \frac{1}{8}x$. Find each value.
   (a) $\Pr(X \le 1)$ (b) $\Pr(2 \le X \le 2.5)$
   (c) $\Pr(3.5 \le X)$

3. If $f(x) = kx^2$, determine the value of k that makes $f(x)$ a probability density function on $0 \le x \le 2$.

4. If $f(x) = k/\sqrt{x}$, determine the value of k that makes $f(x)$ a probability density function on $1 \le x \le 4$.

5. Suppose that the outcomes X of an experiment lie between 0 and ∞, and X has an exponential density function $f(x) = 2e^{-2x}$. Find each value.

   (a) $\Pr(X \le .1)$ (b) $\Pr(.1 \le X \le .5)$
   (c) $\Pr(1 \le X)$ (d) the average value of X

6. Suppose that the outcomes X of an experiment are exponentially distributed, with density function $f(x) = .25e^{-.25x}$. Find each value.
   (a) $\Pr(1 \le X \le 2)$ (b) $\Pr(X \le 3)$
   (c) $\Pr(4 \le X)$ (d) the average value of X

7. An automated machine produces an automobile part every 3 minutes. An inspector arrives at a random time and must wait X minutes for a part.
   (a) Find the probability density function for X.
   (b) find the probability that the inspector must wait at least 1 minute.
   (c) Find the probability that the inspector must wait no more than 1 minute.

8. The annual income of the households in a certain community ranges between \$5000 and \$25,000. Let X represent the annual income (in thousands of dollars) of a household chosen at random in this community, and

suppose that the probability density function for X is $f(x) = kx$, $5 \leq x \leq 25$.

(a) Find the value of k that makes $f(x)$ a density function.

(b) Find the fraction of households that have an annual income between $5000 and $10,000.

(c) What fraction of the households have an income exceeding $20,000?

9. Suppose that in a certain farming region, and in a certain year, the number X of bushels of wheat produced on a given acre has a probability density function $f(x) = (x - 30)/50$, $30 \leq x \leq 40$.

(a) What is the probability that an acre selected at random produced less than 35 bushels of wheat?

(b) If the farming region had 20,000 acres of wheat, how many acres produced less than 35 bushels of wheat?

10. The parent corporation for a franchised chain of fast-food restaurants claims that the fraction X of their new restaurants that make a profit during their first year of operation has the probability density $f(x) = 12x^2 - 12x^3$, $0 \leq x \leq 1$.

(a) What is the likelihood that less than 40% of the restaurants opened this year will make a profit during their first year of operation?

(b) What is the likelihood that more than 50% of the restaurants will make a profit during their first year of operation?

11. Suppose that at a certain supermarket the amount of time one must wait at the express lane is a random variable with density function $f(x) = \frac{11}{10}(x + 1)^{-2}$, $0 \leq x \leq 10$. Find the probability of having to wait less than 4 minutes at the express lane.

12. Suppose that in a certain cell population, cells divide every 10 days, and the age of a cell selected at random is a random variable X with the density function $f(x) = 2ke^{-kx}$, $0 \leq x \leq 10$, $k = (.1)\ln 2$.

(a) Find the probability that a cell is at most 5 days old.

(b) Upon examination of a slide, a microbiologist finds that 10% of the cells are undergoing mitosis (a change in the cell leading to division). Compute the length of time required for mitosis; that is, find the number M such that

$$\int_{10-M}^{10} 2ke^{-kx} \, dx = .10.$$

13. At a certain gas station, it takes an average of 4 minutes to get served. Suppose that the service time X for a car has an exponential probability density function.

(a) What fraction of the cars are served within 2 minutes?

(b) What is the probability that a car will have to wait at least 4 minutes?

14. The emergency flasher on an automobile is guaranteed for the first 12,000 miles that the car is driven. On the average, the flashers last about 50,000 miles. Let X be the time of failure of the flasher (measured in thousands of miles), and suppose X has an exponential probability density function. What percentage of the emergency flashers will have to be replaced during the warranty period?

15. Let X be the number of seconds between successive cars at a toll booth on the Ohio Turnpike on a typical Saturday afternoon. It can be shown that X has an exponential density function. If the average interarrival time is 2 seconds, find the probability that X is at least 3 seconds.

16. Let X be the relief time (in minutes) of an arthritic patient who has taken an analgesic for pain. Suppose that a certain analgesic provides relief within 4 minutes for 75% of a large group of patients, and suppose the density function for X is $f(x) = ke^{-kx}$. (This model has been used by some medical researchers.) Then one estimates that $\Pr(X \leq 4) = .75$. Use this estimate to find an approximate value for k. [*Hint:* First show that $\Pr(X \leq 4) = 1 - e^{-4k}$.]

Upon studying the vacancies occurring in the U.S. Supreme Court, it has been determined that the time elapsed between successive resignations is exponentially distributed with average value 2 years.[*]

17. A new president takes office at the same time a justice retires. Find the probability that the next vacancy on the court will take place during his 4-year term.

18. Find the probability that the composition of the U.S. Supreme Court will remain unchanged for a period of 5 years or more.

19. Consider a group of patients that have been treated for an acute disease such as cancer, and let X be the number of years a person lives after receiving the treatment (the "survival time"). Under suitable conditions, the density function for X will be $f(x) = ke^{-kx}$ for some constant k. The *survival function* $S(x)$ is the probability that a person chosen at random from the group of patients survives until at least time x. Suppose that the probability is .90 that a patient will survive at least 5 years [i.e., $S(5) = .90$]. Find the constant k in the exponential density function $f(x)$.

---

1. Property II is not satisfied:

$$\int_0^1 f(x)\,dx = \int_0^1 x\,dx = \frac{x^2}{2}\Big|_0^1 = \frac{1}{2}.$$

Property II says that this integral should be 1.

2. $\int_{1/2}^1 4x^3\,dx = x^4\Big|_{1/2}^1 = 1 - \frac{1}{16} = \frac{15}{16}.$

Review of Fundamental Concepts of Chapter 6

1. What does it mean to antidifferentiate a function?
2. State the formula for $\int h(x)\,dx$ for each of the following functions.
   (a) $h(x) = x^r$, $r \neq -1$ (b) $h(x) = e^{kx}$
   (c) $h(x) = \dfrac{1}{x}$ (d) $h(x) = f(x) + g(x)$
   (e) $h(x) = kf(x)$
3. In the formula $\Delta x = \dfrac{b-a}{n}$, what do a, b, n, and Δx denote?
4. What is a Riemann sum?
5. Give an interpretation of the area under a rate of change function. Give a concrete example.
6. What is a definite integral?
7. What is the difference between a definite integral and an indefinite integral?
8. State the fundamental theorem of calculus.
9. How is $F(x)\big|_a^b$ calculated and what is it called?
10. Outline a procedure for finding the area of a region bounded by two curves.
11. State the formula for each of the following quantities:
   (a) average value of a function
   (b) consumers' surplus
   (c) future value of an income stream
   (d) volume of a solid of revolution

▶ Chapter 6 Supplementary Exercises

Calculate the following integrals.

1. $\int e^{-x/2}\,dx$
2. $\int \dfrac{5}{\sqrt{x-7}}\,dx$
3. $\int (3x^4 - 4x^3)\,dx$
4. $\int (2x+3)^7\,dx$
5. $\int \sqrt{4-x}\,dx$
6. $\int \left(\dfrac{5}{x} - \dfrac{x}{5}\right)dx$
7. $\int_1^4 \dfrac{1}{x^2}\,dx$
8. $\int_3^6 e^{2-(x/3)}\,dx$
9. $\int_0^5 (5+3x)^{-1}\,dx$

10. Find the area under the curve $y = 1 + \sqrt{x}$ from $x = 1$ to $x = 9$.
11. Find the area under the curve $y = (3x-2)^{-3}$ from $x = 1$ to $x = 2$.
12. Find the area of the region bounded by the curves $y = 16 - x^2$ and $y = 10 - x$.
13. Find the area of the region bounded by the curves $y = x^3 - 3x + 1$ and $y = x + 1$.
14. Find the area of the region between the curves $y = 2x^2 + x$ and $y = x^2 + 2$ from $x = 0$ to $x = 2$.
15. Find the function $f(x)$ for which $f'(x) = (x-5)^2$, $f(8) = 2$.
16. Find the function $f(x)$ for which $f'(x) = e^{-5x}$, $f(0) = 1$.
17. Describe all solutions of the following differential equations, where y represents a function of t.
   (a) $y' = 4t$
   (b) $y' = 4y$
   (c) $y' = e^{4t}$
18. Let k be a constant, and let $y = f(t)$ be a function such that $y' = kty$. Show that $y = Ce^{kt^2/2}$, for some constant C. [*Hint:* Use the product rule to evaluate $\dfrac{d}{dt}[f(t)e^{-kt^2/2}]$, and then apply Theorem II of Section 6.1.]
19. An airplane tire plant finds that its marginal cost of producing tires is $.04x + 150$ dollars at a production level of x tires per day. If fixed costs are \$500 per day, find the cost of producing x tires per day.

20. Suppose that the marginal revenue function for a company is $400 - 3x^2$. Find the additional revenue received from doubling production if currently 10 units are being produced.

21. A drug is injected into a patient at the rate of $f(t)$ cubic centimeters per minute at time t. What does the area under the graph of $y = f(t)$ from $t = 0$ to $t = 4$ represent?

22. A rock thrown straight up into the air has a velocity of $v(t) = -9.8t + 20$ meters per second after t seconds.

   (a) Determine the distance the rock travels during the first 2 seconds.

   (b) Represent the answer to part (a) as an area.

23. Use a Riemann sum with $n = 4$ and left endpoints to estimate the area under the graph in Fig. 1 for $0 \leq x \leq 2$.

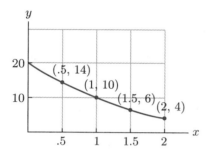

Figure 1.

24. Redo Exercise 23 using right endpoints.

25. Use a Riemann sum with $n = 2$ and midpoints to estimate the area under the graph of

$$f(x) = \frac{1}{x + 2}$$

on the interval $0 \leq x \leq 2$. Then use a definite integral to find the exact value of the area to five decimal places.

26. Use a Riemann sum with $n = 5$ and midpoints to estimate the area under the graph of $f(x) = e^{2x}$ on the interval $0 \leq x \leq 1$. Then use a definite integral to find the exact value of the area to five decimal places.

27. Find the consumers' surplus for the demand curve $p = \sqrt{25 - .04x}$ at the sales level $x = 400$.

28. Three thousand dollars is deposited in the bank at 4% interest compounded continuously. What will be the average value of the money in the account during the next 10 years?

29. Find the average value of $f(x) = 1/x^3$ from $x = \frac{1}{3}$ to $x = \frac{1}{2}$.

30. Suppose that the interval $0 \leq x \leq 1$ is divided into 100 subintervals with a width of $\Delta x = .01$. Show that the sum

$$\left[3e^{-.01}\right] \Delta x + \left[3e^{-.02}\right] \Delta x + \left[3e^{-.03}\right] \Delta x$$
$$+ \cdots + \left[3e^{-1}\right] \Delta x$$

is close to $3(1 - e^{-1})$.

31. In Fig. 2, three regions are labeled with their areas. Determine $\int_a^c f(x)\, dx$ and determine $\int_a^d f(x)\, dx$.

32. Find the volume of the solid of revolution generated by revolving about the x-axis the region under the curve $y = 1 - x^2$ from $x = 0$ to $x = 1$.

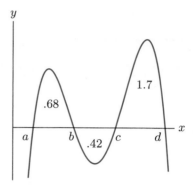

Figure 2.

33. A store has an inventory of Q units of a certain product at time $t = 0$. The store sells the product at the steady rate of Q/A units per week, and exhausts the inventory in A weeks.

   (a) Find a formula $f(t)$ for the amount of product in inventory at time t.

   (b) Find the average inventory level during the period $0 \leq t \leq A$.

34. A retail store sells a certain product at the rate of $g(t)$ units per week at time t, where $g(t) = rt$. At time $t = 0$, the store has Q units of the product in inventory.

   (a) Find a formula $f(t)$ for the amount of product in inventory at time t.

   (b) Determine the value of r in part (a) such that the inventory is exhausted in A weeks.

   (c) Using $f(t)$, with r as in part (b), find the average inventory level during the period $0 \leq t \leq A$.

35. Let x be any positive number, and define $g(x)$ to be the number determined by the definite integral

$$g(x) = \int_0^x \frac{1}{1 + t^2}\, dt.$$

   (a) Give a geometric interpretation of the number $g(3)$.

   (b) Find the derivative $g'(x)$.

36. For each number x satisfying $-1 \leq x \leq 1$, define $h(x)$ by

$$h(x) = \int_{-1}^x \sqrt{1 - t^2}\, dt.$$

(a) Give a geometric interpretation of the values $h(0)$ and $h(1)$.

(b) Find the derivative $h'(x)$.

37. Suppose that the interval $0 \leq t \leq 3$ is divided into 1000 subintervals of width Δt. Let $t_1, t_2, \ldots, t_{1000}$ denote the right endpoints of these subintervals. Suppose that in some problem one needs to estimate the sum

$$5000e^{-.1t_1}\Delta t + 5000e^{-.1t_2}\Delta t$$

$$+ \cdots + 5000e^{-.1t_{1000}}\Delta t.$$

Show that this sum is close to 13,000. (*Note:* A sum such as this would arise if one wanted to compute the present value of a continuous stream of income of $5000 per year for 3 years, with interest compounded continuously at 10%.)

38. What number does

$$\left[e^0 + e^{1/n} + e^{2/n} + e^{3/n} + \cdots + e^{(n-1)/n}\right] \cdot \frac{1}{n}$$

approach as n gets very large?

39. What number does the sum

$$\left[1^3 + \left(1 + \frac{1}{n}\right)^3 + \left(1 + \frac{2}{n}\right)^3 + \left(1 + \frac{3}{n}\right)^3\right.$$

$$\left. + \cdots + \left(1 + \frac{n-1}{n}\right)^3\right] \cdot \frac{1}{n}$$

approach as n gets very large?

40. In Fig. 3, the rectangle has the same area as the region under the graph of $f(x)$. What is the average value of $f(x)$ on the interval $2 \leq x \leq 6$?

41. *True or false:* If $3 \leq f(x) \leq 4$ whenever $0 \leq x \leq 5$, then $3 \leq \frac{1}{5}\int_0^5 f(x)\,dx \leq 4$.

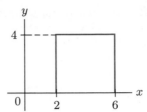 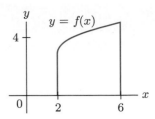

Figure 3.

42. Suppose that water is flowing into a tank at a rate of $r(t)$ gallons per hour, where the rate depends on the time t according to the formula

$$r(t) = 20 - 4t, \qquad 0 \leq t \leq 5.$$

(a) Consider a brief period of time, say from t_1 to t_2. The length of this time period is $\Delta t = t_2 - t_1$. During this period the rate of flow does not change much and is approximately $20 - 4t_1$ (the rate at the beginning of the brief time interval). Approximately how much water flows into the tank during the time from t_1 to t_2?

(b) Explain why the total amount of water added to the tank during the time interval from $t = 0$ to $t = 5$ is given by $\int_0^5 r(t)\,dt$.

43. The annual world rate of water use t years after 1960, for $t \leq 35$, was approximately $860e^{.04t}$ cubic kilometers per year. How much water was used between 1960 and 1995?

44. Suppose that money is deposited steadily into a savings account at the rate of $4500 per year. Determine the balance at the end of 1 year if the account pays 9% interest compounded continuously.

Chapter Project

Table 1 gives the mean annual household income for the years 1990–1996. Let t measure time, with $t = 1$ corresponding to 1990. Let $W(t)$ denote the piecewise-linear function obtained from the data in Table 1.

Table 1	Mean Income of Households in Constant 1996 Dollars

Source: Statistical Abstract of the United States, U.S. Bureau of the Census, 1999, Table 738.

Year	Mean Household Income
1990	44,901
1991	43,685
1992	43,435
1993	44,983
1994	45,665
1995	46,265
1996	47,123

(a) For which values of t is the function $W(t)$ differentiable?

(b) Describe the function $W(t)$ analytically.

(c) Compute $W'(t)$. Be sure to take into account the values of t at which $W(t)$ is not differentiable.

(d) Draw the graph of $W'(t)$. Describe, in words, the relationship between the graphs of $W(t)$ and $W'(t)$.

(e) Express as an integral the area under the graph of $W'(t)$ from $t = 0$ to $t = 6$.

(f) Use the fundamental theorem of calculus to provide an interpretation of the area of part (e).

C H A P T E R

Functions of Several Variables

7

U ntil now, most of our applications of calculus have involved functions of one variable. In real life, however, a quantity of interest often depends on more than one variable. For instance, the sales level of a product may depend not only on its price but also on the prices of competing products, the amount spent on advertising, and perhaps the time of year. The total cost of manufacturing the product depends on the cost of raw materials, labor, plant maintenance, and so on.

This chapter introduces the basic ideas of calculus for functions of more than one variable. Section 7.1 presents two examples that will be used throughout the chapter. Derivatives are treated in Section 7.2 and then used in Sections 7.3 and 7.4 to solve optimization problems more general than those in Chapter 2. The final three sections are devoted to least-squares problems and a brief introduction to integration of functions of two variables.

7.1 Examples of Functions of Several Variables

A function $f(x, y)$ of the two variables x and y is a rule that assigns a number to each pair of values for the variables; for instance,

$$f(x, y) = e^x(x^2 + 2y).$$

An example of a function of three variables is

$$f(x, y, z) = 5xy^2z.$$

▶ Example 1   A store sells butter at \$2.50 per pound and margarine at \$1.40 per pound. The revenue from the sale of x pounds of butter and y pounds of margarine is given by the function

$$f(x, y) = 2.50x + 1.40y.$$

Determine and interpret $f(200, 300)$.

Solution $f(200, 300) = 2.50(200) + 1.40(300) = 500 + 420 = 920$. The revenue from the sale of 200 pounds of butter and 300 pounds of margarine is \$920.   ◆

A function $f(x, y)$ of two variables may be graphed in a manner analogous to that for functions of one variable. It is necessary to use a three-dimensional coordinate system, where each point is identified by three coordinates (x, y, z). For each choice of x, y, the graph of $f(x, y)$ includes the point $(x, y, f(x, y))$. This graph is usually a surface in three-dimensional space, with equation $z = f(x, y)$. See Fig. 1. Three graphs of specific functions are shown in Fig. 2.

Application to Architectural Design   When designing a building, one would like to know, at least approximately, how much heat the building loses per day. The heat loss affects many aspects of the design, such as the size of the heating plant, the size and location of duct work, and so on. A building loses heat through its sides, roof, and floor. How much heat is lost will generally differ for each face of the building and will depend on such factors as insulation, materials used in construction, exposure (north, south, east, or west), and climate. It is possible to estimate how much heat is lost per square foot of each face. Using these data, one can construct a heat-loss function as in the following example.

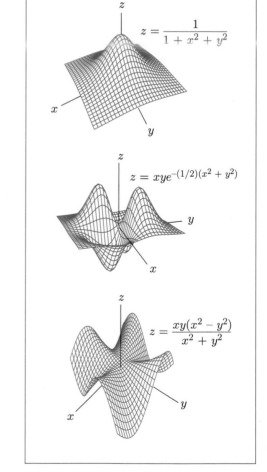

$$z = \frac{1}{1 + x^2 + y^2}$$

$$z = xye^{-(1/2)(x^2 + y^2)}$$

$$z = \frac{xy(x^2 - y^2)}{x^2 + y^2}$$

Surface
$z = f(x, y)$ $(x, y, f(x, y))$

(x, y)

Figure 1. Graph of f(x,y).

Figure 2.

▶ Example 2   A rectangular industrial building of dimensions x, y, and z is shown in Fig. 3(a). In Fig. 3(b) we give the amount of heat lost per day by each side of the building, measured in suitable units of heat per square foot. Let $f(x, y, z)$ be the total daily heat loss for such a building.

(a) Find a formula for $f(x, y, z)$.

(b) Find the total daily heat loss if the building has length 100 feet, width 70 feet, and height 50 feet.

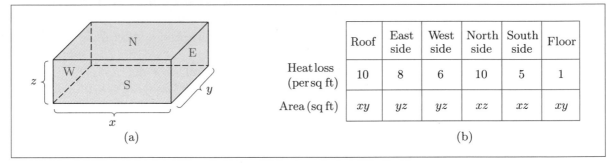

	Roof	East side	West side	North side	South side	Floor
Heat loss (per sq ft)	10	8	6	10	5	1
Area (sq ft)	xy	yz	yz	xz	xz	xy

(a)   (b)

Figure 3. Heat loss from an industrial building.

Solution   (a) The total heat loss is the sum of the amount of heat loss through each face of the building. The heat loss through the roof is

[heat loss per square foot of roof] $\cdot$ [area of roof in square feet] $= 10xy$.

Similarly, the heat loss through the east side is $8yz$. Continuing in this way, we see that the total daily heat loss is

$$f(x, y, z) = 10xy + 8yz + 6yz + 10xz + 5xz + 1 \cdot xy.$$

We collect terms to obtain

$$f(x, y, z) = 11xy + 14yz + 15xz.$$

(b) The amount of heat loss when $x = 100$, $y = 70$, and $z = 50$ is given by $f(100, 70, 50)$, which equals

$$f(100, 70, 50) = 11(100)(70) + 14(70)(50) + 15(100)(50)$$
$$= 77{,}000 + 49{,}000 + 75{,}000 = 201{,}000.$$ ◆

In Section 7.3 we will determine the dimensions x, y, z that minimize the heat loss for a building of specific volume.

Production Functions in Economics   The costs of a manufacturing process can generally be classified as one of two types: cost of labor and cost of capital. The meaning of the cost of labor is clear. By the cost of capital, we mean the cost of buildings, tools, machines, and similar items used in the production process. A manufacturer usually has some control over the relative portions of labor and capital utilized in his production process. He can completely automate production so that labor is at a minimum, or he can utilize mostly labor and little capital. Suppose that x units of labor and y units of capital are used.* Let

---

*Economists normally use L and K, respectively, for labor and capital. However, for simplicity, we use x and y.

$f(x, y)$ denote the number of units of finished product that are manufactured. Economists have found that $f(x, y)$ is often a function of the form

$$f(x, y) = Cx^A y^{1-A},$$

where A and C are constants, $0 < A < 1$. Such a function is called a *Cobb-Douglas production function*.

▶ Example 3   (*Production in a Firm*)   Suppose that during a certain time period the number of units of goods produced when utilizing x units of labor and y units of capital is $f(x, y) = 60x^{3/4}y^{1/4}$.

    (a) How many units of goods will be produced by using 81 units of labor and 16 units of capital?

    (b) Show that whenever the amounts of labor and capital being used are doubled, so is the production. (Economists say that the production function has "constant returns to scale.")

Solution   (a) $f(81, 16) = 60(81)^{3/4} \cdot (16)^{1/4} = 60 \cdot 27 \cdot 2 = 3240$. There will be 3240 units of goods produced.

    (b) Utilization of a units of labor and b units of capital results in the production of $f(a, b) = 60a^{3/4}b^{1/4}$ units of goods. Utilizing $2a$ and $2b$ units of labor and capital, respectively, results in $f(2a, 2b)$ units produced. Set $x = 2a$ and $y = 2b$. Then we see that

$$f(2a, 2b) = 60(2a)^{3/4}(2b)^{1/4}$$

$$= 60 \cdot 2^{3/4} \cdot a^{3/4} \cdot 2^{1/4} \cdot b^{1/4}$$

$$= 60 \cdot 2^{(3/4 + 1/4)} \cdot a^{3/4}b^{1/4}$$

$$= 2^1 \cdot 60a^{3/4}b^{1/4}$$

$$= 2f(a, b). \qquad \blacklozenge$$

Level Curves   It is possible graphically to depict a function $f(x, y)$ of two variables using a family of curves called level curves. Let c be any number. Then the graph of the equation $f(x, y) = c$ is a curve in the xy-plane called the *level curve of height c*. This curve describes all points of height c on the graph of the function $f(x, y)$. As c varies, we have a family of level curves indicating the sets of points on which $f(x, y)$ assumes various values c. In Fig. 4, we have drawn the graph and various level curves for the function $f(x, y) = x^2 + y^2$.

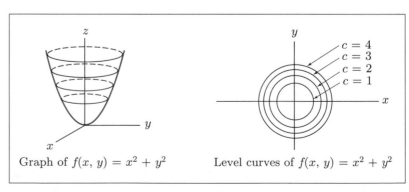

Graph of $f(x, y) = x^2 + y^2$ Level curves of $f(x, y) = x^2 + y^2$

Figure 4. Level curves.

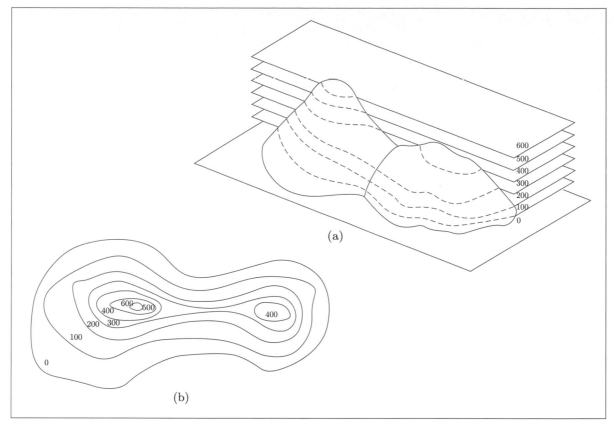

(a)

(b)

Figure 5. Topographic level curves show altitudes.

Level curves often have interesting physical interpretations. For example, surveyors draw *topographic maps* that use level curves to represent points having equal altitude. Here $f(x, y) =$ the altitude at point (x, y). Figure 5(a) shows the graph of $f(x, y)$ for a typical hilly region. Figure 5(b) shows the level curves corresponding to various altitudes. Note that when the level curves are closer together, the surface is steeper.

▶ Example 4   Determine the level curve at height 600 for the production function $f(x, y) = 60x^{3/4}y^{1/4}$ of Example 3.

Solution   The level curve is the graph of $f(x, y) = 600$, or

$$60x^{3/4}y^{1/4} = 600$$

$$y^{1/4} = \frac{10}{x^{3/4}}$$

$$y = \frac{10,000}{x^3}.$$

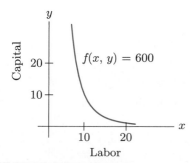

Figure 6. A level curve of a production function.

Of course, since x and y represent quantities of labor and capital, they must both be positive. We have sketched the graph of the level curve in Fig. 6. The points on the curve are precisely those combinations of capital and labor that yield 600 units of production. Economists call this curve an *isoquant*.   ◆

Practice Problems 7.1

1. Let $f(x, y, z) = x^2 + y/(x - z) - 4$. Compute $f(3, 5, 2)$.

2. Suppose that in a certain country the daily demand for coffee is given by $f(p_1, p_2) = 16p_1/p_2$ thousand pounds, where p_1 and p_2 are the respective prices of tea and coffee in dollars per pound. Compute and interpret $f(3, 4)$.

▶ Exercises 7.1

1. Let $f(x, y) = x^2 - 3xy - y^2$. Compute $f(5, 0)$, $f(5, -2)$, and $f(a, b)$.

2. Let $g(x, y) = \sqrt{x^2 + 2y^2}$. Compute $g(-3, 0)$, $g(4, 1)$, and $g(a, b)$.

3. Let $g(x, y, z) = x/(y - z)$. Compute $g(2, 3, 4)$ and $g(7, 46, 44)$.

4. Let $f(x, y, z) = x^2 e^{\sqrt{y^2 + z^2}}$. Compute $f(1, 0, 1)$ and $f(2, 3, -4)$.

5. Let $f(x, y) = xy$. Show that $f(2 + h, 3) - f(2, 3) = 3h$.

6. Let $f(x, y) = xy$. Show that $f(2, 3 + k) - f(2, 3) = 2k$.

7. Find a formula $C(x, y, z)$ that gives the cost of materials for the closed rectangular box in Fig. 7(a), with dimensions in feet, assuming that the material for the top and bottom costs \$3 per square foot and the material for the sides costs \$5 per square foot.

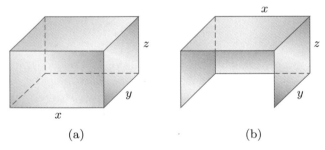

(a)　　　　　　　(b)

Figure 7.

8. Find a formula $C(x, y, z)$ that gives the cost of material for the rectangular enclosure in Fig. 7(b), with dimensions in feet, assuming that the material for the top costs \$3 per square foot and the material for the back and two sides costs \$5 per square foot.

9. Consider the Cobb-Douglas production function $f(x, y) = 20x^{1/3}y^{2/3}$. Compute $f(8, 1)$, $f(1, 27)$, and $f(8, 27)$. Show that for any positive constant k, $f(8k, 27k) = kf(8, 27)$.

10. Let $f(x, y) = 10x^{2/5}y^{3/5}$. Show that $f(3a, 3b) = 3f(a, b)$.

11. The present value of A dollars to be paid t years in the future (assuming a 5% continuous interest rate) is $P(A, t) = Ae^{-.05t}$. Find and interpret $P(100, 13.8)$.

12. Refer to Example 3. Suppose that labor costs \$100 per unit and capital costs \$200 per unit. Express as a function of two variables, $C(x, y)$, the cost of utilizing x units of labor and y units of capital.

13. The value of residential property for tax purposes is usually much lower than its actual market value. If v is the market value, then the *assessed value* for real estate taxes might be only 40% of v. Suppose the property tax, T, in a community is given by the function

$$T = f(r, v, x) = \frac{r}{100}(.40v - x),$$

where v is the estimated market value of a property (in dollars), x is a *homeowner's exemption* (a number of dollars depending on the type of property), and r is the tax rate (stated in dollars per hundred dollars) of net assessed value.

(a) Determine the real estate tax on a property valued at \$200,000 with a homeowner's exemption of \$5000, assuming a tax rate of \$2.50 per hundred dollars of net assessed value.

(b) Determine the tax due if the tax rate increases by 20% to \$3.00 per hundred dollars of net assessed value. Assume the same property value and homeowner's exemption. Does the tax due also increase by 20%?

14. Let $f(r, v, x)$ be the real estate tax function of Exercise 13.

(a) Determine the real estate tax on a property valued at \$100,000 with a homeowner's exemption of \$5000, assuming a tax rate of \$2.20 per hundred dollars of net assessed value.

(b) Determine the real estate tax when the market value rises 20% to \$120,000. Assume the same homeowner's exemption and a tax rate of \$2.20 per hundred dollars of net assessed value. Does the tax due also increase by 20%?

Draw the level curves of heights 0, 1, and 2 for the functions in Exercises 15 and 16.

15. $f(x, y) = 2x + y$　　　16. $f(x, y) = -x^2 + y$

17. Draw the level curve of the function $f(x, y) = x - y$ containing the point $(5, 3)$.

18. Draw the level curve of the function $f(x, y) = xy$ containing the point $(2, \frac{1}{2})$.

19. Find a function $f(x, y)$ that has the line $y = 3x - 4$ as a level curve.

20. Find a function $f(x, y)$ that has the curve $y = 3/x^2$ as a level curve.

21. Suppose that a topographic map is viewed as the graph of a certain function $f(x, y)$. What are the level curves?

22. A certain production process uses labor and capital. If the quantities of these commodities are x and y, respectively, then the total cost is $100x + 200y$ dollars. Draw the level curves of height 600, 800, and 1000 for this function. Explain the significance of these curves. (Economists frequently refer to these lines as *budget lines* or *isocost lines*.)

Match the graphs of the functions in Exercises 23–26 to the systems of level curves shown in Fig. 8(a)–(d).

23.

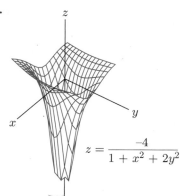

$$z = \frac{-4}{1 + x^2 + 2y^2}$$

24.

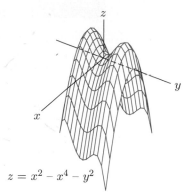

$$z = x^2 - x^4 - y^2$$

25.

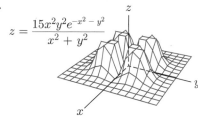

$$z = \frac{15x^2 y^2 e^{-x^2 - y^2}}{x^2 + y^2}$$

26.

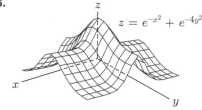

$$z = e^{-x^2} + e^{-4y^2}$$

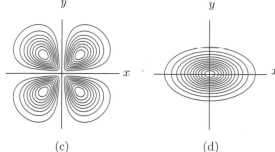

(a)   (b)   (c)   (d)

Figure 8.

Solutions to Practice Problems 7.1	

1. Substitute 3 for x, 5 for y, and 2 for z.

$$f(3, 5, 2) = 3^2 + \frac{5}{3 - 2} - 4 = 10$$

2. To compute $f(3, 4)$, substitute 3 for p_1 and 4 for p_2 into $f(p_1, p_2) = 16p_1/p_2$. Thus

$$f(3, 4) = 16 \cdot \tfrac{3}{4} = 12.$$

Therefore, if the price of tea is \$3 per pound and the price of coffee is \$4 per pound, then 12,000 pounds of coffee will be sold each day. (Notice that as the price of coffee increases, the demand decreases.)

7.2 **Partial Derivatives**

In Chapter 1 we introduced the notion of a derivative to measure the rate at which a function $f(x)$ is changing with respect to changes in the variable x. Let us now study the analog of the derivative for functions of two (or more) variables.

Let $f(x, y)$ be a function of the two variables x and y. Since we want to know how $f(x, y)$ changes with respect to both changes in the variable x and changes in the variable y, we shall define two derivatives of $f(x, y)$ (to be called "partial derivatives"), one with respect to each of the variables. The *partial derivative of $f(x, y)$ with respect to x*, written $\dfrac{\partial f}{\partial x}$, is the derivative of $f(x, y)$, where y is treated as a constant and $f(x, y)$ is considered as a function of x alone. The *partial derivative of $f(x, y)$ with respect to y*, written $\dfrac{\partial f}{\partial y}$, is the derivative of $f(x, y)$, where x is treated as a constant.

▶ **Example 1** Let $f(x, y) = 5x^3 y^2$. Compute $\dfrac{\partial f}{\partial x}$ and $\dfrac{\partial f}{\partial y}$.

Solution To compute $\dfrac{\partial f}{\partial x}$, we think of $f(x, y)$ written as

$$f(x, y) = \left[5y^2\right] x^3,$$

where the brackets emphasize that $5y^2$ is to be treated as a constant. Therefore, when differentiating with respect to x, $f(x, y)$ is just a constant times x^3. Recall that if k is any constant, then

$$\frac{d}{dx}(kx^3) = 3 \cdot k \cdot x^2.$$

Thus

$$\frac{\partial f}{\partial x} = 3 \cdot \left[5y^2\right] \cdot x^2 = 15x^2 y^2.$$

After some practice, it is unnecessary to place the y^2 in front of the x^3 before differentiating.

Now, in order to compute $\dfrac{\partial f}{\partial y}$, we think of

$$f(x, y) = \left[5x^3\right] y^2.$$

When differentiating with respect to y, $f(x, y)$ is simply a constant (namely, $5x^3$) times y^2. Hence

$$\frac{\partial f}{\partial y} = 2 \cdot \left[5x^3\right] \cdot y = 10x^3 y. \qquad \blacklozenge$$

▶ **Example 2** Let $f(x, y) = 3x^2 + 2xy + 5y$. Compute $\dfrac{\partial f}{\partial x}$ and $\dfrac{\partial f}{\partial y}$.

Solution To compute $\dfrac{\partial f}{\partial x}$, we think of

$$f(x, y) = 3x^2 + [2y]x + [5y].$$

Now we differentiate $f(x, y)$ as if it were a quadratic polynomial in x:

$$\frac{\partial f}{\partial x} = 6x + [2y] + 0 = 6x + 2y.$$

Note that $5y$ is treated as a constant when differentiating with respect to x, so the partial derivative of $5y$ with respect to x is zero.

To compute $\dfrac{\partial f}{\partial y}$, we think of

$$f(x, y) = \left[3x^2\right] + [2x]y + 5y.$$

Then

$$\frac{\partial f}{\partial y} = 0 + [2x] + 5 = 2x + 5.$$

Note that $3x^2$ is treated as a constant when differentiating with respect to y, so the partial derivative of $3x^2$ with respect to y is zero. ◆

▶ **Example 3** Compute $\dfrac{\partial f}{\partial x}$ and $\dfrac{\partial f}{\partial y}$ for each of the following.

(a) $f(x, y) = (4x + 3y - 5)^8$

(b) $f(x, y) = e^{xy^2}$

(c) $f(x, y) = y/(x + 3y)$

Solution (a) To compute $\dfrac{\partial f}{\partial x}$, we think of

$$f(x, y) = (4x + [3y - 5])^8.$$

By the general power rule,

$$\frac{\partial f}{\partial x} = 8 \cdot (4x + [3y - 5])^7 \cdot 4 = 32(4x + 3y - 5)^7.$$

Here we used the fact that the derivative of $4x + 3y - 5$ with respect to x is just 4.

To compute $\dfrac{\partial f}{\partial y}$, we think of

$$f(x, y) = ([4x] + 3y - 5)^8.$$

Then

$$\frac{\partial f}{\partial y} = 8 \cdot ([4x] + 3y - 5)^7 \cdot 3 = 24(4x + 3y - 5)^7.$$

(b) To compute $\dfrac{\partial f}{\partial x}$, we observe that

$$f(x, y) = e^{x[y^2]},$$

so that

$$\frac{\partial f}{\partial x} = [y^2] e^{x[y^2]} = y^2 e^{xy^2}.$$

To compute $\dfrac{\partial f}{\partial y}$, we think of

$$f(x,y) = e^{[x]y^2}.$$

Thus

$$\frac{\partial f}{\partial y} = e^{[x]y^2} \cdot 2[x]y = 2xye^{xy^2}.$$

(c) To compute $\dfrac{\partial f}{\partial x}$, we use the general power rule to differentiate $[y](x+[3y])^{-1}$ with respect to x:

$$\frac{\partial f}{\partial x} = (-1) \cdot [y](x + [3y])^{-2} \cdot 1 = -\frac{y}{(x + 3y)^2}.$$

To compute $\dfrac{\partial f}{\partial y}$, we use the quotient rule to differentiate

$$f(x,y) = \frac{y}{[x] + 3y}$$

with respect to y. We find that

$$\frac{\partial f}{\partial y} = \frac{([x] + 3y) \cdot 1 - y \cdot 3}{([x] + 3y)^2} = \frac{x}{(x + 3y)^2}.$$

The use of brackets to highlight constants is helpful initially in order to compute partial derivatives. From now on we shall merely form a mental picture of those terms to be treated as constants and dispense with brackets. ◆

A partial derivative of a function of several variables is also a function of several variables and hence can be evaluated at specific values of the variables. We write

$$\frac{\partial f}{\partial x}(a,b)$$

for $\dfrac{\partial f}{\partial x}$ evaluated at $x = a$, $y = b$. Similarly,

$$\frac{\partial f}{\partial y}(a,b)$$

denotes the function $\dfrac{\partial f}{\partial y}$ evaluated at $x = a$, $y = b$.

▶ **Example 4** Let $f(x,y) = 3x^2 + 2xy + 5y$.

(a) Calculate $\dfrac{\partial f}{\partial x}(1,4)$. (b) Evaluate $\dfrac{\partial f}{\partial y}$ at $(x,y) = (1,4)$.

Solution (a) $\dfrac{\partial f}{\partial x} = 6x + 2y$, $\dfrac{\partial f}{\partial x}(1,4) = 6 \cdot 1 + 2 \cdot 4 = 14$

(b) $\dfrac{\partial f}{\partial y} = 2x + 5$, $\dfrac{\partial f}{\partial y}(1,4) = 2 \cdot 1 + 5 = 7$ ◆

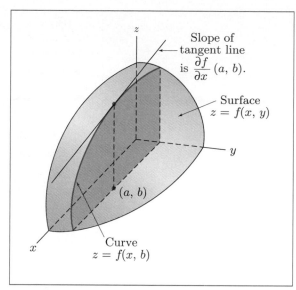

Figure 1. $\frac{\partial f}{\partial x}$ gives the slope of a curve formed by holding y constant.

Geometric Interpretation of Partial Derivatives   Consider the three-dimensional surface $z = f(x, y)$ in Fig. 1. If y is held constant at b and x is allowed to vary, the equation

$$z = f(x, b)$$

describes a curve on the surface. [The curve is formed by cutting the surface $z = f(x, y)$ with a vertical plane parallel to the xz-plane.] The value of $\frac{\partial f}{\partial x}(a, b)$ is the slope of the tangent line to the curve at the point where $x = a$ and $y = b$.

Likewise, if x is held constant at a and y is allowed to vary, the equation

$$z = f(a, y)$$

describes the curve on the surface $z = f(x, y)$ shown in Fig. 2. The value of the partial derivative $\frac{\partial f}{\partial y}(a, b)$ is the slope of this curve at the point where $x = a$ and $y = b$.

Partial Derivatives and Rates of Change   Since $\frac{\partial f}{\partial x}$ is simply the ordinary derivative with y held constant, $\frac{\partial f}{\partial x}$ gives the rate of change of $f(x, y)$ with respect to x for y held constant. In other words, keeping y constant and increasing x by one (small) unit produces a change in $f(x, y)$ that is approximately given by $\frac{\partial f}{\partial x}$. An analogous interpretation holds for $\frac{\partial f}{\partial y}$.

▶ Example 5   Interpret the partial derivatives of $f(x, y) = 3x^2 + 2xy + 5y$ calculated in Example 4.

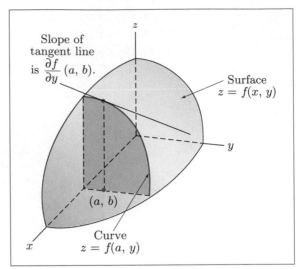

Figure 2. $\frac{\partial f}{\partial y}$ gives the slope of a curve formed by holding x constant.

Solution   We showed in Example 4 that

$$\frac{\partial f}{\partial x}(1,4) = 14, \qquad \frac{\partial f}{\partial y}(1,4) = 7.$$

The fact that

$$\frac{\partial f}{\partial x}(1,4) = 14$$

means that if y is kept constant at 4 and x is allowed to vary near 1, then $f(x,y)$ changes at a rate 14 times the change in x. That is, if x increases by one small unit, then $f(x,y)$ increases by approximately 14 units. If x increases by h units (where h is small), then $f(x,y)$ increases by approximately $14 \cdot h$ units. That is,

$$f(1+h,4) - f(1,4) \approx 14 \cdot h.$$

Similarly, the fact that

$$\frac{\partial f}{\partial y}(1,4) = 7$$

means that if we keep x constant at 1 and let y vary near 4, then $f(x,y)$ changes at a rate equal to seven times the change in y. So for a small value of k, we have

$$f(1,4+k) - f(1,4) \approx 7 \cdot k. \qquad \blacklozenge$$

We can generalize the interpretations of $\frac{\partial f}{\partial x}$ and $\frac{\partial f}{\partial y}$ given in Example 5 to yield the following general fact.

Let $f(x,y)$ be a function of two variables. Then if h and k are small, we have

$$f(a+h,b) - f(a,b) \approx \frac{\partial f}{\partial x}(a,b) \cdot h$$

$$f(a,b+k) - f(a,b) \approx \frac{\partial f}{\partial y}(a,b) \cdot k.$$

Partial derivatives can be computed for functions of any number of variables. When taking the partial derivative with respect to one variable, we treat the other variables as constant.

▶ **Example 6** Let $f(x, y, z) = x^2yz - 3z$.

(a) Compute $\dfrac{\partial f}{\partial x}$, $\dfrac{\partial f}{\partial y}$, and $\dfrac{\partial f}{\partial z}$.

(b) Calculate $\dfrac{\partial f}{\partial z}(2, 3, 1)$.

Solution (a) $\dfrac{\partial f}{\partial x} = 2xyz$, $\dfrac{\partial f}{\partial y} = x^2z$, $\dfrac{\partial f}{\partial z} = x^2y - 3$

(b) $\dfrac{\partial f}{\partial z}(2, 3, 1) = 2^2 \cdot 3 - 3 = 12 - 3 = 9$ ◆

▶ **Example 7** Let $f(x, y, z)$ be the heat-loss function computed in Example 2 of Section 7.1. That is, $f(x, y, z) = 11xy + 14yz + 15xz$. Calculate and interpret $\dfrac{\partial f}{\partial x}(10, 7, 5)$.

Solution We have

$$\frac{\partial f}{\partial x} = 11y + 15z$$

$$\frac{\partial f}{\partial x}(10, 7, 5) = 11 \cdot 7 + 15 \cdot 5 = 77 + 75 = 152.$$

The quantity $\dfrac{\partial f}{\partial x}$ is commonly referred to as the *marginal heat loss with respect to change in x*. Specifically, if x is changed from 10 by h units (where h is small) and the values of y and z remain fixed at 7 and 5, then the amount of heat loss will change by approximately $152 \cdot h$ units. ◆

▶ **Example 8** (*Production*) Consider the production function $f(x, y) = 60x^{3/4}y^{1/4}$, which gives the number of units of goods produced when utilizing x units of labor and y units of capital.

(a) Find $\dfrac{\partial f}{\partial x}$ and $\dfrac{\partial f}{\partial y}$.

(b) Evaluate $\dfrac{\partial f}{\partial x}$ and $\dfrac{\partial f}{\partial y}$ at $x = 81$, $y = 16$.

(c) Interpret the numbers computed in part (b).

Solution (a) $\dfrac{\partial f}{\partial x} = 60 \cdot \dfrac{3}{4}x^{-1/4}y^{1/4} = 45x^{-1/4}y^{1/4} = 45\dfrac{y^{1/4}}{x^{1/4}}$

$\dfrac{\partial f}{\partial y} = 60 \cdot \dfrac{1}{4}x^{3/4}y^{-3/4} = 15x^{3/4}y^{-3/4} = 15\dfrac{x^{3/4}}{y^{3/4}}$

(b) $\dfrac{\partial f}{\partial x}(81, 16) = 45 \cdot \dfrac{16^{1/4}}{81^{1/4}} = 45 \cdot \dfrac{2}{3} = 30$

$\dfrac{\partial f}{\partial y}(81, 16) = 15 \cdot \dfrac{81^{3/4}}{16^{3/4}} = 15 \cdot \dfrac{27}{8} = \dfrac{405}{8} = 50\tfrac{5}{8}$

(c) The quantities $\dfrac{\partial f}{\partial x}$ and $\dfrac{\partial f}{\partial y}$ are referred to as the *marginal productivity of labor* and the *marginal productivity of capital*. If the amount of capital is held fixed at $y = 16$ and the amount of labor increases by 1 unit, then the quantity of goods produced will increase by approximately 30 units. Similarly, an increase in capital of 1 unit (with labor fixed at 81) results in an increase in production of approximately $50\frac{5}{8}$ units of goods. ◆

Just as we formed second derivatives in the case of one variable, we can form second partial derivatives of a function $f(x, y)$ of two variables. Since $\dfrac{\partial f}{\partial x}$ is a function of x and y, we can differentiate it with respect to x or y. The partial derivative of $\dfrac{\partial f}{\partial x}$ with respect to x is denoted by $\dfrac{\partial^2 f}{\partial x^2}$. The partial derivative of $\dfrac{\partial f}{\partial x}$ with respect to y is denoted by $\dfrac{\partial^2 f}{\partial y \partial x}$. Similarly, the partial derivative of the function $\dfrac{\partial f}{\partial y}$ with respect to x is denoted by $\dfrac{\partial^2 f}{\partial x \partial y}$, and the partial derivative of $\dfrac{\partial f}{\partial y}$ with respect to y is denoted by $\dfrac{\partial^2 f}{\partial y^2}$. Almost all functions $f(x, y)$ encountered in applications (and all functions $f(x, y)$ in this text) have the property that

$$\frac{\partial^2 f}{\partial y \partial x} = \frac{\partial^2 f}{\partial x \partial y}.$$

Note that in computing $\dfrac{\partial f}{\partial x}$ and $\dfrac{\partial f}{\partial y}$, verifying the last equation is a check that you have done the differentiation correctly.

▶ **Example 9** Let $f(x, y) = x^2 + 3xy + 2y^2$. Calculate $\dfrac{\partial^2 f}{\partial x^2}, \dfrac{\partial^2 f}{\partial y^2}, \dfrac{\partial^2 f}{\partial x \partial y}$, and $\dfrac{\partial^2 f}{\partial y \partial x}$.

Solution First we compute $\dfrac{\partial f}{\partial x}$ and $\dfrac{\partial f}{\partial y}$.

$$\frac{\partial f}{\partial x} = 2x + 3y, \qquad \frac{\partial f}{\partial y} = 3x + 4y$$

To compute $\dfrac{\partial^2 f}{\partial x^2}$, we differentiate $\dfrac{\partial f}{\partial x}$ with respect to x:

$$\frac{\partial^2 f}{\partial x^2} = 2.$$

Similarly, to compute $\dfrac{\partial^2 f}{\partial y^2}$, we differentiate $\dfrac{\partial f}{\partial y}$ with respect to y:

$$\frac{\partial^2 f}{\partial y^2} = 4.$$

To compute $\dfrac{\partial^2 f}{\partial x \partial y}$, we differentiate $\dfrac{\partial f}{\partial y}$ with respect to x:

$$\frac{\partial^2 f}{\partial x \partial y} = 3.$$

Finally, to compute $\dfrac{\partial^2 f}{\partial y \partial x}$, we differentiate $\dfrac{\partial f}{\partial x}$ with respect to y:

$$\frac{\partial^2 f}{\partial y \partial x} = 3.$$ ◆

Incorporating Technology Graphing calculators can evaluate partial derivatives. See the appropriate calculator appendix for details. The function from Example 4 and its first partial derivatives are specified in Fig. 3(a) and evaluated
in Fig. 3(b).

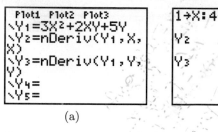

(a)   (b)

Figure 3.

<table>
<tr><td>**Practice Problems
7.2**</td></tr>
</table>

1. The number of TV sets an appliance store sells per week is given by a function
   of two variables, $f(x, y)$, where x is the price per TV set and y is the amount of
   money spent weekly on advertising. Suppose that the current price is $400 per set
   and that currently $2000 per week is being spent for advertising.

   (a) Would you expect $\dfrac{\partial f}{\partial x}(400, 2000)$ to be positive or negative?

   (b) Would you expect $\dfrac{\partial f}{\partial y}(400, 2000)$ to be positive or negative?

2. The monthly mortgage payment for a house is a function of two variables, $f(A, r)$,
   where A is the amount of the mortgage and the interest rate is $r\%$. For a 30-
   year mortgage, $f(92,000, 9) = 740.25$ and $\dfrac{\partial f}{\partial r}(92,000, 9) = 66.20$. What is the
   significance of the number 66.20?

▶ Exercises 7.2

Find $\dfrac{\partial f}{\partial x}$ and $\dfrac{\partial f}{\partial y}$ for each of the following functions.

1. $f(x, y) = 5xy$

2. $f(x, y) = 3x^2 + 2y + 1$

3. $f(x, y) = 2x^2 e^y$

4. $f(x, y) = x + e^{xy}$

5. $f(x, y) = \dfrac{y^2}{x}$

6. $f(x, y) = \dfrac{x}{1 + e^y}$

7. $f(x, y) = (2x - y + 5)^2$

8. $f(x, y) = (9x^2 y + 3x)^{12}$

9. $f(x, y) = x^2 e^{3x} \ln y$

10. $f(x, y) = (x - \ln y) e^{xy}$

11. $f(x, y) = \dfrac{x - y}{x + y}$

12. $f(x, y) = \dfrac{2xy}{e^x}$

13. Let $f(L, K) = 3\sqrt{LK}$. Find $\dfrac{\partial f}{\partial L}$.

14. Let $f(p, q) = e^{q/p}$. Find $\dfrac{\partial f}{\partial q}$ and $\dfrac{\partial f}{\partial p}$.

15. Let $f(x, y, z) = (1 + x^2 y)/z$. Find $\dfrac{\partial f}{\partial x}$, $\dfrac{\partial f}{\partial y}$, and $\dfrac{\partial f}{\partial z}$.

16. Let $f(x, y, z) = x^2 y + 3yz - z^2$. Find $\dfrac{\partial f}{\partial x}$, $\dfrac{\partial f}{\partial y}$, and $\dfrac{\partial f}{\partial z}$.

17. Let $f(x, y, z) = xze^{yz}$. Find $\dfrac{\partial f}{\partial x}$, $\dfrac{\partial f}{\partial y}$, and $\dfrac{\partial f}{\partial z}$.

18. Let $f(x, y, z) = ze^{(x+3y)z}$. Find $\dfrac{\partial f}{\partial x}$, $\dfrac{\partial f}{\partial y}$, and $\dfrac{\partial f}{\partial z}$.

19. Let $f(x, y) = x^2 + 2xy + y^2 + 3x + 5y$. Find $\dfrac{\partial f}{\partial x}(2, -3)$
and $\dfrac{\partial f}{\partial y}(2, -3)$.

20. Let $f(x, y) = xye^{2x-y}$. Evaluate $\dfrac{\partial f}{\partial x}$ and $\dfrac{\partial f}{\partial y}$ at
$(x, y) = (1, 2)$.

21. Let $f(x, y, z) = xy^2 z + 5$. Evaluate $\dfrac{\partial f}{\partial y}$ at $(x, y, z) = (2, -1, 3)$.

22. Let $f(x, y, z) = \dfrac{x}{y - z}$. Compute $\dfrac{\partial f}{\partial y}(2, -1, 3)$.

23. Let $f(x, y) = x^3 y + 2xy^2$. Find $\dfrac{\partial^2 f}{\partial x^2}$, $\dfrac{\partial^2 f}{\partial y^2}$, $\dfrac{\partial^2 f}{\partial x\, \partial y}$, and
$\dfrac{\partial^2 f}{\partial y\, \partial x}$.

24. Let $f(x, y) = xe^y + x^4 y + y^3$. Find $\dfrac{\partial^2 f}{\partial x^2}$, $\dfrac{\partial^2 f}{\partial y^2}$, $\dfrac{\partial^2 f}{\partial x\, \partial y}$,
and $\dfrac{\partial^2 f}{\partial y\, \partial x}$.

25. A farmer can produce $f(x, y) = 200\sqrt{6x^2 + y^2}$ units of produce by utilizing x units of labor and y units of capital. (The capital is used to rent or purchase land, materials, and equipment.)

(a) Calculate the marginal productivities of labor and capital when $x = 10$ and $y = 5$.

(b) Let h be a small number. Use the result of part (a) to determine the approximate effect on production of changing labor from 10 to $10 + h$ units while keeping capital fixed at 5 units.

(c) Use part (b) to estimate the change in production when labor decreases from 10 to 9.5 units and capital stays fixed at 5 units.

26. The productivity of a country is given by $f(x, y) = 300x^{2/3}y^{1/3}$, where x and y are the amount of labor and capital.

(a) Compute the marginal productivities of labor and capital when $x = 125$ and $y = 64$.

(b) Use part (a) to determine the approximate effect on productivity of increasing capital from 64 to 66 units, while keeping labor fixed at 125 units.

(c) What would be the approximate effect of decreasing labor from 125 to 124 units while keeping capital fixed at 64 units?

27. In a certain suburban community, commuters have the choice of getting into the city by bus or train. The demand for these modes of transportation varies with their cost. Let $f(p_1, p_2)$ be the number of people who will take the bus when p_1 is the price of the bus ride and p_2 is the price of the train ride. For example, if $f(4.50, 6) = 7000$, then 7000 commuters will take the bus when the price of a bus ticket is \$4.50 and the price of a train ticket is \$6.00. Explain why $\dfrac{\partial f}{\partial p_1} < 0$ and $\dfrac{\partial f}{\partial p_2} > 0$.

28. Refer to Exercise 27. Let $g(p_1, p_2)$ be the number of people who will take the train when p_1 is the price of the bus ride and p_2 is the price of the train ride. Would you expect $\dfrac{\partial g}{\partial p_1}$ to be positive or negative? How about $\dfrac{\partial g}{\partial p_2}$?

29. Let p_1 be the average price of VCRs, p_2 the average price of video tape, $f(p_1, p_2)$ the demand for VCRs, and $g(p_1, p_2)$ the demand for video tape. Explain why $\dfrac{\partial f}{\partial p_2} < 0$ and $\dfrac{\partial g}{\partial p_1} < 0$.

30. The demand for a certain gas-guzzling car is given by $f(p_1, p_2)$, where p_1 is the price of the car and p_2 is the price of gasoline. Explain why $\dfrac{\partial f}{\partial p_1} < 0$ and $\dfrac{\partial f}{\partial p_2} < 0$.

31. The volume (V) of a certain amount of a gas is determined by the temperature (T) and the pressure (P)

by the formula $V = .08(T/P)$. Calculate and interpret $\dfrac{\partial V}{\partial P}$ and $\dfrac{\partial V}{\partial T}$ when $P = 20$, $T = 300$.

32. Using data collected from 1929 to 1941, Richard Stone[*] determined that the yearly quantity Q of beer consumed in the United Kingdom was approximately given by the formula $Q = f(m, p, r, s)$, where

$$f(m, p, r, s) = (1.058)m^{.136}p^{-.727}r^{.914}s^{.816}$$

and m is the aggregate real income (personal income after direct taxes, adjusted for retail price changes), p is the average retail price of the commodity (in this case, beer), r is the average retail price level of all other consumer goods and services, and s is a measure of the strength of the beer. Determine which partial derivatives are positive and which are negative and give interpretations. (For example, since $\dfrac{\partial f}{\partial r} > 0$, people buy more beer when the prices of other goods increase and the other factors remain constant.)

33. Richard Stone (see Exercise 32) determined that the yearly consumption of food in the United States was given by

$$f(m, p, r) = (2.186)m^{.595}p^{-.543}r^{.922}.$$

Determine which partial derivatives are positive and which are negative and give interpretations of these facts.

34. For the production function $f(x, y) = 60x^{3/4}y^{1/4}$ considered in Example 8, think of $f(x, y)$ as the revenue when utilizing x units of labor and y units of capital. Under actual operating conditions, say $x = a$ and $y = b$, $\dfrac{\partial f}{\partial x}(a, b)$ is referred to as the *wage per unit of labor* and $\dfrac{\partial f}{\partial y}(a, b)$ is referred to as the *wage per unit of capital*. Show that

$$f(a, b) = a \cdot \left[\frac{\partial f}{\partial x}(a, b) \right] + b \cdot \left[\frac{\partial f}{\partial y}(a, b) \right].$$

(This equation shows how the revenue is distributed between labor and capital.)

35. Compute $\dfrac{\partial^2 f}{\partial x^2}$ where $f(x, y) = 60x^{3/4}y^{1/4}$, a production function (where x is units of labor). Explain why $\dfrac{\partial^2 f}{\partial x^2}$ is always negative.

36. Compute $\dfrac{\partial^2 f}{\partial y^2}$ where $f(x, y) = 60x^{3/4}y^{1/4}$, a production function (where y is units of capital). Explain why $\dfrac{\partial^2 f}{\partial y^2}$ is always negative.

---

37. Let $f(x, y) = 3x^2 + 2xy + 5y$, as in Example 5. Show that

$$f(1 + h, 4) - f(1, 4) = 14h + 3h^2.$$

Thus the error in approximating $f(1 + h, 4) - f(1, 4)$ by $14h$ is $3h^2$. (If $h = .01$, for instance, the error is only $.0003$.)

38. Physicians, particularly pediatricians, sometimes need to know the body surface area of a patient. For instance, the surface area is used to adjust the results of certain tests of kidney performance. Tables are available that give the approximate body surface area A in square meters of a person who weighs W kilograms and is H centimeters tall. The following empirical formula* is also used:

$$A = .007W^{.425}H^{.725}.$$

Evaluate $\frac{\partial A}{\partial W}$ and $\frac{\partial A}{\partial H}$ when $W = 54$, $H = 165$, and give a physical interpretation of your answers. You may use the approximations $(54)^{.425} \approx 5.4$, $(54)^{-.575} \approx .10$, $(165)^{.725} \approx 40.5$, $(165)^{-.275} \approx .25$.

Solutions to Practice Problems 7.2	

1. (a) Negative. $\frac{\partial f}{\partial x}(400, 2000)$ is approximately the change in sales due to a \$1 increase in x (price). Since raising prices lowers sales, we would expect $\frac{\partial f}{\partial x}(400, 2000)$ to be negative.

(b) Positive. $\frac{\partial f}{\partial y}(400, 2000)$ is approximately the change in sales due to a \$1 increase in advertising. Since spending more money on advertising brings in more customers, we would expect sales to increase; that is, $\frac{\partial f}{\partial y}(400, 2000)$ is most likely positive.

2. If the interest rate is raised from 9% to 10%, then the monthly payment will increase by about \$66.20. [An increase to $9\frac{1}{2}$% causes an increase in the monthly payment of about $\frac{1}{2} \cdot (66.20)$ or \$33.10, and so on.]

7.3 Maxima and Minima of Functions of Several Variables

Previously, we studied how to determine the maxima and minima of functions of a single variable. Let us extend that discussion to functions of several variables.

If $f(x, y)$ is a function of two variables, then we say that $f(x, y)$ has a *relative maximum* when $x = a$, $y = b$ if $f(x, y)$ is at most equal to $f(a, b)$ whenever x is near a and y is near b. Geometrically, the graph of $f(x, y)$ has a peak at $(x, y) = (a, b)$. [See Fig. 1(a).] Similarly, we say that $f(x, y)$ has a *relative minimum* when $x = a$, $y = b$ if $f(x, y)$ is at least equal to $f(a, b)$ whenever x is near a and y is near b. Geometrically, the graph of $f(x, y)$ has a pit whose bottom occurs at $(x, y) = (a, b)$. [See Fig. 1(b).]

Suppose the function $f(x, y)$ has a relative minimum at $(x, y) = (a, b)$, as in Fig. 2. When y is held constant at b, $f(x, y)$ is a function of x with a relative minimum at $x = a$. Therefore, the tangent line to the curve $z = f(x, b)$ is horizontal at $x = a$ and hence has slope 0. That is,

$$\frac{\partial f}{\partial x}(a, b) = 0.$$

Likewise, when x is held constant at a, $f(x, y)$ is a function of y with a relative minimum at $y = b$. Therefore, its derivative with respect to y is zero at $y = b$.

---

*See J. Routh, *Mathematical Preparation for Laboratory Technicians* (Philadelphia, Pa.: W. B. Saunders Co., 1971), p. 92.

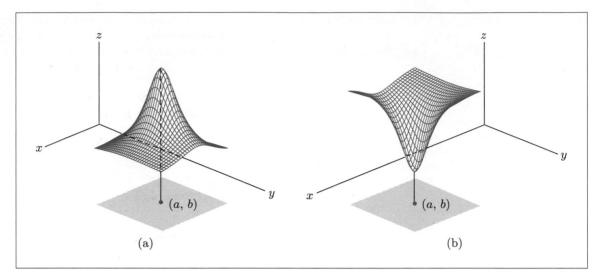

Figure 1. Maximum and minimum points.

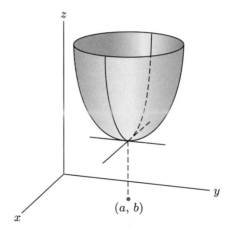

Figure 2. Horizontal tangent lines at a relative minimum.

That is,

$$\frac{\partial f}{\partial y}(a, b) = 0.$$

Similar considerations apply when $f(x, y)$ has a relative maximum at $(x, y) = (a, b)$.

First-Derivative Test for Functions of Two Variables   If $f(x, y)$ has either a relative maximum or minimum at $(x, y) = (a, b)$, then

$$\frac{\partial f}{\partial x}(a, b) = 0$$

and

$$\frac{\partial f}{\partial y}(a, b) = 0.$$

A relative maximum or minimum may or may not be an absolute maximum or minimum. However, to simplify matters in this text, the examples and exercises have been chosen so that if an absolute extremum of $f(x, y)$ exists, it will occur at a point where $f(x, y)$ has a relative extremum.

▶ Example 1   The function $f(x, y) = 3x^2 - 4xy + 3y^2 + 8x - 17y + 30$ has the graph pictured in Fig. 2. Find the point (a, b) at which $f(x, y)$ attains its minimum value.

Solution   We look for those values of x and y at which both partial derivatives are zero. The partial derivatives are

$$\frac{\partial f}{\partial x} = 6x - 4y + 8$$

$$\frac{\partial f}{\partial y} = -4x + 6y - 17.$$

Setting $\dfrac{\partial f}{\partial x} = 0$ and $\dfrac{\partial f}{\partial y} = 0$, we obtain

$$6x - 4y + 8 = 0 \quad \text{or} \quad y = \frac{6x + 8}{4}$$

$$-4x + 6y - 17 = 0 \quad \text{or} \quad y = \frac{4x + 17}{6}.$$

By equating these two expressions for y, we have

$$\frac{6x + 8}{4} = \frac{4x + 17}{6}.$$

Cross-multiplying, we see that

$$36x + 48 = 16x + 68$$
$$20x = 20$$
$$x = 1.$$

When we substitute this value for x into our first equation for y in terms of x, we obtain

$$y = \frac{6x + 8}{4} = \frac{6 \cdot 1 + 8}{4} = \frac{7}{2}.$$

If $f(x, y)$ has a minimum, it must occur where $\dfrac{\partial f}{\partial x} = 0$ and $\dfrac{\partial f}{\partial y} = 0$. We have determined that the partial derivatives are zero only when $x = 1$, $y = \frac{7}{2}$. From Fig. 2 we know that $f(x, y)$ has a minimum, so it must be at $(x, y) = (1, \frac{7}{2})$. ◆

▶ Example 2   (*Price Discrimination*)   A monopolist markets his product in two countries and can charge different amounts in each country. Let x be the number of units to be sold in the first country and y the number of units to be sold in the second country. Due to the laws of demand, the monopolist must set the price at $97 - (x/10)$ dollars in the first country and $83 - (y/20)$ dollars in the second country in order to sell all the units. The cost of producing these units is $20{,}000 + 3(x + y)$. Find the values of x and y that maximize the profit.

Solution   Let $f(x, y)$ be the profit derived from selling x units in the first country and y in the second. Then

$$f(x, y) = [\text{revenue from first country}] + [\text{revenue from second country}] - [\text{cost}]$$

$$= \left(97 - \frac{x}{10}\right)x + \left(83 - \frac{y}{20}\right)y - [20{,}000 + 3(x + y)]$$

$$= 97x - \frac{x^2}{10} + 83y - \frac{y^2}{20} - 20{,}000 - 3x - 3y$$

$$= 94x - \frac{x^2}{10} + 80y - \frac{y^2}{20} - 20{,}000.$$

To find where $f(x, y)$ has its maximum value, we look for those values of x and y at which both partial derivatives are zero.

$$\frac{\partial f}{\partial x} = 94 - \frac{x}{5}$$

$$\frac{\partial f}{\partial y} = 80 - \frac{y}{10}$$

We set $\dfrac{\partial f}{\partial x} = 0$ and $\dfrac{\partial f}{\partial y} = 0$ to obtain

$$94 - \frac{x}{5} = 0 \quad \text{or} \quad x = 470$$

$$80 - \frac{y}{10} = 0 \quad \text{or} \quad y = 800.$$

Therefore, the firm should adjust its prices to levels where it will sell 470 units in the first country and 800 units in the second country. That is, it should charge $50 in the first country and $43 in the second.   ◆

▶ **Example 3**   Suppose that we want to design a rectangular building having volume 147,840 cubic feet. Assuming that the daily loss of heat is given by

$$w = 11xy + 14yz + 15xz,$$

where x, y, and z are, respectively, the length, width, and height of the building, find the dimensions of the building for which the daily heat loss is minimal.

Solution   We must minimize the function

$$w = 11xy + 14yz + 15xz, \tag{1}$$

where x, y, z satisfy the constraint equation (refer to Section 2.5)

$$xyz = 147{,}840.$$

For simplicity, let us denote 147,840 by V. Then $xyz = V$, so that $z = V/xy$. We substitute this expression for z into the objective function (1) to obtain a heat-loss function $g(x, y)$ of two variables—namely,

$$g(x, y) = 11xy + 14y\frac{V}{xy} + 15x\frac{V}{xy} = 11xy + \frac{14V}{x} + \frac{15V}{y}.$$

To minimize this function, we first compute the partial derivatives with respect to x and y; then we equate them to zero.

$$\frac{\partial g}{\partial x} = 11y - \frac{14V}{x^2} = 0$$

$$\frac{\partial g}{\partial y} = 11x - \frac{15V}{y^2} = 0$$

These two equations yield

$$y = \frac{14V}{11x^2} \tag{2}$$

$$11xy^2 = 15V. \tag{3}$$

If we substitute the value of y from (2) into (3), we see that

$$11x \left(\frac{14V}{11x^2} \right)^2 = 15V$$

$$\frac{14^2 V^2}{11x^3} = 15V$$

$$x^3 = \frac{14^2 \cdot V^2}{11 \cdot 15 \cdot V} = \frac{14^2 \cdot V}{11 \cdot 15}$$

$$= \frac{14^2 \cdot 147{,}840}{11 \cdot 15}$$

$$= 175{,}616.$$

Therefore, we see (using a calculator) that

$$x = 56.$$

From equation (2) we find that

$$y = \frac{14 \cdot V}{11x^2} = \frac{14 \cdot 147{,}840}{11 \cdot 56^2} = 60.$$

Finally,

$$z = \frac{V}{xy} = \frac{147{,}840}{56 \cdot 60} = 44.$$

Thus the building should be 56 feet long, 60 feet wide, and 44 feet high in order to minimize the heat loss.* ◆

When considering a function of two variables, we find points (x, y) at which $f(x, y)$ has a potential relative maximum or minimum by setting $\dfrac{\partial f}{\partial x}$ and $\dfrac{\partial f}{\partial y}$ equal to zero and solving for x and y. However, if we are given no additional information about $f(x, y)$, it may be difficult to determine whether we have found a maximum or a minimum (or neither). In the case of functions of one variable, we studied concavity and deduced the second-derivative test. There is an analog of the second derivative test for functions of two variables, but it is much more complicated than the one-variable test. We state it without proof.

---

*For further discussion of this heat-loss problem, as well as other examples of optimization in architectural design, see L. March, "Elementary Models of Built Forms," Chapter 3 in *Urban Space and Structures*, L. Martin and L. March, eds. (Cambridge: Cambridge University Press, 1972).

Second-Derivative Test for Functions of Two Variables Suppose $f(x, y)$ is a function and (a, b) is a point at which

$$\frac{\partial f}{\partial x}(a, b) = 0 \quad \text{and} \quad \frac{\partial f}{\partial y}(a, b) = 0,$$

and let

$$D(x, y) = \frac{\partial^2 f}{\partial x^2} \cdot \frac{\partial^2 f}{\partial y^2} - \left(\frac{\partial^2 f}{\partial x \, \partial y}\right)^2.$$

1. If

$$D(a, b) > 0 \quad \text{and} \quad \frac{\partial^2 f}{\partial x^2}(a, b) > 0,$$

then $f(x, y)$ has a relative minimum at (a, b).

2. If

$$D(a, b) > 0 \quad \text{and} \quad \frac{\partial^2 f}{\partial x^2}(a, b) < 0,$$

then $f(x, y)$ has a relative maximum at (a, b).

3. If

$$D(a, b) < 0,$$

then $f(x, y)$ has neither a relative maximum nor a relative minimum at (a, b).

4. If $D(a, b) = 0$, then no conclusion can be drawn from this test.

The saddle-shaped graph in Fig. 3 illustrates a function $f(x, y)$ for which $D(a, b) < 0$. Both partial derivatives are zero at $(x, y) = (a, b)$ and yet the function has neither a relative maximum nor a relative minimum there. (Observe that the function has a relative maximum with respect to x when y is held constant and a relative minimum with respect to y when x is held constant.)

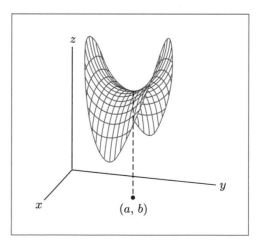

Figure 3.

▶ Example 4   Let $f(x, y) = x^3 - y^2 - 12x + 6y + 5$. Find all possible relative maximum and minimum points of $f(x, y)$. Use the second-derivative test to determine the nature of each such point.

Solution   Since

$$\frac{\partial f}{\partial x} = 3x^2 - 12, \qquad \frac{\partial f}{\partial y} = -2y + 6,$$

we find that $f(x, y)$ has a potential relative extreme point when

$$3x^2 - 12 = 0$$
$$-2y + 6 = 0.$$

From the first equation, $3x^2 = 12$, $x^2 = 4$, and $x = \pm 2$. From the second equation, $y = 3$. Thus, $\frac{\partial f}{\partial x}$ and $\frac{\partial f}{\partial y}$ are both zero when $(x, y) = (2, 3)$ and when $(x, y) = (-2, 3)$. To apply the second-derivative test, compute

$$\frac{\partial^2 f}{\partial x^2} = 6x, \qquad \frac{\partial^2 f}{\partial y^2} = -2, \qquad \frac{\partial^2 f}{\partial x\, \partial y} = 0,$$

and

$$D(x, y) = \frac{\partial^2 f}{\partial x^2} \cdot \frac{\partial^2 f}{\partial y^2} - \left(\frac{\partial^2 f}{\partial x\, \partial y} \right)^2 = (6x)(-2) - 0^2 = -12x. \qquad (4)$$

Since $D(2, 3) = -12(2) = -24$, which is negative, case 3 of the second-derivative test says that $f(x, y)$ has neither a relative maximum nor a relative minimum at $(2, 3)$. However, $D(-2, 3) = -12(-2) = 24$. Since $D(-2, 3)$ is positive, the function $f(x, y)$ has either a relative maximum or a relative minimum at $(-2, 3)$. To determine which, we compute

$$\frac{\partial^2 f}{\partial x^2}(-2, 3) = 6(-2) = -12 < 0.$$

By case 2 of the second-derivative test, the function $f(x, y)$ has a relative maximum at $(-2, 3)$. ◆

In this section we have restricted ourselves to functions of two variables, but the case of three or more variables is handled in a similar fashion. For instance, here is the first-derivative test for a function of three variables.

If $f(x, y, z)$ has a relative maximum or minimum at $(x, y, z) = (a, b, c)$, then

$$\frac{\partial f}{\partial x}(a, b, c) = 0$$

$$\frac{\partial f}{\partial y}(a, b, c) = 0$$

$$\frac{\partial f}{\partial z}(a, b, c) = 0.$$

Practice Problems 7.3

1. Find all points (x, y) where $f(x, y) = x^3 - 3xy + \frac{1}{2}y^2 + 8$ has a possible relative maximum or minimum.

2. Apply the second-derivative test to the function $g(x, y)$ of Example 3 to confirm that a relative minimum actually occurs when $x = 56$ and $y = 60$.

▶ Exercises 7.3

Find all points (x, y) where $f(x, y)$ has a possible relative maximum or minimum.

1. $f(x, y) = x^2 - 3y^2 + 4x + 6y + 8$
2. $f(x, y) = \frac{1}{2}x^2 + y^2 - 3x + 2y - 5$
3. $f(x, y) = x^2 - 5xy + 6y^2 + 3x - 2y + 4$
4. $f(x, y) = -3x^2 + 7xy - 4y^2 + x + y$
5. $f(x, y) = x^3 + y^2 - 3x + 6y$
6. $f(x, y) = x^2 - y^3 + 5x + 12y + 1$
7. $f(x, y) = \frac{1}{3}x^3 - 2y^3 - 5x + 6y - 5$
8. $f(x, y) = x^4 - 8xy + 2y^2 - 3$
9. The function $f(x, y) = 2x + 3y + 9 - x^2 - xy - y^2$ has a maximum at some point (x, y). Find the values of x and y where this maximum occurs.
10. The function $f(x, y) = \frac{1}{2}x^2 + 2xy + 3y^2 - x + 2y$ has a minimum at some point (x, y). Find the values of x and y where this minimum occurs.

In Exercises 11–16, both first partial derivatives of the function $f(x, y)$ are zero at the given points. Use the second-derivative test to determine the nature of $f(x, y)$ at each of these points. If the second-derivative test is inconclusive, so state.

11. $f(x, y) = 3x^2 - 6xy + y^3 - 9y$; $(3, 3)$, $(-1, -1)$
12. $f(x, y) = 6xy^2 - 2x^3 - 3y^4$; $(0, 0)$, $(1, 1)$, $(1, -1)$
13. $f(x, y) = 2x^2 - x^4 - y^2$; $(-1, 0)$, $(0, 0)$, $(1, 0)$
14. $f(x, y) = x^4 - 4xy + y^4$; $(0, 0)$, $(1, 1)$, $(-1, -1)$
15. $f(x, y) = ye^x - 3x - y + 5$; $(0, 3)$
16. $f(x, y) = \dfrac{1}{x} + \dfrac{1}{y} + xy$; $(1, 1)$

Find all points (x, y) where $f(x, y)$ has a possible relative maximum or minimum. Then use the second-derivative test to determine, if possible, the nature of $f(x, y)$ at each of these points. If the second-derivative test is inconclusive, so state.

17. $f(x, y) = x^2 - 2xy + 4y^2$
18. $f(x, y) = 2x^2 + 3xy + 5y^2$
19. $f(x, y) = -2x^2 + 2xy - y^2 + 4x - 6y + 5$
20. $f(x, y) = -x^2 - 8xy - y^2$
21. $f(x, y) = x^2 + 2xy + 5y^2 + 2x + 10y - 3$
22. $f(x, y) = x^2 - 2xy + 3y^2 + 4x - 16y + 22$
23. $f(x, y) = x^3 - y^2 - 3x + 4y$
24. $f(x, y) = x^3 - 2xy + 4y$
25. $f(x, y) = 2x^2 + y^3 - x - 12y + 7$
26. $f(x, y) = x^2 + 4xy + 2y^4$
27. Find the possible values of x, y, z at which
$$f(x, y, z) = 2x^2 + 3y^2 + z^2 - 2x - y - z$$
assumes its minimum value.

28. Find the possible values of x, y, z at which
$$f(x, y, z) = 5 + 8x - 4y + x^2 + y^2 + z^2$$
assumes its minimum value.

29. U.S. postal rules require that the length plus the girth of a package cannot exceed 84 inches in order to be mailed. Find the dimensions of the rectangular package of greatest volume that can be mailed. [*Note:* From Fig. 4 we see that $84 = $ (length) + (girth) = $l + (2x + 2y)$.]

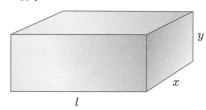

Figure 4.

30. Find the dimensions of the rectangular box of least surface area that has a volume of 1000 cubic inches.
31. A company manufactures and sells two products, call them I and II, that sell for $10 and $9 per unit, respectively. The cost of producing x units of product I and y units of product II is
$$400 + 2x + 3y + .01(3x^2 + xy + 3y^2).$$
Find the values of x and y that maximize the company's profits. [*Note:* Profit = (revenue) − (cost).]
32. A monopolist manufactures and sells two competing products, call them I and II, that cost $30 and $20 per unit, respectively, to produce. The revenue from marketing x units of product I and y units of product II is $98x + 112y - .04xy - .1x^2 - .2y^2$. Find the values of x and y that maximize the monopolist's profits.
33. A company manufactures and sells two products, call them I and II, that sell for p_I and p_{II} per unit, respectively. Let $C(x, y)$ be the cost of producing x units of product I and y units of product II. Show that if the company's profit is maximized when $x = a$, $y = b$, then
$$\frac{\partial C}{\partial x}(a, b) = p_I \quad \text{and} \quad \frac{\partial C}{\partial y}(a, b) = p_{II}.$$
34. A monopolist manufactures and sells two competing products, call them I and II, that cost p_I and p_{II} per unit, respectively, to produce. Let $R(x, y)$ be the revenue from marketing x units of product I and y units of product II. Show that if the monopolist's profit is maximized when $x = a$, $y = b$, then
$$\frac{\partial R}{\partial x}(a, b) = p_I \quad \text{and} \quad \frac{\partial R}{\partial y}(a, b) = p_{II}.$$

Solutions to Practice Problems 7.3

1. Compute the first partial derivatives of $f(x, y)$ and solve the system of equations that results from setting the partials equal to zero.

$$\frac{\partial f}{\partial x} = 3x^2 - 3y = 0$$

$$\frac{\partial f}{\partial y} = -3x + y = 0$$

Solve each equation for y in terms of x.

$$\begin{cases} y = x^2 \\ y = 3x \end{cases}$$

Equate expressions for y and solve for x.

$$x^2 = 3x$$
$$x^2 - 3x = 0$$
$$x(x - 3) = 0$$
$$x = 0 \quad \text{or} \quad x = 3$$

When $x = 0$, $y = 0^2 = 0$. When $x = 3$, $y = 3^2 = 9$. Therefore, the possible relative maximum or minimum points are $(0, 0)$ and $(3, 9)$.

2. We have

$$g(x, y) = 11xy + \frac{14V}{x} + \frac{15V}{y},$$

$$\frac{\partial g}{\partial x} = 11y - \frac{14V}{x^2}, \quad \text{and} \quad \frac{\partial g}{\partial y} = 11x - \frac{15V}{y^2}.$$

Now,

$$\frac{\partial^2 g}{\partial x^2} = \frac{28V}{x^3}, \quad \frac{\partial^2 g}{\partial y^2} = \frac{30V}{y^3}, \quad \text{and} \quad \frac{\partial^2 g}{\partial x \, \partial y} = 11.$$

Therefore,

$$D(x, y) = \frac{28V}{x^3} \cdot \frac{30V}{y^3} - (11)^2$$

$$D(56, 60) = \frac{28(147{,}840)}{(56)^3} \cdot \frac{30(147{,}840)}{(60)^3} - 121$$

$$= 484 - 121 = 363 > 0,$$

and

$$\frac{\partial^2 g}{\partial x^2}(56, 60) = \frac{28(147{,}840)}{(56)^3} > 0.$$

It follows that $g(x, y)$ has a relative minimum at $x = 56$, $y = 60$.

7.4 Lagrange Multipliers and Constrained Optimization

We have seen a number of optimization problems in which we were required to minimize (or maximize) an objective function where the variables were subject to a constraint equation. For instance, in Example 4 of Section 2.5, we minimized the cost of a rectangular enclosure by minimizing the objective function $21x + 14y$, where x and y were subject to the constraint equation $600 - xy = 0$. In the preceding section (Example 3) we minimized the daily heat loss from a building by minimizing the objective function $11xy + 14yz + 15xz$, subject to the constraint equation $147{,}840 - xyz = 0$.

Figure 1 gives a graphical illustration of what happens when an objective function is maximized subject to a constraint. The graph of the objective function is the cone-shaped surface $z = 36 - x^2 - y^2$, and the colored curve on that surface consists of those points whose x- and y-coordinates satisfy the constraint equation $x + 7y - 25 = 0$. The constrained maximum is at the highest point on this curve. Of course, the surface itself has a higher "unconstrained maximum" at $(x, y, z) = (0, 0, 36)$, but these values of x and y do not satisfy the constraint equation.

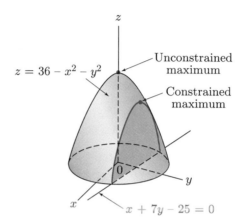

Figure 1. A constrained optimization problem.

In this section we introduce a powerful technique for solving problems of this type. Let us begin with the following general problem, which involves two variables.

Problem   Let $f(x, y)$ and $g(x, y)$ be functions of two variables. Find values of x and y that maximize (or minimize) the objective function $f(x, y)$ and that also satisfy the constraint equation $g(x, y) = 0$.

Of course, if we can solve the equation $g(x, y) = 0$ for one variable in terms of the other and substitute the resulting expression into $f(x, y)$, we arrive at a function of a single variable that can be maximized (or minimized) by using the methods of Chapter 2. However, this technique can be unsatisfactory for two reasons. First, it may be difficult to solve the equation $g(x, y) = 0$ for x or for y. For example, if $g(x, y) = x^4 + 5x^3y + 7x^2y^3 + y^5 - 17 = 0$, then it is difficult to write y as a function of x or x as a function of y. Second, even if $g(x, y) = 0$ can be solved for one variable in terms of the other, substitution of the result into $f(x, y)$ may yield a complicated function.

One clever idea for handling the preceding problem was discovered by the eighteenth-century mathematician Lagrange, and the technique that he pioneered today bears his name—the method of *Lagrange multipliers*. The basic idea of this method is to replace $f(x, y)$ by an auxiliary function of three variables $F(x, y, \lambda)$, defined as

$$F(x, y, \lambda) = f(x, y) + \lambda g(x, y).$$

The new variable λ (lambda) is called a *Lagrange multiplier* and always multiplies the constraint function $g(x, y)$. The following theorem is stated without proof.

Theorem Suppose that, subject to the constraint $g(x, y) = 0$, the function $f(x, y)$ has a relative maximum or minimum at $(x, y) = (a, b)$. Then there is a value of λ, say $\lambda = c$, such that the partial derivatives of $F(x, y, \lambda)$ all equal zero at $(x, y, \lambda) = (a, b, c)$.

The theorem implies that if we locate all points (x, y, λ) where the partial derivatives of $F(x, y, \lambda)$ are all zero, then among the corresponding points (x, y) we will find all possible places where $f(x, y)$ may have a constrained relative maximum or minimum. Thus the first step in the method of Lagrange multipliers is to set the partial derivatives of $F(x, y, \lambda)$ equal to zero and solve for x, y, and λ:

$$\frac{\partial F}{\partial x} = 0 \tag{L-1}$$

$$\frac{\partial F}{\partial y} = 0 \tag{L-2}$$

$$\frac{\partial F}{\partial \lambda} = 0. \tag{L-3}$$

From the definition of $F(x, y, \lambda)$, we see that $\dfrac{\partial F}{\partial \lambda} = g(x, y)$. Thus the third equation (L-3) is just the original constraint equation $g(x, y) = 0$. So when we find a point (x, y, λ) that satisfies (L-1), (L-2), and (L-3), the coordinates x and y will automatically satisfy the constraint equation.

The first example applies this method to the problem described in Fig. 1.

▶ **Example 1** Maximize $36 - x^2 - y^2$ subject to the constraint $x + 7y - 25 = 0$.

Solution Here $f(x, y) = 36 - x^2 - y^2$, $g(x, y) = x + 7y - 25$, and

$$F(x, y, \lambda) = 36 - x^2 - y^2 + \lambda(x + 7y - 25).$$

Equations (L-1) to (L-3) read

$$\frac{\partial F}{\partial x} = -2x + \lambda = 0 \tag{1}$$

$$\frac{\partial F}{\partial y} = -2y + 7\lambda = 0 \tag{2}$$

$$\frac{\partial F}{\partial \lambda} = x + 7y - 25 = 0. \tag{3}$$

We solve the first two equations for λ:

$$\lambda = 2x$$

$$\lambda = \tfrac{2}{7}y. \tag{4}$$

If we equate these two expressions for λ, we obtain

$$2x = \tfrac{2}{7}y$$

$$x = \tfrac{1}{7}y. \tag{5}$$

Substituting this expression for x into equation (3), we have

$$\tfrac{1}{7}y + 7y - 25 = 0$$

$$\tfrac{50}{7}y = 25$$

$$y = \tfrac{7}{2}.$$

With this value for y, equations (4) and (5) produce the values of x and λ:

$$x = \tfrac{1}{7}y = \tfrac{1}{7}\left(\tfrac{7}{2}\right) = \tfrac{1}{2}$$

$$\lambda = \tfrac{2}{7}y = 1.$$

Therefore, the partial derivatives of $F(x, y, \lambda)$ are zero when $x = \tfrac{1}{2}$, $y = \tfrac{7}{2}$, and $\lambda = 1$. So the maximum value of $36 - x^2 - y^2$ subject to the constraint $x + 7y - 25 = 0$ is

$$36 - \left(\tfrac{1}{2}\right)^2 - \left(\tfrac{7}{2}\right)^2 = \tfrac{47}{2}. \qquad \blacklozenge$$

The preceding technique for solving three equations in the three variables x, y, and λ can usually be applied to solve Lagrange multiplier problems. Here is the basic procedure.

1. Solve (L-1) and (L-2) for λ in terms of x and y; then equate the resulting expressions for λ.

2. Solve the resulting equation for one of the variables.

3. Substitute the expression so derived into the equation (L-3) and solve the resulting equation of one variable.

4. Use the one known variable and the equations of steps 1 and 2 to determine the other two variables.

In most applications we know that an absolute (constrained) maximum or minimum exists. In the event that the method of Lagrange multipliers produces exactly one possible relative extreme value, we will assume that it is indeed the sought-after absolute extreme value. For instance, the statement of Example 1 is meant to imply that there is an absolute maximum value. Since we determined that there was just one possible relative extreme value, we concluded that it was the absolute maximum value.

▶ Example 2 Using Lagrange multipliers, minimize $42x + 28y$, subject to the constraint $600 - xy = 0$, where x and y are restricted to positive values. (This problem arose in Example 4 of Section 2.5, where $42x + 28y$ was the cost of building a 600-square-foot enclosure having dimensions x and y.)

Solution We have $f(x, y) = 42x + 28y$, $g(x, y) = 600 - xy$, and

$$F(x, y, \lambda) = 42x + 28y + \lambda(600 - xy).$$

The equations (L-1) to (L-3), in this case, are

$$\frac{\partial F}{\partial x} = 42 - \lambda y = 0$$

$$\frac{\partial F}{\partial y} = 28 - \lambda x = 0$$

$$\frac{\partial F}{\partial \lambda} = 600 - xy = 0.$$

From the first two equations we see that

$$\lambda = \frac{42}{y} = \frac{28}{x}. \hspace{3cm} \text{(step 1)}$$

Therefore,

$$42x = 28y$$

and

$$x = \frac{2}{3}y. \hspace{3cm} \text{(step 2)}$$

Substituting this expression for x into the third equation, we derive

$$600 - \left(\frac{2}{3}y\right)y = 0$$

$$y^2 = \frac{3}{2} \cdot 600 = 900$$

$$y = \pm 30. \hspace{3cm} \text{(step 3)}$$

We discard the case $y = -30$ because we are interested only in positive values of x and y. Using $y = 30$, we find that

$$\left.\begin{array}{l} x = \dfrac{2}{3}(30) = 20 \\[2mm] \lambda = \dfrac{28}{20} = \dfrac{7}{5}. \end{array}\right\} \hspace{2cm} \text{(step 4)}$$

So the minimum value of $42x + 28y$ with x and y subject to the constraint occurs when $x = 20$, $y = 30$, and $\lambda = \frac{7}{5}$. That minimum value is

$$42 \cdot (20) + 28 \cdot (30) = 1680. \hspace{2cm} \blacklozenge$$

▶ **Example 3** (*Production*) Suppose that x units of labor and y units of capital can produce $f(x, y) = 60x^{3/4}y^{1/4}$ units of a certain product. Also suppose that each unit of labor costs \$100, whereas each unit of capital costs \$200. Assume that \$30,000 is available to spend on production. How many units of labor and how many units of capital should be utilized in order to maximize production?

Solution The cost of x units of labor and y units of capital equals $100x + 200y$. Therefore, since we want to use all the available money (\$30,000), we must satisfy the constraint equation

$$100x + 200y = 30{,}000$$

or

$$g(x, y) = 30{,}000 - 100x - 200y = 0.$$

The objective function is $f(x, y) = 60x^{3/4}y^{1/4}$. In this case, we have

$$F(x, y, \lambda) = 60x^{3/4}y^{1/4} + \lambda(30{,}000 - 100x - 200y).$$

The equations (L-1) to (L-3) read

$$\frac{\partial F}{\partial x} = 45x^{-1/4}y^{1/4} - 100\lambda = 0 \qquad \text{(L-1)}$$

$$\frac{\partial F}{\partial y} = 15x^{3/4}y^{-3/4} - 200\lambda = 0 \qquad \text{(L-2)}$$

$$\frac{\partial F}{\partial \lambda} = 30{,}000 - 100x - 200y = 0. \qquad \text{(L-3)}$$

By solving the first two equations for λ, we see that

$$\lambda = \frac{45}{100}x^{-1/4}y^{1/4} = \frac{9}{20}x^{-1/4}y^{1/4}$$

$$\lambda = \frac{15}{200}x^{3/4}y^{-3/4} = \frac{3}{40}x^{3/4}y^{-3/4}.$$

Therefore, we must have

$$\frac{9}{20}x^{-1/4}y^{1/4} = \frac{3}{40}x^{3/4}y^{-3/4}.$$

To solve for y in terms of x, let us multiply both sides of this equation by $x^{1/4}y^{3/4}$:

$$\frac{9}{20}y = \frac{3}{40}x$$

or

$$y = \frac{1}{6}x.$$

Inserting this result in (L-3), we find that

$$100x + 200\left(\frac{1}{6}x\right) = 30{,}000$$

$$\frac{400x}{3} = 30{,}000$$

$$x = 225.$$

Hence

$$y = \frac{225}{6} = 37.5.$$

So maximum production is achieved by using 225 units of labor and 37.5 units of capital. ◆

In Example 3 it turns out that, at the optimum value of x and y,

$$\lambda = \frac{9}{20}x^{-1/4}y^{1/4} = \frac{9}{20}(225)^{-1/4}(37.5)^{1/4} \approx .2875$$

$$\frac{\partial f}{\partial x} = 45x^{-1/4}y^{1/4} = 45(225)^{-1/4}(37.5)^{1/4} \qquad \text{(6)}$$

$$\frac{\partial f}{\partial y} = 15x^{3/4}y^{-3/4} = 15(225)^{3/4}(37.5)^{-3/4}. \qquad \text{(7)}$$

It can be shown that the Lagrange multiplier λ can be interpreted as the *marginal productivity of money*. That is, if one additional dollar is available, then approximately .2875 additional units of the product can be produced.

Recall that the partial derivatives $\dfrac{\partial f}{\partial x}$ and $\dfrac{\partial f}{\partial y}$ are called the marginal productivity of labor and capital, respectively. From (6) and (7) we have

$$\frac{[\text{marginal productivity of labor}]}{[\text{marginal productivity of capital}]} = \frac{45(225)^{-1/4}(37.5)^{1/4}}{15(225)^{3/4}(37.5)^{-3/4}}$$

$$= \frac{45}{15}(225)^{-1}(37.5)^{1}$$

$$= \frac{3(37.5)}{225} = \frac{37.5}{75} = \frac{1}{2}.$$

On the other hand,

$$\frac{[\text{cost per unit of labor}]}{[\text{cost per unit of capital}]} = \frac{100}{200} = \frac{1}{2}.$$

This result illustrates the following law of economics. *If labor and capital are at their optimal levels, then the ratio of their marginal productivities equals the ratio of their unit costs.*

The method of Lagrange multipliers generalizes to functions of any number of variables. For instance, we can maximize $f(x, y, z)$, subject to the constraint equation $g(x, y, z) = 0$, by considering the Lagrange function

$$F(x, y, z, \lambda) = f(x, y, z) + \lambda g(x, y, z).$$

The analogs of equations (L-1) to (L-3) are

$$\frac{\partial F}{\partial x} = 0$$

$$\frac{\partial F}{\partial y} = 0$$

$$\frac{\partial F}{\partial z} = 0$$

$$\frac{\partial F}{\partial \lambda} = 0.$$

Let us now show how we can solve the heat-loss problem of Section 7.3 by using this method.

▶ **Example 4** Use Lagrange multipliers to find the values of x, y, z that minimize the objective function

$$f(x, y, z) = 11xy + 14yz + 15xz,$$

subject to the constraint

$$xyz = 147{,}840.$$

Solution The Lagrange function is

$$F(x, y, z, \lambda) = 11xy + 14yz + 15xz + \lambda(147{,}840 - xyz).$$

The conditions for a relative minimum are

$$\frac{\partial F}{\partial x} = 11y + 15z - \lambda yz = 0$$

$$\frac{\partial F}{\partial y} = 11x + 14z - \lambda xz = 0$$

$$\frac{\partial F}{\partial z} = 14y + 15x - \lambda xy = 0$$

$$\frac{\partial F}{\partial \lambda} = 147{,}840 - xyz = 0. \tag{8}$$

From the first three equations we have

$$\left. \begin{array}{l} \lambda = \dfrac{11y + 15z}{yz} = \dfrac{11}{z} + \dfrac{15}{y} \\[2mm] \lambda = \dfrac{11x + 14z}{xz} = \dfrac{11}{z} + \dfrac{14}{x} \\[2mm] \lambda = \dfrac{14y + 15x}{xy} = \dfrac{14}{x} + \dfrac{15}{y} \end{array} \right\}. \tag{9}$$

Let us equate the first two expression for λ:

$$\frac{11}{z} + \frac{15}{y} = \frac{11}{z} + \frac{14}{x}$$

$$\frac{15}{y} = \frac{14}{x}$$

$$x = \frac{14}{15}y.$$

Next, we equate the second and third expressions for λ in (9):

$$\frac{11}{z} + \frac{14}{x} = \frac{14}{x} + \frac{15}{y}$$

$$\frac{11}{z} = \frac{15}{y}$$

$$z = \frac{11}{15}y.$$

We now substitute the expressions for x and z into the constraint equation (8) and obtain

$$\frac{14}{15}y \cdot y \cdot \frac{11}{15}y = 147{,}840$$

$$y^3 = \frac{(147{,}840)(15)^2}{(14)(11)} = 216{,}000$$

$$y = 60.$$

From this, we find that

$$x = \frac{14}{15}(60) = 56 \quad \text{and} \quad z = \frac{11}{15}(60) = 44.$$

We conclude that the heat loss is minimized when $x = 56$, $y = 60$, and $z = 44$.

◆

In the solution of Example 4, we found that at the optimal values of x, y, and z,

$$\frac{14}{x} = \frac{15}{y} = \frac{11}{z}.$$

Referring to Example 2 of Section 7.1, we see that 14 is the combined heat loss through the east and west sides of the building, 15 is the heat loss through the north and south sides of the building, and 11 is the heat loss through the floor and roof. Thus we have that under optimal conditions

$$\frac{[\text{heat loss through east and west sides}]}{[\text{distance between east and west sides}]}$$

$$= \frac{[\text{heat loss through north and south sides}]}{[\text{distance between north and south sides}]}$$

$$= \frac{[\text{heat loss through floor and roof}]}{[\text{distance between floor and roof}]}.$$

This is a principle of optimal design: Minimal heat loss occurs when the distance between each pair of opposite sides is some fixed constant times the heat loss from the pair of sides.

The value of λ in Example 4 corresponding to the optimal values of x, y, and z is

$$\lambda = \frac{11}{z} + \frac{15}{y} = \frac{11}{44} + \frac{15}{60} = \frac{1}{2}.$$

One can show that the Lagrange multiplier λ is the marginal heat loss with respect to volume. That is, if a building of volume slightly more than 147,840 cubic feet is optimally designed, then $\frac{1}{2}$ unit of additional heat will be lost for each additional cubic foot of volume.

Practice Problems 7.4

1. Let $F(x, y, \lambda) = 2x + 3y + \lambda(90 - 6x^{1/3}y^{2/3})$. Find $\dfrac{\partial F}{\partial x}$.

2. Refer to Exercise 29 of Section 7.3. What is the function $F(x, y, \lambda)$ when the exercise is solved using the method of Lagrange multipliers?

▶ Exercises 7.4

Solve the following exercises by the method of Lagrange multipliers.

1. Minimize the function $x^2 + 3y^2 + 10$, subject to the constraint $8 - x - y = 0$.

2. Maximize the function $x^2 - y^2$, subject to the constraint $2x + y - 3 = 0$.

3. Maximize $x^2 + xy - 3y^2$, subject to the constraint $2 - x - 2y = 0$.

4. Minimize $\frac{1}{2}x^2 - 3xy + y^2 + \frac{1}{2}$, subject to the constraint $3x - y - 1 = 0$.

5. Find the values of x, y that maximize the function

$$-2x^2 - 2xy - \tfrac{3}{2}y^2 + x + 2y,$$

subject to the constraint $x + y - \frac{5}{2} = 0$.

6. Find the values of x, y that minimize the function

$$x^2 + xy + y^2 - 2x - 5y,$$

subject to the constraint $1 - x + y = 0$.

7. Find the two positive numbers whose product is 25 and whose sum is as small as possible.

8. Four hundred eighty dollars are available to fence in a rectangular garden. The fencing for the north and south sides of the garden costs $10 per foot and the fencing for the east and west sides costs $15 per foot. Find the dimensions of the largest possible garden.

9. Three hundred square inches of material are available to construct an open rectangular box with a square

base. Find the dimensions of the box that maximize the volume.

10. The amount of space required by a particular firm is $f(x, y) = 1000\sqrt{6x^2 + y^2}$, where x and y are, respectively, the number of units of labor and capital utilized. Suppose that labor costs \$480 per unit and capital costs \$40 per unit and that the firm has \$5000 to spend. Determine the amounts of labor and capital that should be utilized in order to minimize the amount of space required.

11. Find the dimensions of the rectangle of maximum area that can be inscribed in the unit circle. [See Fig. 2(a).]

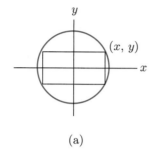

(a)

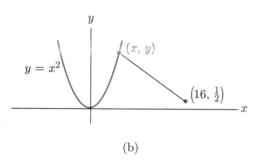

(b)

Figure 2.

12. Find the point on the parabola $y = x^2$ that has minimal distance from the point $\left(16, \frac{1}{2}\right)$. [See Fig. 2(b).] [*Suggestion*: If d denotes the distance from (x, y) to $\left(16, \frac{1}{2}\right)$, then $d^2 = (x - 16)^2 + (y - \frac{1}{2})^2$. If d^2 is minimized, then d will be minimized.]

13. Suppose that a firm makes two products A and B that use the same raw materials. Given a fixed amount of raw materials and a fixed amount of manpower, the firm must decide how much of its resources should be allocated to the production of A and how much to B. If x units of A and y units of B are produced, suppose that x and y must satisfy

$$9x^2 + 4y^2 = 18{,}000.$$

The graph of this equation (for $x \geq 0$, $y \geq 0$) is called a *production possibilities curve* (Fig. 3). A point (x, y) on this curve represents a *production schedule* for the firm, committing it to produce x units of A and y units of B. The reason for the relationship between x and y

involves the limitations on personnel and raw materials available to the firm. Suppose that each unit of A yields a \$3 profit, whereas each unit of B yields a \$4 profit. Then the profit of the firm is

$$P(x, y) = 3x + 4y.$$

Find the production schedule that maximizes the profit function $P(x, y)$.

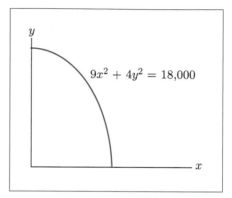

Figure 3. A production possibilities curve.

14. A firm makes x units of product A and y units of product B and has a production possibilities curve given by the equation $4x^2 + 25y^2 = 50{,}000$ for $x \geq 0$, $y \geq 0$. (See Exercise 13.) Suppose profits are \$2 per unit for product A and \$10 per unit for product B. Find the production schedule that maximizes the total profit.

15. The production function for a firm is $f(x, y) = 64x^{3/4}y^{1/4}$, where x and y are the number of units of labor and capital utilized. Suppose that labor costs \$96 per unit and capital costs \$162 per unit and that the firm decides to produce 3456 units of goods.

   (a) Determine the amounts of labor and capital that should be utilized in order to minimize the cost. That is, find the values of x, y that minimize $96x + 162y$, subject to the constraint $3456 - 64x^{3/4}y^{1/4} = 0$.

   (b) Find the value of λ at the optimal level of production.

   (c) Show that, at the optimal level of production, we have

$$\frac{[\text{marginal productivity of labor}]}{[\text{marginal productivity of capital}]}$$

$$= \frac{[\text{unit price of labor}]}{[\text{unit price of capital}]}.$$

16. Consider the monopolist of Example 2, Section 7.3, who sells his goods in two countries. Suppose that he must set the same price in each country. That is, $97 - (x/10) = 83 - (y/20)$. Find the values of x and y that maximize profits under this new restriction.

17. Find the values of x, y, and z that maximize the function xyz subject to the constraint $36 - x - 6y - 3z = 0$.

18. Find the values of x, y, and z that maximize the function $xy + 3xz + 3yz$ subject to the constraint $9 - xyz = 0$.

19. Find the values of x, y, z that maximize the function

$$3x + 5y + z - x^2 - y^2 - z^2,$$

subject to the constraint $6 - x - y - z = 0$.

20. Find the values of x, y, z that minimize the function

$$x^2 + y^2 + z^2 - 3x - 5y - z,$$

subject to the constraint $20 - 2x - y - z = 0$.

21. The material for a closed rectangular box costs \$2 per square foot for the top and \$1 per square foot for the sides and bottom. Using Lagrange multipliers, find the dimensions for which the volume of the box is 12 cubic feet and the cost of the materials is minimized. [Referring to Fig. 4(a), the cost will be $3xy + 2xz + 2yz$.]

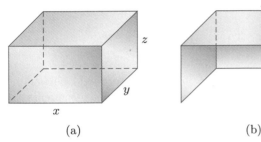

(a)   (b)

Figure 4.

22. Use Lagrange multipliers to find the three positive numbers whose sum is 15 and whose product is as large as possible.

23. Find the dimensions of an open rectangular glass tank of volume 32 cubic feet for which the amount of material needed to construct the tank is minimized. [See Fig. 4(a).]

24. A shelter for use at the beach has a back, two sides, and a top made of canvas. [See Fig. 4(b).] Find the dimensions that maximize the volume and require 96 square feet of canvas.

25. Let $f(x, y)$ be any production function where x represents labor (costing \$$a$ per unit) and y represents capital (costing \$$b$ per unit). Assuming that \$$c$ is available, show that, at the values of x, y that maximize production,

$$\frac{\frac{\partial f}{\partial x}}{\frac{\partial f}{\partial y}} = \frac{a}{b}.$$

Note: Let $F(x, y, \lambda) = f(x, y) + \lambda(c - ax - by)$. The result follows from (L-1) and (L-2).

26. By applying the result in Exercise 25 to the production function $f(x, y) = kx^\alpha y^\beta$, show that, for the values of x, y that maximize production, we have

$$\frac{y}{x} = \frac{a\beta}{b\alpha}.$$

(This tells us that the ratio of capital to labor does not depend on the amount of money available nor on the level of production but only on the numbers a, b, α, and β.)

Solutions to Practice Problems 7.4

1. The function can be written as

$$F(x, y, \lambda) = 2x + 3y + \lambda \cdot 90 - \lambda \cdot 6x^{1/3}y^{2/3}.$$

When differentiating with respect to x, both y and λ should be treated as constants (so $\lambda \cdot 90$ and $\lambda \cdot 6$ are also regarded as constants).

$$\frac{\partial F}{\partial x} = 2 - \lambda \cdot 6 \cdot \frac{1}{3}x^{-2/3} \cdot y^{2/3}$$

$$= 2 - 2\lambda x^{-2/3}y^{2/3}$$

(*Note*: It is not necessary to write out the multiplication by λ as we did. Most people just do this mentally and then differentiate.)

2. The quantity to be maximized is the volume xyl. The constraint is that length plus girth is 84. This translates to $84 = l + 2x + 2y$ or $84 - l - 2x - 2y = 0$. Therefore,

$$F(x, y, l, \lambda) = xyl + \lambda(84 - l - 2x - 2y).$$

7.5 The Method of Least Squares

Nowadays, people compile graphs of literally thousands of different quantities: the purchasing value of the dollar as a function of time; the pressure of a fixed volume of air as a function of temperature; the average income of people as a function of their years of formal education; or the incidence of strokes as a function of blood pressure. The observed points on such graphs tend to be irregularly distributed due to the complicated nature of the phenomena underlying them as well as to errors made in observation. (For example, a given procedure for measuring average income may not count certain groups.)

In spite of the imperfect nature of the data, we are often faced with the problem of making assessments and predictions based on them. Roughly speaking, this problem amounts to filtering the sources of errors in the data and isolating the basic underlying trend. Frequently, on the basis of a suspicion or a working hypothesis, we may suspect that the underlying trend is linear—that is, the data should lie on a straight line. But which straight line? This is the problem that the *method of least squares* attempts to answer. To be more specific, let us consider the following problem:

Problem of Fitting a Straight Line to Data   Given observed data points $(x_1, y_1), (x_2, y_2), \ldots, (x_N, y_N)$ on a graph, find the straight line that "best" fits these points.

In order to completely understand the statement of the problem being considered, we must define what it means for a line to "best" fit a set of points. If (x_i, y_i) is one of our observed points, then we will measure how far it is from a given line $y = Ax + B$ by the vertical distance from the point to the line. Since the point on the line with x-coordinate x_i is $(x_i, Ax_i + B)$, this vertical distance is the distance between the y-coordinates $Ax_i + B$ and y_i. (See Fig. 1.) If $E_i = (Ax_i + B) - y_i$, then either E_i or $-E_i$ is the vertical distance from (x_i, y_i) to the line. To avoid this ambiguity, we work with the square of this vertical distance, namely,

$$E_i^2 = (Ax_i + B - y_i)^2.$$

The total error in approximating the data points $(x_1, y_1), \ldots, (x_N, y_N)$ by the line $y = Ax + B$ is usually measured by the sum E of the squares of the vertical

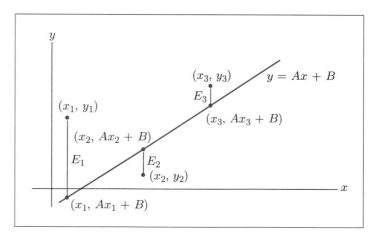

Figure 1. Fitting a line to data points.

distances from the points to the line,

$$E = E_1^2 + E_2^2 + \cdots + E_N^2.$$

E is called the *least-squares error* of the observed points with respect to the line. If all the observed points lie on the line $y = Ax + B$, then all E_i are zero and the error E is zero. If a given observed point is far away from the line, the corresponding E_i^2 is large and hence makes a large contribution to the error E.

In general, we cannot expect to find a line $y = Ax + B$ that fits the observed points so well that the error E is zero. Actually, this situation will occur only if the observed points lie on a straight line. However, we can rephrase our original problem as follows:

Problem Given observed data points $(x_1, y_1), (x_2, y_2), \ldots, (x_N, y_N)$, find a straight line $y = Ax + B$ for which the error E is as small as possible. This line is called the *least-squares line* or *regression line*.

It turns out that this problem is a minimization problem in the two variables A and B and so can be solved by using the methods of Section 7.3. Let us consider an example.

▶ **Example 1** Find the straight line that minimizes the least-squares error for the points $(1, 4)$, $(2, 5)$, $(3, 8)$.

Solution Let the straight line be $y = Ax + B$. When $x = 1, 2, 3$, the y-coordinate of the corresponding point of the line is $A + B$, $2A + B$, $3A + B$, respectively. Therefore, the squares of the vertical distances from the points $(1, 4)$, $(2, 5)$, $(3, 8)$ are, respectively,

$$E_1^2 = (A + B - 4)^2$$
$$E_2^2 = (2A + B - 5)^2$$
$$E_3^2 = (3A + B - 8)^2.$$

(See Fig. 2.) Thus the least-squares error is

$$E = E_1^2 + E_2^2 + E_3^2 = (A + B - 4)^2 + (2A + B - 5)^2 + (3A + B - 8)^2.$$

This error obviously depends on the choice of A and B. Let $f(A, B)$ denote this least-squares error. We want to find values of A and B that minimize $f(A, B)$. To do so, we take partial derivatives with respect to A and B and set the partial derivatives equal to zero:

$$\frac{\partial f}{\partial A} = 2(A + B - 4) + 2(2A + B - 5) \cdot 2 + 2(3A + B - 8) \cdot 3$$
$$= 28A + 12B - 76 = 0$$

$$\frac{\partial f}{\partial B} = 2(A + B - 4) + 2(2A + B - 5) + 2(3A + B - 8)$$
$$= 12A + 6B - 34 = 0.$$

In order to find A and B, we must solve the system of simultaneous linear equations

$$28A + 12B = 76$$
$$12A + 6B = 34.$$

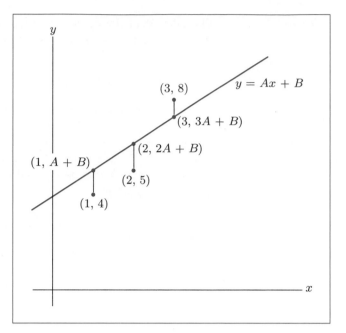

Figure 2.

Multiplying the second equation by 2 and subtracting from the first equation, we have $4A = 8$, or $A = 2$. Therefore, $B = \frac{5}{3}$, and the straight line that minimizes the least-squares error is $y = 2x + \frac{5}{3}$. ◆

The minimization process used in Example 1 can be applied to a general set of data points $(x_1, y_1), \ldots, (x_N, y_N)$ to obtain the following algebraic formula for A and B.

$$A = \frac{N \cdot \Sigma xy - \Sigma x \cdot \Sigma y}{N \cdot \Sigma x^2 - (\Sigma x)^2}$$

$$B = \frac{\Sigma y - A \cdot \Sigma x}{N}$$

where

$\Sigma x =$ sum of the x-coordinates of the data points

$\Sigma y =$ sum of the y-coordinates of the data points

$\Sigma xy =$ sum of the products of the coordinates of the data points

$\Sigma x^2 =$ sum of the squares of the x-coordinates of the data points

$N =$ number of data points.

That is,

$$\Sigma x = x_1 + x_2 + \cdots + x_N$$

$$\Sigma y = y_1 + y_2 + \cdots + y_N$$

$$\Sigma xy = x_1 \cdot y_1 + x_2 \cdot y_2 + \cdots + x_N \cdot y_N$$

$$\Sigma x^2 = x_1^2 + x_2^2 + \cdots + x_N^2.$$

▶ Example 2   The following table* gives the crude male death rate for lung cancer in 1950 and the per capita consumption of cigarettes in 1930 in various countries.

Country	Cigarette Consumption (Per Capita)	Lung Cancer Deaths (Per Million Males)
Norway	250	95
Sweden	300	120
Denmark	350	165
Australia	470	170

(a) Use the preceding formulas to obtain the straight line that best fits this data.

(b) In 1930 the per capita cigarette consumption in Finland was 1100. Use the straight line found in part (a) to estimate the male lung cancer death rate in Finland in 1950.

Solution   (a) The points are plotted in Fig. 3. The sums are calculated in Table 1 and then used to determine the values of A and B.

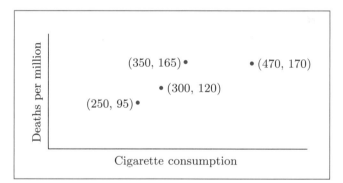

Figure 3. Lung cancer data for least-squares analysis.

Table 1	Least-Squares Calculation for Smoking Data		
x	y	xy	x^2
250	95	23,750	62,500
300	120	36,000	90,000
350	165	57,750	122,500
470	170	79,900	220,900
$\Sigma x = 1370$	$\Sigma y = 550$	$\Sigma xy = 197,400$	$\Sigma x^2 = 495,900$

*These data were obtained from *Smoking and Health*, Report of the Advisory Committee to the Surgeon General of the Public Health Service, U.S. Department of Health, Education, and Welfare, Washington, D.C., Public Health Service Publication No. 1103, p. 176.

$$A = \frac{4 \cdot 197{,}400 - 1370 \cdot 550}{4 \cdot 495{,}900 - 1370^2} = \frac{36{,}100}{106{,}700} = \frac{361}{1067} \approx .338$$

$$B = \frac{550 - \frac{361}{1067} \cdot 1370}{4} = \frac{1067 \cdot 550 - 361 \cdot 1370}{1067 \cdot 4}$$

$$= \frac{92{,}280}{4268} \approx 21.621$$

Therefore, the equation of the least-squares line is $y = .338x + 21.621$.

(b) We use the straight line to estimate the lung cancer death rate in Finland by setting $x = 1100$. Then we get

$$y = .338(1100) + 21.621 = 393.421 \approx 393.$$

Therefore, we estimate the lung cancer death rate in Finland to be 393 deaths per million males. (*Note:* The actual rate was 350.) ◆

Incorporating Technology After the x- and y-values are placed into lists on a graphing calculator [see Fig. 4(a)], the statistical routine LinReg calculates the coefficients of the least-squares line [see Fig. 4(b)]. If desired, the equation for the line can automatically be assigned to a function and graphed along with the original points [see Fig. 4(c)]. The details for these tasks appear in Appendices A–D.

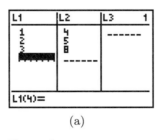
(a)

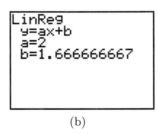
(b)

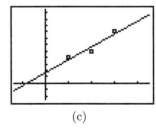
(c)

Figure 4.

Practice Problems 7.5

1. Let $E = (A + B + 2)^2 + (3A + B)^2 + (6A + B - 8)^2$. What is $\dfrac{\partial E}{\partial A}$?

2. Find the formula (of the type in Problem 1) that gives the least-squares error for the points $(1, 10)$, $(5, 8)$, and $(7, 0)$.

▶ Exercises 7.5

1. Find the least-squares error E for the least-squares line fit to the four points in Fig. 5.

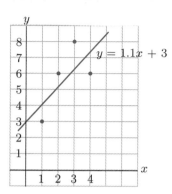

$$y = 1.1x + 3$$

Figure 5.

2. Find the least-squares error E for the least-squares line fit to the five points in Fig. 6.

3. Find the formula (of the type in Practice Problem 1) that gives the least-squares error for the points $(2, 6)$, $(5, 10)$, and $(9, 15)$.

4. Find the formula (of the type in Practice Problem 1) that gives the least-squares error for the points $(8, 4)$, $(9, 2)$, and $(10, 3)$.

In Exercises 5–8, use partial derivatives to obtain the formula for the best least-squares fit to the data points.

5. $(1, 2)$, $(2, 5)$, $(3, 11)$

6. $(1, 8)$, $(2, 4)$, $(4, 3)$

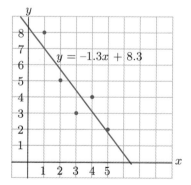

$$y = -1.3x + 8.3$$

Figure 6.

7. $(1, 9), (2, 8), (3, 6), (4, 3)$

8. $(1, 5), (2, 7), (3, 6), (4, 10)$

9. Complete Table 2 and find the values of A and B for the straight line that provides the best least-squares fit to the data.

Table 2

x	y	xy	x^2
1	7		
2	6		
3	4		
4	3		
$\Sigma x =$	$\Sigma y =$	$\Sigma xy =$	$\Sigma x^2 =$

10. Complete Table 3 and find the values of A and B for the straight line that provides the best least-squares fit to the data.

Table 3

x	y	xy	x^2
1	2		
2	4		
3	7		
4	9		
5	12		
$\Sigma x =$	$\Sigma y =$	$\Sigma xy =$	$\Sigma x^2 =$

In the remaining exercises, use one or more of the three methods discussed in this section (partial derivatives, formulas, or graphing utilities) to obtain the formula for the least-squares line.

11. Table 4* gives the U.S. per capita health care expenditures for the years 1990–1994.

   (a) Find the least-squares line for these data.

   (b) Use the least-squares line to predict the per capita health care expenditures for the year 2000.

   (c) Use the least-squares line to predict when per capita health care expenditures will reach $6000.

Table 4 — **U.S. Per Capita Health Care Expenditures**

Years (after 1990)	Dollars (in thousands)
0	2.688
1	2.902
2	3.144
3	3.331
4	3.510

*U.S. Health Care Financing Administration, *Health Care Financing Review*, Spring 1996.

12. Table 5 shows the 1994 price of a gallon (in U.S. dollars) of fuel and the average miles driven per automobile for several countries.[†]

    (a) Find the straight line that provides the best least-squares fit to these data.

    (b) In 1994, the price of gas in Japan was $4.14 per gallon. Use the straight line of part (a) to estimate the average number of miles automobiles were driven in Japan.

Table 5	Effect of Gas Prices on Miles Driven	
Country	Price per Gallon	Average Miles per Auto
Canada	$1.57	10,371
England	$2.86	10,186
France	$3.31	8740
Germany	$3.34	7674
Sweden	$3.44	7456
United States	$1.24	11,099

13. Table 6 gives the percent of persons 25 years and over who have completed four or more years of college.[‡]

Table 6	College Completion Rates					
Year	1970	1975	1980	1985	1990	1995
Percent	10.7	13.9	16.2	19.4	21.3	23.0

    (a) Use the method of least squares to obtain the straight line that best fits these data. (*Hint*: First convert *Year* to *Years after 1970*.)

    (b) Estimate the percent for the year 1993.

    (c) If the trend determined by the straight line in part (a) continues, when will the percent reach 27.1?

14. Table 7 gives the number (in millions) of cars in use in the United States[§] for certain years.

Table 7	Automobile Population		
Year	Cars	Year	Cars
1980	104.6	1991	123.3
1985	114.7	1992	120.3
1989	122.8	1993	121.1
1990	123.3	1994	122

    (a) Use the method of least squares to obtain the straight line that best fits these data. (*Hint*: First convert *Years* to *Years after 1980*.)

    (b) Estimate the number of cars in use in 1983.

    (c) If the trend determined by the straight line in part (a) continues, when will the number of cars in use reach 130 million?

15. An ecologist wished to know whether certain species of aquatic insects have their ecological range limited by temperature. He collected the data in Table 8, relating the average daily temperature at different portions of a creek with the elevation of that portion of the creek (above sea level).[*]

    (a) Find the straight line that provides the best least-squares fit to these data.

    (b) Use the linear function to estimate the average daily temperature for this creek at altitude 3.2 kilometers.

Table 8	Relationship Between Elevation and Temperature in a Creek
Elevation (kilometers)	Average Temperature (degrees Celsius)
2.7	11.2
2.8	10
3.0	8.5
3.5	7.5

---

[*]The authors express their thanks to Dr. J. David Allen, formerly of the Department of Zoology at the University of Maryland, for providing the data for this exercise.

[†]Source: Energy Information Administration, *International Energy Annual*. U.S. Highway Administration, *Highway Statistics*, 1994.

[‡]U.S. Bureau of the Census.

[§]American Automobile Manufacturers Association, Inc., Detroit, Mich., *Motor Vehicle Facts and Figures*.

**Solutions to
Practice Problems
7.5**

1. $\dfrac{\partial E}{\partial A} = 2(A + B + 2)\cdot 1 + 2(3A + B)\cdot 3 + 2(6A + B - 8)\cdot 6$

$= (2A + 2B + 4) + (18A + 6B) + (72A + 12B - 96)$

$= 92A + 20B - 92$

(Notice that we used the general power rule when differentiating and so had to always multiply by the derivative of the quantity inside the parentheses. Also, you might be tempted to first square the terms in the expression for E and then differentiate. We recommend that you resist that temptation.)

2. $E = (A + B - 10)^2 + (5A + B - 8)^2 + (7A + B)^2$. In general, E is a sum of squares, one for each point being fitted. The point (a, b) gives rise to the term $(aA + B - b)^2$.

7.6 Nonlinear Regression

In the last section, we learned how to approximate a set of data points with a linear function $y = Ax + B$. The parameters m and b were chosen to minimize the sum of the squares of the errors. Statisticians and laboratory experimentalists call this method *linear regression* rather than the *method of least squares*, the latter being the name that mathematicians prefer. The linear function $y = Ax + B$ is called the *regression function* corresponding to the given set of data.

Unfortunately, not all data sets can be approximated by regression functions as simple as the linear function $y = Ax + B$. Some sets of data require approximation by an exponential function $y = Ae^{kx}$, others by a quadratic function $y = Ax^2 + Bx + C$, and yet others by a linear function of two variables $z = Ax + By + C$. In this section, we discuss these more general forms of regression and how to apply them to real-world data.

Terminology and Notation   Suppose we have a set of data that gives the value of the dependent variable y for various values of the single independent variable x. This data can be represented as a list of data points

$$(x_1, y_1), (x_2, y_2), \ldots, (x_N, y_N).$$

Suppose that we seek to approximate this data by a regression function $\hat{y} = \hat{f}(x)$, where $\hat{f}(x)$ is chosen from a specific family of functions (linear, quadratic, exponential, etc.). For each data point, we may compute

$$\hat{y}_i = \hat{f}(x_i) \qquad (1 \leq i \leq N).$$

That is, $\hat{y}_i$ is the value of the regression function at the ith x-value x_i and is called the *predicted value* of x_i. If the regression function fits the data perfectly, then the predicted value would exactly match the observed value y_i. However, this is not usually the case, and the predicted and observed values are different. The difference

$$\hat{y}_i - y_i \qquad (1 \leq i \leq N)$$

measures by how much the predicted value differs from the observed value and is called a *residual*. The sum of the squares of the residuals gives a measure of how close the function $\hat{f}$ approximates the data set and is denoted *SSE* (shorthand for "sum of the squares of the errors"). Using this terminology, we may state the following:

Problem of Regression Analysis   Given a set of data points and a specific family of possible regression functions, select the function for which *SSE* is minimized.

In case the family of functions is the set of linear functions $y = mx + b$, this is just the problem considered in the last section. In this section, we will consider the problem for other families of functions.

Exponential Regression   Consider the set of data shown in Table 1. It appears that as the value of x increases, the value of y increases at an increasing rate. (See Fig. 1.) This is not characteristic of a linear function. However, it does suggest an exponential function $\hat{y} = Ae^{kx}$, where A and k are parameters whose values we seek. In this case, SSE is given by the expression

$$
\begin{aligned}
SSE &= (\hat{y}_1 - y_1)^2 + (\hat{y}_2 - y_2)^2 + (\hat{y}_3 - y_3)^2 + (\hat{y}_4 - y_4)^2 + (\hat{y}_5 - y_5)^2 \\
&= (Ae^{k \cdot 0} - y_1)^2 + (Ae^{k \cdot 1} - y_2)^2 + (Ae^{k \cdot 2} - y_3)^2 + (Ae^{k \cdot 3} - y_4)^2 \\
&\quad + (Ae^{k \cdot 4} - y_5)^2 \\
&= (A - 3.1)^2 + (Ae^{k} - 5.5)^2 + (Ae^{2k} - 17.6)^2 + (Ae^{3k} - 57.3)^2 \\
&\quad + (Ae^{4k} - 199.7)^2.
\end{aligned}
$$

Table 1

x	y
0	3.1
1	5.5
2	17.6
3	57.3
4	199.7

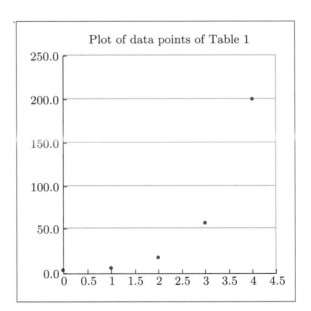

Figure 1.

If SSE has a local minimum, then we have

$$
\frac{\partial SSE}{\partial A} = 0, \qquad \frac{\partial SSE}{\partial k} = 0.
$$

By carrying out the differentiation, we get a system of two equations in unknowns A and k. By solving for A and k, we can determine the regression function $\hat{y} = Ae^{kx}$. We will omit these computations, which are quite tedious. Instead, let's consider two other approaches to determining the regression function.

For the first approach, take natural logarithms of both sides of the regression equation:

$$
\ln \hat{y} = \ln(Ae^{kx}) = \ln A + kx \ln e = \ln A + kx.
$$

Introduce the new variables $\hat{Y} = \ln \hat{y}$, $b = \ln A$. Then the relationship between $\hat{Y}$ and x is

$$
\hat{Y} = kx + b,
$$

which is a linear equation. This suggests that we transform the original problem into a linear regression problem with data points $(x_i, \hat{Y}_i) = (x_i, \ln y_i)$ for $1 \le i \le 5$, which we can solve using the methods of the preceding section. Table 2 shows the computation of the new data set. The exponential regression function is

$$\hat{Y} = 0.8749 + 1.6074x$$
$$\ln \hat{y} = 0.8749 + 1.6074x$$
$$\hat{y} = e^{0.8749 + 1.6074x}$$
$$= 2.3987 e^{1.6074x}.$$

Figure 2 shows the graph of the regression function along with the given data points.

Table 2

x	$\ln y$
0	1.131402
1	1.704748
2	2.867899
3	4.048301
4	5.296816

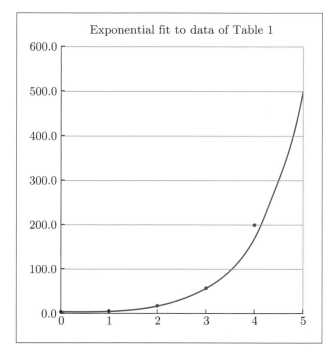

Figure 2.

The second method for finding an exponential regression function is to use a graphing calculator or spreadsheet. For example, the TI-83 has many different varieties of regression, accessed through the STAT|CALC menu. (See Fig. 3.) Select the command **ExpReg** from this menu and specify the lists containing the values of x_i and y_i, as shown in Fig. 4. (We are assuming that you entered the data into lists L1 and L2, respectively.)

Figure 3.

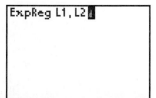

Figure 4.

Pressing **ENTER** gives the regression function as shown in Fig. 5. Note that the regression is written in the form $a \cdot b^x$, where a and b are specified on the screen. To write this as an exponential function, note that

$$a \cdot b^x = ae^{(\ln b)x}$$

so that the desired exponential function is $y = Ae^{kx}$, with $A = a$ and $k = \ln b$.

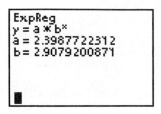

Figure 5.

Plot1 Plot2 Plot3
\Y1 = 2.3987722312*2.9079200871^X
\Y2 =
\Y3 =
\Y4 =
\Y5 =
\Y6 =
\Y7 =

Figure 6.

Moreover, the regression function is stored as Y1, as shown in Fig. 6. This makes it easy to graph the regression function along with the statistical plot showing the data points. (See Fig. 7.) Note that the TI-83 works with the original y-values rather than the transformed values $\hat{Y}_i$.

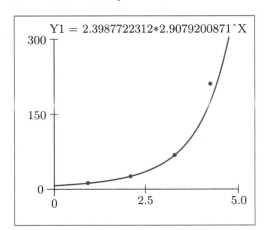

Figure 7.

Power and Logarithmic Regression   Power functions of the form $\hat{y} = Ax^b$ are often used for regression fits. For example, if you have a set of data that increases, but not so rapidly as an exponential function, a power function is a good bet. Finding the power function that minimizes SSE can be accomplished by taking partial derivatives with respect to A and b, setting them equal to 0, and solving the resulting system of equations. However, an elementary change of variables can be used to reduce a power regression problem to an elementary linear regression. Indeed, suppose that

$$\hat{y} = Ax^b.$$

Taking natural logarithms of both sides gives us

$$\ln \hat{y} = \ln A + b \ln x.$$

Make the change of variables $\hat{Y} = \ln \hat{y}$, $a = \ln A$, $\hat{X} = \ln x$; then the power regression formula has the form

$$\hat{Y} = a + b\hat{X},$$

which is a simple linear regression equation for the data

$$(\ln x_i, \ln y_i) \qquad (1 \le i \le N).$$

Using the methods of the preceding section, we can solve for a and b. Then we may determine the parameters A and b of the power regression formula.

In some applications, the data shows an increase, but an extremely slow increase. For such data, a logarithmic regression function may be used. We may fit a logarithmic function of the form $\hat{y} = A \ln x + B$ to a set of data as follows. Introduce the variable $X = \ln x$. Then the logarithmic regression function may be written in the form

$$\hat{y} = AX + B,$$

another linear equation. So we may reduce a logarithmic regression to a linear regression for the data:

$$(\ln x_i, y_i) \qquad (1 \le i \le N).$$

The parameters A and B for the linear regression are the same as the parameters for the logarithmic regression.

Note that the TI-83 has built-in commands for power regression and logarithmic regression. When using these commands, it is only necessary to enter the original data. No changes of variables are required.

In the exercises, we will show various applied situations in which power and logarithmic regression are used.

Multiple Regression   Consider the data in Table 3. The dependent variable z is a function of *two* independent variables x and y. A regression problem with more than one independent variable is called a *multiple regression*. One common form of multiple regression problem is to fit a linear function to data that is a function of several variables. For instance, in case of the data in Table 3, the linear function $z = ax + by + c$ that minimizes SSE must satisfy the three equations

$$\frac{\partial SSE}{\partial a} = 0$$

$$\frac{\partial SSE}{\partial b} = 0$$

$$\frac{\partial SSE}{\partial c} = 0.$$

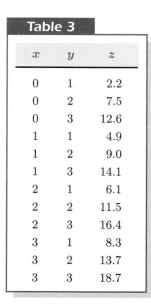

Table 3

x	y	z
0	1	2.2
0	2	7.5
0	3	12.6
1	1	4.9
1	2	9.0
1	3	14.1
2	1	6.1
2	2	11.5
2	3	16.4
3	1	8.3
3	2	13.7
3	3	18.7

Writing down these equations is mildly tedious, since SSE is a sum of twelve squares. However, the partial derivatives are easy to calculate and give rise to a system of three equations in the three variables a, b, and c. Solving this system gives us the multiple regression function $z = ax + by + c$. We leave the computations to the chapter project.

Solving large systems of linear equations can be tedious. For this reason, multiple regression problems are rarely solved manually. Instead, computer programs are used to perform the necessary computations.

Here is a sketch of how to use Excel to solve the preceding multiple regression problem. First enter the data of Table 3 into three columns of a spreadsheet. At the head of each column, include a label (e.g., "x", "y", "z"). Next give the command **Tools|Data Analysis|Regression**.* You will see a dialog box in which to insert the following information: For the X-variable range input the range of the independent variable; that is, the range containing x- and y-values. For the Y-variable range, input the range of the dependent variable; that is, the z-range. Always check the **Labels** box and include in your ranges the labels x, y, and z at the tops of the columns. Finally, give a cell address, which can serve as the upper-left corner of the report that is to be generated. The dialog box has a number of other parameters, but ignore them for now.

Once the dialog box is filled out, press ENTER and your spreadsheet will contain the report shown in Fig. 8. Note that the column labels for the independent variables are included in the last two rows of the report. This helps in interpreting the output.

Summary Output

Regression	Statistics
Multiple R	0.998419
R Square	0.99684
Adjusted F	0.996138
Standard E	0.305763
Observation	12

ANOVA

	df	SS	MS	F	Significance F
Regression	2	265.4353	132.7176	1419.58	5.6E-12
Residual	9	0.841417	0.093491		
Total	11	266.2767			

	Coefficient	standard Err	t Stat	P-value	Lower 95%	Upper 95%	Lower 95.0%	Upper 95.0%
Intercept	−2.71833	0.261839	-10.3817	2.62E-06	−3.31066	−2.12601	−3.31066	−2.12601
x	2.04	0.078948	25.83994	9.39E-10	1.861408	2.218592	1.861408	2.218592
y	5.0375	0.108103	46.59891	4.83E-12	4.792953	5.282047	4.792953	5.282047

Figure 8.

For our purposes, the important information is contained in the highlighted cells, which give the coefficients of the regression function:

$$z = 2.04x + 5.0375y - 2.71833.$$

The remaining cells of the report give data that statisticians use to determine how well the regression function fits the data. We will not discuss this information other than to say that you will definitely meet it again if you take a course in business statistics.

*Note that the Regression tool is an Excel add-in for Both Office 97 and later versions. The add-in is part of the **Data Analysis Toolpack** and needs to be loaded. Here's how. Pull down the menu **Tools|Add-ins** and check **Analysis Toolpack**. Whenever you subsequently start Excel, the Regression tool will be loaded. You can tell if you have loaded the **Data Analysis Toolpack** if the entry **Data Analysis** appears in the **Tools** menu.

Note that most graphing calculators (including the TI-83) don't support multiple regression at this time. However, as the capabilities of graphing calculators expand, it is likely that multiple regression will come within the set of their capabilities.

Polynomial Regression   In many applications, it is necessary to fit a quadratic, cubic, or higher-degree polynomial to a set of data. Take, for example, the quadratic case. Here the regression function is a quadratic function

$$y = ax^2 + bx + c$$

with three parameters a, b, c, which must be determined. To solve this problem, regard the variable x^2 as the independent variable w. Then the desired regression function is the linear function

$$y = aw + bx + c.$$

We can determine this function using multiple regression, as just discussed.

The following example illustrates the computations using Excel. Suppose that a set of x-y pairs of data are as given in Table 4. (Ignore the last column of the table for now.) Figure 9 shows a scatter plot of the data points. The picture suggests that a quadratic function might be used to fit the data.

Table 4		
y	x	x^2
0	0	0
3.1	1	1
8.5	2	4
12.4	3	9
18.9	4	16
26.8	5	25
35	6	36
43.7	7	49
53.6	8	64
62.9	9	81
75.3	10	100

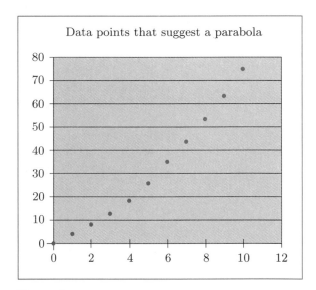

Figure 9.

Here is how to find the regression function using Excel. Since we are looking for a quadratic regression function, we use independent variables x and x^2. To tabulate the values of the second of these, we add an x^2 column to our data table and compute the entries, as shown in Table 4. The data for the independent variables includes both the columns for x and x^2. The data for the dependent variable is the y-column as before. The regression report is shown in Fig. 10. Note that the highlighted cells give us the coefficients of the regression function, namely,

$$y = 0.435548x^2 + 3.187249x - 0.25315.$$

Summary Output	
Regression	Statistics
Multiple R	0.999782
R Square	0.999565
Adjusted F	0.999456
Standard E	0.591269
Observation	11

ANOVA

	df	SS	MS	F	Significance F
Regression	2	6420.965	3210.483	9183.327	3.59E-14
Residual	8	2.796793	0.349599		
Total	11	6423.762			

	Coefficient	standard Err	t Stat	P-value	Lower 95%	Upper 95%	Lower 95.0%	Upper 95.0%
Intercept	−0.25315	0.45046	-0.56197	0.589527	−1.29191	0.785616	−1.29191	0.785616
x	3.187249	0.20958	15.20776	3.469E-07	2.703956	3.670543	2.703956	3.670543
x^2	0.435548	0.020186	21.57716	2.24E-08	0.389	0.482096	0.389	0.482096

Figure 10.

The regression function is graphed along with the data points in Fig. 11. We see that the choice of a quadratic regression function was definitely a good one. After you have learned some statistics, you can give a quantitative meaning to the word "good."

The TI-83 has quadratic, cubic, and quartic regression built in. These are accessed on the menu **STAT|CALC** using the commands **QuadReg**, **CubicReg**,

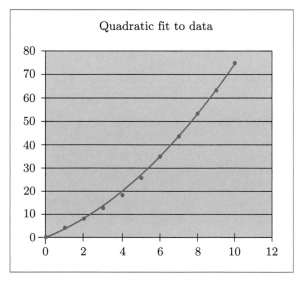

Figure 11.

Figure 12.

and **QuartReg**, respectively. Note that on the TI-83, it is not necessary to enter the data column for x^2; the calculator computes the regression function using only the x- and y-lists as input. Let's rework the preceding quadratic regression problem using the TI-83.

First, we enter the data into lists as shown in Fig. 12. Then we give the command **STAT|CALC|QuadReg**, as shown in Fig. 13. This will cause the command **QuadReg** to be displayed on the main window. We complete the command by adding the lists containing the x- and y-variables (x in L1 and y in L2), as shown in Fig. 14. We press ENTER and the calculator shows the regression report shown in Fig. 15.

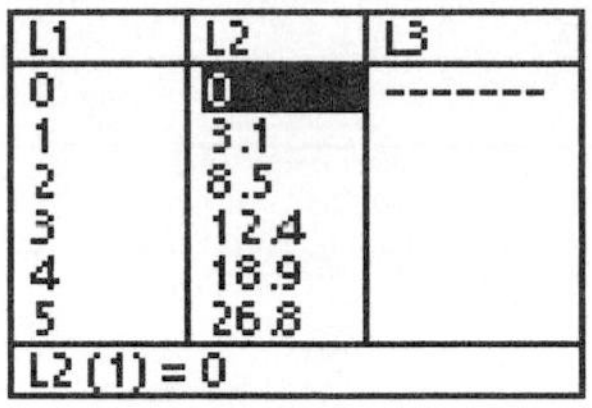

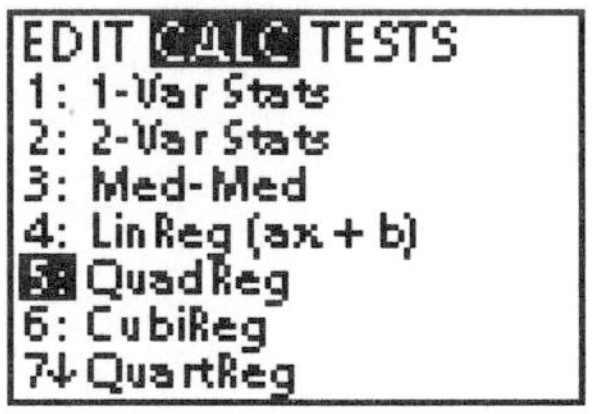

Figure 13.

Choosing a Model In our discussions, we showed how to fit various types of functions to data. The possible regression functions include linear and polynomial functions, as well as exponential and logarithmic functions. However, we sidestepped the issue of how to decide which sort of function you should choose for a particular set of data. This question is considered in detail in a course on mathematical modeling or a course in statistics. For our purposes, we will rely on an intuitive analysis of a scatter plot of the data to suggest the shape of a curve fitting the data. Based on the general shape, we can then choose an appropriate regression function.

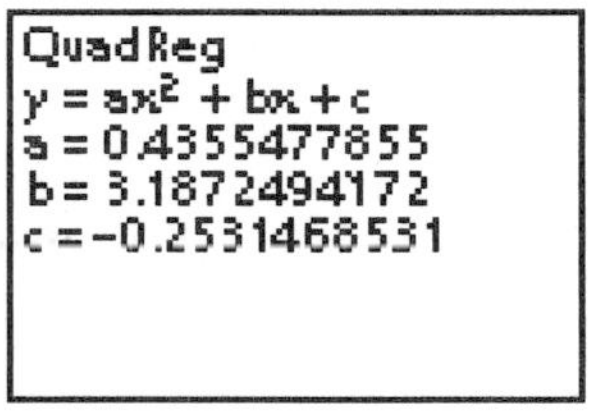

Figure 14.

An intuitive analysis is not always definitive. For example, consider data leading to a curve as shown in Fig. 16. The graph shows one definite local maximum and a probable local minimum, as well as an inflection point. A cubic curve can have all of these characteristics. However, so can a curve of fourth, fifth, or higher degree. Which type of regression function should you choose? For our purposes, use the following principle: All other things being equal, choose the simplest regression function. So if going to a higher-degree polynomial doesn't buy you a significantly better fit to the data, stick with the cubic curve. In judging the fit to the data, just let your eyes be your guide, at least until you learn about quantitative measure of fit in a more advanced course.

Figure 15.

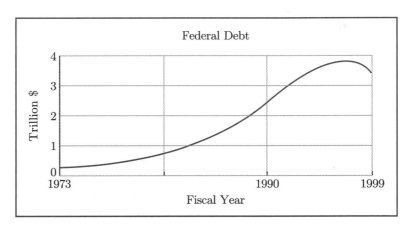

Figure 16.

Practice Problem 7.6	1. Suppose that $SSE = 0$ for a particular regression function fitting a set of data. What conclusion can you draw?

▶ Exercises 7.6

In Exercises 1–8, do the following.

(a) Determine the regression function of the specified type that best fits the given data.

(b) In the case of simple regression, graph the data points and the regression function on the same coordinate system.

1. Exponential regression

x	y
0	2.4
1	3.6
2	5
3	7.2
4	12.1
5	17.5

2. Exponential regression

x	y
0	2009
1	856
2	366
3	156.8
4	66.8
5	28.5

3. Multiple linear regression

x	y	z
−1	−1	7.9
−1	0	6.9
−1	1	5.6
0	−1	11.7
0	0	11
0	1	9.4
1	−1	15.6
1	0	14.3
1	1	13.4

4. Multiple linear regression

x	y	z
−1	−1	−15.1
−1	0	−9.5
−1	1	−5.1
0	−1	−4.7
0	0	0.1
0	1	5.9
1	−1	6.1
1	0	11.6
1	1	15.5

5. Quadratic regression

x	y
1	7.7
2	17
3	32.1
4	48.5
5	71.1
6	98.1
7	127.7
8	160.2
9	198.5

6. Quadratic regression

x	y
1	0.7
2	4.7
3	0.2
4	−0.6
5	−6.5
6	−9
7	−14.4
8	−19
9	−30.2

7. Quadratic regression

x	y	x	y
−5	0.5	0	3.6
−4	3.6	1	1.1
−3	5.1	2	−2.3
−2	5.7	3	−7.1
−1	5.2	4	−12.8
		5	−19.3

8. Exponential regression

x	y	x	y
0	2	3	0.45
0.5	1.56	3.5	0.35
1	1.21	4	0.27
1.5	0.94	4.5	0.21
2	0.74	5	0.16
2.5	0.57	5.5	0.13

In Exercises 9–12, determine the regression function of the specified type that best fits the given data. You may use any suitable technology. The data is available in Excel spreadsheet form on the Web site (**http://www.prenhall.com/goldstein**).

9. Exponential regression

x = Age (years)	y = Total Expected Deaths per 1000 Alive at Specified Age (for 1995)	x = Age (years)	y = Total Expected Deaths per 1000 Alive at Specified Age (for 1995)	x = Age (years)	y = Total Expected Deaths per 1000 Alive at Specified Age (for 1995)
0	75.8	24	53.1	48	31.1
1	75.4	25	52.2	49	30.2
2	74.4	26	51.2	50	29.3
3	73.4	27	50.3	51	28.5
4	72.5	28	49.3	52	27.6
5	71.5	29	48.4	53	26.8
6	70.5	30	47.5	54	25.9
7	69.5	31	46.5	55	25.1
8	68.5	32	45.6	56	24.3
9	67.5	33	44.7	57	23.5
10	66.6	34	43.8	58	22.7
11	65.6	35	42.8	59	21.9
12	64.6	36	41.9	60	21.1
13	63.6	37	41	61	20.4
14	62.6	38	40.1	62	19.6
15	61.6	39	39.2	63	18.9
16	60.7	40	38.3	64	18.2
17	59.7	41	37.3	65	17.4
18	58.8	42	36.4	70	14.1
19	57.8	43	35.5	75	11
20	56.9	44	34.6	80	8.3
21	55.9	45	33.8	85	6
22	55	46	32.9		
23	54.1	47	32		

10. Exponential regression

Year	x = Elapsed Time Since 1970 (years)	y = Shares Traded on the New York Stock Exchange (millions)
1970	0	3,124
1975	5	4,839
1980	10	11,562
1981	11	12,049
1982	12	16,669
1983	13	21,846
1984	14	23,309
1985	15	27,774
1986	16	36,009
1987	17	48,143
1988	18	41,118
1989	19	42,022
1990	20	39,946
1991	21	45,599
1992	22	51,826
1993	23	67,461
1994	24	74,003
1995	25	87,873
1996	26	105,477
1997	27	134,404

11. Quadratic regression

x = Year	y = Annual Production of Audio Cassettes
1982	182.3
1983	236.8
1984	332
1985	339.1
1986	344.5
1987	410
1988	450.1
1989	446.2
1990	442.2
1991	360.1
1992	366.4
1993	339.5
1994	345.4
1995	272.6
1996	225.3
1997	172.6

12. Quadratic regression

x = Year	y = Total Number of Pieces of U.S. Mail (millions)
1990	3340
1992	3497
1993	3672
1994	3781
1995	3865
1996	4020
1997	4115

Solution to Practice Problem 7.6	**1.** Each of the terms $(\hat{y}_i - y_i)^2$ must equal 0, since each is nonnegative and their sum is 0. Therefore, $\hat{y}_i = y_i$ for $1, 2, \ldots, N$. That is, the regression curve passes through each of the data points.

7.7 Double Integrals

Up to this point, our discussion of calculus of several variables has been confined to the study of differentiation. Let us now take up the topic of integration of functions of several variables. As has been the case throughout most of this chapter, we restrict our discussion to functions $f(x, y)$ of two variables.

Let us begin with some motivation. Before we define the concept of an integral for functions of several variables, let us review the essential features of the integral in one variable.

Consider the definite integral $\int_a^b f(x)\, dx$. To write down this integral takes two pieces of information. The first is the function $f(x)$. The second is the interval over which the integration is to be performed. In this case, the interval is the portion of the x-axis from $x = a$ to $x = b$. The value of the definite integral is a number. In case the function $f(x)$ is nonnegative throughout the interval from $x = a$ to $x = b$, this number equals the area under the graph of $f(x)$ from $x = a$ to $x = b$. (See Fig. 1.) If $f(x)$ is negative for some values of x in the interval, then the integral still equals the area bounded by the graph, but areas below the x-axis are counted as negative.

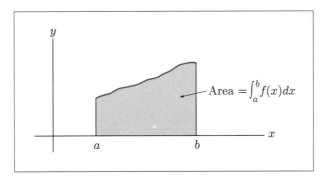

Figure 1.

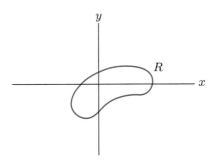

Figure 2. A region in the xy-plane.

Let us generalize the foregoing ingredients to a function $f(x, y)$ of two variables. First, we must provide a two-dimensional analog of the interval from $x = a$ to $x = b$. This is easy. We take a two-dimensional region R of the plane, such as the region shown in Fig. 2. As our generalization of $f(x)$, we take a function $f(x, y)$ of two variables. Our generalization of the definite integral is denoted

$$\iint_R f(x, y)\, dx\, dy$$

and is called the *double integral of* $f(x, y)$ *over the region* R. The value of the double integral is a number defined as follows. For the sake of simplicity, let us begin by assuming that $f(x, y) \geq 0$ for all points (x, y) in the region R. [This is the analog of the assumption that $f(x) \geq 0$ for all x in the interval from $x = a$ to $x = b$.] This means that the graph of f lies above the region R in three-dimensional space. (See Fig. 3.) The portion of the graph over R determines a solid figure. (See Fig. 4.) This figure is called the *solid bounded by* $f(x, y)$ *over the region* R. We define the double integral $\iint_R f(x, y)\, dx\, dy$ to be the volume of this solid. In case the graph of $f(x, y)$ lies partially above the region R and partially below, we define the double integral to be the volume of the solid above the region minus the volume of the solid below the region. That is, we count volumes below the xy-plane as negative.

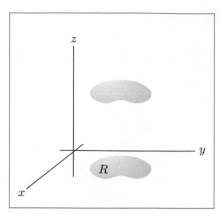

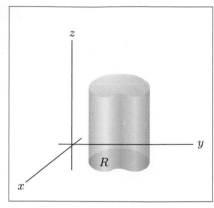

Figure 3. Graph of f(x,y) above the region R.

Figure 4. Solid bounded by f(x,y) over R.

Now that we have defined the notion of a double integral, we must learn how to calculate its value. To do so, let us introduce the notion of an iterated integral. Let $f(x, y)$ be a function of two variables, let $g(x)$ and $h(x)$ be two functions of x alone, and let a and b be numbers. Then an *iterated integral* is an expression of the form

$$\int_a^b \left(\int_{g(x)}^{h(x)} f(x, y)\, dy \right) dx.$$

To explain the meaning of this collection of symbols, we proceed from the inside out. We evaluate the integral

$$\int_{g(x)}^{h(x)} f(x, y)\, dy$$

by considering $f(x, y)$ as a function of y alone. This is indicated by the dy in the inner integral. We treat x as a constant in this integration. So we evaluate the integral by first finding an antiderivative $F(x, y)$ with respect to y. The integral above is then evaluated as

$$F(x, h(x)) - F(x, g(x)).$$

That is, we evaluate the antiderivative between the limits $y = g(x)$ and $y = h(x)$. This gives us a function of x alone. To complete the evaluation of the integral, we integrate this function from $x = a$ to $x = b$. The next two examples illustrate the procedure for evaluating iterated integrals.

▶ Example 1   Evaluate the iterated integral $\int_1^2 \left(\int_3^4 (y - x)\, dy \right) dx.$

Solution   Here $g(x)$ and $h(x)$ are constant functions: $g(x) = 3$ and $h(x) = 4$. We evaluate the inner integral first. The variable in this integral is y, so we treat x as a constant.

$$\int_3^4 (y - x)\, dy = \left(\frac{1}{2}y^2 - xy\right)\Big|_3^4$$

$$= \left(\frac{1}{2}\cdot 16 - x\cdot 4\right) - \left(\frac{1}{2}\cdot 9 - x\cdot 3\right)$$

$$= 8 - 4x - \frac{9}{2} + 3x$$

$$= \frac{7}{2} - x$$

Now we carry out the integration with respect to x:

$$\int_1^2 \left(\frac{7}{2} - x\right) dx = \frac{7}{2}x - \frac{1}{2}x^2\Big|_1^2$$

$$= \left(\frac{7}{2}\cdot 2 - \frac{1}{2}\cdot 4\right) - \left(\frac{7}{2} - \frac{1}{2}\cdot 1\right)$$

$$= (7 - 2) - (3) = 2.$$

So the value of the iterated integral is 2. ◆

▶ **Example 2**   Evaluate the iterated integral $\int_0^1 \left(\int_{\sqrt{x}}^{x+1} 2xy\, dy\right) dx$.

Solution   We evaluate the inner integral first.

$$\int_{\sqrt{x}}^{x+1} 2xy\, dy = xy^2\Big|_{\sqrt{x}}^{x+1} = x(x+1)^2 - x(\sqrt{x})^2$$

$$= x(x^2 + 2x + 1) - x\cdot x$$

$$= x^3 + 2x^2 + x - x^2$$

$$= x^3 + x^2 + x$$

Now we evaluate the outer integral.

$$\int_0^1 (x^3 + x^2 + x)\, dx = \frac{1}{4}x^4 + \frac{1}{3}x^3 + \frac{1}{2}x^2\Big|_0^1 = \frac{1}{4} + \frac{1}{3} + \frac{1}{2} = \frac{13}{12}$$

So the value of the iterated integral is $\frac{13}{12}$. ◆

Let us now return to the discussion of the double integral $\iint_R f(x, y)\, dx\, dy$. When the region R has a special form, the double integral may be expressed as an iterated integral. Namely, let us suppose that R is bounded by the graphs of $y = g(x)$, $y = h(x)$ and by the vertical lines $x = a$ and $x = b$. (See Fig. 5.) In this case, we have the following fundamental result, which we cite without proof.

Let R be the region in the xy-plane bounded by the graphs of $y = g(x)$, $y = h(x)$, and the vertical lines $x = a$, $x = b$. Then

$$\iint_R f(x, y)\, dx\, dy = \int_a^b \left(\int_{g(x)}^{h(x)} f(x, y)\, dy\right) dx.$$

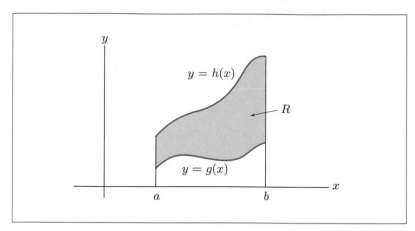

Figure 5.

Since the value of the double integral gives the volume of the solid bounded by the graph of $f(x, y)$ over the region R, the preceding result may be used to calculate volumes, as the next two examples show.

▶ **Example 3**   Calculate the volume of the solid bounded above by the function $y - x$ and lying over the rectangular region R: $1 \leq x \leq 2$, $3 \leq y \leq 4$. (See Fig. 6.)

Solution   The desired volume is given by the double integral $\iint\limits_{R} (y - x)\, dx\, dy$. By the result just cited, this double integral is equal to the iterated integral

$$\int_1^2 \left(\int_3^4 (y - x)\, dy \right) dx.$$

The value of this iterated integral was shown in Example 1 to be 2, so the volume of the solid shown in Fig. 6 is 2.   ◆

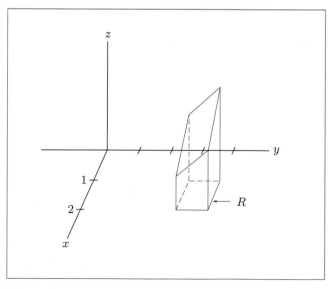

Figure 6.

▶ Example 4   Calculate $\iint\limits_R 2xy\,dx\,dy$, where R is the region shown in Fig. 7.

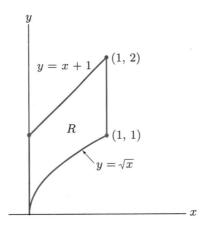

Figure 7.

Solution   The region R is bounded below by $y = \sqrt{x}$, above by $y = x + 1$, on the left by $x = 0$, and on the right by $x = 1$. Therefore,

$$\iint\limits_R 2xy\,dx\,dy = \int_0^1 \left(\int_{\sqrt{x}}^{x+1} 2xy\,dy \right) dx = \frac{13}{12} \quad \text{(by Example 2).} \qquad \blacklozenge$$

In our discussion, we have confined ourselves to iterated integrals in which the inner integral was with respect to y. In a completely analogous manner, we may treat iterated integrals in which the inner integral is with respect to x. Such iterated integrals may be used to evaluate double integrals over regions R bounded by curves of the form $x = g(y)$, $x = h(y)$, and horizontal lines $y = a$, $y = b$. The computations are analogous to those given in this section.

Practice Problems 7.7

1. Calculate the iterated integral

$$\int_0^2 \left(\int_0^{x/2} e^{2y-x}\,dy \right) dx.$$

2. Calculate

$$\iint\limits_R e^{2y-x}\,dx\,dy$$

where R is the region in Fig. 8.

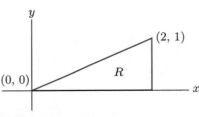

Figure 8.

▶ Exercises 7.7

Calculate the following iterated integrals.

1. $\displaystyle\int_0^1 \left(\int_0^1 e^{x+y}\, dy \right) dx$

2. $\displaystyle\int_{-1}^1 \left(\int_0^2 (3x^3 + y^2)\, dy \right) dx$

3. $\displaystyle\int_{-2}^0 \left(\int_{-1}^1 xe^{xy}\, dy \right) dx$

4. $\displaystyle\int_1^3 \left(\int_2^5 (2y - 3x)\, dy \right) dx$

5. $\displaystyle\int_1^4 \left(\int_x^{x^2} xy\, dy \right) dx$

6. $\displaystyle\int_0^2 \left(\int_x^{5x} y\, dy \right) dx$

7. $\displaystyle\int_{-1}^1 \left(\int_x^{2x} (x + y)\, dy \right) dx$

8. $\displaystyle\int_0^1 \left(\int_0^x (x + e^y)\, dy \right) dx$

Let R be the rectangle consisting of all points (x, y) such that $0 \le x \le 2$, $2 \le y \le 3$. Calculate the following double integrals. Interpret each as a volume.

9. $\displaystyle\iint_R xy^2\, dx\, dy$

10. $\displaystyle\iint_R (xy + y^2)\, dx\, dy$

11. $\displaystyle\iint_R e^{-x-y}\, dx\, dy$

12. $\displaystyle\iint_R e^{y-x}\, dx\, dy$

Calculate the volumes over the following regions R bounded above by the graph of $f(x, y) = x^2 + y^2$.

13. R is the rectangle bounded by the lines $x = 1$, $x = 3$, $y = 0$, $y = 1$.

14. R is the region bounded by the lines $x = 0$, $x = 1$ and the curves $y = 0$ and $y = \sqrt[3]{x}$.

Solutions to Practice Problems 7.7	

1. $\displaystyle\int_0^2 \left(\int_0^{x/2} e^{2y-x}\, dy \right) dx = \int_0^2 \left(\frac{1}{2} e^{2y-x} \Big|_0^{x/2} \right) dx$

$\displaystyle = \int_0^2 \left(\frac{1}{2} e^{2(x/2)-x} - \frac{1}{2} e^{2(0)-x} \right) dx$

$\displaystyle = \int_0^2 \left(\frac{1}{2} - \frac{1}{2} e^{-x} \right) dx$

$\displaystyle = \frac{1}{2} x + \frac{1}{2} e^{-x} \Big|_0^2$

$\displaystyle = \frac{1}{2} \cdot 2 + \frac{1}{2} e^{-2} - \left(\frac{1}{2} \cdot 0 + \frac{1}{2} e^{-0} \right)$

$\displaystyle = 1 + \frac{1}{2} e^{-2} - 0 - \frac{1}{2}$

$\displaystyle = \frac{1}{2} + \frac{1}{2} e^{-2}$

2. The line passing through the points $(0, 0)$ and $(2, 1)$ has equation $y = x/2$. Hence the region R is bounded below by $y = 0$, above by $y = x/2$, on the left by $x = 0$, and on the right by $x = 2$. Therefore,

$$\iint_R e^{2y-x}\, dx\, dy = \int_0^2 \left(\int_0^{x/2} e^{2y-x}\, dy \right) dx = \frac{1}{2} + \frac{1}{2} e^{-2}$$

by Problem 1.

Review of Fundamental Concepts of Chapter 7

1. Give an example of a level curve of a function of two variables.

2. Explain how to find a first partial derivative of a function of two variables.

3. Explain how to find a second partial derivative of a function of two variables.

4. What expression involving a partial derivative gives an approximation to $f(a+h,b) - f(a,b)$?

5. Interpret $\dfrac{\partial f}{\partial y}(2,3)$ as a rate of change.

6. Give an example of a Cobb-Douglas production function. What is the marginal productivity of labor? Of capital?

7. Explain how to find possible relative extreme points for a function of several variables.

8. State the second-derivative test for functions of two variables.

9. Outline how the method of Lagrange multipliers is used to solve an optimization problem.

10. What is the least-squares line approximation to a set of data points? How is the line determined?

11. Give a geometric interpretation to $\iint\limits_{R} f(x,y)\,dx\,dy$, where $f(x,y) \geq 0$.

12. Give a formula for evaluating a double integral in terms of an iterated integral.

▶ Chapter 7 Supplementary Exercises

1. Let $f(x,y) = x\sqrt{y}/(1+x)$. Compute $f(2,9)$, $f(5,1)$, and $f(0,0)$.

2. Let $f(x,y,z) = x^2 e^{y/z}$. Compute $f(-1,0,1)$, $f(1,3,3)$, and $f(5,-2,2)$.

3. If A dollars are deposited in a bank (assuming a 6% continuous interest rate), the amount in the account after t years is $f(A,t) = Ae^{.06t}$. Find and interpret $f(10, 11.5)$.

4. Let $f(x,y,\lambda) = xy + \lambda(5 - x - y)$. Find $f(1,2,3)$.

5. Let $f(x,y) = 3x^2 + xy + 5y^2$. Find $\dfrac{\partial f}{\partial x}$ and $\dfrac{\partial f}{\partial y}$.

6. Let $f(x,y) = 3x - \frac{1}{2}y^4 + 1$. Find $\dfrac{\partial f}{\partial x}$ and $\dfrac{\partial f}{\partial y}$.

7. Let $f(x,y) = e^{x/y}$. Find $\dfrac{\partial f}{\partial x}$ and $\dfrac{\partial f}{\partial y}$.

8. Let $f(x,y) = x/(x-2y)$. Find $\dfrac{\partial f}{\partial x}$ and $\dfrac{\partial f}{\partial y}$.

9. Let $f(x,y,z) = x^3 - yz^2$. Find $\dfrac{\partial f}{\partial x}$, $\dfrac{\partial f}{\partial y}$, and $\dfrac{\partial f}{\partial z}$.

10. Let $f(x,y,\lambda) = xy + \lambda(5 - x - y)$. Find $\dfrac{\partial f}{\partial x}$, $\dfrac{\partial f}{\partial y}$, and $\dfrac{\partial f}{\partial \lambda}$.

11. Let $f(x,y) = x^3 y + 8$. Compute $\dfrac{\partial f}{\partial x}(1,2)$ and $\dfrac{\partial f}{\partial y}(1,2)$.

12. Let $f(x,y,z) = (x+y)z$. Evaluate $\dfrac{\partial f}{\partial y}$ at $(x,y,z) = (2,3,4)$.

13. Let $f(x,y) = x^5 - 2x^3 y + \frac{1}{2}y^4$. Find $\dfrac{\partial^2 f}{\partial x^2}$, $\dfrac{\partial^2 f}{\partial y^2}$, $\dfrac{\partial^2 f}{\partial x\,\partial y}$, and $\dfrac{\partial^2 f}{\partial y\,\partial x}$.

14. Let $f(x,y) = 2x^3 + x^2 y - y^2$. Compute $\dfrac{\partial^2 f}{\partial x^2}$, $\dfrac{\partial^2 f}{\partial y^2}$, and $\dfrac{\partial^2 f}{\partial x\,\partial y}$ at $(x,y) = (1,2)$.

15. A dealer in a certain brand of electronic calculator finds that (within certain limits) the number of calculators she can sell per week is given by $f(p,t) = -p + 6t - .02pt$, where p is the price of the calculator and t is the number of dollars spent on advertising. Compute $\dfrac{\partial f}{\partial p}(25, 10{,}000)$ and $\dfrac{\partial f}{\partial t}(25, 10{,}000)$ and interpret these numbers.

16. Suppose that the crime rate in a certain city can be approximated by a function $f(x,y,z)$, where x is the unemployment rate, y is the amount of social services available, and z is the size of the police force. Explain why $\dfrac{\partial f}{\partial x} > 0$, $\dfrac{\partial f}{\partial y} < 0$, and $\dfrac{\partial f}{\partial z} < 0$.

In Exercises 17–20, find all points (x,y) where $f(x,y)$ has a possible relative maximum or minimum.

17. $f(x,y) = -x^2 + 2y^2 + 6x - 8y + 5$

18. $f(x,y) = x^2 + 3xy - y^2 - x - 8y + 4$

19. $f(x,y) = x^3 + 3x^2 + 3y^2 - 6y + 7$

20. $f(x,y) = \frac{1}{2}x^2 + 4xy + y^3 + 8y^2 + 3x + 2$

In Exercises 21–23, find all points (x,y) where $f(x,y)$ has a possible relative maximum or minimum. Then use the second-derivative test to determine, if possible, the nature of $f(x,y)$ at each of these points. If the second-derivative test is inconclusive, so state.

21. $f(x,y) = x^2 + 3xy + 4y^2 - 13x - 30y + 12$

22. $f(x,y) = 7x^2 - 5xy + y^2 + x - y + 6$

23. $f(x, y) = x^3 + y^2 - 3x - 8y + 12$

24. Find the values of x, y, z at which

$$f(x, y, z) = x^2 + 4y^2 + 5z^2 - 6x + 8y + 3$$

assumes its minimum value.

25. Using the method of Lagrange multipliers, maximize the function $3x^2 + 2xy - y^2$, subject to the constraint $5 - 2x - y = 0$.

26. Using the method of Lagrange multipliers, find the values of x, y that minimize the function $-x^2 - 3xy - \frac{1}{2}y^2 + y + 10$, subject to the constraint $10 - x - y = 0$.

27. Using the method of Lagrange multipliers, find the values of x, y, z that minimize the function $3x^2 + 2y^2 + z^2 + 4x + y + 3z$, subject to the constraint $4 - x - y - z = 0$.

28. Using the method of Lagrange multipliers, find the dimensions of the rectangular box of volume 1000 cubic inches for which the sum of the dimensions is minimized.

29. A person wants to plant a rectangular garden along one side of a house and put a fence on the other three sides of the garden. (See Fig. 1.) Using the method of Lagrange multipliers, find the dimensions of the garden of greatest area that can be enclosed by using 40 feet of fencing.

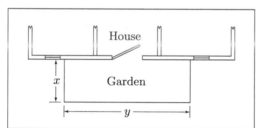

Figure 1. A garden.

30. The solution to Exercise 29 is $x = 10$, $y = 20$, $\lambda = 10$. Suppose that one additional foot of fencing becomes available. Compute the new optimal dimensions and the new area, and show that the increase in area (compared with the area in Exercise 29) is approximately equal to 10 (the value of λ).

In Exercises 31–33, find the straight line that best fits the following data points, where "best" is meant in the sense of least squares.

31. $(1, 1)$, $(2, 3)$, $(3, 6)$

32. $(1, 1)$, $(3, 4)$, $(5, 7)$

33. $(0, 1)$, $(1, -1)$, $(2, -3)$, $(3, -5)$

In Exercises 34 and 35, calculate the iterated integral.

34. $\int_0^1 \left(\int_0^4 (x\sqrt{y} + y)\, dy \right) dx$

35. $\int_0^5 \left(\int_1^4 (2xy^4 + 3)\, dy \right) dx$

In Exercises 36 and 37, let R be the rectangle consisting of all points (x, y) such that $0 \le x \le 4$, $1 \le y \le 3$, and calculate the double integral.

36. $\iint_R (2x + 3y)\, dx\, dy$ **37.** $\iint_R 5\, dx\, dy$

38. The present value of y dollars after x years at 15% continuous interest is $f(x, y) = ye^{-.15x}$. Sketch some sample level curves. (Economists call this collection of level curves a *discount system*.)

Chapter Project

In this project, we will explore multiple regression from a calculus viewpoint. The accompanying table gives data points (x, y, z), where z is a function of x and y.

Table 1	Data on which to Perform Multiple Regression	
x	y	z
0	1	2.2
0	2	7.5
0	3	12.6
1	1	4.9
1	2	9.0
1	3	14.1
2	1	6.1
2	2	11.5
2	3	16.4
3	1	8.3
3	2	13.7
3	3	18.7

Assume that we wish to fit a linear regression function of two variables to this data. The general form of such a function is

$$z = ax + by + c.$$

(a) Determine a formula for SSE for this regression problem, in terms of a, b, c.

(b) Compute the partial derivatives $\dfrac{\partial SSE}{\partial a}, \dfrac{\partial SSE}{\partial b}, \dfrac{\partial SSE}{\partial c}$.

(c) By setting the partial derivatives equal to 0, show that for SSE to be a minimum, the parameters a, b, c must satisfy a certain system of three linear equations. Explicitly determine this system of equations.

(d) Solve the system of equations of part (c). You may do this manually or using any technology that is available to you.

(e) Use the results of part (d) to determine the regression function z.

(f) Derive the regression function z, using the regression package in Excel. Check that Excel gives the same regression function as obtained in part (e).

CHAPTER

The Trigonometric Functions

8

I n this chapter we expand the collection of functions to which we can apply calculus by introducing the trigonometric functions. As we shall see, these functions are *periodic*. That is, after a certain point their graphs repeat themselves. This repetitive phenomenon is not displayed by any of the functions that we have considered until now. Yet many natural phenomena are repetitive or cyclical—for example, the motion of the planets in our solar system, earthquake vibrations, and the natural rhythm of the heart. Thus the functions introduced in this chapter add considerably to our capacity to describe physical processes.

8.1 Radian Measure of Angles

The ancient Babylonians introduced angle measurement in terms of degrees, minutes, and seconds, and these units are still generally used today for navigation and practical measurements. In calculus, however, it is more convenient to measure angles in terms of *radians*, for in this case the differentiation formulas for the trigonometric functions are easier to remember and use. Also, the radian is becoming more widely used today in scientific work because it is the unit of angle measurement in the international metric system (Système Internationale des Unités).

In order to define a radian, we consider a circle of radius 1 and measure angles in terms of distances around the circumference. The central angle determined by an arc of length 1 along the circumference is said to have a measure of *one radian*. (See Fig. 1.) Since the circumference of the circle of radius 1 has length 2π, there are 2π radians in one full revolution of the circle. Equivalently,

$$360° = 2\pi \text{ radians.} \tag{1}$$

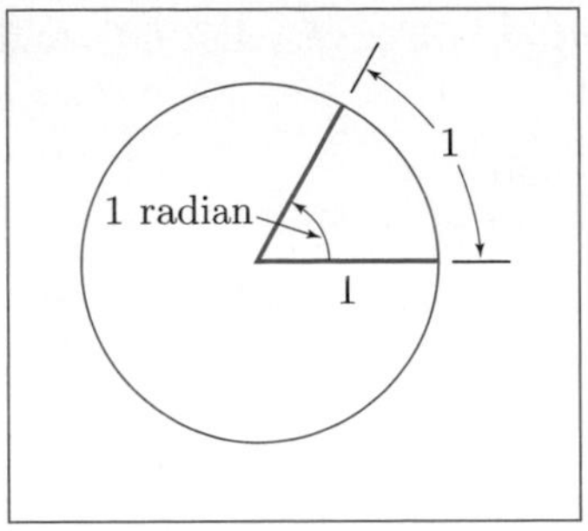

Figure 1.

The following important relations should be memorized (see Fig. 2):

$$90° = \frac{\pi}{2} \text{ radians} \qquad \text{(one quarter-revolution)}$$

$$180° = \pi \text{ radians} \qquad \text{(one half-revolution)}$$

$$270° = \frac{3\pi}{2} \text{ radians} \qquad \text{(three quarter-revolutions)}$$

$$360° = 2\pi \text{ radians} \qquad \text{(one full revolution)}.$$

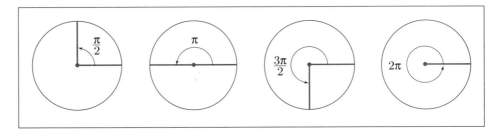

Figure 2.

From formula (1) we see that

$$1° = \frac{2\pi}{360} \text{ radians} = \frac{\pi}{180} \text{ radians}.$$

If d is any number, then

$$d° = d \times \frac{\pi}{180} \text{ radians.} \qquad (2)$$

That is, in order to convert degrees to radians, multiply the number of degrees by $\pi/180$.

▶ Example 1   Convert 45°, 60°, and 135° to radians.

Solution

$$45° = \overset{1}{\cancel{45}} \times \frac{\pi}{\underset{4}{\cancel{180}}} \text{ radians} = \frac{\pi}{4} \text{ radians}$$

$$60° = \overset{1}{\cancel{60}} \times \frac{\pi}{\underset{3}{\cancel{180}}} \text{ radians} = \frac{\pi}{3} \text{ radians}$$

$$135° = \overset{3}{\cancel{135}} \times \frac{\pi}{\underset{4}{\cancel{180}}} \text{ radians} = \frac{3\pi}{4} \text{ radians}$$

These three angles are shown in Fig. 3.   ◆

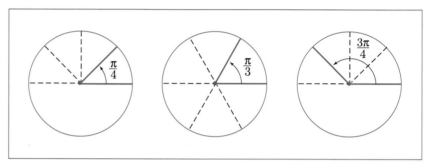

Figure 3.

We usually omit the word "radian" when measuring angles because all our angle measurements will be in radians unless degrees are specifically indicated.

For our purposes, it is important to be able to speak of negative as well as positive angles, so let us define what we mean by a negative angle. We shall usually consider angles that are in *standard position* on a coordinate system, with the vertex of the angle at $(0, 0)$ and one side, called the "initial side," along the positive x-axis. We measure such an angle from the initial side to the "terminal side," where a *counterclockwise angle is positive* and a *clockwise angle is negative*. Some examples are given in Fig. 4.

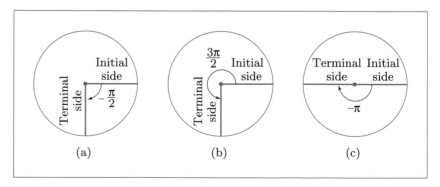

Figure 4.

Notice in Fig. 4(a) and (b) how essentially the same picture can describe more than one angle.

By considering angles formed from more than one revolution (in the positive or negative direction), we can construct angles whose measure is of arbitrary size (i.e., not necessarily between -2π and 2π). Some examples are illustrated in Fig. 5.

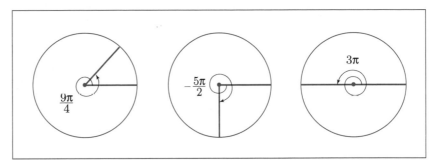

Figure 5.

▶ Example 2   (a) What is the radian measure of the angle in Fig. 6?

            (b) Construct an angle of $5\pi/2$ radians.

Solution   (a) The angle described in Fig. 6 consists of one full revolution (2π radians) plus three quarter-revolutions [$3 \times (\pi/2)$ radians]. That is,

$$t = 2\pi + 3 \times \frac{\pi}{2} = 4 \times \frac{\pi}{2} + 3 \times \frac{\pi}{2} = \frac{7\pi}{2}.$$

            (b) Think of $5\pi/2$ radians as $5 \times (\pi/2)$ radians—that is, five quarter-revolutions of the circle. This is one full revolution plus one quarter-revolution. An angle of $5\pi/2$ radians is shown in Fig. 7.   ◆

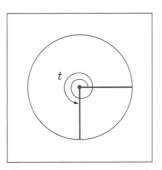

Figure 6.

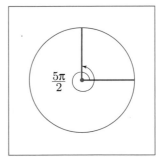

Figure 7.

Practice Problems 8.1

1. A right triangle has one angle of $\pi/3$ radians. What are the other angles?

2. How many radians are there in an angle of $-780°$? Draw the angle.

▶ Exercises 8.1

Convert the following to radian measure.

1. 30°, 120°, 315°

2. 18°, 72°, 150°

3. 450°, −210°, −90°

4. 990°, −270°, −540°

Give the radian measure of each angle described.

5.
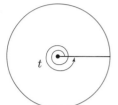

6.
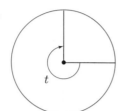 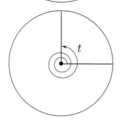

7.
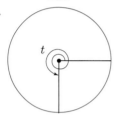

8.
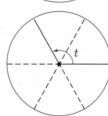

9.
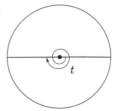

10.
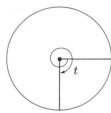

11.

12.
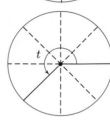

Construct angles with the following radian measure.

13. $3\pi/2$, $3\pi/4$, 5π

14. $\pi/3$, $5\pi/2$, 6π

15. $-\pi/3$, $-3\pi/4$, $-7\pi/2$

16. $-\pi/4$, $-3\pi/2$, -3π

17. $\pi/6$, $-2\pi/3$, $-\pi$

18. $2\pi/3$, $-\pi/6$, $7\pi/2$

Solutions to Practice Problems 8.1

1. The sum of the angles of a triangle is 180° or π radians. Since a right angle is $\pi/2$ radians and one angle is $\pi/3$ radians, the remaining angle is $\pi - (\pi/2 + \pi/3) = \pi/6$ radians.

2. $-780° = -780 \times (\pi/180)$ radians $= -13\pi/3$ radians. Since $-13\pi/3 = -4\pi - \pi/3$, we draw the angle by first making two revolutions in the negative direction and then a rotation of $\pi/3$ in the negative direction. See Fig. 8.

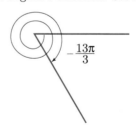

$-\dfrac{13\pi}{3}$

Figure 8.

8.2 The Sine and the Cosine

Given a number t, we consider an angle of t radians placed in standard position, as in Fig. 1, and we let P be a point on the terminal side of this angle. Denote the coordinates of P by (x, y) and let r be the length of the segment OP; that is, $r = \sqrt{x^2 + y^2}$. The *sine* and *cosine* of t, denoted by $\sin t$ and $\cos t$, respectively, are defined by the ratios

$$\sin t = \frac{y}{r}$$

$$\cos t = \frac{x}{r}.$$

(1)

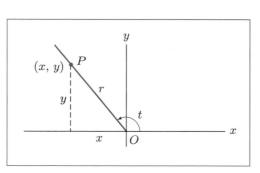

Figure 1. Diagram for definitions of sine and cosine.

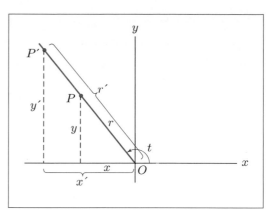

Figure 2.

It does not matter which point on the ray through P we use to define $\sin t$ and $\cos t$. If $P' = (x', y')$ is another point on the same ray and if r' is the length of OP' (Fig. 2), then, by properties of similar triangles, we have

$$\frac{y'}{r'} = \frac{y}{r} = \sin t$$

$$\frac{x'}{r'} = \frac{x}{r} = \cos t.$$

Three examples that illustrate the definition of $\sin t$ and $\cos t$ are shown in Fig. 3.

When $0 < t < \pi/2$, the values of $\sin t$ and $\cos t$ may be expressed as ratios of the lengths of the sides of a right triangle. Indeed, if we are given a right triangle as in Fig. 4, then we have

$$\sin t = \frac{\text{opposite}}{\text{hypotenuse}}, \qquad \cos t = \frac{\text{adjacent}}{\text{hypotenuse}}. \tag{2}$$

A typical application of (2) appears in Example 1.

▶ Example 1   The hypotenuse of a right triangle is four units and one angle is .7 radian. Determine the length of the side opposite this angle.

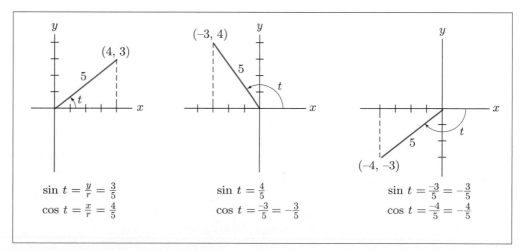

$$\sin t = \frac{y}{r} = \frac{3}{5}$$
$$\cos t = \frac{x}{r} = \frac{4}{5}$$

$$\sin t = \frac{4}{5}$$
$$\cos t = \frac{-3}{5} = -\frac{3}{5}$$

$$\sin t = \frac{-3}{5} = -\frac{3}{5}$$
$$\cos t = \frac{-4}{5} = -\frac{4}{5}$$

Figure 3. Calculation of sin t and cos t.

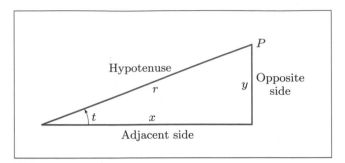

Figure 4.

Solution   See Fig. 5. Since $y/4 = \sin .7$, we have

$$y = 4\sin .7$$
$$y = 4(.64422) = 2.57688. \qquad \blacklozenge$$

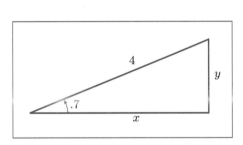

Figure 5.

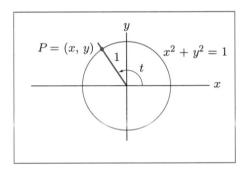

Figure 6.

Another way to describe the sine and cosine functions is to choose the point P in Fig. 1 so that $r = 1$. That is, choose P on the unit circle. (See Fig. 6.) In this case

$$\sin t = \frac{y}{1} = y$$

$$\cos t = \frac{x}{1} = x.$$

So the y-coordinate of P is $\sin t$ and the x-coordinate of P is $\cos t$. Thus we have the following result.

Alternative Definition of Sine and Cosine Functions   We can think of $\cos t$ and $\sin t$ as the x- and y-coordinates of the point P on the unit circle that is determined by an angle of t radians (Fig. 7).

▶ Example 2   Find a value of t such that $0 < t < \pi/2$ and $\cos t = \cos(-\pi/3)$.

Solution   On the unit circle we locate the point P that is determined by an angle of $-\pi/3$ radians. The x-coordinate of P is $\cos(-\pi/3)$. There is another point Q on the unit circle with the same x-coordinate. (See Fig. 8.) Let t be the radian measure of the angle determined by Q. Then

$$\cos t = \cos\left(-\frac{\pi}{3}\right)$$

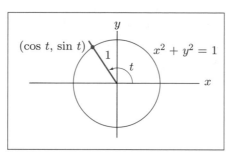

Figure 7.

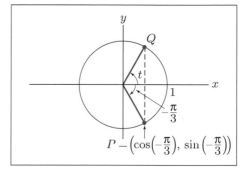

Figure 8.

because Q and P have the same x-coordinate. Also, $0 < t < \pi/2$. From the symmetry of the diagram it is clear that $t = \pi/3$. ◆

Properties of the Sine and Cosine Functions   Each number t determines a point $(\cos t, \sin t)$ on the unit circle $x^2 + y^2 = 1$ as in Fig. 7. Therefore, $(\cos t)^2 + (\sin t)^2 = 1$. It is convenient (and traditional) to write $\sin^2 t$ instead of $(\sin t)^2$ and $\cos^2 t$ instead of $(\cos t)^2$. Thus we can write the last formula as follows:

$$\cos^2 t + \sin^2 t = 1. \tag{3}$$

The numbers t and $t \pm 2\pi$ determine the same point on the unit circle (because 2π represents a full revolution of the circle). But $t + 2\pi$ and $t - 2\pi$ correspond to the points $(\cos(t+2\pi), \sin(t+2\pi))$ and $(\cos(t-2\pi), \sin(t-2\pi))$, respectively. Hence

$$\cos(t \pm 2\pi) = \cos t, \qquad \sin(t \pm 2\pi) = \sin t. \tag{4}$$

Figure 9(a) illustrates another property of the sine and cosine—namely,

$$\cos(-t) = \cos t, \qquad \sin(-t) = -\sin t. \tag{5}$$

Figure 9(b) shows that the points P and Q corresponding to t and to $\pi/2 - t$ are reflections of each other through the line $y = x$. Consequently, the coordinates of Q are obtained by interchanging the coordinates of P. This means that

$$\cos\left(\frac{\pi}{2} - t\right) = \sin t, \qquad \sin\left(\frac{\pi}{2} - t\right) = \cos t. \tag{6}$$

The equations in (3) to (6) are called *identities* because they hold for all values of t. Another identity that holds for all numbers s and t is

$$\sin(s + t) = \sin s \cos t + \cos s \sin t. \tag{7}$$

A proof of (7) may be found in any introductory book on trigonometry. There are a number of other identities concerning trigonometric functions, but we shall not need them here.

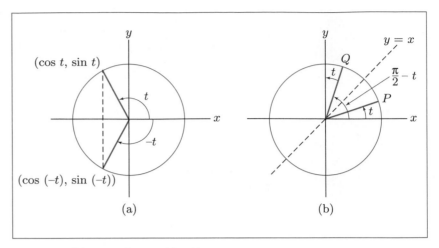

Figure 9. Diagrams for two identities.

The Graph of sin t   Let us analyze what happens to $\sin t$ as t increases from 0 to π. When $t = 0$, the point $P = (\cos t, \sin t)$ is at $(1, 0)$, as in Fig. 10(a). As t increases, P moves counterclockwise around the unit circle. The y-coordinate of P—that is, $\sin t$—increases until $t = \pi/2$, where $P = (0, 1)$. [See Fig. 10(c).] As t increases from $\pi/2$ to π, the y-coordinate of P—that is, $\sin t$—decreases from 1 to 0. [See Fig. 10(d) and (e).]

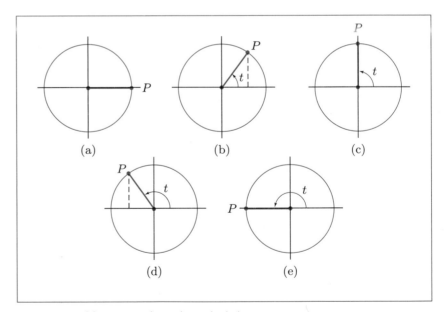

Figure 10. Movement along the unit circle.

Part of the graph of $\sin t$ is sketched in Fig. 11. Notice that, for t between 0 and π, the values of $\sin t$ increase from 0 to 1 and then decrease back to 0, just as we predicted from Fig. 10. For t between π and 2π, the values of $\sin t$ are negative. Can you explain why? The graph of $y = \sin t$ for t between 2π and 4π is exactly the same as the graph for t between 0 and 2π. This result follows from formula (4). We say that the sine function is *periodic with period* 2π because the graph repeats itself every 2π units. We can use this fact to make a quick sketch of part of the graph for negative values of t (Fig. 12).

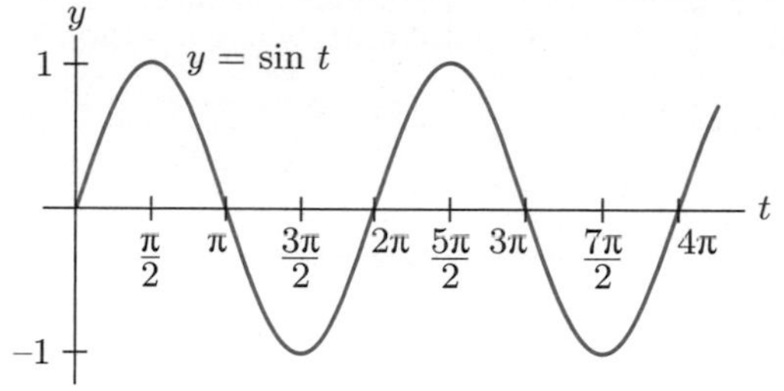

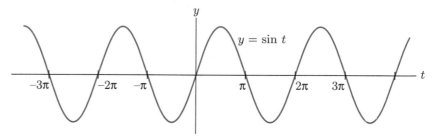

Figure 11.

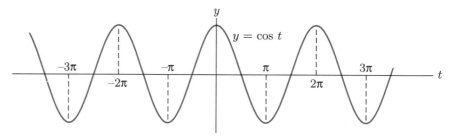

Figure 12. Graph of the *sine* function.

The Graph of cos t   By analyzing what happens to the first coordinate of the point $(\cos t, \sin t)$ as t varies, we obtain the graph of $\cos t$. Note from Fig. 13 that the graph of the cosine function is also periodic with period 2π.

Figure 13. Graph of the cosine function.

A Remark About Notation   The sine and cosine functions assign to each number t the values $\sin t$ and $\cos t$, respectively. There is nothing special, however, about the letter t. Although we chose to use the letters t, x, y, and r in the *definition* of the sine and cosine, other letters could have been used as well. Now that the sine and cosine of every number are defined, we are free to use *any* letter to represent the independent variable.

Incorporating Technology   Many graphing calculators have a special window setting for graphing trigonometric functions. TI calculators have a ZTrig window $[-2\pi, 2\pi]$ *by* $[-4, 4]$ with x-scale $\pi/2$. See Fig. 14.

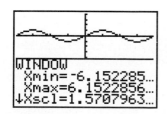

Figure 14.

<table>
<tr><td>

Practice Problems 8.2

</td><td>

1. Find $\cos t$, where t is the radian measure of the angle shown in Fig. 15.

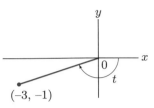

Figure 15.

2. Assume that $\cos(1.17) = .390$. Use properties of cosine and sine to determine $\sin(1.17)$, $\cos(1.17 + 4\pi)$, $\cos(-1.17)$, and $\sin(-1.17)$.

</td></tr>
</table>

▶ Exercises 8.2

In Exercises 1–12, give the values of $\sin t$ and $\cos t$, where t is the radian measure of the angle shown.

1.

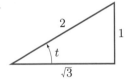

2.

3.

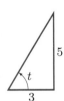

4.

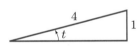

5.

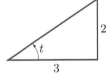

6.

7. $(-2, 1)$

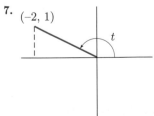

8.

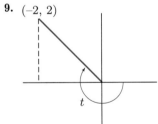

$(2, -3)$

9. $(-2, 2)$

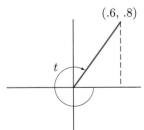

10. $(.6, .8)$

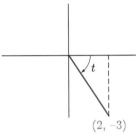

11.

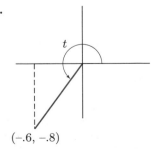

$(-.6, -.8)$

12.

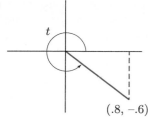

(.8, −.6)

Exercises 13–20 refer to various right triangles whose sides and angles are labeled as in Fig. 16. Round off all lengths of sides to one decimal place.

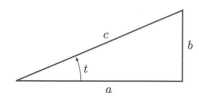

Figure 16.

13. Estimate t if $a = 12$, $b = 5$, and $c = 13$.

14. If $t = 1.1$ and $c = 10.0$, find b.

15. If $t = 1.1$ and $b = 3.2$, find c.

16. If $t = .4$ and $c = 5.0$, find a.

17. If $t = .4$ and $a = 10.0$, find c.

18. If $t = .9$ and $c = 20.0$, find a and b.

19. If $t = .5$ and $a = 2.4$, find b and c.

20. If $t = 1.1$ and $b = 3.5$, find a and c.

Find t such that $0 \leq t \leq \pi$ and t satisfy the stated condition.

21. $\cos t = \cos(-\pi/6)$ **22.** $\cos t = \cos(3\pi/2)$

23. $\cos t = \cos(5\pi/4)$ **24.** $\cos t = \cos(-4\pi/6)$

25. $\cos t = \cos(-5\pi/8)$ **26.** $\cos t = \cos(-3\pi/4)$

Find t such that $-\pi/2 \leq t \leq \pi/2$ and t satisfy the stated condition.

27. $\sin t = \sin(3\pi/4)$ **28.** $\sin t = \sin(7\pi/6)$

29. $\sin t = \sin(-4\pi/3)$ **30.** $\sin t = -\sin(3\pi/8)$

31. $\sin t = -\sin(\pi/6)$ **32.** $\sin t = -\sin(-\pi/3)$

33. $\sin t = \cos t$ **34.** $\sin t = -\cos t$

35. By referring to Fig. 10, describe what happens to $\cos t$ as t increases from 0 to π.

36. Use the unit circle to describe what happens to $\sin t$ as t increases from π to 2π.

37. Determine the value of $\sin t$ when $t = 5\pi$, -2π, $17\pi/2$, $-13\pi/2$.

38. Determine the value of $\cos t$ when $t = 5\pi$, -2π, $17\pi/2$, $-13\pi/2$.

39. Assume that $\cos(.19) = .98$. Use properties of cosine and sine to determine $\sin(.19)$, $\cos(.19-4\pi)$, $\cos(-.19)$, and $\sin(-.19)$.

40. Assume that $\sin(.42) = .41$. Use properties of cosine and sine to determine $\sin(-.42)$, $\sin(6\pi - .42)$, and $\cos(.42)$.

Technology Exercises

41. In any given locality, tap water temperature varies during the year. In Dallas, Texas, the tap water temperature (in degrees Fahrenheit) t days after the beginning of a year is given approximately by the formula*

$$T = 59 + 14 \cos\left[\frac{2\pi}{365}(t - 208)\right], \qquad 0 \leq t \leq 365.$$

(a) Graph the function in the window $[0, 365]$ *by* $[-10, 73]$.

(b) What is the temperature on February 14, that is, when $t = 45$?

(c) Use the fact that the value of the cosine function ranges from -1 to 1 to find the coldest and warmest tap water temperatures during the year.

(d) Use the TRACE feature or the MINIMUM command to estimate the day during which the tap water temperature is coldest. Find the exact day algebraically by using the fact that $\cos(-\pi) = -1$.

(e) Use the TRACE feature or the MAXIMUM command to estimate the day during which the tap water temperature is warmest. Find the exact day algebraically by using the fact that $\cos(0) = 1$.

(f) The average tap water temperature during the year is $59°$. Find the two days during which the average temperature is achieved. (*Note:* Answer this question both graphically and algebraically.)

42. In any given locality, the length of daylight varies during the year. In Des Moines, Iowa, the number of minutes of daylight in a day t days after the beginning of a year is given approximately by the formula†

$$D = 720 + 200 \sin\left[\frac{2\pi}{365}(t - 79.5)\right], \qquad 0 \leq t \leq 365.$$

(a) Graph the function in the window $[0, 365]$ *by* $[-100, 940]$.

(b) How many minutes of daylight are there on February 14, that is, when $t = 45$?

---

*See D. Rapp, *Solar Energy* (Englewood Cliffs, N.J.: Prentice-Hall, Inc., 1981), p. 171.

†See D. R. Duncan et al., "Climate Curves," *School Science and Mathematics*, Vol. LXXVI (January 1976), pp. 41–49.

(c) Use the fact that the value of the sine function ranges from -1 to 1 to find the shortest and longest amounts of daylight during the year.

(d) Use the TRACE feature or the MINIMUM command to estimate the day with the shortest amount of daylight. Find the exact day algebraically by using the fact that $\sin(3\pi/2) = -1$.

(e) Use the TRACE feature or the MAXIMUM command to estimate the day with the longest amount of daylight. Find the exact day algebraically by using the fact that $\sin(\pi/2) = 1$.

(f) Find the two days during which the amount of daylight equals the amount of darkness. (These days are called *equinoxes*.) (*Note*: Answer this question both graphically and algebraically.)

Solutions to Practice Problems 8.2

1. Here $P = (x, y) = (-3, -1)$. The length of the line segment OP is

$$r = \sqrt{x^2 + y^2} = \sqrt{(-3)^2 + (-1)^2} = \sqrt{10}.$$

Then

$$\cos t = \frac{x}{r} = \frac{-3}{\sqrt{10}} \approx -.94868.$$

2. Given $\cos(1.17) = .390$, use the relation $\cos^2 t + \sin^2 t = 1$ with $t = 1.17$ to solve for $\sin(1.17)$:

$$\cos^2(1.17) + \sin^2(1.17) = 1$$
$$\sin^2(1.17) = 1 - \cos^2(1.17)$$
$$= 1 - (.390)^2 = .8479.$$

So

$$\sin(1.17) = \sqrt{.8479} \approx .921.$$

Also, from properties (4) and (5),

$$\cos(1.17 + 4\pi) = \cos(1.17) = .390$$
$$\cos(-1.17) = \cos(1.17) = .390$$
$$\sin(-1.17) = -\sin(1.17) = -.921.$$

8.3 Differentiation and Integration of sin t and cos t

In this section we study the two differentiation rules

$$\frac{d}{dt} \sin t = \cos t \tag{1}$$

$$\frac{d}{dt} \cos t = -\sin t. \tag{2}$$

It is not difficult to see why these rules might be true. Formula (1) says that the slope of the curve $y = \sin t$ at a particular value of t is given by the corresponding value of $\cos t$. To check it, we draw a careful graph of $y = \sin t$ and estimate the slope at various points, as indicated in Fig. 1. Let us plot the slope as a function of t (Fig. 2). As can be seen, the "slope function" (i.e., the derivative) of $\sin t$ has a graph similar to the curve $y = \cos t$. Thus formula (1) seems to be reasonable. A similar analysis of the graph of $y = \cos t$ would show why (2) might be true.

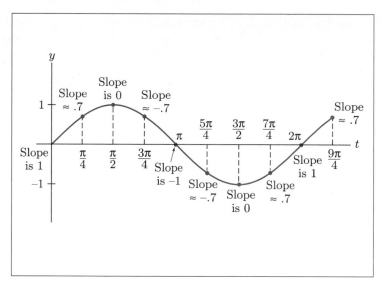

Figure 1. Slope estimates along the graph of y = sin t.

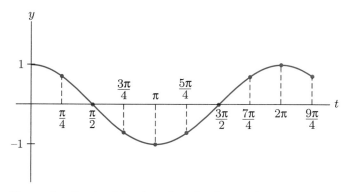

Figure 2. Graph of the slope function for y = sin t.

Proofs of these differentiation rules are outlined in an appendix at the end of this section.

Combining (1), (2), and the chain rule, we obtain the following general rules.

$$\frac{d}{dt}\sin g(t) = [\cos g(t)]\, g'(t) \tag{3}$$

$$\frac{d}{dt}\cos g(t) = [-\sin g(t)]\, g'(t) \tag{4}$$

▶ **Example 1**   Differentiate.

(a) $\sin 3t$ 　　　　　　　　　　(b) $(t^2 + 3\sin t)^5$

Solution   (a) $\dfrac{d}{dt}(\sin 3t) = (\cos 3t)\dfrac{d}{dt}(3t) = (\cos 3t)\cdot 3 = 3\cos 3t$

(b) $\dfrac{d}{dt}(t^2 + 3\sin t)^5 = 5(t^2 + 3\sin t)^4 \cdot \dfrac{d}{dt}(t^2 + 3\sin t)$

$$= 5(t^2 + 3\sin t)^4(2t + 3\cos t) \qquad ◆$$

▶ **Example 2**   Differentiate.

(a) $\cos(t^2 + 1)$ (b) $\cos^2 t$

Solution   (a) $\dfrac{d}{dt}\cos(t^2 + 1) = -\sin(t^2 + 1)\dfrac{d}{dt}(t^2 + 1) = -\sin(t^2 + 1)\cdot(2t)$

$$= -2t\sin(t^2 + 1)$$

(b) Recall that the notation $\cos^2 t$ means $(\cos t)^2$.

$$\frac{d}{dt}\cos^2 t = \frac{d}{dt}(\cos t)^2 = 2(\cos t)\frac{d}{dt}\cos t = -2\cos t\sin t \qquad ◆$$

▶ **Example 3**   Differentiate.

(a) $t^2\cos 3t$ (b) $(\sin 2t)/t$

Solution   (a) From the product rule we have

$$\frac{d}{dt}(t^2\cos 3t) = t^2\frac{d}{dt}\cos 3t + (\cos 3t)\frac{d}{dt}t^2$$

$$= t^2(-3\sin 3t) + (\cos 3t)(2t)$$

$$= -3t^2\sin 3t + 2t\cos 3t.$$

(b) From the quotient rule we have

$$\frac{d}{dt}\left(\frac{\sin 2t}{t}\right) = \frac{t\dfrac{d}{dt}\sin 2t - (\sin 2t)\cdot 1}{t^2} = \frac{2t\cos 2t - \sin 2t}{t^2}. \qquad ◆$$

▶ **Example 4**   A V-shaped trough is to be constructed with sides that are 200 centimeters long and 30 centimeters wide (Fig. 3). Find the angle t between the sides that maximizes the capacity of the trough.

Solution   The volume of the trough is its length times its cross-sectional area. Since the length is constant, it suffices to maximize the cross-sectional area. Let us rotate the diagram of a cross section so that one side is horizontal (Fig. 4). Note that $h/30 = \sin t$, so $h = 30\sin t$. Thus the area A of the cross section is

$$A = \tfrac{1}{2}\cdot\text{base}\cdot\text{height}$$

$$A = \tfrac{1}{2}(30)(h) = 15(30\sin t) = 450\sin t.$$

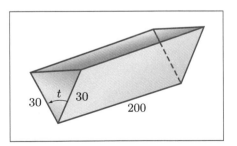

Figure 3.

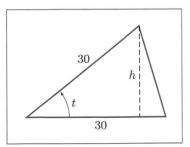

Figure 4.

To find where A is a maximum, we set the derivative equal to zero and solve for t.

$$\frac{dA}{dt} = 0$$

$$450 \cos t = 0$$

Physical considerations force us to consider only values of t between 0 and π. From the graph of $y = \cos t$ we see that $t = \pi/2$ is the only value of t between 0 and π that makes $\cos t = 0$. So in order to maximize the volume of the trough, the two sides should be perpendicular to one another. ◆

▶ **Example 5** Calculate the following indefinite integrals.

(a) $\displaystyle\int \sin t \, dt$

(b) $\displaystyle\int \sin 3t \, dt$

Solution (a) Since $\dfrac{d}{dt}(-\cos t) = \sin t$, we have

$$\int \sin t \, dt = -\cos t + C$$

where C is an arbitrary constant.

(b) From part (a) we guess that an antiderivative of $\sin 3t$ should resemble the function $-\cos 3t$. However, if we differentiate, we find that

$$\frac{d}{dt}(-\cos 3t) = (\sin 3t) \cdot \frac{d}{dt}(3t) = 3 \sin 3t,$$

which is three times too much. So we multiply this last equation by $\frac{1}{3}$ to derive that

$$\frac{d}{dt}\left(-\tfrac{1}{3}\cos 3t\right) = \sin 3t,$$

so

$$\int \sin 3t \, dt = -\tfrac{1}{3}\cos 3t + C. \qquad ◆$$

▶ **Example 6** Find the area under the curve $y = \sin 3t$ from $t = 0$ to $t = \pi/3$.

Solution The area is shaded in Fig. 5.

$$[\text{shaded area}] = \int_0^{\pi/3} \sin 3t \, dt$$

$$= -\frac{1}{3}\cos 3t \Big|_0^{\pi/3}$$

$$= -\frac{1}{3}\cos\left(3 \cdot \frac{\pi}{3}\right) - \left(-\frac{1}{3}\cos 0\right)$$

$$= -\frac{1}{3}\cos \pi + \frac{1}{3}\cos 0$$

$$= \frac{1}{3} + \frac{1}{3} = \frac{2}{3}. \qquad ◆$$

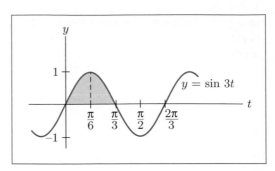

Figure 5.

As we mentioned earlier, the trigonometric functions are required to model situations that are repetitive (or periodic). The next example illustrates such a situation.

▶ Example 7 In many mathematical models used to study the interaction between predators and prey, both the number of predators and the number of prey are described by periodic functions. Suppose that in one such model, the number of predators (in a particular geographical region) at time t is given by the equation

$$N(t) = 5000 + 2000\cos(2\pi t/36),$$

where t is measured in months from June 1, 1990.

(a) At what rate is the number of predators changing on August 1, 1990?

(b) What is the average number of predators during the time interval from June 1, 1990, to June 1, 1993?

Solution (a) The date August 1, 1990, corresponds to $t = 2$. The rate of change of $N(t)$ is given by the derivative $N'(t)$:

$$N'(t) = \frac{d}{dt}\left[5000 + 2000\cos\left(\frac{2\pi t}{36}\right)\right]$$

$$= 2000\left[-\sin\left(\frac{2\pi t}{36}\right)\cdot\left(\frac{2\pi}{36}\right)\right]$$

$$= -\frac{1000\pi}{9}\sin\left(\frac{2\pi t}{36}\right),$$

$$N'(2) = -\frac{1000\pi}{9}\sin\left(\frac{\pi}{9}\right)$$

$$\approx -119.$$

Thus, on August 1, 1990, the number of predators is decreasing at the rate of 119 per month.

(b) The time interval from June 1, 1990, to June 1, 1993, corresponds to $t = 0$

to $t = 36$. The average value of $N(t)$ over this interval is

$$\frac{1}{36-0} \int_0^{36} N(t)\, dt = \frac{1}{36} \int_0^{36} \left[5000 + 2000 \cos\left(\frac{2\pi t}{36}\right)\right] dt$$

$$= \frac{1}{36} \left[5000t + \frac{2000}{2\pi/36} \sin\left(\frac{2\pi t}{36}\right)\right]\Bigg|_0^{36}$$

$$= \frac{1}{36} \left[5000 \cdot 36 + \frac{2000}{2\pi/36} \sin(2\pi)\right]$$

$$- \frac{1}{36} \left[5000 \cdot 0 + \frac{2000}{2\pi/36} \sin(0)\right]$$

$$= 5000.$$

We have sketched the graph of $N(t)$ in Fig. 6. Note how $N(t)$ oscillates around 5000, the average value. ◆

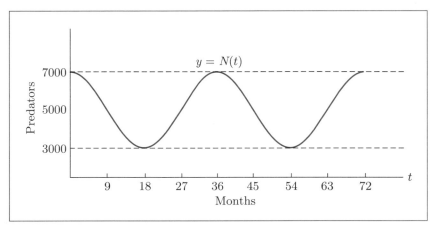

Figure 6. Periodic fluctuation of a predator population.

APPENDIX Informal Justification of the Differentiation Rules for sin t and cos t

First, let us examine the derivatives of $\cos t$ and $\sin t$ at $t = 0$. The function $\cos t$ has a maximum at $t = 0$; consequently, its derivative there must be zero. [See Fig. 7(a).] If we approximate the tangent line at $t = 0$ by a secant line, as in Fig. 7(b), then the slope of the secant line must approach 0 as $h \to 0$. Since the slope of the secant line is $(\cos h - 1)/h$, we conclude that

$$\lim_{h \to 0} \frac{\cos h - 1}{h} = 0. \tag{5}$$

It appears from the graph of $y = \sin t$ that the tangent line at $t = 0$ has slope 1 [Fig. 8(a)]. If it does, then the slope of the approximating secant line [Fig. 8(b)] must approach 1. Since the slope of the secant line in Fig. 8(b) is $(\sin h)/h$, this would imply that

$$\lim_{h \to 0} \frac{\sin h}{h} = 1. \tag{6}$$

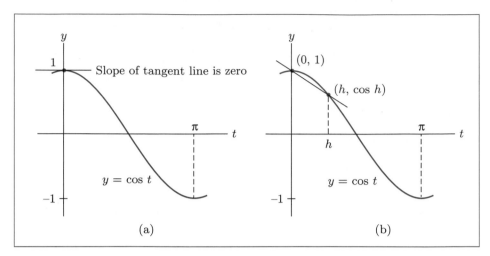

Figure 7.

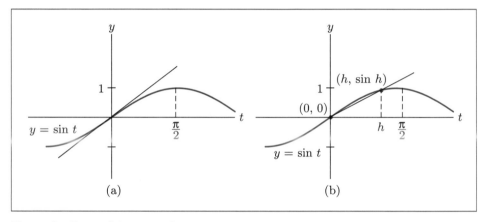

Figure 8. Slope of sin t at t = 0.

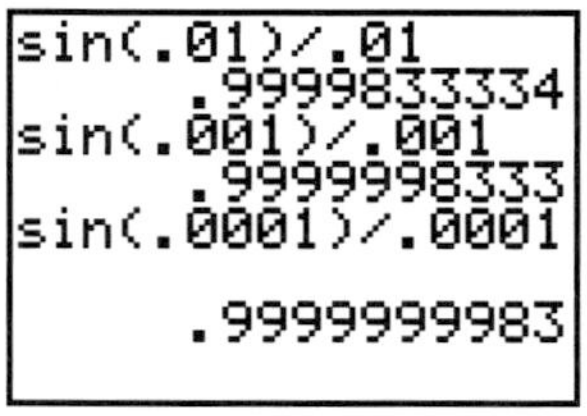

We can evaluate $\sin h/h$ for small values of h with a calculator. See Fig. 9. The numerical evidence does not prove (6), but should be sufficiently convincing for our purposes.

To obtain the differentiation formula for $\sin t$, we approximate the slope of a tangent line by the slope of a secant line. (See Fig. 10.) The slope of a secant line is

$$\frac{\sin(t + h) - \sin t}{h}.$$

From Formula (7) of Section 8.2, we note that $\sin(t+h) = \sin t \cos h + \cos t \sin h$. Thus,

$$[\text{slope of secant line}] = \frac{(\sin t \cos h + \cos t \sin h) - \sin t}{h}$$

$$= \frac{\sin t (\cos h - 1) + \cos t \sin h}{h}$$

$$= (\sin t)\frac{\cos h - 1}{h} + (\cos t)\frac{\sin h}{h}.$$

```
sin(.01)/.01
        .9999833334
sin(.001)/.001
        .9999998333
sin(.0001)/.0001
        .9999999983
```

Figure 9.

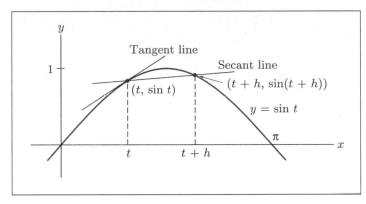

Figure 10. Secant line approximation for y = sin t.

From (5) and (6) it follows that

$$\frac{d}{dt}\sin t = \lim_{h \to 0} \left[(\sin t)\frac{\cos h - 1}{h} + (\cos t)\frac{\sin h}{h} \right]$$

$$= (\sin t) \lim_{h \to 0} \frac{\cos h - 1}{h} + (\cos t) \lim_{h \to 0} \frac{\sin h}{h}$$

$$= (\sin t) \cdot 0 + (\cos t) \cdot 1$$

$$= \cos t.$$

A similar argument may be given to verify the formula for the derivative of $\cos t$. Here is a shorter proof that uses the chain rule and the two identities

$$\cos t = \sin \left(\frac{\pi}{2} - t \right), \qquad \sin t = \cos \left(\frac{\pi}{2} - t \right).$$

[See Formula (6) of Section 8.2.] We have

$$\frac{d}{dt}\cos t = \frac{d}{dt}\sin \left(\frac{\pi}{2} - t \right)$$

$$= \cos \left(\frac{\pi}{2} - t \right) \cdot \frac{d}{dt} \left(\frac{\pi}{2} - t \right)$$

$$= \cos \left(\frac{\pi}{2} - t \right) \cdot (-1)$$

$$= -\sin t.$$

Practice Problems 8.3

1. Differentiate $y = 2\sin[t^2 + (\pi/6)]$.
2. Differentiate $y = e^t \sin 2t$.

▶ Exercises 8.3

Differentiate (with respect to t or x).

1. $\sin 4t$

2. $-3\cos t$

3. $4\sin t$

4. $\cos(-3t)$

5. $2\cos 3t$

6. $2\sin \pi t$

7. $t + \cos \pi t$

8. $t^2 - 2\sin 4t$

9. $\sin(\pi - t)$

10. $2\cos(t + \pi)$

11. $\cos^3 t$

12. $\sin t^3$

13. $\sin \sqrt{x-1}$

14. $\cos(e^x)$

15. $\sqrt{\sin(x-1)}$

16. $e^{\cos x}$

17. $(1+\cos t)^8$

18. $\sqrt[3]{\sin \pi t}$

19. $\cos^2 x^3$

20. $\sin^3 x + 4\sin^2 x$

21. $e^x \sin x$

22. $x\sqrt{\cos x}$

23. $\sin 2x \cos 3x$

24. $\sin^3 x \cos x$

25. $\dfrac{\sin t}{\cos t}$

26. $\dfrac{e^t}{\cos 2t}$

27. $\ln(\cos t)$

28. $\ln(\sin 2t)$

29. $\sin(\ln t)$

30. $(\cos t)\ln t$

31. Find the slope of the line tangent to the graph of $y = \cos 3x$ at $x = 13\pi/6$.

32. Find the slope of the line tangent to the graph of $y = \sin 2x$ at $x = 5\pi/4$.

33. Find the equation of the line tangent to the graph of $y = 3\sin x + \cos 2x$ at $x = \pi/2$.

34. Find the equation of the line tangent to the graph of $y = 3\sin 2x - \cos 2x$ at $x = 3\pi/4$.

Find the following indefinite integrals.

35. $\displaystyle\int \cos 2x\, dx$

36. $\displaystyle\int \sin \frac{x}{3}\, dx$

37. $\displaystyle\int \sin(4x+1)\, dx$

38. $\displaystyle\int \cos(5-x)\, dx$

39. Suppose that a person's blood pressure P at time t (in seconds) is given by $P = 100 + 20\cos 6t$.

   (a) Find the maximum value of P (called the systolic pressure) and the minimum value of P (called the diastolic pressure) and give one or two values of t where these maximum and minimum values of P occur.

   (b) If time is measured in seconds, approximately how many heartbeats per minute are predicted by the equation for P?

40. The *basal metabolism* (BM) of an organism over a certain time period may be described as the total amount of heat in kilocalories (kcal) the organism produces during that period, assuming that the organism is at rest and not subject to stress. The *basal metabolic rate* (BMR) is the rate in kcal per hour at which the organism produces heat. The BMR of an animal such as a desert rat fluctuates in response to changes in temperature and other environmental factors. The BMR generally follows a *diurnal* cycle—rising at night during low temperatures and decreasing during the warmer daytime temperatures. Find the BM for 1 day if $\mathrm{BMR}(t) = .4 + .2\cos(\pi t/12)$ kcal per hour ($t = 0$ corresponds to 3 A.M.). See Fig. 11.

41. As h approaches 0, what value is approached by

$$\frac{\sin\left(\dfrac{\pi}{2}+h\right)-1}{h}?$$

$$\left[\textit{Hint: } \sin\frac{\pi}{2} = 1.\right]$$

42. As h approaches 0, what value is approached by

$$\frac{\cos(\pi+h)+1}{h}?$$

$$[\textit{Hint: } \cos\pi = -1.]$$

43. The average weekly temperature in Washington, D.C. t weeks after the beginning of the year is

$$f(t) = 54 + 23\sin\left[\frac{2\pi}{52}(t-12)\right].$$

The graph of this function is sketched in Fig. 12.

   (a) What is the average weekly temperature at week 18?

   (b) At week 20, how fast is the temperature changing?

   (c) When is the average weekly temperature 39 degrees?

   (d) When is the average weekly temperature falling at the rate of 1 degree per week?

   (e) When is the average weekly temperature greatest? Least?

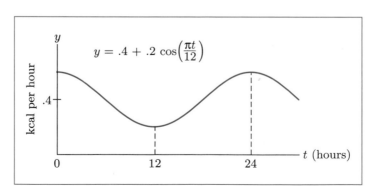

Figure 11. Diurnal cycle of the basal metabolic rate.

(f) When is the average weekly temperature increasing fastest? Decreasing fastest?

44. The number of hours of daylight per day in Washington, D.C. t weeks after the beginning of the year is

$$f(t) - 12.18 + 2.725 \sin\left[\frac{2\pi}{52}(t-12)\right].$$

The graph of this function is sketched in Fig. 13.

(a) How many hours of daylight are there after 42 weeks?

(b) After 32 weeks, how fast is the number of hours of daylight decreasing?

(c) When is there 14 hours of daylight per day?

(d) When is the number of hours of daylight increasing at the rate of 15 minutes per week?

(e) When are the days longest? Shortest?

(f) When is the number of hours of daylight increasing fastest? Decreasing fastest?

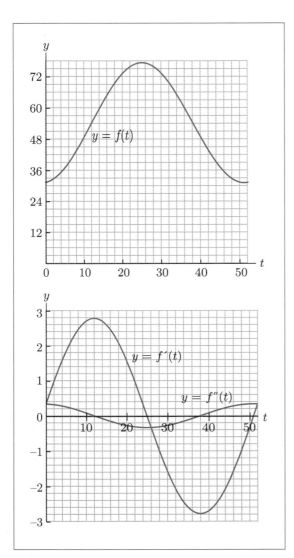

Figure 12.

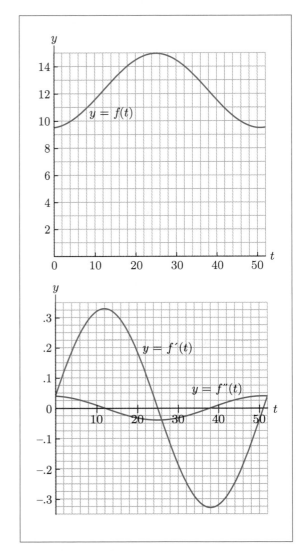

Figure 13.

Solutions to Practice Problems 8.3

1. By the chain rule,

$$y' = 2\cos\left(t^2 + \frac{\pi}{6}\right) \cdot \frac{d}{dt}\left(t^2 + \frac{\pi}{6}\right)$$

$$= 2\cos\left(t^2 + \frac{\pi}{6}\right) \cdot 2t$$

$$= 4t\cos\left(t^2 + \frac{\pi}{6}\right).$$

2. By the product rule,

$$y' = e^t \frac{d}{dt}(\sin 2t) + (\sin 2t)\frac{d}{dt}e^t = 2e^t \cos 2t + e^t \sin 2t.$$

8.4 The Tangent and Other Trigonometric Functions

Certain functions involving the sine and cosine functions occur so frequently in applications that they have been given special names. The *tangent* (tan), *cotangent* (cot), *secant* (sec), and *cosecant* (csc) are such functions and are defined as follows:

$$\tan t = \frac{\sin t}{\cos t}, \qquad \cot t = \frac{\cos t}{\sin t},$$

$$\sec t = \frac{1}{\cos t}, \qquad \csc t = \frac{1}{\sin t}.$$

They are defined only for t such that the denominators in the preceding quotients are not zero. These four functions, together with the sine and cosine, are called the *trigonometric functions*. Our main interest in this section is with the tangent function. Some properties of the cotangent, secant, and cosecant are developed in the exercises.

Many identities involving the trigonometric functions can be deduced from the identities given in Section 8.2. We shall mention just one:

$$\tan^2 t + 1 = \sec^2 t. \tag{1}$$

[Here $\tan^2 t$ means $(\tan t)^2$ and $\sec^2 t$ means $(\sec t)^2$.] This identity follows from the identity $\sin^2 t + \cos^2 t = 1$ when we divide everything by $\cos^2 t$.

An important interpretation of the tangent function can be given in terms of the diagram used to define the sine and cosine. For a given t, let us construct an angle of t radians, as in Fig. 1. Since $\sin t = y/r$ and $\cos t = x/r$, we have

$$\frac{\sin t}{\cos t} = \frac{y/r}{x/r} = \frac{y}{x},$$

where this formula holds provided that $x \neq 0$. Thus

$$\tan t = \frac{y}{x}. \tag{2}$$

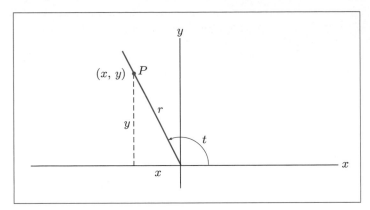

Figure 1.

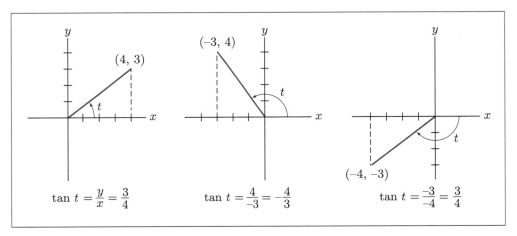

Figure 2.

Three examples that illustrate this property of the tangent appear in Fig. 2.

When $0 < t < \pi/2$, the value of $\tan t$ is a ratio of the lengths of the sides of a right triangle. In other words, suppose that we are given a triangle as in Fig. 3. Then we have

$$\tan t = \frac{\text{opposite}}{\text{adjacent}}. \tag{3}$$

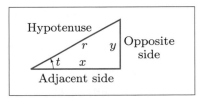

Figure 3.

▶ Example 1   The angle of elevation from an observer to the top of a building is 29° (Fig. 4). If the observer is 100 meters from the base of the building, how high is the building?

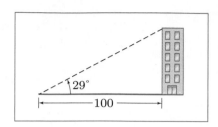

Figure 4.

Solution   Let h denote the height of the building. Then (3) implies that

$$\frac{h}{100} = \tan 29°$$

$$h = 100 \tan 29°.$$

We convert $29°$ into radians. We find that $29° = (\pi/180) \cdot 29$ radians $\approx .5$ radian, and $\tan .5 = .54630$. Hence

$$h \approx 100(.54630) = 54.63 \text{ meters.} \qquad \blacklozenge$$

The Derivative of tan t   Since $\tan t$ is defined in terms of $\sin t$ and $\cos t$, we can compute the derivative of $\tan t$ from our rules of differentiation. That is, by applying the quotient rule for differentiation, we have

$$\frac{d}{dt}(\tan t) = \frac{d}{dt}\left(\frac{\sin t}{\cos t}\right) = \frac{(\cos t)(\cos t) - (\sin t)(-\sin t)}{(\cos t)^2}$$

$$= \frac{\cos^2 t + \sin^2 t}{\cos^2 t} = \frac{1}{\cos^2 t}.$$

Now

$$\frac{1}{\cos^2 t} = \frac{1}{(\cos t)^2} = \left(\frac{1}{\cos t}\right)^2 = (\sec t)^2 = \sec^2 t.$$

So the derivative of $\tan t$ can be expressed in two equivalent ways:

$$\frac{d}{dt}(\tan t) = \frac{1}{\cos^2 t} = \sec^2 t. \qquad (4)$$

Combining (4) with the chain rule, we have

$$\frac{d}{dt}(\tan g(t)) = [\sec^2 g(t)]g'(t). \qquad (5)$$

▶ Example 2   Differentiate.

    (a) $\tan(t^3 + 1)$ (b) $\tan^3 t$

Solution   (a) From (5) we find that

$$\frac{d}{dt}[\tan(t^3 + 1)] = \sec^2(t^3 + 1) \cdot \frac{d}{dt}(t^3 + 1)$$

$$= 3t^2 \sec^2(t^3 + 1).$$

(b) We write $\tan^3 t$ as $(\tan t)^3$ and use the chain rule (in this case, the general power rule):

$$\frac{d}{dt}(\tan t)^3 = (3\tan^2 t)\cdot\frac{d}{dt}\tan t = 3\tan^2 t\,\sec^2 t. \qquad ◆$$

The Graph of tan t   Recall that $\tan t$ is defined for all t except where $\cos t = 0$. (We cannot have zero in the denominator of $\sin t/\cos t$.) The graph of $\tan t$ is sketched in Fig. 5. Note that $\tan t$ is periodic with period π.

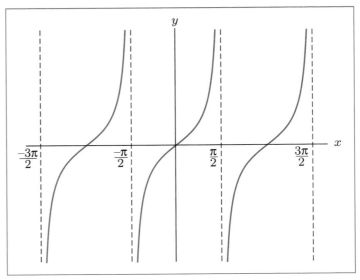

Figure 5. Graph of the tangent function.

1. Show that the slope of a straight line is equal to the tangent of the angle that the line makes with the x-axis.

2. Calculate $\int_0^{\pi/4} \sec^2 t\,dt$.

▶ Exercises 8.4

1. Suppose that $0 < t < \pi/2$ and use Fig. 3 to describe $\sec t$ as a ratio of the lengths of the sides of a right triangle.

2. Describe $\cot t$ for $0 < t < \pi/2$ as a ratio of the lengths of the sides of a right triangle.

In Exercises 3–10, give the values of $\tan t$ and $\sec t$, where t is the radian measure of the angle shown.

3.

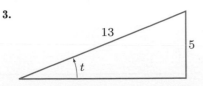

4.

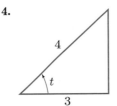

5. (−2, 1)

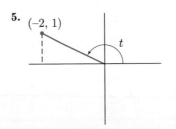

6.

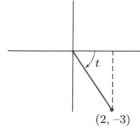

(2, −3)

7. (−2, 2)

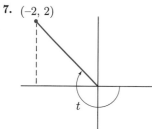

8. (.6, .8)

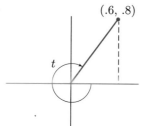

9.

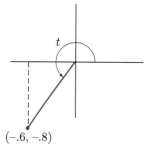

(−.6, −.8)

10.

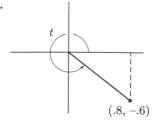

(.8, −.6)

11. Find the width of a river at points A and B if the angle BAC is $90°$, the angle ACB is $40°$, and the distance from A to C is 75 feet. See Fig. 6.

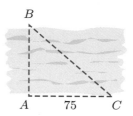

B

A 75 C

Figure 6.

12. The angle of elevation from an observer to the top of a church is .3 radian, while the angle of elevation from the observer to the top of the church spire is .4 radian. If the observer is 70 meters from the church, how tall is the spire on top of the church?

Differentiate (with respect to t or x).

13. $f(t) = \sec t$ **14.** $f(t) = \csc t$

15. $f(t) = \cot t$ **16.** $f(t) = \cot 3t$

17. $f(t) = \tan 4t$ **18.** $f(t) = \tan \pi t$

19. $f(x) = 3\tan(\pi - x)$ **20.** $f(x) = 5\tan(2x + 1)$

21. $f(x) = 4\tan(x^2 + x + 3)$ **22.** $f(x) = 3\tan(1 - x^2)$

23. $y = \tan \sqrt{x}$ **24.** $y = 2\tan \sqrt{x^2 - 4}$

25. $y = x\tan x$ **26.** $y = e^{3x}\tan 2x$

27. $y = \tan^2 x$ **28.** $y = \sqrt{\tan x}$

29. $y = (1 + \tan 2t)^3$ **30.** $y = \tan^4 3t$

31. $y = \ln(\tan t + \sec t)$ **32.** $y = \ln(\tan t)$

Solutions to Practice Problems 8.4

1. A line of positive slope m is shown in Fig. 7(a). Here, $\tan\theta = m/1 = m$. Suppose that $y = mx + b$ where the slope m is negative. The line $y = mx$ has the same slope and makes the same angle with the x-axis. [See Fig. 7(b).] We see that $\tan\theta = -m/-1 = m$.

2. $\displaystyle\int_0^{\pi/4} \sec^2 t \, dt = \tan t \Big|_0^{\pi/4} = \tan\frac{\pi}{4} - \tan 0 = 1 - 0 = 1$

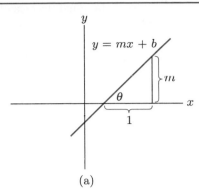
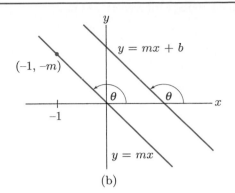

(a)   (b)

Figure 7.

Solutions to Practice Problems 8.4 (Continued)

Review of Fundamental Concepts of Chapter 8

1. Explain the radian measure of an angle.
2. Give the formula for converting degree measure to radian measure.
3. Give the triangle interpretation of $\sin t$, $\cos t$, and $\tan t$ for t between 0 and $\pi/2$.
4. Define $\sin t$, $\cos t$, and $\tan t$ for an angle of measure t for any t.
5. What does it mean when we say that the sine and cosine functions are periodic with period 2π?

6. Give verbal descriptions of the graphs of $\sin t$, $\cos t$, and $\tan t$.
7. State as many identities involving the sine and cosine functions as you can recall.
8. Define $\cot t$, $\sec t$, and $\csc t$ for an angle of measure t.
9. State an identity involving $\tan t$ and $\sec t$.
10. What are the derivatives of $\sin g(t)$, $\cos g(t)$, and $\tan g(t)$?

▶ Chapter 8 Supplementary Exercises

Determine the radian measure of the angles shown in Exercises 1–3.

1.

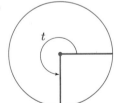

2.

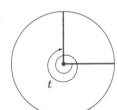

3.

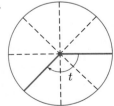

Construct angles with the following radian measure.

4. $-\pi$ **5.** $\dfrac{5\pi}{4}$ **6.** $-\dfrac{9\pi}{2}$

In Exercises 7–10, the point with the given coordinates determines an angle of t radians, where $0 \le t \le 2\pi$. Find $\sin t$, $\cos t$, $\tan t$.

7. $(3, 4)$ **8.** $(-.6, .8)$
9. $(-.6, -.8)$ **10.** $(3, -4)$

11. If $\sin t = \frac{1}{5}$, what are the possible values for $\cos t$?

12. If $\cos t = -\frac{2}{3}$, what are the possible values for $\sin t$?

13. Find the four values of t between -2π and 2π at which $\sin t = \cos t$.

14. Find the four values of t between -2π and 2π at which $\sin t = -\cos t$.

15. When $-\pi/2 < t < 0$, is $\tan t$ positive or negative?

16. When $\pi/2 < t < \pi$, is $\sin t$ positive or negative?

17. A gabled roof is to be built on a house that is 30 feet wide, so that the roof rises at a pitch of $23°$. Determine the length of the rafters needed to support the roof.

18. A tree casts a 60-foot shadow when the angle of elevation of the sun (measured from the horizontal) is $53°$. How tall is the tree?

Differentiate (with respect to t or x).

19. $f(t) = 3\sin t$

20. $f(t) = \sin 3t$

21. $f(t) = \sin\sqrt{t}$

22. $f(t) = \cos t^3$

23. $g(x) = x^3 \sin x$

24. $g(x) = \sin(-2x)\cos 5x$

25. $f(x) = \dfrac{\cos 2x}{\sin 3x}$

26. $f(x) = \dfrac{\cos x - 1}{x^3}$

27. $f(x) = \cos^3 4x$

28. $f(x) = \tan^3 2x$

29. $y = \tan(x^4 + x^2)$

30. $y = \tan e^{-2x}$

31. $y = \sin(\tan x)$

32. $y = \tan(\sin x)$

33. $y = \sin x \tan x$

34. $y = \ln x \cos x$

35. $y = \ln(\sin x)$

36. $y = \ln(\cos x)$

37. $y = e^{3x}\sin^4 x$

38. $y = \sin^4 e^{3x}$

39. $f(t) = \dfrac{\sin t}{\tan 3t}$

40. $f(t) = \dfrac{\tan 2t}{\cos t}$

41. $f(t) = e^{\tan t}$

42. $f(t) = e^t \tan t$

43. If $f(t) = \sin^2 t$, find $f''(t)$.

44. Show that $y = 3\sin 2t + \cos 2t$ satisfies the differential equation $y'' = -4y$.

45. If $f(s, t) = \sin s \cos 2t$, find $\dfrac{\partial f}{\partial s}$ and $\dfrac{\partial f}{\partial t}$.

46. If $z = \sin wt$, find $\dfrac{\partial z}{\partial w}$ and $\dfrac{\partial z}{\partial t}$.

47. If $f(s, t) = t\sin st$, find $\dfrac{\partial f}{\partial s}$ and $\dfrac{\partial f}{\partial t}$.

48. The identity

$$\sin(s + t) = \sin s \cos t + \cos s \sin t$$

was given in Section 8.2. Compute the partial derivative of each side with respect to t and obtain an identity involving $\cos(s + t)$.

49. Find the equation of the line tangent to the graph of $y = \tan t$ at $t = \pi/4$.

50. Sketch the graph of $f(t) = \sin t + \cos t$ for $-2\pi \le t \le 2\pi$, using the following steps:

(a) Find all t (between -2π and 2π) such that $f'(t) = 0$. Plot the corresponding points on the graph of $y = f(t)$.

(b) Check the concavity of $f(t)$ at the points in part (a). Make sketches of the graph near these points.

(c) Determine any inflection points and plot them. Then complete the sketch of the graph.

51. Sketch the graph of $y = t + \sin t$ for $0 \le t \le 2\pi$.

52. Find the area under the curve $y = 2 + \sin 3t$ from $t = 0$ to $t = \pi/2$.

53. Find the area of the region between the curve $y = \sin t$ and the t-axis from $t = 0$ to $t = 2\pi$.

54. Find the area of the region between the curve $y = \cos t$ and the t-axis from $t = 0$ to $t = 3\pi/2$.

55. Find the area of the region bounded by the curves $y = x$ and $y = \sin x$ from $x = 0$ to $x = \pi$.

A spirogram is a device that records on a graph the volume of air in a person's lungs as a function of time. If a person undergoes spontaneous hyperventilation, the spirogram trace will closely approximate a sine curve. A typical trace is given by

$$V(t) = 3 + .05\sin\left(160\pi t - \frac{\pi}{2}\right),$$

where t is measured in minutes and $V(t)$ is the lung volume in liters. (See Fig. 1.) Exercises 56–58 refer to this function.

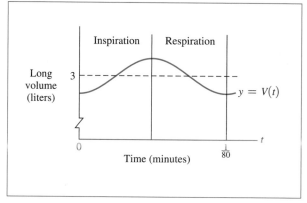

Figure 1. Spirogram trace.

56. (a) Compute $V(0)$, $V(\frac{1}{320})$, $V(\frac{1}{160})$, and $V(\frac{1}{80})$.

(b) What is the maximum lung volume?

57. (a) Find a formula for the rate of flow of air into the lungs at time t.

(b) Find the maximum rate of flow of air during inspiration (i.e., breathing in). This is called the *peak inspiratory flow*.

(c) Inspiration occurs during the time from $t = 0$ to $t = 1/160$. Find the average rate of flow of air during inspiration. This quantity is called the *mean inspiratory flow*.

58. The *minute volume* is defined to be the total amount of air inspired (breathed in) during one minute. According to a standard text on respiratory physiology, when a person undergoes spontaneous hyperventilation, the peak inspiratory flow equals π times the minute volume, and the mean inspiratory flow equals twice the minute volume.[*] Verify these assertions using the data from Exercise 57.

[*]J. F. Nunn, *Applied Respiratory Physiology*, 2nd ed. (London: Butterworths, 1977), p. 122.

Chapter Project

On day t of the year (January 1 corresponds to $t = 1$), the length of the day, $D(t, \lambda)$, at latitude λ is given by the formula

$$D(t, \lambda) = 12 + A(\lambda) \sin\left[\frac{2\pi}{365}(t - 79.5)\right],$$

where $A(\lambda)$ depends on the latitude.

(a) Show that the function is periodic in the variable t. Determine the period and explain its significance.

(b) Using either on-line or library reference materials, determine the latitude of your location.

(c) Consult your local newspaper to determine the length of the current day.

(d) Use the result of (c) to determine the value of $A(\lambda)$ for your latitude.

(e) Graph the function $D(t, \lambda)$ for your latitude λ.

(f) For your latitude, determine a formula for the rate at which the length of the day is changing. Determine this rate for Jan. 1, Mar. 1, Sept. 15, and the current date.

(g) Consult your local newspaper to determine the times for sunrise and sunset for today and tomorrow. From these, determine the rate at which the length of the day is changing. Compare this result with the rate computed in part (f).

(h) Using the data from your local newspaper, determine functions that give the time of sunrise and sunset at your latitude as a function of t.

(i) When is the length of the day the longest? How long is the day at this time?

(j) When is the length of the day the shortest? How long is the day at this time?

Appendices

Appendix A
Calculus and the TI-82 Calculator

Functions

A. Define a function
- Press $\boxed{Y=}$ to obtain **Y=** edit screen.
- Cursor down to the function to be defined. (Press $\boxed{\text{CLEAR}}$ if the function is to be redefined.)
- Type in the expression for the function. (*Note*: Press $\boxed{X,T,\theta}$ to display X.)

B. Select or deselect a function
(Functions with highlighted equal signs are said to be *selected*. Pressing $\boxed{\text{GRAPH}}$ instructs the calculator to graph all selected functions and pressing $\boxed{\text{2nd}}$ $[\text{TABLE}]$ creates a column of the table for each selected function.)
- Press $\boxed{Y=}$ to obtain the **Y=** edit screen.
- Cursor down to the function to be selected or deselected.
- Move the cursor to the equal sign and press $\boxed{\text{ENTER}}$ to toggle the state of the function on and off.

C. Display a function name, that is, $Y_1, Y_2, Y_3, \ldots$
- Press $\boxed{\text{2nd}}$ $\boxed{\text{Y-VARS}}$ **1** to invoke the list of function names.
- Press the number of the function name, or cursor down to the function and press $\boxed{\text{ENTER}}$.

D. Combine functions
Suppose Y_1 is $f(x)$ and Y_2 is $g(x)$.
- If $Y_3 = Y_1 + Y_2$, then Y_3 is $f(x) + g(x)$. (Similarly for $-$, $\times$, and $\div$.)
- If $Y_3 = Y_1(Y_2)$, then Y_3 is $f(g(x))$.

Specific Window Settings

A. Customize a window
- Press $\boxed{\text{WINDOW}}$ to open the window-setting screen, and edit the following values as desired.
- Xmin = *the leftmost value on the x-axis*

- Xmax = *the rightmost value on the x-axis*
- Xscl = *the distance between tick marks on the x-axis*
- Ymin = *the bottom value on the y-axis*
- Ymax = *the top value on the y-axis*
- Yscl = *the distance between tick marks on the y-axis*

  Note 1: The notation $[a, b]$ *by* $[c, d]$ stands for the window settings Xmin $= a$, Xmax $= b$, Ymin $= c$, Ymax $= d$.

  Note 2: The default values of Xscl and Yscl are 1. The value of Xscl should be made large (small) if the difference between Xmax and Xmin is large (small). For instance, with the window settings $[0, 100]$ *by* $[-1, 1]$, good scale settings are Xscl $= 10$ and Yscl $= .1$.

B. Use a predefined window setting
- Press $\boxed{\text{ZOOM}}$ to display the list of predefined settings.
- Either press a number or move the cursor down to an item and press $\boxed{\text{ENTER}}$.
- Select ZStandard to obtain $[-10, 10]$ *by* $[-10, 10]$, Xscl $=$ Yscl $= 1$.
- Select ZDecimal to obtain $[-4.7, 4.7]$ *by* $[-3.1, 3.1]$, Xscl $=$ Yscl $= 1$. (When TRACE is used with this setting, points have nice x-coordinates.)
- Select ZSquare to obtain a true-aspect window. (With such a window setting, lines that should be perpendicular actually look perpendicular, and the graph of $y = \sqrt{1 - x^2}$ looks like the top half of a circle.)
- Select ZTrig to obtain $[-2\pi, 2\pi]$ *by* $[-4, 4]$, Xscl $= \pi/2$, Yscl $= 1$, a good setting for the graphs of trigonometric functions.

C. Some nice window settings
(With these settings, one unit on the x-axis has the same length as one unit on the y-axis, and tracing progresses over simple values.)

- $[-4.7, 4.7]$ by $[-3.1, 3.1]$
- $[-2.35, 2.35]$ by $[-1.55, 1.55]$
- $[-7.05, 7.05]$ by $[-4.65, 4.65]$
- $[-9.4, 9.4]$ by $[-6.2, 6.2]$
- $[0, 9.4]$ by $[0, 6.2]$
- $[0, 18.8]$ by $[0, 12.4]$
- $[0, 47]$ by $[0, 31]$
- $[0, 94]$ by $[0, 62]$

General principle: (Xmax − Xmin) should be a number of the form $k \cdot 9.4$, where k is a whole number or $\frac{1}{2}, \frac{3}{2}, \frac{5}{2}, \ldots$, then (Ymax − Ymin) should be $(31/47) \cdot$ (Xmax − Xmin).

Derivative, Slopes, and Tangent Lines

A. Compute $f'(a)$ from the Home screen using nDeriv($f(x)$, X, a)

- Press $\boxed{\text{MATH}}$ **8** to display **nDeriv(**.
- Enter either $\mathbf{Y_1}, \mathbf{Y_2}, \ldots$ or an expression for $f(x)$.
- Type in the remaining items and press $\boxed{\text{ENTER}}$.

B. Define derivatives of the function $\mathbf{Y_1}$

- Set $\mathbf{Y_2} = \mathbf{nDeriv(Y_1, X, X)}$ to obtain the 1st derivative.
- Set $\mathbf{Y_3} = \mathbf{nDeriv(Y_2, X, X)}$ to obtain the 2nd derivative.

  Note: **nDeriv(** is obtained by pressing $\boxed{\text{MATH}}$ **8**.

C. Compute the slope of graph at a point

- Press $\boxed{\text{2nd}}$ [CALC] **6**.
- Use the arrow keys to move to the point of the graph.
- Press $\boxed{\text{ENTER}}$.

  Note: This process usually works best with a nice window setting.

D. Draw a tangent line to a graph

- Press $\boxed{\text{2nd}}$ [DRAW] **5** to select **Tangent**.
- Use the arrow keys to move to a point of the graph.
- Press $\boxed{\text{ENTER}}$.

  Note 1: This process usually works best with a nice window setting.

  Note 2: To remove all tangent lines, press $\boxed{\text{2nd}}$ [DRAW] **1** to execute **ClrDraw**.

Special Points on the Graph of $\mathbf{Y_1}$

A. Find a point of intersection with the graph of $\mathbf{Y_2}$, from the Home screen

Evaluate **solve($\mathbf{Y_1} - \mathbf{Y_2}$, X, g)**, where g is a guess, as follows:

- Press $\boxed{\text{MATH}}$ **0** to display **solve(**.
- Enter either $\mathbf{Y_1} - \mathbf{Y_2}$ or an expression for the difference of the two functions.
- Type in the remaining items, where g is (hopefully) close to a value of x for which $\mathbf{Y_1} = \mathbf{Y_2}$. Press $\boxed{\text{ENTER}}$.
- A value of x for which $\mathbf{Y_1} = \mathbf{Y_2}$ will be displayed.

B. Find intersection points, with graphs displayed

- Press $\boxed{\text{GRAPH}}$ to display the graphs of all selected functions.
- Press $\boxed{\text{2nd}}$ [CALC] **5** to select **intersect**.
- Reply to **"First curve?"** by using $\boxed{\triangle}$ or $\boxed{\triangledown}$ (if necessary) to move the cursor to one of the two curves and then pressing $\boxed{\text{ENTER}}$.
- Reply to **"Second curve?"** by using $\boxed{\triangle}$ or $\boxed{\triangledown}$ (if necessary) to move the cursor to the other curves and then pressing $\boxed{\text{ENTER}}$.
- Reply to **"Guess?"** by moving the cursor near the point of intersection and pressing $\boxed{\text{ENTER}}$.

C. Find the second coordinate of point with first coordinate specified, call it a

From the Home screen

- Display $\mathbf{Y_1}(a)$ and press $\boxed{\text{ENTER}}$.
  or
- Press a $\boxed{\text{STO} \triangleright}$ $\boxed{\text{X,T,}\theta}$ $\boxed{\text{ENTER}}$ to assign the value a to the variable X.
- Display $\mathbf{Y_1}$ and press $\boxed{\text{ENTER}}$.

  From the Home screen or with the graph displayed

- Press $\boxed{\text{2nd}}$ [CALC] **1** to select **value**.
- Type in the value of a and press $\boxed{\text{ENTER}}$.
- If desired, press the up-arrow key to move to points on graphs of other selected functions.

  With the graph displayed

- Press $\boxed{\text{TRACE}}$.
- Move cursor with $\boxed{\triangleright}$ and/or $\boxed{\triangleleft}$ until the x-coordinate of the cursor is close as possible to a. (*Note*: This process usually works best if one of the nice window settings discussed above is used.)

D. Find the first coordinate of a point with second coordinate specified, call it b

From the Home screen

- Press $\boxed{\text{MATH}}$ **0** to display **solve(**.
- Continue typing to obtain **solve($\mathbf{Y_1} - b$, X, c)** where c is a guess for the value of the first coordinate. (*Note*: The expression for $\mathbf{Y_1}$ can be used in place of $\mathbf{Y_1}$.)
- Press $\boxed{\text{ENTER}}$.

  With the graph displayed

- Set $\mathbf{Y_2} = b$.
- Find the point of intersection of $\mathbf{Y_1}$ and $\mathbf{Y_2}$ as in part B above.

E. Find an x-intercept (that is, a root of $\mathbf{Y_1} = 0$)

- Graph $\mathbf{Y_1}$.
- Press $\boxed{\text{2nd}}$ [CALC] **2** to select **root**.
- Move the cursor to a point just to the left of a root and press $\boxed{\text{ENTER}}$.
- Move the cursor to a point just to the right of the root and press $\boxed{\text{ENTER}}$.

- Move the cursor to a point near the root and press ENTER.

F. Find a relative extreme point

- Set $Y_2 = \text{nDeriv}(Y_1, X, X)$ or set Y_2 equal to the exact expression for the derivative of Y_1. [To display **nDeriv(** press MATH **8**.]
- Select Y_2 and deselect all other functions.
- Graph Y_2.
- Find an x-intercept of Y_2, call it r, at which the graph of Y_2 crosses the x-axis.
- The point $(r, Y_1(r))$ will be a possible relative extreme point of Y_1.

G. Find an inflection point

- Set $Y_2 = \text{nDeriv}(Y_1, X, X)$ and $Y_3 = \text{nDeriv}(Y_2, X, X)$, or set Y_3 equal to the exact expression for the second derivative of Y_1.
- Select Y_3 and deselect all other functions.
- Graph Y_3.
- Find an x-intercept of Y_3, call it r, at which the graph of Y_3 crosses the x-axis.
- The point $(r, Y_1(r))$ will be a possible inflection point of Y_1.

Tables

A. Display values of $f(x)$ for evenly spaced values of x

- Press Y=, assign the function $f(x)$ to Y_1.
- Press 2nd [TblSet].
- Set **TblMin** =*first value of x.*
- Set **ΔTbl** =*increment for values of x.*
- Set both **Indpnt** and **Depend** to **Auto**.
- Press 2nd [TABLE].

Note 1: You can use the down- and up-arrow keys to look at function values for other values of x.

Note 2: The table can display values of more than one function. For example, in the first step you can assign the function $g(x)$ to Y_2 and also select Y_2 to obtain a table with columns for X, Y_1, and Y_2.

B. Display values of $f(x)$ for arbitrary values of x

- Press Y= and assign the function $f(x)$ to Y_1.
- Press 2nd [TblSet].
- Set **Indpnt** to **Ask** by moving the cursor to Ask and pressing ENTER.
- Leave **Depend** set to **Auto**.
- Press 2nd [TABLE].
- Type in any value for X and press ENTER.
- Repeat the previous step for any other values of X.

Riemann Sums

Suppose that Y_1 is $f(x)$, and c, d, and Δx are numbers, then **sum(seq(Y_1, X, c, d, Δx))** computes

$$f(c) + f(c + \Delta x) + f(c + 2\Delta x) + \cdots + f(d).$$

The function **sum** is in the LIST/MATH menu and the function **seq** is in the LIST/OPS menu.

A. Compute $[f(x_1) + f(x_2) + \cdots + f(x_n)] \cdot \Delta x$

On the Home screen, evaluate sum(seq($f(x)$, X, x_1, x_n, Δx)) ∗ Δx as follows:

- Press 2nd [LIST] and move the cursor right to MATH.
- Press **5** (to display **sum(**.
- Press 2nd [LIST] **5** to display **seq(**.
- Enter either Y_1 or an expression for $f(x)$.
- Type in the remaining items and press ENTER.

Definite Integrals and Antiderivatives

A. Compute $\int_a^b f(x)\, dx$

On the Home screen, evaluate fnInt($f(x)$, x, a, b) as follows:

- Press MATH **9** to display **fnInt(**.
- Enter either Y_1 or an expression for $f(x)$.
- Type in the remaining items and press ENTER.

B. Shade a region under the graph of a function and find its area

- Press 2nd [CALC] **7** to select $\int \mathbf{f(x)}\, \mathbf{dx}$.
- If necessary, use △ or ▽ to move the cursor to the graph.
- In response to the request for a Lower Limit, move the cursor to the left endpoint of the region and press ENTER.
- In response to the request for an Upper Limit, move the cursor to the right endpoint of the region and press ENTER.

Note 1: This process often works best with a nice window setting.

Note 2: To remove the shading, press 2nd [DRAW] **1** to execute the **ClrDraw** command.

C. Obtain the graph of the solution to the differential equation $y' = g(x)$, $y(a) = b$

[That is, obtain the graph of the function $f(x)$ that is an antiderivative of $g(x)$ and satisfies the additional condition $f(a) = b$.]

- Set $Y_1 = \mathbf{g(X)}$.
- Set $Y_2 = \text{fnInt}(Y_1, X, a, X) + b$. [Press MATH **9** to display **fnInt(**.] The function Y_2 is an antiderivative of $g(x)$ and can be evaluated and graphed.

D. Shade the region between two curves

Suppose the graph of Y_1 lies below the graph of Y_2 for $a \leq x \leq b$ and both functions have been selected. To shade the region between these two curves execute the instructions **Shade(Y₁,Y₂,1,**a**,**b**)** as follows.

- Press 2nd [DRAW] **7** to display **Shade(**.
- Type in the remaining items and press ENTER.

Note 1: Replace 1 by 2 for a striped shading.

Note 2: To remove the shading, press 2nd [DRAW] **1** ENTER to execute the **ClrDraw** command.

Functions of Several Variables

A. Specify a function of several variables and its derivatives

- In the **Y=** edit screen, set $Y_1 = f(X, Y)$. (The letter Y is entered by pressing ALPHA [Y].)
- Set $Y_2 = $ **nDeriv(Y₁,X,X)**. Y_2 will be $\frac{\partial f}{\partial x}$.
- Set $Y_3 = $ **nDeriv(Y₁,Y,Y)**. Y_3 will be $\frac{\partial f}{\partial y}$.
- Set $Y_4 = $ **nDeriv(Y₂,X,X)**. Y_4 will be $\frac{\partial^2 f}{\partial x^2}$.
- Set $Y_5 = $ **nDeriv(Y₃,Y,Y)**. Y_5 will be $\frac{\partial^2 f}{\partial y^2}$.
- Set $Y_6 = $ **nDeriv(Y₃,X,X)**. Y_6 will be $\frac{\partial^2 f}{\partial x \partial y}$.

B. Evaluate one of the functions in part A at $x = a$ and $y = b$

- In the Home screen, assign the value a to the variable X with a STO ▷ X,T,θ.
- Press b STO ▷ ALPHA [Y] to assign the value b to the variable Y.
- Display the name of one of the functions, such as $Y_1, Y_2, \ldots$, and press ENTER.

Least-Squares Approximations

A. Obtain the equation of the least-squares line

Assume the points are $(x_1, y_1), \ldots, (x_n, y_n)$.

- Press STAT **1** for the EDIT screen, and obtain a table used for entering the data.
- If necessary, clear data from columns L_1 and L_2 as follows. Move the cursor to the top of column L_1 and press CLEAR ENTER. Repeat for column L_2.
- To enter the x-coordinates of the points, move the cursor to the first blank row of column L_1, enter the value of x_1, and press ENTER. Repeat with $x_2, \ldots, x_n$.

- Move the cursor to the first blank row of column L_2, and enter the values of $y_1, \ldots, y_n$.
- Press STAT ▷ for the CALC menu, and press **5** to place **LinReg(ax + b)** on the Home screen.
- Press ENTER to obtain the slope and y-intercept of the least-squares line.

B. Assign the least-squares line to a function

- Press Y=, move the cursor to the function, and press CLEAR to erase the current expression.
- Press VARS **5** to select the **Statistics** variables.
- Press ▷ ▷ to select the EQ menu, and press **7** for **RegEQ** (Regression Equation), which then assigns the least-squares line equation to the function.

C. Display the points from part A

- Press Y= and deselect all functions.
- Press 2nd [STAT PLOT] ENTER to select **Plot1**, press ENTER to turn **Plot1 ON**.
- Select the first plot from the four icons for the types of plots. This icon corresponds to a scatter plot.
- Select L_1 from Xlist and select L_2 from Ylist.
- Press GRAPH to display the data points.

Note 1: Press ZOOM **9** to ensure that the current window setting is large enough to display the points.

Note 2: When you finish using the point-plotting feature, turn it off. Press 2nd [STAT PLOT] ENTER to select **Plot1**. Then press ▷ to move to **OFF** and press ENTER.

D. Display the line and the points from part A

- Press Y= and deselect all functions except for the function containing the equation of the least-squares line.
- Carry out all but the first step of part C.

Miscellaneous Items and Tips

A. From the Home screen, if you plan to reuse a recently entered line with some minor changes, press 2nd [ENTRY] until the previous line appears. You can then make alterations to the line and press ENTER to execute the line.

B. If you plan to use TRACE to examine the values of various points on a graph, set Ymin to a value that is lower than is actually necessary for the graph. Then, the values of x and y will not obliterate the graph while you trace.

C. To clear the Home screen, press CLEAR twice.

Appendix B
Calculus and the TI-83/TI-83 Plus Calculators

Functions

A. Define a function

- Press $\boxed{\text{Y=}}$ to obtain the **Y=** editor.

- Cursor down to the function to be defined. (Press $\boxed{\text{CLEAR}}$ if the function is to be redefined.)

- . Type in the expression for the function. (*Note*: Press $\boxed{\text{X,T,}\theta\text{,}n}$ to display X.)

B. Select or deselect a function
(Functions with highlighted equal signs are said to be *selected*. Pressing $\boxed{\text{GRAPH}}$ instructs the calculator to graph all selected functions and pressing $\boxed{\text{2nd}}$ [TABLE] creates a column of the table for each selected function.)

- Press $\boxed{\text{Y=}}$ to obtain the **Y=** editor.

- Cursor down to the function to be selected or deselected.

- Move the cursor to the equal sign and press $\boxed{\text{ENTER}}$ to toggle the state of the function on and off.

C. Display a function name, that is, $Y_1, Y_2, Y_3, \ldots$

- Press $\boxed{\text{VARS}}$ $\boxed{\triangleright}$ and then press **1** or $\boxed{\text{ENTER}}$ to display the list of function names.

- Press the number for the desired function, or cursor down to the function and press $\boxed{\text{ENTER}}$.

D. Select a style [such as Line (\\), Thick (\\), or Dot (∵.)] for the graph of a function

- Move the cursor to the symbol at the left of a function in the **Y=** editor.

- Press $\boxed{\text{ENTER}}$ repeatedly to select one of the seven styles.

E. Combine functions
Suppose Y_1 is $f(x)$ and Y_2 is $g(x)$.

- If $Y_3 = Y_1 + Y_2$, then Y_3 is $f(x) + g(x)$. (Similarly for $-$, $\times$, and $\div$.)

- If $Y_3 = Y_1(Y_2)$, then Y_3 is $f(g(x))$.

Specific Window Settings

A. Customize a window

- Press $\boxed{\text{WINDOW}}$ to open the window-setting screen, and edit the following values as desired:

- Xmin = *the leftmost value on the x-axis*

- Xmax = *the rightmost value on the x-axis*

- Xscl = *the distance between tick marks on the x-axis*

- Ymin = *the bottom value on the y-axis*

- Ymax = *the top value on the y-axis*

- Yscl = *the distance between tick marks on the y-axis*

- Xres = *a whole number from* 1 *to* 8
  Usually Xres = 1. Higher values speed up graphing, but with a loss of resolution.

  Note 1: The notation $[a, b]$ *by* $[c, d]$ stands for the window settings Xmin = a, Xmax = b, Ymin = c, Ymax = d.

  Note 2: The default values of Xscl and Yscl are 1. The value of Xscl should be made large (small) if the difference between Xmax and Xmin is large (small). For instance, with the window settings $[0, 100]$ *by* $[-1, 1]$, good scale settings are Xscl = 10 and Yscl = .1.

B. Use a predefined window setting

- Press $\boxed{\text{ZOOM}}$ to display the list of predefined settings.

- Either press a number or move the cursor down to an item and press $\boxed{\text{ENTER}}$.

- Select ZStandard to obtain $[-10, 10]$ *by* $[-10, 10]$, Xscl = Yscl = 1.

- Select ZDecimal to obtain $[-4.7, 4.7]$ *by* $[-3.1, 3.1]$, Xscl = Yscl = 1. (When TRACE is used with this setting, points have nice x-coordinates.)

- Select ZSquare to obtain a true-aspect window. (With such a window setting, lines that should be perpendicular actually look perpendicular, and the graph of $y = \sqrt{1 - x^2}$ looks like the top half of a circle.)

- Select ZTrig to obtain almost $[-2\pi, 2\pi]$ *by* $[-4, 4]$, Xscl = $\pi/2$, Yscl = 1, a good setting for the graphs of trigonometric functions.

C. Some nice window settings
(With these settings, one unit on the x-axis has the same length as one unit on the y-axis, and tracing progresses over simple values.)

- $[-4.7, 4.7]$ *by* $[-3.1, 3.1]$
- $[-2.35, 2.35]$ *by* $[-1.55, 1.55]$
- $[-7.05, 7.05]$ *by* $[-4.65, 4.65]$
- $[-9.4, 9.4]$ *by* $[-6.2, 6.2]$
- $[0, 9.4]$ *by* $[0, 6.2]$
- $[0, 18.8]$ *by* $[0, 12.4]$
- $[0, 47]$ *by* $[0, 31]$
- $[0, 94]$ *by* $[0, 62]$

General principle: (Xmax − Xmin) should be a number of the form $k \cdot 9.4$, where k is a whole number or $\frac{1}{2}, \frac{3}{2}, \frac{5}{2}, \ldots$, then (Ymax − Ymin) should be $(31/47) \cdot (\text{Xmax} - \text{Xmin})$.

Derivative, Slopes, and Tangent Lines

A. Compute $f'(a)$ from the home screen using nDeriv($f(x)$, X, a)

- Press $\boxed{\text{MATH}}$ **8** to display **nDeriv(**.

- Enter either $Y_1, Y_2, \ldots$ or an expression for $f(x)$.

- Type in the remaining items and press $\boxed{\text{ENTER}}$.

B. Define derivatives of the function Y_1

- Set $Y_2 = \mathbf{nDeriv}(Y_1, X, X)$ to obtain the 1st derivative.

- Set $Y_3 = \mathbf{nDeriv}(Y_2, X, X)$ to obtain the 2nd derivative.

  Note: $\mathbf{nDeriv(}$ is obtained by pressing [MATH] **8**.

C. Compute the slope of a graph at a point

- Press [2nd] [CALC] **6** to obtain $\mathbf{dy/dx}$.

- Use the arrow keys to move to the point of the graph or type in the x-coordinate of the point.

- Press [ENTER].

D. Draw a tangent line to a graph

- Press [2nd] [DRAW] **5** to select **Tangent**.

- Use the arrow keys to move to a point of the graph or type in the x-coordinate of a point.

- Press [ENTER].

  Note: To remove all tangent lines, press [2nd] [DRAW] **1** to execute **ClrDraw**.

Special Points on the Graph of Y_1

A. Find a point of intersection with the graph of Y_2, from the home screen

- Press [MATH] **0** for the (equation) **Solver**.

- Press [△] [CLEAR] to clear the edit screen of the **EQUATION SOLVER**.

- Enter either $Y_1 - Y_2$ or an expression for the difference of the two functions to the right of **"eqn:0="**.

- Press [▽] [CLEAR]. The equation to be solved will be on the first line of the screen and the cursor will be just to the right of **"X="**.

- Type in a guess for the x-coordinate of a point of intersection and then press [ALPHA] [SOLVE]. (The word SOLVE is above the [ENTER] key.) After a delay, the value you typed in will be replaced by the value of the x-coordinate of a point of intersection.

  Note 1: You needn't be concerned with the last two lines displayed.

  Note 2: If you press [ENTER] by mistake, the cursor will move to the third line of the display. Press [△] [ALPHA] [SOLVE] to continue.

- You can now insert a different guess to the right of **"X="** and press [ALPHA] [SOLVE] again to obtain the x-coordinate of another point of intersection.

B. Find intersection points, with graphs displayed

- Press [GRAPH] to display the graphs of all selected functions.

- Press [2nd] [CALC] **5** to select **intersect**.

- Reply to **"First curve?"** by using [△] or [▽] (if necessary) to move the cursor to one of the two curves and then pressing [ENTER].

- Reply to **"Second curve?"** by using [△] or [▽] (if necessary) to move the cursor to the other curves and then pressing [ENTER].

- Reply to **"Guess?"** by moving the cursor near the point of intersection (or typing in an approximate value of the x-coordinate of the point of intersection) and pressing [ENTER].

C. Find the second coordinate of the point whose first coordinate is a

  From the home screen

- Display $Y_1(a)$ and press [ENTER].
  or

- Press a [STO ▷] [X,T,θ,n] [ENTER] to assign the value a to the variable X.

- Display Y_1 and press [ENTER].

  From the home screen or with the graph displayed

- Press [2nd] [CALC] **1** to select **value**.

- Type in the value of a and press [ENTER].

  Note: The value of a must be between Xmin and Xmax.

- If desired, press [△] to move to points on graphs of other selected functions.

  With the graph displayed

- Press [TRACE].

- Type in the value of a and press [ENTER].

D. Find the first coordinate of a point whose second coordinate is b

  From the home screen

- Select the **EQUATION SOLVER** screen as in part A above, and set **"eqn:0"** $= Y_1 - b$.

- Press [▽] [CLEAR] to switch to another screen and clear the initial value of X.

- Type in a guess for the value of X and then press [ALPHA] [SOLVE].

  With the graphs displayed

- Set $Y_2 = b$.

- Find the point of intersection of the graphs of Y_1 and Y_2 as in part B above.

E. Find an x-intercept (that is, a *zero* of Y_1)

- Graph Y_1.

- Press [2nd] [CALC] **2** to select **zero**.

  Note: If more than one graph is displayed press [△] until the expression for Y_1 appears at the top of the screen.

- Move the cursor to a point just to the left of a zero (or type in a number less than a zero) and press [ENTER].

- Move the cursor to a point just to the right of the zero (or type in a number greater than a zero) and press [ENTER].

- Move the cursor to a point near the zero (or type in a number near the zero) and press ENTER.

F. Find a relative extreme point

- Set $\mathtt{Y_2 = nDeriv(Y_1, X, X)}$ or set Y_2 equal to the exact expression for the derivative of Y_1. [To display **nDeriv(** press MATH 8.]

- Select Y_2 and deselect all other functions.

- Graph Y_2.

- Find an x-intercept of Y_2, call it r, at which the graph of Y_2 crosses the x-axis.

- The point $(r, Y_1(r))$ will be a possible relative extreme point of Y_1.

G. Find an inflection point

- Set $\mathtt{Y_2 = nDeriv(Y_1, X, X)}$ and $\mathtt{Y_3 = nDeriv(Y_2, X, X)}$, or set Y_3 equal to the exact expression for the second derivative of Y_1.

- Select Y_3 and deselect all other functions.

- Graph Y_3.

- Find an x-intercept of Y_3, call it r, at which the graph of Y_3 crosses the x-axis.

- The point $(r, Y_1(r))$ will be a possible inflection point of Y_1.

Tables

A. Display values of $f(x)$ for evenly spaced values of x

- Press Y=, assign the function $f(x)$ to Y_1.

- Press 2nd [TblSet].

- Set **TblStart** =*first value of x*.

- Set Δ**Tbl** =*increment for values of x*.

- Set both **Indpnt** and **Depend** to **Auto**.

- Press 2nd [TABLE].

  Note 1: You can use △ or ▽ to look at function values for other values of x.

  Note 2: The table can display values of more than one function. For example, in the first step you can assign the function $g(x)$ to Y_2 and also select Y_2 to obtain a table with columns for X, Y_1, and Y_2.

B. Display values of $f(x)$ for arbitrary values of x

- Press Y= and assign the function $f(x)$ to Y_1, and deselect all other functions.

- Press 2nd [TblSet].

- Set **Indpnt** to **Ask** by moving the cursor to Ask and pressing ENTER.

- Leave **Depend** set to **Auto**.

- Press 2nd [TABLE].

- Type in any value for X and press ENTER.

- Repeat the previous step for any other values of X.

Riemann Sums

Suppose Y_1 is $f(x)$, and c, d, and Δx are numbers. Then **sum(seq($Y_1, X, c, d, \Delta x$))** computes

$$f(c) + f(c + \Delta x) + f(c + 2\Delta x) + \cdots + f(d).$$

The function **sum** is in the LIST/MATH menu and the function **seq** is in the LIST/OPS menu.

A. Compute $[f(x_1) + f(x_2) + \cdots + f(x_n)] \cdot \Delta x$

On the home screen, evaluate sum(seq($f(x)$, X, x_1, x_n, Δx)) $*$ Δx as follows.

- Press 2nd [LIST] and move the cursor right to MATH

- Press 5 (to display **sum(**.

- Press 2nd [LIST] and move the cursor right to OPS.

- Press 5 (to display **sum(**.

- Enter either Y_1 or an expression for $f(x)$.

- Type in the remaining items and press ENTER.

Definite Integrals and Antiderivatives

A. Compute $\int_a^b f(x)\, dx$

On the home screen, evaluate fnInt($f(x)$, x, a, b) as follows:

- Press MATH 9 to display **fnInt(**.

- Enter either Y_1 or an expression for $f(x)$.

- Type in the remaining items and press ENTER.

B. Shade a region under the graph of a function and find its area

- Press 2nd [CALC] 7 to select $\int \mathbf{f(x)}\, \mathbf{dx}$.

- If necessary, use △ or ▽ to move the cursor to the graph.

- In response to the request for a Lower Limit, move the cursor to the left endpoint of the region (or type in the value of a) and press ENTER.

- In response to the request for an Upper Limit, move the cursor to the right endpoint of the region or type in the value of b) and press ENTER.

  Note: To remove the shading, press 2nd [DRAW] 1 to execute the **ClrDraw** command.

- Press 2nd [LIST] and move the cursor to **OPS**.

- Press 5 to display **seq(**.

C. Obtain the graph of the solution to the differential equation $y' = g(x)$, $y(a) = b$

[That is, obtain the graph of the function $f(x)$ that is an antiderivative of $g(x)$ and satisfies the additional condition $f(a) = b$.]

- Set $\mathtt{Y_1 = g(X)}$.

- Set $Y_2 = \mathtt{fnInt}(Y_1, X, a, X) + b$. [Press MATH 9 to display $\mathtt{fnInt}$(.] The function Y_2 is an antiderivative of $g(x)$ and can be evaluated and graphed.

  Note: The graphing of Y_2 proceeds very slowly. Graphing can be speeded up by setting xRes (in the WINDOW screen) to a high value.

D. Shade the region between two curves

  Suppose the graph of Y_1 lies below the graph of Y_2 for $a \le x \le b$ and both functions have been selected. To shade the region between these two curves execute the instructions $\mathtt{Shade(Y_1, Y_2}, a, b)$ as follows:

- Press 2nd [DRAW] 7 to display $\mathtt{Shade}$(.

- Type in the remaining items and press ENTER.

  Note: To remove the shading, press 2nd [DRAW] 1 ENTER to execute the $\mathtt{ClrDraw}$ command.

Functions of Several Variables

A. Specify a function of several variables and its derivatives

- In the **Y=** editor, set $Y_1 = f(X, Y)$. (The letter Y is entered by pressing ALPHA [Y].)

- Set $Y_2 = \mathtt{nDeriv}(Y_1, X, X)$. Y_2 will be $\frac{\partial f}{\partial x}$.

- Set $Y_3 = \mathtt{nDeriv}(Y_1, Y, Y)$. Y_3 will be $\frac{\partial f}{\partial y}$.

- Set $Y_4 = \mathtt{nDeriv}(Y_2, X, X)$. Y_4 will be $\frac{\partial^2 f}{\partial x^2}$.

- Set $Y_5 = \mathtt{nDeriv}(Y_3, Y, Y)$. Y_5 will be $\frac{\partial^2 f}{\partial y^2}$.

- Set $Y_6 = \mathtt{nDeriv}(Y_3, X, X)$. Y_6 will be $\frac{\partial^2 f}{\partial x \, \partial y}$.

B. Evaluate one of the functions in part A at $x = a$ and $y = b$

- In the home screen, assign the value a to the variable X with a STO ▷ X,T,θ,n.

- Press b STO ▷ ALPHA [Y] to assign the value b to the variable Y.

- Display the name of one of the functions, such as $Y_1, Y_2, \ldots$, and press ENTER.

Least-Squares Approximations

A. Obtain the equation of the least-squares line

  Assume the points are $(x_1, y_1), \ldots, (x_n, y_n)$.

- Press STAT 1 for the EDIT screen, and obtain a table used for entering the data.

- If necessary, clear data from columns L_1 and L_2 as follows. Move the cursor to the top of column L_1 and press CLEAR ENTER. Repeat for column L_2.

- To enter the x-coordinates of the points, move the cursor to the first blank row of column L_1, enter the value of x_1, and press ENTER. Repeat with $x_2, \ldots, x_n$.

- Move the cursor to the first blank row of column L_2, and enter the values of $y_1, \ldots, y_n$.

- Press STAT ▷ for the CALC menu, and press 4 to place $\mathtt{LinReg(ax + b)}$ on the home screen.

- Press ENTER to obtain the slope and y-intercept of the least-squares line.

B. Assign the least-squares line to a function

- Press Y=, move the cursor to the function, and press CLEAR to erase the current expression.

- Press VARS 5 to select the **Statistics** variables.

- Press ▷ ▷ to select the EQ menu, and press 1 for **RegEQ** (Regression Equation), which then assigns the least-squares line equation to the function.

C. Display the points from part A

- Press Y= and deselect all functions.

- Press 2nd [STAT PLOT] ENTER to select **Plot1**, press ENTER to turn **Plot1 ON**.

- Select the first plot from the six icons for the types of plots. This icon corresponds to a scatter plot.

- Press GRAPH to display the data points.

  Note 1: Press ZOOM 9 to ensure that the current window setting is large enough to display the points.

  Note 2: When you finish using the point-plotting feature, turn it off. Press 2nd [STAT PLOT] ENTER to select **Plot1**. Then press ▷ to move to **OFF** and press ENTER.

D. Display the line and the points from part A

- Press Y= and deselect all functions except for the function containing the equation of the least-squares line.

- Carry out all but the first step of part C.

Miscellaneous Items and Tips

A. From the home screen, if you plan to reuse a recently entered line with some minor changes, press 2nd [ENTRY] until the previous line appears. You can then make alterations to the line and press ENTER to execute the line.

B. If you plan to use TRACE to examine the values of various points on a graph, set Ymin to a value that is lower than is actually necessary for the graph. Then, the values of x and y will not obliterate the graph while you trace.

C. To clear the home screen, press CLEAR twice.

Appendix C
Calculus and the TI-85 Calculator

Functions

A. Define a function, say y1, from the Home screen

- Press [2nd] [QUIT] to invoke the Home screen.
- Press [2nd] [alpha] [Y] **1** [ALPHA] [=] followed by an expression for the function, and press [ENTER].

B. Define a function from the function editor

- Press [GRAPH] [F1] to select $y(x) =$ from the GRAPH menu and obtain the screen for defining functions; that is, the function editor.
- Cursor down to a function. (To delete an existing expression, press [CLEAR]. To create an additional function, cursor down to the last function and press [ENTER].)
- Type in an expression for the function. (Press [F1] or [x-VAR] to display x. Press [F2] or [2nd] [alpha] [Y] to display y.)

C. Select or deselect a function in the function editor

(Functions with highlighted equal signs are said to be *selected*. The graph screen displays the graphs of all selected functions.)

- Press [GRAPH] [F1] to invoke the function editor.
- Cursor down to the function to be selected or deselected.
- Press [F5], that is, SELECT, to toggle the state of the function on and off.

D. Display a function name, that is, $y1, y2, y3, \ldots$

- Press [2nd] [alpha] [Y] followed by the number.

or

- Press [2nd] [VARS] [MORE] [F3] to invoke a list containing the function names.
- Cursor down to the desired function.
- Press [ENTER] to display the selected function name.

E. Combine functions

Suppose y1 is $f(x)$ and y2 is $g(x)$.

- If y3 = y1 + y2, then y3 is $f(x) + g(x)$. (Similarly for $-$, $\times$, and $\div$.)
- If y3 = evalF(y1, x, y2), then y3 is $f(g(x))$. (To display evalF(, press [2nd] [CALC] [F1].)

Specify Window Settings

A. Customize a viewing rectangle window

- Press [GRAPH] [F2] to invoke the RANGE screen and edit the following values as desired.
- xMin = *the leftmost value on the x-axis*
- xMax = *the rightmost value on the x-axis*
- xScl = *the distance between tick marks on the x-axis*
- yMin = *the bottom value on the y-axis*
- yMax = *the top value on the y-axis*
- yScl = *the distance between tick marks on the y-axis*

Note 1: The notation $[a, b]$ *by* $[c, d]$ stands for the range settings xMin = a, xMax = b, yMin = c, yMax = d.

Note 2: The default values of xScl and yScl are 1. The value of xScl should be made large (small) if the difference between xMax and xMin is large (small). For instance, with the window settings $[0, 100]$ *by* $[-1, 1]$, good scale settings are xScl = 10 and yScl = .1.

B. Use a predefined range setting

- Press [GRAPH] [F3] to invoke the ZOOM menu.
- Press [F4], that is, ZSTD, to obtain $[-10, 10]$ *by* $[-10, 10]$, xScl = yScl = 1.
- Press [MORE] [F2], that is, ZSQR, to obtain a true-aspect viewing rectangle. (With such a viewing rectangle, lines that should be perpendicular actually look perpendicular, and the graph of $y = \sqrt{1 - x^2}$ looks like the top half of a circle.)
- Press [MORE] [F4], that is ZDECM, to obtain $[-6.3, 6.3]$ *by* $[-3.1, 3.1]$, xScl = yScl = 1. (When TRACE is used with this viewing rectangle, points have nice x-coordinates.)
- Press [MORE] [F3], that is ZTRIG, to obtain $[-21\pi/8, 21\pi/8]$ *by* $[-4, 4]$, xScl = $\pi/2$, yScl = 1, a good setting for the graphs of trigonometric functions.

C. Some nice range settings

With these settings, one unit on the x-axis has the same length as one unit on the y-axis, and tracing progresses over simple values.

- $[-6.3, 6.3]$ *by* $[-3.7, 3.7]$
- $[-3.15, 3.15]$ *by* $[-1.85, 1.85]$
- $[-9.45, 9.45]$ *by* $[-5.55, 5.55]$
- $[-12.6, 12.6]$ *by* $[-7.4, 7.4]$
- $[0, 12.6]$ *by* $[0, 7.4]$
- $[0, 25.2]$ *by* $[0, 14.8]$
- $[0, 63]$ *by* $[0, 37]$
- $[0, 126]$ *by* $[0, 74]$

General principle: (xMax − xMin) should be a number of the form $k \cdot 6.3$, where k is a whole number or $\frac{1}{2}, \frac{3}{2}, \frac{5}{2}, \ldots$, then (yMax − yMin) should be $(37/63) \cdot$ (xMax − xMin).

Derivative, Slopes, and Tangent Lines

A. Compute $f'(a)$ from the home screen using der1($f(x)$, x, a)

- Press [2nd] [CALC] [F3] to display **der1(**.
- Enter either y1, y2, ... or an expression for $f(x)$.

- Type in the remaining items and press ENTER.

B. **Define derivatives of the function y5**

- Set **y1 = der1(y5, x, x)** to obtain the 1st derivative.
- Set **y2 = der2(y5, x, x)** to obtain the 2nd derivative.
- Set **y3 = nDer(y2, x, x)** to obtain the 3rd derivative.
- Set **y4 = nDer(y3, x, x)** to obtain the 4th derivative.

    Note: **der1**, **der2**, and **nDer** are found on the menu obtained by pressing 2nd [CALC].

C. **Compute the slope of graph at a point**

- Press GRAPH F5 to display the graph of the function.
- Press MORE F1 F4 to select dy/dx from the GRAPH/MATH menu.
- Use the arrow keys to move to the point of the graph.
- Press ENTER.

    Note: This process usually works best with a nice range setting.

D. **Draw a tangent line to a graph**

- Press GRAPH F5 to display the graph of the function.
- Press MORE F1 MORE MORE F3 to select TANLN from the GRAPH/MATH menu.
- Move the cursor to any point on the graph.
- Press ENTER to draw the tangent line through the point and display the slope of the curve at that point. The slope is displayed at the bottom of the screen as $dy/dx = slope$.
- To draw another tangent line, press GRAPH and then repeat the previous three steps.

    Note 1: To remove all tangent lines, press GRAPH MORE F2 MORE F5.

    Note 2: This process usually works best with a nice range or a range in which the x-coordinate of the point is halfway between xMin and xMax.

Special Points on the Graph of y1

A. **Find a point of intersection with the graph of y2, from the Home screen**

- Press 2nd [SOLVER].
- To the right of **"eqn:"** enter **y1 − y2 = 0** and press ENTER. (y1 and y2 can be entered via F keys and the equal sign is entered with ALPHA [=].)
- To the right of **"x="** type in a guess for the x-coordinate of the point of intersection, and then press F5. After a little delay, a value of x for which y1 = y2 will be displayed in place of your guess.

B. **Find intersection points, with graphs displayed**

- Press GRAPH F5 to display the graphs of all selected functions.

- Press MORE F1 MORE F5 to select ISECT from the GRAPH/MATH menu.
- If necessary, use △ or ▽ to place the cursor on one of the two curves.
- Move the cursor close to the point of intersection and then press ENTER.
- If necessary, use △ or ▽ to place the cursor on the other curve.
- Press ENTER to display the coordinates of the point of intersection.

C. **Find the second coordinate of the point whose first coordinate is a**

    From the Home screen
    Compute **evalF(y1, x, a)** as follows:

- Press 2nd [CALC] F1 to display **evalF(**.
- Enter either y1 or an expression for the function.
- Type in the remaining items and press ENTER.
    or
- Press a STO ▷ X-VAR ENTER to assign the value a to the variable x.
- Display y1 and press ENTER.

    From the Home screen or with the graph displayed

- Press GRAPH MORE MORE F1.
- Type in the value of a and press ENTER. (The value of a must be between xMin and xMax.)
- If desired, press the up-arrow key, △, to move to points on graphs of other selected functions.

    With the graph displayed

- Press F4; that is, TRACE.
- Move cursor with ▷ and/or ◁ until the x-coordinate of the cursor is as close as possible to a. *Note*: Usually works best if one of the nice range settings discussed above is used.

D. **Find the first coordinate of a point whose second coordinate is b**

- Set **y2** = b.
- Find the point of intersection of the graphs of y1 and y2 as in part B above.

E. **Find an x-intercept of a graph of a function**

- Press GRAPH MORE F1 F3 to select ROOT from the GRAPH/MATH menu.
- Move the cursor along the graph of the function close to an x-intercept, and press ENTER.

F. **Find a relative extreme point**

- Set **y2 = der1(y1, x, x)** or set y2 equal to the exact expression for the derivative of y1. [To display **der1(** press 2nd [CALC] F3.]
- Select y2 and deselect all other functions.
- Graph y2.
- Find an x-intercept of y2, call it r, at which the graph of y2 crosses the x-axis.

- The point $(r, y1(r))$ will be a possible relative extreme point of Y_1.

G. Find an inflection point

- Set **y2 = der2(y1, x, x)** or set y2 equal to the exact expression for the second derivative of y1. (To display der2, press $\boxed{\text{2nd}}$ [CALC] $\boxed{\text{F4}}$.)

- Select y2 and deselect all other functions.

- Graph y2.

- Find an x-intercept of y2, call it r, at which the graph of y2 crosses the x-axis.

- The point whose first coordinate is r will be a possible inflection point of y1.

Riemann Sums

Suppose that y1 is $f(x)$, and c, d, and Δx are numbers, then **sum(seq(y1,x,c,d,Δx))** computes

$$f(c) + f(c + \Delta x) + f(c + 2\Delta x) + \cdots + f(d).$$

The functions **sum** and **seq** are found in the LIST/OPS menu.

A. Compute $[f(x_1) + f(x_2) + \cdots + f(x_n)] \cdot \Delta x$

- On the Home screen, evaluate sum(seq($f(x)$, x, x_1, x_n, Δx)) $* \Delta x$ as follows:

- Press $\boxed{\text{2nd}}$ [LIST] $\boxed{\text{F5}}$ $\boxed{\text{MORE}}$ $\boxed{\text{F1}}$ $\boxed{(}$ $\boxed{\text{F3}}$ to display sum(seq(.

- Enter either y1 or an expression for $f(x)$.

- Type in the remaining items and press $\boxed{\text{ENTER}}$.

Definite Integrals and Antiderivatives

A. Compute $\int_a^b f(x)\, dx$

- On the Home screen, evaluate fnInt($f(x)$, x, a, b) as follows.

- Press $\boxed{\text{2nd}}$ [CALC] $\boxed{\text{F5}}$ to display **fnInt(**.

- Enter either y1 or an expression for $f(x)$.

- Type in the remaining items and press $\boxed{\text{ENTER}}$.

B. Find the area of a region under the graph of a function

- Press $\boxed{\text{GRAPH}}$ $\boxed{\text{MORE}}$ $\boxed{\text{F1}}$ $\boxed{\text{F5}}$ to select $\int f(x)$ from the GRAPH/MATH menu.

- If necessary, use $\boxed{\triangle}$ or $\boxed{\triangledown}$ to move the cursor to the graph.

- Move the cursor to the left endpoint of the region and press $\boxed{\text{ENTER}}$.

- Move the cursor to the right endpoint of the region and press $\boxed{\text{ENTER}}$.

Note: This process usually works best with a nice range setting.

C. Obtain the graph of the solution to the differential equation $y' = g(x)$, $y(a) = b$

That is, obtain the graph of the function $f(x)$ that is an antiderivative of $g(x)$ and satisfies the additional condition $f(a) = b$.

- Set **y1 = g(x)**.

- Set **y2 = fnInt(y1, x, a, x) + b**. (To display **fnInt(**, press $\boxed{\text{2nd}}$ [CALC] $\boxed{\text{F5}}$.) The function y2 is an antiderivative of $g(x)$ and can be evaluated and graphed.

Note: The graphing of y2 proceeds very slowly.

D. Shade the region between two curves

Suppose the graph of y1 lies below the graph of y2 for $a \leq x \leq b$ and both functions have been selected. To shade the region between these two curves, execute the instruction **Shade(y1,y2,a,b)** as follows.

- Press $\boxed{\text{GRAPH}}$ $\boxed{\text{MORE}}$ $\boxed{\text{F2}}$ $\boxed{\text{F1}}$ to display **Shade(** from the GRAPH/DRAW menu.

- Type in the remaining items and press $\boxed{\text{ENTER}}$.

Note: To remove the shading, press $\boxed{\text{GRAPH}}$ $\boxed{\text{MORE}}$ $\boxed{\text{F2}}$ $\boxed{\text{MORE}}$ $\boxed{\text{F5}}$ to execute **ClrDraw** from the MATH/DRAW menu.

Functions of Several Variables

A. Specify a function of several variables and its derivatives

- In the $y(x) =$ function editor, set y1 $= f(x, y)$. (The letters x and y can be entered by pressing $\boxed{\text{F1}}$ and $\boxed{\text{F2}}$.)

- Set **y2 = der1(y1, x, x)**. y2 will be $\frac{\partial f}{\partial x}$.

- Set **y3 = der1(y1, y, y)**. y3 will be $\frac{\partial f}{\partial y}$.

- Set **y4 = der2(y1, x, x)**. y4 will be $\frac{\partial^2 f}{\partial x^2}$.

- Set **y5 = der2(y1, y, y)**. y5 will be $\frac{\partial^2 f}{\partial y^2}$.

- Set **y6 = nDer(y3, x, x)**. y6 will be $\frac{\partial^2 f}{\partial x\, \partial y}$.

B. Evaluate one of the functions in part A at $x = a$ and $y = b$

- In the Home screen, assign the value a to the variable x with a $\boxed{\text{STO} \triangleright}$ $\boxed{\text{x-VAR}}$.

- Press b $\boxed{\text{STO} \triangleright}$ $\boxed{\text{2nd}}$[alpha] [Y] to assign the value b to the variable y.

- Display the name of one of the functions, such as y1, y2, ..., and press $\boxed{\text{ENTER}}$.

Least-Squares Approximations

A. Obtain the equation of the least-squares line

Assume the points are $(x_1, y_1), \ldots, (x_n, y_n)$.

- Press $\boxed{\text{STAT}}$ $\boxed{\text{F2}}$ $\boxed{\text{ENTER}}$ $\boxed{\text{ENTER}}$ to obtain a list used for entering the data.

- Press $\boxed{\text{F5}}$ to clear all previous data.

- Enter the data for the points by pressing x_1 $\boxed{\text{ENTER}}$ y_1 $\boxed{\text{ENTER}}$ x_2 $\boxed{\text{ENTER}}$ y_2 $\boxed{\text{ENTER}}$ ... x_n $\boxed{\text{ENTER}}$ y_n.

- Press $\boxed{\text{STAT}}$ $\boxed{\text{F1}}$ $\boxed{\text{ENTER}}$ $\boxed{\text{ENTER}}$ $\boxed{\text{F2}}$ to obtain the values of a and b where the least-squares line has equation $y = bx + a$.

- If desired, the straight line (and the points) can be graphed with the following steps:
  - (a) First press $\boxed{\text{GRAPH}}$ $\boxed{\text{F1}}$ and deselect all functions.
  - (b) Press $\boxed{\text{STAT}}$ $\boxed{\text{F3}}$ to invoke the statistical DRAW menu. (If any graphs appear, press $\boxed{\text{F5}}$ to clear them.)
  - (c) Press $\boxed{\text{F4}}$ to draw the least-squares line and press $\boxed{\text{F2}}$ to draw the n points.

B. Assign the least-squares line to a function

- Press $\boxed{\text{GRAPH}}$ $\boxed{\text{F1}}$, move the cursor to the function, and press $\boxed{\text{CLEAR}}$ to erase the current expression for the function.

- Press $\boxed{\text{STAT}}$ $\boxed{\text{F5}}$ $\boxed{\text{MORE}}$ $\boxed{\text{MORE}}$ $\boxed{\text{F2}}$ to assign the equation (known as RegEq) to the function.

C. Display the points from part A

- Press $\boxed{\text{GRAPH}}$ $\boxed{\text{F1}}$ and deselect all functions.

- Press $\boxed{\text{STAT}}$ $\boxed{\text{F3}}$ $\boxed{\text{F2}}$ to select SCAT from the STAT/DRAW menu.

Note 1: Make sure the current window setting is large enough to display the points.

Note 2: To also draw the least-squares line, press $\boxed{\text{STAT}}$ $\boxed{\text{F3}}$ $\boxed{\text{F4}}$ to select DRREG from the STAT/DRAW menu.

Note 3: To erase the points, press $\boxed{\text{STAT}}$ $\boxed{\text{F3}}$ $\boxed{\text{F5}}$ to select CLDRW from the STAT/DRAW menu.

Miscellaneous Items and Tips

A. From the Home screen, if you plan to reuse the most recently entered line with some minor changes, press $\boxed{\text{2nd}}$ [ENTRY] to display the previous line. You can then make alterations to the line and press $\boxed{\text{ENTER}}$ to execute the line.

B. If you plan to use TRACE to examine the values of various points on a graph, set yMin to a value that is lower than is actually necessary for the graph. Then, the values of x and y will not obliterate the graph while you trace.

C. To clear the Home screen, press $\boxed{\text{CLEAR}}$ twice.

D. When two menus are displayed at the same time, you can remove the top menu by pressing $\boxed{\text{EXIT}}$. (After that, you can remove the remaining menu with $\boxed{\text{CLEAR}}$.)

E. To obtain the solutions of a quadratic equation, or of any equation of the form $p(x) = 0$, where $p(x)$ is a polynomial of degree ≤ 30, press $\boxed{\text{2nd}}$ [POLY], enter the degree of the polynomial, enter the coefficients of the polynomial, and press $\boxed{\text{F5}}$. Some of the solutions might be complex numbers.

Appendix D
Calculus and the TI-86 Calculator

Functions

A. Define a function, say y1, from the home screen

- Press $\boxed{\text{2nd}}$ $\lfloor$QUIT$\rfloor$ to invoke the home screen.
- Press $\boxed{\text{2nd}}$ [alpha] [Y] **1** $\boxed{\text{ALPHA}}$ [=] followed by an expression for the function, and press $\boxed{\text{ENTER}}$.

B. Define a function from the function editor

- Press $\boxed{\text{GRAPH}}$ $\boxed{\text{F1}}$ to select $y(x) =$ from the GRAPH menu and obtain the screen for defining functions; that is, the function editor.
- Cursor down to a function. (To delete an existing expression, press $\boxed{\text{CLEAR}}$. To create an additional function, cursor down to the last function and press $\boxed{\text{ENTER}}$.)
- Type in an expression for the function. (*Note*: Press $\boxed{\text{F1}}$ or $\boxed{\text{x-VAR}}$ to display x. Press $\boxed{\text{F2}}$ or $\boxed{\text{2nd}}$ [alpha] [Y] to display y.)

C. Select or deselect a function in the function editor

Functions with highlighted equal signs are said to be *selected*. The graph screen displays the graphs of all selected functions, and tables contain a column for each selected function.

- Press $\boxed{\text{GRAPH}}$ $\boxed{\text{F1}}$ to invoke the function editor.
- Cursor down to the function to be selected or deselected.
- Press $\boxed{\text{F5}}$, that is, SELCT, to toggle the state of the function on and off.

D. Select a style [such as Line (\), Thick (\), or Dot (⋅⋅)] for the graph of a function

- Move the cursor to a function in the function editor.
- Press $\boxed{\text{MORE}}$ and then press $\boxed{\text{F3}}$ repeatedly to select one of the seven styles.

E. Display a function name, that is, y1, y2, y3, …

- Press $\boxed{\text{2nd}}$ [alpha] [Y] followed by the number.
  or
- Press $\boxed{\text{2nd}}$ [CATLG/VARS] $\boxed{\text{MORE}}$ $\boxed{\text{F4}}$ to invoke a list containing the function names.
- Cursor down to the desired function.
- Press $\boxed{\text{ENTER}}$ to display the selected function name.

F. Combine functions

Suppose y1 is $f(x)$ and y2 is $g(x)$.

- If y3 = y1 + y2, then y3 is $f(x) + g(x)$. (Similarly for $-$, $\times$, and $\div$.)
- If y3 = y1(y2), then y3 is $f(g(x))$.

Specify Window Settings

A. Customize a window

- Press $\boxed{\text{GRAPH}}$ $\boxed{\text{F2}}$ to invoke the window editor and edit the following values as desired.
- xMin = *the leftmost value on the x-axis*
- xMax = *the rightmost value on the x-axis*
- xScl = *the distance between tick marks on the x-axis*
- yMin = *the bottom value on the y-axis*
- yMax = *the top value on the y-axis*
- yScl = *the distance between tick marks on the y-axis*
  Usually xRes = 1. Higher values speed up graphing, but with a loss of resolution.

  Note 1: The notation $[a, b]$ *by* $[c, d]$ stands for the window settings xMin = a, xMax = b, yMin = c, yMax = d.

  Note 2: The default values of xScl and yScl are 1. The value of xScl should be made large (small) if the difference between xMax and xMin is large (small). For instance, with the window settings $[0, 100]$ *by* $[-1, 1]$, good scale settings are xScl = 10 and yScl = .1.

B. Use a predefined window setting

- Press $\boxed{\text{GRAPH}}$ $\boxed{\text{F3}}$ to invoke the ZOOM menu.
- Press $\boxed{\text{F4}}$, that is ZSTD, to obtain $[-10, 10]$ *by* $[-10, 10]$, xScl = yScl = 1.
- Press $\boxed{\text{MORE}}$ $\boxed{\text{F2}}$, that is ZSQR, to obtain a true-aspect window. (With such a window setting, lines that should be perpendicular actually look perpendicular, and the graph of $y = \sqrt{1 - x^2}$ looks like the top half of a circle.)
- Press $\boxed{\text{MORE}}$ $\boxed{\text{F4}}$, that is ZDECM, to obtain $[-6.3, 6.3]$ *by* $[-3.1, 3.1]$, xScl = yScl = 1. (When TRACE is used with this setting, points have nice x-coordinates.)
- Press $\boxed{\text{MORE}}$ $\boxed{\text{F3}}$, that is ZTRIG, to obtain $[-21\pi/8, 21\pi/8]$ *by* $[-4, 4]$, xScl = $\pi/2$, yScl = 1, a good setting for the graphs of trigonometric functions.

C. Some nice window settings

With these settings, one unit on the x-axis has the same length as one unit on the y-axis, and tracing progresses over simple values.

- $[-6.3, 6.3]$ *by* $[-3.7, 3.7]$
- $[-3.15, 3.15]$ *by* $[-1.85, 1.85]$
- $[-9.45, 9.45]$ *by* $[-5.56, 5.56]$
- $[-12.6, 12.6]$ *by* $[-7.4, 7.4]$
- $[0, 12.6]$ *by* $[0, 7.4]$
- $[0, 25.2]$ *by* $[0, 14.8]$
- $[0, 63]$ *by* $[0, 37]$
- $[0, 126]$ *by* $[0, 74]$

General principle: (xMax $-$ xMin) should be a number of the form $k \cdot 12.6$, where k is a whole number or $\frac{1}{2}, \frac{3}{2}, \frac{5}{2}, \ldots$, then (yMax $-$ yMin) should be $(37/63) \cdot (\text{xMax} - \text{xMin})$.

Derivative, Slopes, and Tangent Lines

A. Compute $f'(a)$ from the home screen using der1($f(x), x, a$)

- Press $\boxed{\text{2nd}}$ $\boxed{\text{CALC}}$ $\boxed{\text{F3}}$ to display **der1(**.
- Enter either $y1, y2, \ldots$ or an expression for $f(x)$.
- Type in the remaining items and press $\boxed{\text{ENTER}}$.

B. Define derivatives of the function y5

- Set **y1 = der1(y5, x, x)** to obtain the 1st derivative.
- Set **y2 = der2(y5, x, x)** to obtain the 2nd derivative.
- Set **y3 = nDer(y2, x, x)** to obtain the 3rd derivative.
- Set **y4 = nDer(y3, x, x)** to obtain the 4th derivative.

  Note 1: **der1**, **der2**, and **nDer** are found on the menu obtained by pressing $\boxed{\text{2nd}}$ [CALC].

  Note 2: To speed up the display of the 3rd and 4th derivatives, set the value of xRes in the WINDOW screen to at least 5.

C. Compute the slope of a graph at a point

- Press $\boxed{\text{GRAPH}}$ $\boxed{\text{F5}}$ to display the graph of the function.
- Press $\boxed{\text{MORE}}$ $\boxed{\text{F1}}$ $\boxed{\text{F2}}$ to select dy/dx from the GRAPH/MATH menu.
- Use the arrow keys to move to the point of the graph and then press $\boxed{\text{ENTER}}$. (This process usually works best with a nice window setting.) Or, type in the value of the x-coordinate of a point (any number between xMin and xMax), and press $\boxed{\text{ENTER}}$.

D. Draw a tangent line to a graph

- Press $\boxed{\text{GRAPH}}$ $\boxed{\text{F5}}$ to display the graph of the function.
- Press $\boxed{\text{MORE}}$ $\boxed{\text{F1}}$ $\boxed{\text{MORE}}$ $\boxed{\text{MORE}}$ $\boxed{\text{F1}}$ to select TANLN from the GRAPH/MATH menu.
- Move the cursor to any point on the graph or type in the first coordinate of a point.
- Press $\boxed{\text{ENTER}}$ to draw the tangent line through the point and display the slope of the curve at that point. The slope is displayed at the bottom of the screen as $dy/dx = slope$.
- To draw another tangent line, press $\boxed{\text{GRAPH}}$ and then repeat the previous three steps.

  Note: To remove all tangent lines, press $\boxed{\text{GRAPH}}$ $\boxed{\text{MORE}}$ $\boxed{\text{F2}}$ $\boxed{\text{MORE}}$ $\boxed{\text{MORE}}$ $\boxed{\text{F1}}$ to select CLDRW from the GRAPH/DRAW menu.

Special Points on the Graph of y1

A. Find a point of intersection with the graph of y2, from the home screen

- Press $\boxed{\text{2nd}}$ [SOLVER].
- To the right of **"eqn:"** enter **y1 − y2 = 0** and press $\boxed{\text{ENTER}}$. (*Note*: y1 and y2 can be entered via F keys and the equal sign is entered with $\boxed{\text{ALPHA}}$ $\boxed{[=]}$.)

- To the right of **"x="** type in a guess for the x-coordinate of the point of intersection, and then press $\boxed{\text{F5}}$. After a little delay, a value of x for which y1 = y2 will be displayed in place of your guess.

B. Find intersection points, with graphs displayed

- Press $\boxed{\text{GRAPH}}$ $\boxed{\text{F5}}$ to display the graphs of all selected functions.
- Press $\boxed{\text{MORE}}$ $\boxed{\text{F1}}$ $\boxed{\text{MORE}}$ $\boxed{\text{F3}}$ to select ISECT from the GRAPH/MATH menu.
- Reply to **"First curve?"** by using $\boxed{\triangle}$ or $\boxed{\triangledown}$ (if necessary) to place the cursor on one of the two curves and then pressing $\boxed{\text{ENTER}}$.
- Reply to **"Second curve?"** by using $\boxed{\triangle}$ or $\boxed{\triangledown}$ (if necessary) to place the cursor on the other curve and then pressing $\boxed{\text{ENTER}}$.
- Reply to **"Guess?"** by moving the cursor close to the desired point of intersection (or typing in a guess for the first coordinate of the desired point) and pressing $\boxed{\text{ENTER}}$.

C. Find the second coordinate of the point whose first coordinate is a

From the home screen

- Display **y1(**a**)** and press $\boxed{\text{ENTER}}$.
  or
- Press a $\boxed{\text{STO} \triangleright}$ $\boxed{\text{X-VAR}}$ $\boxed{\text{ENTER}}$ to assign the value a to the variable x.
- Display y1 and press $\boxed{\text{ENTER}}$.

  From the home screen or with the graph displayed

- Press $\boxed{\text{GRAPH}}$ $\boxed{\text{MORE}}$ $\boxed{\text{MORE}}$ $\boxed{\text{F1}}$ to display **"Eval x="**.
- Type in the value of a and press $\boxed{\text{ENTER}}$. (The value of a must be between xMin and xMax.)
- If desired, press the up-arrow key, $\boxed{\triangle}$, to move to points on graphs of other selected functions.

  With the graph displayed

- Press $\boxed{\text{F4}}$; that is, TRACE.
- Type in the value of a and press $\boxed{\text{ENTER}}$.

D. Find the first coordinate of a point whose second coordinate is b

- Set **y2 = b**.
- Find the point of intersection of the graphs of y1 and y2 as in part B above.

E. Find an x-intercept of a graph of a function

- Press $\boxed{\text{GRAPH}}$ $\boxed{\text{MORE}}$ $\boxed{\text{F1}}$ $\boxed{\text{F1}}$ to select ROOT from the GRAPH/MATH menu.
- If necessary, use $\boxed{\triangle}$ or $\boxed{\triangledown}$ to place the cursor on the desired graph.
- In response to the request for a Left Bound, move the cursor to a point whose first coordinate is less than the desired x-intercept or type in a value less than the desired x-intercept. Then press the $\boxed{\text{ENTER}}$ key.

- In response to the request for a Right Bound, move the cursor to a point whose first coordinate is greater than the desired x-intercept or type in a value greater than the desired x-intercept. Then press the ENTER key.
- In response to the request for a Guess, move the cursor to a point near the desired x-intercept or type in a value close to the desired x-intercept. Then press the ENTER key.

F. Find a relative extreme point

- Set $\mathbf{y2 = der1(y1,x,x)}$ or set y2 equal to the exact expression for the derivative of y1. (To display **der1(**, press 2nd [CALC] F3.)
- Select y2 and deselect all other functions.
- Graph y2.
- Find an x-intercept of y2, call it r, at which the graph of y2 crosses the x-axis.
- The point $(r, y1(r))$ will be a possible relative extreme point of y1.

G. Find an inflection point

- Set $\mathbf{y2 = der2(y1,x,x)}$ or set y2 equal to the exact expression for the second derivative of y1. (To display **der2**, press 2nd [CALC] F4.)
- Select y2 and deselect all other functions.
- Graph y2.
- Find an x-intercept of y2, call it r, at which the graph of y2 crosses the x-axis.
- The point $(r, y1(r))$ will be a possible inflection point of y1.

Tables

A. Display values of $f(x)$ for evenly spaced values of x

- Press GRAPH F1 and assign the function $f(x)$ to y1.
- Press TABLE F2 to invoke the TABLE SETUP window.
- Set **TblStart** = *first value of* x.
- Set Δ**Tbl** = *increment for values of* x.
- set **Indpnt** to **Auto**.
- Press F1 to display the table.

  Note 1: You can use △ or ▽ to look at function values for other values of x.

  Note 2: The table can display values of more than one function. For example, in the first step you can assign the function $g(x)$ to another function and also select that other function to obtain a table with columns for x and each of the selected functions.

B. Display values of $f(x)$ for arbitrary values of x

- Press GRAPH F1, assign the function $f(x)$ to a function, and deselect all other functions.

- Press TABLE F2.
- Set **Indpnt** to **Ask** by moving the cursor to **Ask** and pressing ENTER.
- Press F1.
- Type in any value for x and press ENTER.
- Repeat the previous step for as many values as you like.

Riemann Sums

Suppose that y1 is $f(x)$, and c, d, and Δx are numbers, then $\mathbf{sum(seq(y1,x,}c,d,\Delta x\mathbf{))}$ computes

$$f(c) + f(c + \Delta x) + f(c + 2\Delta x) + \cdots + f(d).$$

The functions **sum** and **seq** are found in the LIST/OPS menu.

A. Compute $[f(x_1) + f(x_2) + \cdots + f(x_n)] \cdot \Delta x$

On the home screen, evaluate sum(seq($f(x)$, x, x_1, x_n, Δx)) $* \Delta x$ as follows:

- Press 2nd [LIST] F5 MORE F1 (F3 to display sum(seq(.
- Enter either y1 or an expression for $f(x)$.
- Type in the remaining items and press ENTER.

Definite Integrals and Antiderivatives

A. Compute $\int_a^b f(x)\,dx$

On the home screen, evaluate fnInt($f(x)$, x, a, b) as follows:

- Press 2nd [CALC] F5 to display **fnInt(**.
- Enter either y1 or an expression for $f(x)$.
- Type in the remaining items and press ENTER.

B. Shade a region under the graph of a function from $x = a$ to $x = b$

- Press GRAPH MORE F1 F3 to select $\int f(x)$ from the GRAPH/MATH menu.
- If necessary, use △ or ▽ to move the cursor to the graph.
- In response to the request for a Lower Limit, move the cursor to the left endpoint of the region (or type in the value of a) and press ENTER.
- In response to the request for an Upper Limit, move the cursor to the right endpoint of the region (or type in the value of b) and press ENTER.

  Note: To remove the shading press GRAPH MORE F2 MORE MORE F1 to select CLDRW from the GRAPH/DRAW menu.

C. Obtain the graph of the solution to the differential equation $y' = g(x)$, $y(a) = b$

[That is, obtain the function $f(x)$ that is an antiderivative of $g(x)$ and satisfies the additional condition $f(a) = b$.]

- Set $\mathbf{y1} = \mathbf{g(x)}$.
- Set $\mathbf{y2} = \mathbf{fnInt(y1, x, a, x) + b}$. (To display $\mathbf{fnInt}$, press $\boxed{\text{2nd}}$ $\boxed{\text{CALC}}$ $\boxed{\text{F5}}$.) The function y2 is an antiderivative of $g(x)$ and can be evaluated and graphed.

  Note: The graphing of y2 proceeds very slowly. Graphing can be sped up by setting xRes (in the GRAPH/WIND menu) to a high value.

D. Shade the region between two curves

Suppose the graph of y1 lies below the graph of y2 for $a \leq x \leq b$ and both functions have been selected. To shade the region between these two curves, execute the instruction $\mathbf{Shade(y1, y2, a, b)}$ as follows.

- Press $\boxed{\text{GRAPH}}$ $\boxed{\text{MORE}}$ $\boxed{\text{F2}}$ $\boxed{\text{F1}}$ to display $\mathbf{Shade(}$ from the GRAPH/DRAW menu.
- Type in the remaining items and press $\boxed{\text{ENTER}}$.

  Note: To remove the shading, press $\boxed{\text{GRAPH}}$ $\boxed{\text{MORE}}$ $\boxed{\text{F2}}$ $\boxed{\text{MORE}}$ $\boxed{\text{MORE}}$ $\boxed{\text{F1}}$ to execute $\mathbf{ClrDraw}$ from the GRAPH/DRAW menu.

Functions of Several Variables

A. Specify a function of several variables and its derivatives

- In the $y(x) =$ function editor, set $\mathbf{y1} = f(x, y)$. (The letters x and y can be entered by pressing $\boxed{\text{F1}}$ and $\boxed{\text{F2}}$.)
- Set $\mathbf{y2} = \mathbf{der1(y1, x, x)}$. y2 will be $\frac{\partial f}{\partial x}$.
- Set $\mathbf{y3} = \mathbf{der1(y1, y, y)}$. y3 will be $\frac{\partial f}{\partial y}$.
- Set $\mathbf{y4} = \mathbf{der2(y1, x, x)}$. y4 will be $\frac{\partial^2 f}{\partial x^2}$.
- Set $\mathbf{y5} = \mathbf{der2(y1, y, y)}$. y5 will be $\frac{\partial^2 f}{\partial y^2}$.
- Set $\mathbf{y6} = \mathbf{nDer(y3, x, x)}$. y6 will be $\frac{\partial^2 f}{\partial x \, \partial y}$.

B. Evaluate one of the functions in part A at $x = a$ and $y = b$

- In the home screen, assign the value a to the variable x with a $\boxed{\text{STO ▷}}$ $\boxed{\text{X-VAR}}$.
- Press b $\boxed{\text{STO ▷}}$ $\boxed{\text{2nd}}$[alpha] [Y] to assign the value b to the variable y.
- Display the name of one of the functions, such as $y1, y2, \ldots$, and press $\boxed{\text{ENTER}}$.

Least-Squares Approximations

A. Obtain the equation of the least-squares line

Assume the points are $(x_1, y_1), \ldots, (x_n, y_n)$.

- Press $\boxed{\text{2nd}}$ $\boxed{\text{STAT}}$ $\boxed{\text{F2}}$ to obtain a table for entering the data.
- If necessary, clear data from columns xStat and yStat as follows. Move the cursor to xStat and press $\boxed{\text{CLEAR}}$ $\boxed{\text{ENTER}}$. Repeat for the yStat column.

- To enter the x-coordinates of the points, move the cursor to the first blank row the xStat column, type in the value of x_1, and press $\boxed{\text{ENTER}}$. Repeat with $x_2, \ldots, x_n$.
- Move the cursor to the first blank row of the yStat column and enter the values of $y_1, \ldots, y_n$.
- Place a 1 in each of the first n entries of the fStat column. (The other entries should be blank.)
- Press $\boxed{\text{2nd}}$ $\boxed{\text{QUIT}}$ to display the home screen.
- Press $\boxed{\text{2nd}}$ $\boxed{\text{STAT}}$ $\boxed{\text{F1}}$ $\boxed{\text{F3}}$ $\boxed{\text{ENTER}}$ to obtain the values of a and b where the least-squares line has equation $y = a + bx$.
- If desired, the straight line (and the points) can be graphed with the following steps:
  (a) First press $\boxed{\text{GRAPH}}$ $\boxed{\text{F1}}$ and deselect all functions.
  (b) Press $\boxed{\text{2nd}}$ $\boxed{\text{STAT}}$ $\boxed{\text{F4}}$ to invoke the statistical DRAW menu. (If any graphs appear, press $\boxed{\text{MORE}}$ $\boxed{\text{F2}}$ to clear them.)
  (c) Press the button for DRREG (either $\boxed{\text{F1}}$ or $\boxed{\text{MORE}}$ $\boxed{\text{F1}}$) to draw the least-squares line and press $\boxed{\text{MORE}}$ $\boxed{\text{F2}}$ to draw the n points with SCAT.

B. Assign the least-squares line to a function

- Press $\boxed{\text{GRAPH}}$ $\boxed{\text{F1}}$, move the cursor to the function, and press $\boxed{\text{CLEAR}}$ to erase the current expression for the function.
- Press $\boxed{\text{2nd}}$ $\boxed{\text{STAT}}$ $\boxed{\text{F5}}$ $\boxed{\text{MORE}}$ $\boxed{\text{MORE}}$ $\boxed{\text{F2}}$ to assign the equation (known as RegEq) to the function.

C. Display the points from part A

- Press $\boxed{\text{GRAPH}}$ $\boxed{\text{F1}}$ and deselect all functions.
- Press $\boxed{\text{2nd}}$ $\boxed{\text{STAT}}$ $\boxed{\text{F3}}$ to select **STAT PLOTS** menu.
- Press $\boxed{\text{F1}}$ to select PLOT1.
- Move the cursor to **ON** and press $\boxed{\text{ENTER}}$.
- Select SCAT as the Type, select xStat as the Xlist Name, select yStat as the Ylist Name, and select any symbol for Mark.
- Press $\boxed{\text{GRAPH}}$ $\boxed{\text{F5}}$ to display the data points.

  Note 1: Make sure the current window setting is large enough to display the points.

  Note 2: When you finish using the point-plotting feature, turn it off. Press $\boxed{\text{2nd}}$ $\boxed{\text{STAT}}$ $\boxed{\text{F3}}$ $\boxed{\text{F1}}$, move the cursor to **OFF** and press $\boxed{\text{ENTER}}$.

D. Display the line and the points from part A

- Press $\boxed{\text{GRAPH}}$ $\boxed{\text{F1}}$ and deselect all functions except for the function containing the equation of the least-squares line.
- Carry out all but the first step of part C.

Miscellaneous Items and Tips

A. From the home screen, if you plan to reuse a recently entered line with some minor changes, press $\boxed{\text{2nd}}$ [ENTRY] until the previous line appears. You can then make alterations to the line and press $\boxed{\text{ENTER}}$ to execute the line.

B. If you plan to use TRACE to examine the values of various points on a graph, set yMin to a value that is lower than is actually necessary for the graph. Then, the values of x and y will not obliterate the graph while you trace.

C. To clear the home screen, press $\boxed{\text{CLEAR}}$ twice.

D. When two menus are displayed at the same time, you can remove the top menu by pressing $\boxed{\text{EXIT}}$. (The remaining menu can be removed by pressing $\boxed{\text{EXIT}}$ again.)

E. To obtain the solutions of a quadratic equation, or of any equation of the form $p(x) = 0$, where $p(x)$ is a polynomial of degree ≤ 30, press $\boxed{\text{2nd}}$ [POLY], enter the degree of the polynomial, enter the coefficients of the polynomial, and press $\boxed{\text{F5}}$. (Some of the solutions might be complex numbers.)

Answers

EXERCISES 0.1, PAGE 15

1. [diagram: filled dot at −1, filled dot at 4, labeled −1 0 4]   3.  [diagram: filled dot at −2, open dot at √2, labeled −2 0 √2]

5.  [diagram: filled mark at 0, open dot at 3, labeled 0 3]   **7.** $[2, 3)$ **9.** $[-1, 0)$

11. $(-\infty, 3)$ **13.** $0, 10, 0, 70$ **15.** $0, 0, -\frac{9}{8}, a^3 + a^2 - a - 1$ **17.** $\frac{1}{3}, 3, \dfrac{a+1}{a+2}$

19. $a^2 - 1, a^2 + 2a$ **21. (a)** 1990 sales   **(b)** 60   **23.** $x \neq 1, 2$

25. $x < 3$ **27.** Function   **29.** Not a function   **31.** Not a function

33. 1   **35.** 3   **37.** Positive   **39.** Positive   **41.** $-1, 5, 9$ **43.** .03

45. .04   **47.** No   **49.** Yes   **51.** $(a+1)^3$ **53.** $1, 3, 4$ **55.** $\pi, 3, 12$

57. $f(x) = \begin{cases} .06x & \text{for } 50 \leq x \leq 300 \\ .02x + 12 & \text{for } 300 < x \leq 600 \\ .015x + 15 & \text{for } 600 < x \end{cases}$ **59.**

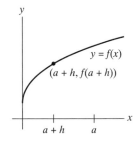

EXERCISES 0.2, PAGE 25

1.

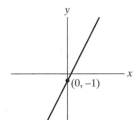

3.

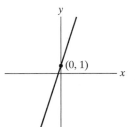

5.

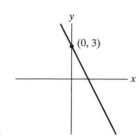

7. $(-\frac{1}{3}, 0), (0, 3)$ **9.** y-intercept $(0, 5)$ **11.** $(12, 0), (0, 3)$

13. **(a)** $K = \frac{1}{250}, V = \frac{1}{50}$ **(b)** $\left(-\frac{1}{K}, 0\right), \left(0, \frac{1}{V}\right)$

15. **(a)** \$58   **(b)** $f(x) = .20x + 18$ **17.** $300x + 1500, x = $ number of days

19. The cost for another 5% is \$25 million. The cost for the final 5% is 21 times as much.

21. $a = 3, b = -4, c = 0$ **23.** $a = -2, b = 3, c = 1$ **25.** $a = -1, b = 0, c = 1$

27.

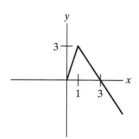

29.

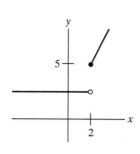

31.

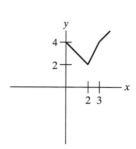

33. 1
35. 10^{-2}
37. 2.5

39. $Y_1 = (3X - 1) * (X \le 2) + (X^2 + 1) * (X > 2)$ **41.** $Y_1 = (X^3 + 2X) * (X \ge -2) * (X < 2)$
$Y_2 = (\sqrt{(X - 2)}) * (X \ge 2)$

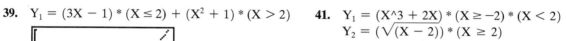

[0, 5] by [0, 25]

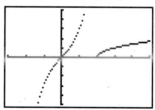

[-3, 5] by [-5, 5]

43.

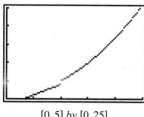

[0, 4] by [-10, 10]

45.

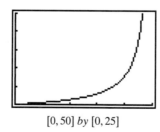

[0, 50] by [0, 25]

47.

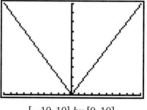

[-10, 10] by [0, 10]

EXERCISES 0.3, PAGE 31

1. $x^2 + 9x + 1$ **3.** $9x^3 + 9x$ **5.** $\dfrac{t}{9} + \dfrac{1}{9t}$ **7.** $\dfrac{3x + 1}{x^2 - x - 6}$ **9.** $\dfrac{4x}{x^2 - 12x + 32}$ **11.** $\dfrac{2x^2 + 5x + 50}{x^2 - 100}$

13. $\dfrac{2x^2 - 2x + 10}{x^2 + 3x - 10}$ **15.** $\dfrac{-x^2 + 5x}{x^2 + 3x - 10}$ **17.** $\dfrac{x^2 + 5x}{-x^2 + 7x - 10}$ **19.** $\dfrac{-x^2 + 3x + 4}{x^2 + 5x - 6}$ **21.** $\dfrac{-x^2 - 3x}{x^2 + 15x + 50}$

23. $\dfrac{5u - 1}{5u + 1}, u \ne 0$ **25.** $\left(\dfrac{x}{1 - x}\right)^6$ **27.** $\left(\dfrac{x}{1 - x}\right)^3 - 5\left(\dfrac{x}{1 - x}\right)^2 + 1$ **29.** $\dfrac{t^3 - 5t^2 + 1}{-t^3 + 5t^2}$ **31.** $2xh + h^2$

33. $4 - 2t - h$ **35.** **(a)** $C(A(t)) = 3000 + 1600t - 40t^2$ **(b)** \$6040

37. $\triangle x = .1, \triangle y = -.3$ **39.** Terminal value $= 3.01, \triangle y = -.03$

41. Terminal value $= 2, \dfrac{\triangle y}{\triangle x} = \dfrac{1}{3}$; terminal value $= 1.5, \dfrac{\triangle y}{\triangle x} = \dfrac{1}{3}$; terminal value $= .9, \dfrac{\triangle y}{\triangle x} = \dfrac{1}{3}$ **43.** Miles per hour

45. Yes, since the values of $\triangle y$ are constant, equal to -2. The function is $y = -2x + 7$.

47. $h(x) = x + \dfrac{1}{8}$; $h(x)$ converts from British sizes to U.S. sizes.

49. The graph of $f(x) + c$ is the graph of $f(x)$ shifted up (if $c > 0$) or down (if $c < 0$) by $|c|$ units.

51.

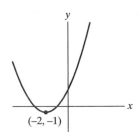

(−2, −1)

53. $f(f(x)) = x, x \neq 1$

EXERCISES 0.4, PAGE 41

1. $2, \frac{3}{2}$ **3.** $\frac{3}{2}$ **5.** No zeros **7.** $1, -\frac{1}{5}$ **9.** $5, 4$ **11.** $2 + \dfrac{\sqrt{6}}{3}, 2 - \dfrac{\sqrt{6}}{3}$ **13.** $(x + 5)(x + 3)$

15. $(x - 4)(x + 4)$ **17.** $3(x + 2)^2$ **19.** $-2(x - 3)(x + 5)$ **21.** $x(3 - x)$ **23.** $-2x(x - \sqrt{3})(x + \sqrt{3})$

25. $(-1, 1), (5, 19)$ **27.** $(-1, 9), (4, 4)$ **29.** $(0, 0), (2, -2)$

31. $(0, 5), (2 - \sqrt{3}, 25 - 23\sqrt{3}/2), (2 + \sqrt{3}, 25 + 23\sqrt{3}/2)$ **33.** $-7, 3$ **35.** $-2, 3$ **37.** -7

39. 16,667 and 78,571 subscribers **41.** $-1, 2$ **43.** ≈ 4.56 **45.** $\approx(-.41, -1.83), (2.41, 3.83)$

47. $\approx(2.14, -25.73), (4.10, -21.80)$ **49.** $[-5, 22]$ by $[-1400, 100]$ **51.** $[-20, 4]$ by $[-500, 2500]$

EXERCISES 0.5, PAGE 47

1. 27 **3.** 1 **5.** .0001 **7.** -16 **9.** 4 **11.** .01 **13.** $\frac{1}{6}$ **15.** 100 **17.** 16 **19.** 125

21. 1 **23.** 4 **25.** $\frac{1}{2}$ **27.** 1000 **29.** 10 **31.** 6 **33.** 16 **35.** 18 **37.** $\frac{4}{9}$ **39.** 7 **41.** $x^6 y^6$

43. $x^3 y^3$ **45.** $\dfrac{1}{\sqrt{x}}$ **47.** $\dfrac{x^{12}}{y^6}$ **49.** $x^{12} y^{20}$ **51.** $x^2 y^6$ **53.** $16x^4$ **55.** x^2 **57.** $\dfrac{1}{x^7}$ **59.** x

61. $\dfrac{27x^6}{8y^3}$ **63.** $2\sqrt{x}$ **65.** $\dfrac{1}{8x^6}$ **67.** $\dfrac{1}{32x^2}$ **69.** $9x^3$ **71.** $x - 1$ **73.** $1 + 6\sqrt{x}$ **77.** 16

79. $\frac{1}{4}$ **81.** 8 **83.** $\frac{1}{32}$ **85.** \$709.26 **87.** \$127,857.61 **89.** \$164.70 **91.** \$1592.75 **93.** \$3268.00

95. $\frac{125}{64}r^4 + \frac{125}{4}r^3 + \frac{375}{2}r^2 + 500r + 500$ **99.** .0008103 **101.** .00000823

EXERCISES 0.6, PAGE 57

1. **3.** 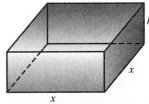 **5.** 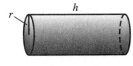 **7.** $P = 8x; 3x^2 = 25$

9. $A = \pi r^2; 2\pi r = 15$ **11.** $V = x^2 h; x^2 + 4xh = 65$ **13.** $\pi r^2 h = 100; C = 11\pi r^2 + 14\pi rh$

15. $2x + 3h = 5000; A = xh$ **17.** $C = 36x + 20h$ **19.** 75 cm^2 **21.** **(a)** 38 **(b)** \$40

23. **(a)** 200 **(b)** 275 **(c)** 25

25. **(a)** $P(x) = 12x - 800$ **(b)** \$640 **(c)** \$3150 **27.** 270 cents

29. A 100-in.^{3} cylinder of radius 3 in. costs \$1.62 to construct. **31.** \$1.08

33. Revenue: \$1800; cost: \$1200 **35.** 40 **37.** $C(1000) = 4000$

39. Find the y-coordinate of the point on the graph whose x-coordinate is 400.

41. The greatest profit, $52,500, occurs when 2500 units of goods are produced.   **43.** Find the x-coordinate of the point on the graph whose y-coordinate is 30,000.   **45.** Find $h(3)$. Find the y-coordinate of the point on the graph whose t-coordinate is 3.   **47.** Find the maximum value of $h(t)$. Find the y-coordinate of the highest point of the graph.   **49.** Solve $h(t) = 100$. Find the t-coordinates of the points whose y-coordinate is 100.

51. **(a)**
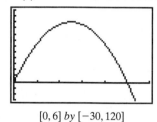
[0, 6] by [−30, 120]

(b) 96 feet   **(c)** 1 and 4 seconds   **(d)** 5 seconds
(e) 2.5 seconds; 100 feet

53. **(a)**
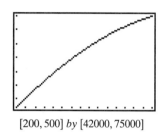
[200, 500] by [42000, 75000]

(b) 350 bicycles per year   **(c)** $68,000   **(d)** $5000
(e) No, revenue would only increase by $4000.

CHAPTER 0: SUPPLEMENTARY EXERCISES, PAGE 61

1. $2, 27\frac{1}{3}, -2, -2\frac{1}{8}, \dfrac{5\sqrt{2}}{2}$ **3.** $a^2 - 4a + 2$ **5.** $x \neq 0, -3$ **7.** All x **9.** Yes   **11.** $5x(x - 1)(x + 4)$

13. $(-1)(x - 6)(x + 3)$ **15.** $-\frac{2}{5}, 1$ **17.** $\left(\dfrac{5 + 3\sqrt{5}}{10}, \dfrac{3\sqrt{5}}{5}\right), \left(\dfrac{5 - 3\sqrt{5}}{10}, -\dfrac{3\sqrt{5}}{5}\right)$ **19.** $x^2 + x - 1$

21. $x^{5/2} - 2x^{3/2}$ **23.** $x^{3/2} - 2x^{1/2}$ **25.** $\dfrac{x^2 - x + 1}{x^2 - 1}$ **27.** $-\dfrac{3x^2 + 1}{3x^2 + 4x + 1}$ **29.** $\dfrac{-3x^2 + 9x - 10}{3x^2 - 5x - 8}$

31. $\dfrac{1}{x^4} - \dfrac{2}{x^2} + 4$ **33.** $(\sqrt{x} - 1)^2$ **35.** $\dfrac{1}{(\sqrt{x} - 1)^2} - \dfrac{2}{\sqrt{x} - 1} + 4$ **37.** $27, 32, 4$ **39.** $301 + 10t + .04t^2$

41. $x^2 + 2x + 1$ **43.** x

CHAPTER 0: APPENDIX, PAGE 66

1.
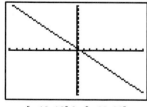
[−10, 10] by [−20, 20]

3.
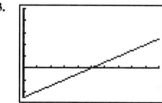
[0, 100] by [−5000, 10000]

5.
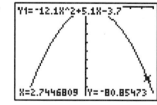
[−3, 3] by [−100, 0]

7.

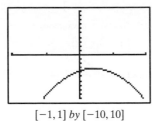

$[-1, 1]$ *by* $[-10, 10]$

9.

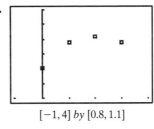

11.

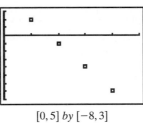

13.

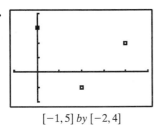

$[-1, 5]$ *by* $[-2, 4]$

15.

$[-1, 4]$ *by* $[0.8, 1.1]$

17.

$[0, 5]$ *by* $[-8, 3]$

19.

$[0, 5]$ *by* $[0, 1]$

21. A graphing utility interprets
$Y_1 = 1/X + 1$ as $\dfrac{1}{x} + 1$, not $\dfrac{1}{x + 1}$.

23.

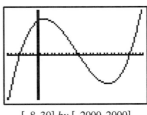

$[-8, 30]$ *by* $[-2000, 2000]$

25.

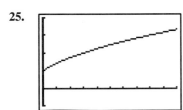

$[0, 10]$ *by* $[-1, 4]$

27.

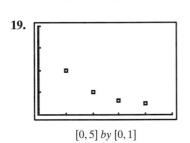

CHAPTER 1

EXERCISES 1.1, PAGE 81

1. -5 **3.** 0 **5.** $\frac{2}{7}$ **7.** $-\frac{2}{3}$ **9.** $y = 3x - 1$ **11.** $y = x + 1$ **13.** $y = 35 - 7x$ **15.** $y = 4$

17. $y = \dfrac{x}{2}$ **19.** $y = -2x$ **21.** $y = 6 - 2x$ **23.** $y = \frac{1}{2}x - \frac{1}{2}$

25. **(a)** C **(b)** B **(c)** D **(d)** A **27.** 2 **29.** $-.75$ **31.** $(2, 5); (3, 7); (0, 1)$

33. $(0, -\frac{5}{4})$; $(1, -\frac{3}{2})$; $(-2, -\frac{3}{4})$ **35.** l_1 **37.**

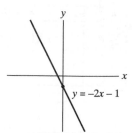

$y = -2x - 1$

39.

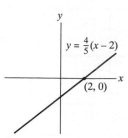

$y = \frac{4}{5}(x - 2)$

$(2, 0)$

41. $y = 5$ **43.** $4x + 5y = 7$ **45.** $y = -2x$ **47.** $y - 1 = 2(x - 1)$ **49.** $y - \frac{1}{4} = -(x + \frac{1}{2})$

51. If the monopolist wants to sell one more unit of goods, then the price per unit must be lowered by 2¢. No one is willing to pay $7 or more for a unit of goods.

53. **(a)–(c)**

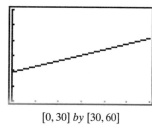

3 $3 + h$

(d) $\dfrac{f(3 + h) - f(3)}{h}$

57. **(a)** $y = .38x + 39.5$ **(b)**

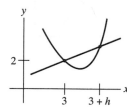

$[0, 30]$ by $[30, 60]$

(c) Every year, .38% more of the world becomes urban.   **(d)** 43.3%   **(e)** 2007   **(f)** 1.9%

EXERCISES 1.2, PAGE 87

1.

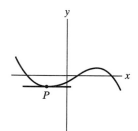

P

3.

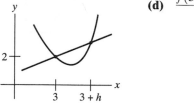

P

5.

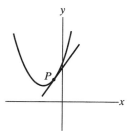

P

7. 1   **9.** -3

11. $\frac{2}{3}$ **13.** Small positive slope   **15.** Zero slope   **17.** Zero slope   **19.** -4; $y - 4 = -4(x + 2)$
21. $\frac{8}{3}$; $y - \frac{16}{9} = \frac{8}{3}(x - \frac{4}{3})$ **23.** $y - 2.25 = 3(x - 1.5)$ **25.** $(\frac{5}{6}, \frac{25}{36})$ **27.** $(-\frac{1}{4}, \frac{1}{16})$ **29.** 12   **31.** $\frac{3}{4}$
33. $y + 1 = 3(x + 1)$ **35.** **(a)** 3, 9   **(b)** increase   **37.** $y - 2.45 = .2(x - 4)$
39. $y - 48 = 2.9(x - 3.5)$

EXERCISES 1.3, PAGE 98

1. 2   **3.** $8x^7$ **5.** $\frac{5}{2}x^{3/2}$ **7.** $\frac{1}{3}x^{-2/3}$ **9.** $-2x^{-3}$ **11.** $-\frac{1}{4}x^{-5/4}$ **13.** 0   **15.** $-3x^{-4}$ **17.** -192
19. $-\frac{1}{9}$ **21.** -1 **23.** $\frac{9}{2}$ **25.** 108   **27.** $\frac{1}{6}$ **29.** 25, -10 **31.** $\frac{1}{32}, -\frac{5}{64}$ **33.** 16, $\frac{8}{3}$
35. 48, $y - 64 = 48(x - 4)$ **37.** $8x^7$ **39.** $\frac{3}{4}x^{-1/4}$ **41.** 0   **43.** $\frac{1}{5}x^{-4/5}$ **45.** 4, $\frac{1}{3}$ **47.** $a = 4, b = 1$
49. 1, 1.5; 2   **51.** 6   **53.** $2x + 5$ **57.** $y - 5 = \dfrac{1}{2}(x - 4)$ **59.** .69315   **61.** .70711   **63.** .11111

65.

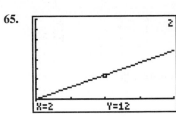

$[0, 4]$ *by* $[-5, 40]$

67. $y = \dfrac{x}{6} + \dfrac{3}{2}$ **69.** $y = -16x + 12$ **71.** 2   **73.** 4   **75.** .11

77. .1004   **79.** 3   **81.** -0.990099 **83.** -1.1 **85.** -1.001

87. $-\triangle x - 1$

89.

x	y	$\dfrac{\triangle y}{\triangle x}$
0	$838.92	
1	$890.20	$51.28
2	$1,020.02	$129.82
3	$850.86	–$169.16
4	$616.24	–$234.62
5	$852.41	$236.17
6	$1,004.65	$152.24
7	$831.17	–$173.48
8	$805.01	–$26.16
9	$838.74	$33.73
10	$963.99	$125.25
11	$875.00	–$88.99
12	$1,046.54	$171.54
13	$1,258.64	$212.10
14	$1,211.57	–$47.07
15	$1,546.47	$334.90
16	$1,895.95	$349.48
17	$1,938.83	$42.88
18	$2,168.57	$229.74
19	$2,753.20	$584.63
20	$2,633.66	–$119.54
21	$3,168.83	$535.17
22	$3,301.11	$132.28
23	$3,754.09	$452.98
24	$3,834.44	$80.35
25	$5,117.12	$1,282.68
26	$6,448.27	$1,331.15
27	$7,908.25	$1,459.98

91. Decreasing at a rate of $119.54 per year.

EXERCISES 1.4, PAGE 110

1. No limit   **3.** 1   **5.** No limit   **7.** -5 **9.** 5   **11.** No limit   **13.** 288   **15.** 0   **17.** 3

19. -4 **21.** -8 **23.** $\frac{6}{7}$ **25.** No limit   **27.** $-\frac{2}{11}$ **29.** 6   **31.** 3   **33.** $-\frac{2}{121}$ **35.** $-\dfrac{\sqrt{3}}{6}$ **37.** 0

39. 3   **41.** $f(x) = \sqrt{x}; a = 9$ **43.** $f(x) = \dfrac{1}{x}; a = 10$ **45.** $f(x) = 3x^2 + 4; a = 1$ **47.** 0   **49.** 0

51. 2 **53.** 0 **55.** .5

EXERCISES 1.5, PAGE 116

1. No **3.** Yes **5.** No **7.** No **9.** Yes **11.** No **13.** Continuous, differentiable
15. Continuous, not differentiable **17.** Continuous, not differentiable **19.** Not continuous, not differentiable
21. $f(5) = 3$ **23.** Not possible **25.** $f(0) = 12$

EXERCISES 1.6, PAGE 122

1. $3x^2 + 2x$ **3.** $2x + 3$ **5.** $5x^4 - \dfrac{1}{x^2}$ **7.** $4x^3 + 3x^2 + 1$ **9.** $6x$ **11.** $3x^2 + 14x$ **13.** $-\dfrac{8}{x^3}$

15. $3 + \dfrac{1}{x^2}$ **17.** $x^2 - x$ **19.** $\dfrac{1}{x^6}$ **21.** $-\dfrac{1}{2\sqrt{x}}$ **23.** $30(3x + 1)^9$ **25.** $\dfrac{45x^2 + 5}{2\sqrt{3x^3 + x}}$

27. $6(4x - 1)(2x^2 - x + 4)^5$ **29.** $\frac{1}{3} - 3x^{-2}$ **31.** $10(1 - 5x)^{-2}$ **33.** $4x^3(1 - x^4)^{-2}$

35. $-2(x^2 + x)^{-3/2}(2x + 1)$ **37.** $\dfrac{3}{2}\left(\dfrac{\sqrt{x}}{2} + 1\right)^{1/2}\left(\dfrac{1}{4}x^{-1/2}\right)$ or $\dfrac{3}{8\sqrt{x}}\left(\dfrac{\sqrt{x}}{2} + 1\right)^{1/2}$ **39.** 4 **41.** 15

43. $f'(4) = 48, y = 48x - 191$ **45.** $f'(x) = 2(3x^2 + x - 2)(6x + 1) = 36x^3 + 18x^2 - 22x - 4$ **47.** 4.8; 1.8
49. 14; 11 **51.** $10; \frac{15}{4}$ **53.** $(5, \frac{161}{3}); (3, 49)$ **55.** $f(4) = 5, f'(4) = \frac{1}{2}$

EXERCISES 1.7, PAGE 127

1. $10t(t^2 + 1)^4$ **3.** $(2t - 1)^{-1/2}$ **5.** $\frac{2}{3}(T^3 + 5T)^{-1/3}(3T^2 + 5)$ **7.** $6P - \frac{1}{2}$ **9.** $2a^2t + b^2$
11. $f'(x) = x - 7, f''(x) = 1$ **13.** $y' = \frac{1}{2}x^{-1/2}, y'' = -\frac{1}{4}x^{-3/2}$ **15.** $f'(r) = 2\pi(hr + 1), f''(r) = 2\pi h$
17. $g'(x) = -5, g''(x) = 0$ **19.** $f'(P) = 15(3P + 1)^4, f''(P) = 180(3P + 1)^3$ **21.** 20 **23.** 54

25. 34 **27.** $8k(2P - 1)^{-3}$ **29.** $f'(3) = -\dfrac{1}{2}, f''(3) = -\dfrac{1}{8}$ **31.** 20

33. **(a)** $f'''(x) = 60x^2 - 24x$ **(b)** $f'''(x) = \dfrac{15}{2\sqrt{x}}$ **35.** **37.** 127.913

$[-4, 4]$ by $[-2, 2]$

EXERCISES 1.8, PAGE 136

1. **(a)** 12; 10; 8.4 **(b)** 8 **3.** **(a)** 14 **(b)** 13 **5.** **(a)** $-\frac{7}{4}$ **(b)** $\frac{5}{6}$ **(c)** January 1, 1980
7. .9 gm/week; .8 gm/week **9.** 63 units/hour **11.** **(a)** 28 km/hour **(b)** 96 km **(c)** $\frac{1}{2}$ hr
13. **(a)** 160 ft/sec **(b)** 96 ft/sec **(c)** -32 ft/sec^2 **(d)** 10 sec **(e)** -160 ft/sec
15. A–b; B–d; C–f; D–e; E–a; F–c; G–g
17. **(a)** 15 ft/sec **(b)** No; positive velocity indicates the object is moving away from the reference point **(c)** 5 ft/sec
19. **(a)** 5010 **(b)** 5005 **(c)** 4990 **(d)** 4980 **(e)** 4997.5
21. Four minutes after it has been poured, the coffee is 120°. At that time, its temperature is decreasing by 5°/min. 119.5°.
23. When the price of a car is $10,000, 200,000 cars are sold. At that price, the number of cars sold decreases by 3 for each dollar increase in price. **25.** When 50 bicycles are manufactured, the cost is $5000. For every additional bicycle manufactured, there is an additional cost of $45. $5090. **27.** The profit from manufacturing and selling 100 luxury cars is $90,000. Each additional car made and sold creates an additional profit of $1200. $88,800. **29.** **(a)** $2.60 **(b)** 100 or 200 units
31. **(a)** $500 billion **(b)** $50 billion/yr **(c)** 1994 **(d)** 1994

33. **(a)**

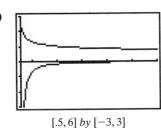

$[.5, 6]\ by\ [-3, 3]$

(b) .85 sec   **(c)** 5 days   **(d)** $-.05$ sec/day   **(e)** 3 days

CHAPTER 1: SUPPLEMENTARY EXERCISES, PAGE 141

1.

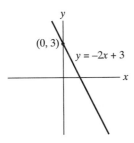

$(0, 3)$ $y = -2x + 3$

3.

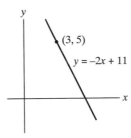

$y = 5x - 10$ $(2, 0)$

5.

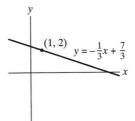

$(3, 5)$ $y = -2x + 11$

7.

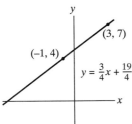

$(3, 7)$ $(-1, 4)$ $y = \frac{3}{4}x + \frac{19}{4}$

9.

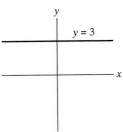

$(1, 2)$ $y = -\frac{1}{3}x + \frac{7}{3}$

11.

$y = 3$

13.

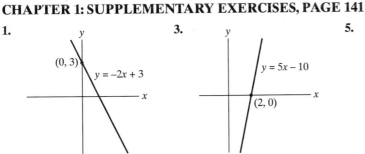

$x = 0$

15. $7x^6 + 3x^2$ **17.** $\dfrac{3}{\sqrt{x}}$ **19.** $-\dfrac{3}{x^2}$ **21.** $48x(3x^2 - 1)^7$ **23.** $-\dfrac{5}{(5x - 1)^2}$

25. $\dfrac{x}{\sqrt{x^2 + 1}}$ **27.** $-\dfrac{1}{4x^{5/4}}$ **29.** 0   **31.** $10[x^5 - (x - 1)^5]^9[5x^4 - 5(x - 1)^4]$

33. $\dfrac{3}{2}t^{-1/2} + \dfrac{3}{2}t^{-3/2}$ **35.** $\dfrac{2(9t^2 - 1)}{(t - 3t^3)^2}$ **37.** $\dfrac{9}{4}x^{1/2} - 4x^{-1/3}$ **39.** 28

41. $14; 3$ **43.** $\frac{15}{2}$ **45.** 33   **47.** $4x^3 - 4x$ **49.** $-\frac{3}{2}(1 - 3P)^{-1/2}$ **51.** 29

53. $300(5x + 1)^2$ **55.** -2 **57.** $3x^{-1/2}$ **59.** Slope -4; tangent $y = -4x + 6$

61.

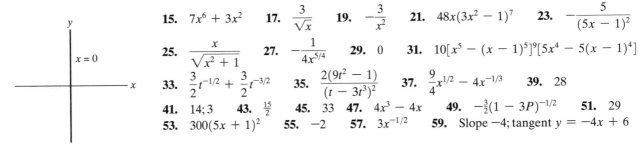

$y = 3x - \frac{9}{4}$ $\left(\frac{3}{2}, \frac{9}{4}\right)$ $y = x^2$ $\frac{3}{2}$

63. $y = 2$ **65.** $f(2) = 3, f'(2) = -1$ **67.** 96 ft/s   **69.** 11 ft

71. $\frac{5}{3}$ ft/s   **73.** **(a)** \$16.10   **(b)** \$16   **75.** $\frac{3}{4}$ in.   **77.** 4

79. Does not exist   **81.** $-\frac{1}{50}$ **83.** The slope of a secant line at $(3, 9)$

CHAPTER 2

EXERCISES 2.1, PAGE 153

1. (a), (e), (f)   **3.** (b), (c), (d)   **5.** Decreasing for $x < -2$, relative minimum point at $x = -2$, minimum value $= -2$, increasing for $x > -2$, concave up, y-intercept $(0, 0)$, x-intercepts $(0, 0)$ and $(-3.6, 0)$.   **7.** Decreasing for $x < 0$, relative minimum point at $x = 0$, increasing for $0 < x < 2$, relative maximum point at $x = 2$, decreasing for $x > 2$, concave up for $x < 1$, concave down for $x > 1$, inflection point at $(1, 3)$, y-intercept $(0, 2)$, x-intercept $(3.6, 0)$.
9. Decreasing for $x < 2$, relative minimum at $x = 2$, minimum value $= 3$, increasing for $x > 2$, concave up for all x, no inflection points, defined for $x > 0$, the line $y = x$ is an asymptote, the y-axis is an asymptote.
11. Decreasing for $1 \le x < 3.2$, relative minimum point at $x = 3.2$, increasing for $x > 3.2$, maximum value $= 6$ (at $x = 1$), minimum value $= .9$ (at $x = 3.2$), inflection point at $x = 4$, concave up for $1 \le x < 4$, concave down for $x > 4$, the line $y = 4$ is an asymptote.   **13.** Slope increases for all x.   **15.** Slope decreases for $x < 3$, increases for $x > 3$. Minimum slope occurs at $x = 3$.   **17. (a)** C, F **(b)** A, B, F **(c)** C

19.

21.

23.

25.

27.

29. Oxygen content decreases until time a, at which time it reaches a minimum. After a, oxygen content steadily increases. The rate of increase increases until b, and then decreases. Time b is the time when oxygen content is increasing fastest.
31. 1960
33. The parachutist's speed levels off to 15 ft/sec.

35.

37.

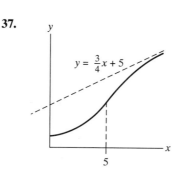

39. (a) Yes   **(b)** Yes   **41.** Relatively low
43. $x = 2$ **45.** $\frac{1}{6}$ unit apart

EXERCISES 2.2, PAGE 163

1. (b), (c), (f)   **3.** (d), (e), (f)   **5.** (d)

7.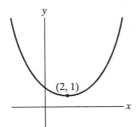

9.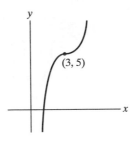

11.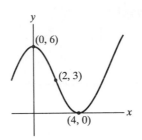

13.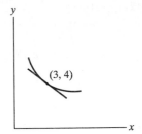

15.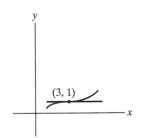

17.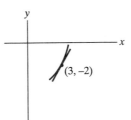

19.

	f	f'	f''
A	POS	POS	NEG
B	0	NEG	0
C	NEG	0	POS

21. $t = 1$ **23. (a)** decreasing   **(b)** The function $f(x)$ is increasing for $1 \leq x < 2$ because the values of $f'(x)$ are positive. The function $f(x)$ is decreasing for $2 < x \leq 3$ because the values of $f'(x)$ are negative. Therefore, $f(x)$ has a relative maximum at $x = 2$. Coordinates: $(2, 9)$ **(c)** The function $f(x)$ is decreasing for $9 \leq x < 10$ because the values of $f'(x)$ are negative. The function $f(x)$ is increasing for $10 < x \leq 11$ because the values of $f'(x)$ are positive. Therefore, $f(x)$ has a relative minimum at $x = 10$.   **(d)** concave down   **(e)** at $x = 6$; coordinates: $(6, 5)$ **(f)** $x = 15$
25. The slope is positive because $f'(6) = 2$, a positive number.   **27.** The slope is 0 because $f'(3) = 0$. Also, $f'(x)$ is positive for x slightly less than 3, and $f'(x)$ is negative for x slightly greater than 3. Hence $f(x)$ changes from increasing to decreasing at $x = 3$.   **29.** $f'(x)$ is increasing at $x = 0$, so the graph of $f(x)$ is concave up.
31. At $x = 1$, $f'(x)$ changes from increasing to decreasing, so the slope of the graph of $f(x)$ changes from increasing to decreasing.   **33.** $y - 3 = 2(x - 6)$ **35.** 3.25   **37. (a)** $\frac{1}{6}$ in.   **(b)** (ii), because the water level is falling.
39. II   **41.** I   **43. (a)** 2 million   **(b)** 30,000 farms per year   **(c)** 1940
(d) 1945 and 1978   **(e)** 1960   **45.** rel. max: $x \approx -2.34$; rel. min: $x \approx 2.34$; inflection point: $x = 0, x \approx \pm 1.41$

EXERCISES 2.3, PAGE 172

1.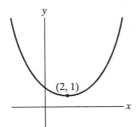

3.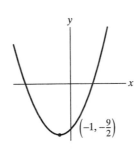

5.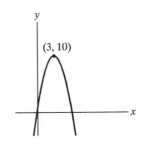

7.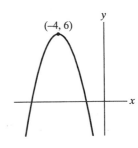

9.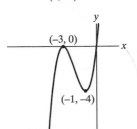

11.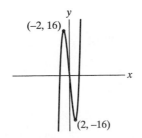

13.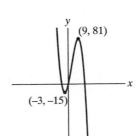

15.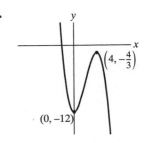

17.

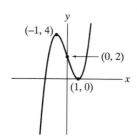

19.

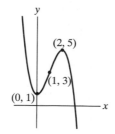

21.

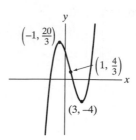

23.

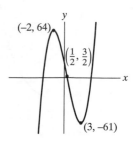

25. No, $f''(x) = 2a \neq 0$ **27.** $(4, 3)$ min   **29.** $(1, 5)$ max   **31.** $(-.1, -3.05)$ min   **33.** $f(x) = g'(x)$
35. **(a)** f has a relative minimum   **(b)** f has an inflection point   **37.** **(a)** 1984   **(b)** 70%   **(c)** 1987; 40%
39. $\left(5, \frac{10}{3}\right)$ **41.** $(5, 5)$

EXERCISES 2.4, PAGE 179

1. $\left(\dfrac{3 \pm \sqrt{5}}{2}, 0\right)$ **3.** $(-2, 0), \left(-\dfrac{1}{2}, 0\right)$ **5.** $\left(\dfrac{1}{2}, 0\right)$ **7.** The derivative $x^2 - 4x + 5$ has no zeros.

9.

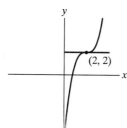

11.

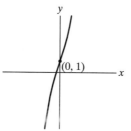

13.

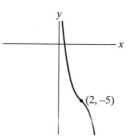

15.

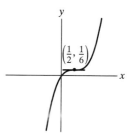

17.

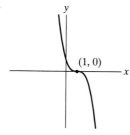

19.

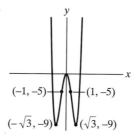

21.

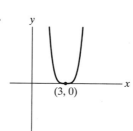

23.

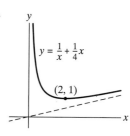

25.

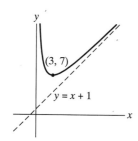

27.

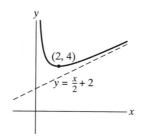

29.

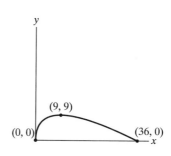

31. $g(x) = f'(x)$

33. **(a)**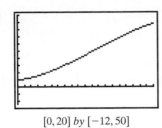

$[0, 20]$ *by* $[-12, 50]$

(b) 15.0 g   **(c)** after 12.0 days   **(d)** 1.6 g/day
(e) after 6.0 days and after 17.6 days   **(f)** after 11.8 days

EXERCISES 2.5, PAGE 188

1. 20   **3.** $t = 4, f(4) = 8$ **5.** **(a)** Objective: $A = xy$, constraint: $8x + 4y = 320$
(b) $A = -2x^2 + 80x$ **(c)** $x = 20$ ft, $y = 40$ ft

7. **(a)**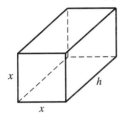

(b) $h + 4x$ **(c)** Objective: $V = x^2h$; constraint: $h + 4x = 84$
(d) $V = -4x^3 + 84x^2$ **(e)** $x = 14$ in., $h = 28$ in.

9. Let x be the length of the fence and y the other dimension.
Objective: $C = 15x + 20y$; constraint: $xy = 75$; $x = 10$ ft, $y = 7.5$ ft
11. Let x be the length of each edge of the base and h the height.
Objective: $A = 2x^2 + 4xh$; constraint: $x^2h = 8000$; 20 cm by 20 cm by 20 cm

13. Let x be the length of the fence parallel to the river and y the length of each section perpendicular to the river.
Objective: $A = xy$; constraint: $6x + 15y = 1500$; $x = 125$ ft, $y = 50$ ft
15. Objective: $P = xy$; constraint: $x + y = 100$; $x = 50$, $y = 50$

17. Objective: $A = \dfrac{\pi x^2}{2} + 2xh$; constraint: $(2 + \pi)x + 2h = 14$; $x = \dfrac{14}{4 + \pi}$ ft

19.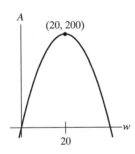

$w = 20$ ft, $x = 10$ ft   **21.** $\left(\dfrac{3}{2}, \sqrt{\dfrac{3}{2}}\right)$

EXERCISES 2.6, PAGE 196

1. Let x be the number of prints and p the price per print. Objective: $R = px$; constraint: $p = 650 - 5x$; 65 prints
3. Let x be the number of tables and p the profit per table. Objective: $P = px$; constraint: $p = 16 - (x/2)$; 16 tables
5. Let x be the number of cases per order and r the number of orders per year.
Objective: $C = 80r + 5x$; constraint: $rx = 10,000$ **(a)** \$4100   **(b)** 400 cases
7. Let r be the number of production runs and x the number of microscopes manufactured per run.
Objective: $C = 2500r + 25x$; constraint: $rx = 1600$; 4 runs
11. Objective: $A = (100 + x)w$; constraint: $2x + 2w = 300$; $x = 25$ ft, $w = 125$ ft
13. Objective: $F = 2x + 3w$; constraint: $xw = 54$; $x = 9$ m, $w = 6$ m
15. Let x be the number of people and c the cost. Objective: $R = xc$; constraint: $c = 1040 - 20x$; 25 people
17. Let x be the length of each edge of the base and h the height.
Objective: $C = 6x^2 + 10xh$; constraint: $x^2h = 150$; 5 ft by 5 ft by 6 ft
19. Let x be the length of each edge of the end and h the length. Objective: $V = x^2h$; constraint: $2x + h = 120$;
40 cm by 40 cm by 40 cm   **21.** Objective: $V = w^2x$; constraint: $2x + w = 16$; $\frac{8}{3}$ in.   **23.** After 20 days

25. $2\sqrt{3}$ by 6   **27.** 10 in. by 10 in. by 4 in.   **29.** ≈ 3.77 cm

EXERCISES 2.7, PAGE 207

1. $1   **3.** 32   **5.** 5   **7.** $x = 20$ units, $p = \$133.33$ **9.** 2 million tons, $156 per ton   **11.** **(a)** $2.00
(b) $2.30   **13.** **(a)** $x = 15 \cdot 10^5$, $p = \$45$.   **(b)** No. Profit is maximized when price is increased to $50.   **15.** 5%
17. **(a)** $75,000   **(b)** $3200 per unit   **(c)** 15 units   **(d)** 32.5 units   **(e)** 35 units

CHAPTER 2: SUPPLEMENTARY EXERCISES, PAGE 210

1. **(a)** increasing: $-3 < x < 1, x > 5$; decreasing: $x < -3, 1 < x < 5$
(b) concave up: $x < -1, x > 3$; concave down: $-1 < x < 3$

3.

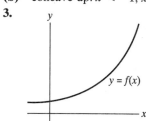

5.

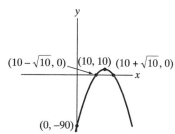

7. d, e   **9.** c, d   **11.** e
13. Graph goes through $(1, 2)$, increasing at $x = 1$.
15. Increasing and concave up at $x = 3$.
17. $(10, 2)$ is a relative minimum point.
19. Graph goes through $(5, -1)$, decreasing at $x = 5$.
21. **(a)** after 2 hours   **(b)** .8   **(c)** after 3 hours
     (d) $-.02$ units per hour

23.

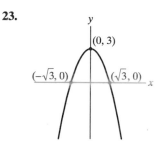

25.

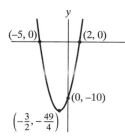

27.

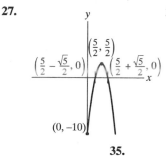

29.

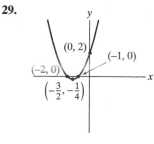

31.

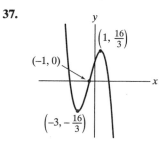

33.

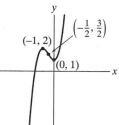

35.

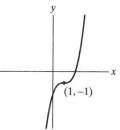

37.

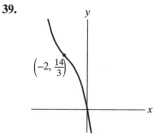

39.

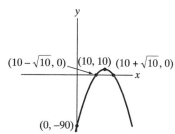

41.

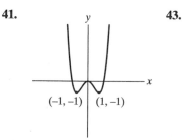

43.

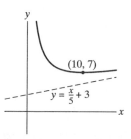

45. $f'(x) = 3x(x^2 + 2)^{1/2}$, $f'(0) = 0$ **47.** $f''(x) = -2x(1 + x^2)^{-2}$, $f''(x)$ is positive for $x < 0$ and negative for $x > 0$.
49. A—c, B—e, C—f, D—b, E—a, F—d   **51.** **(a)** The number of people living between $10 + h$ and 10 mi from the center of the city.   **(b)** If so, $f(x)$ would be decreasing at $x = 10$.

53. The endpoint maximum value of 2 occurs at $x = 0$.　　**55.** Let x be the width and h the height.
Objective: $A = 4x + 2xh + 8h$; constraint: $4xh = 200$; 4 ft by 10 ft by 5 ft　　**57.** $\frac{15}{2}$ in.
59. Let r be the number of production runs and x the number of books manufactured per run.
Objective: $C = 1000r + (.25)x$; constraint: $rx = 400{,}000$; $x = 40{,}000$　　**61.** $x = 3500$

CHAPTER 3

EXERCISES 3.1, PAGE 220

1. $(x + 1) \cdot (3x^2 + 5) + (x^3 + 5x + 2) \cdot 1$, or $4x^3 + 3x^2 + 10x + 7$
3. $(3x^2 - x + 2) \cdot 4x + (2x^2 - 1) \cdot (6x - 1)$, or $24x^3 - 6x^2 + 2x + 1$
5. $(2x - 7) \cdot 5(x - 1)^4(1) + (x - 1)^5 \cdot 2$, or $(x - 1)^4(12x - 37)$
7. $(x^2 + 3) \cdot 10(x^2 - 3)^9(2x) + (x^2 - 3)^{10} \cdot 2x$, or $2x(x^2 - 3)^9(11x^2 + 27)$
9. $\frac{1}{3}(4 - x)^3 \cdot 3(4 + x)^2(1) + (4 + x)^3 \cdot (4 - x)^2(-1)$, or $-2x(4 - x)^2(4 + x)^2$
11. $\dfrac{(4 + x) \cdot (-1) - (4 - x) \cdot 1}{(4 + x)^2}$, or $-\dfrac{8}{(4 + x)^2}$　　**13.** $\dfrac{(x^2 + 1) \cdot 2x - (x^2 - 1) \cdot 2x}{(x^2 + 1)^2}$, or $\dfrac{4x}{(x^2 + 1)^2}$
15. $(-1)(5x^2 + 2x + 5)^{-2}(10x + 2)$　　**17.** $\dfrac{(x + 1) \cdot (2x + 2) - (x^2 + 2x) \cdot 1}{(x + 1)^2}$, or $\dfrac{x^2 + 2x + 2}{(x + 1)^2}$
19. $\dfrac{(3 - x^2) \cdot (6x + 5) - (3x^2 + 5x + 1) \cdot (-2x)}{(3 - x^2)^2}$, or $\dfrac{5(x + 1)(x + 3)}{(3 - x^2)^2}$
21. $\dfrac{(x^2 + 1)^2 \cdot 1 - x \cdot 2(x^2 + 1)(2x)}{(x^2 + 1)^4}$, or $\dfrac{1 - 3x^2}{(x^2 + 1)^3}$
23. $\dfrac{(x + 2)^2 \cdot 4(x - 1)^3(1) - (x - 1)^4 \cdot 2(x + 2)(1)}{(x + 2)^4}$, or $\dfrac{2(x - 1)^3(x + 5)}{(x + 2)^3}$
25. $\dfrac{3x^4 - 4x^2 - 3}{x^2}$　　**27.** $2x^{1/2} \cdot 3(3x^2 - 1)^2(6x) + (3x^2 - 1)^3 \cdot x^{-1/2}$, or $x^{-1/2}(3x^2 - 1)^2(39x^2 - 1)$
29. $(x + 3) \cdot \frac{1}{2}(2x - 3)^{-1/2}(2) + (2x - 3)^{1/2} \cdot 1$, or $3x(2x - 3)^{-1/2}$
31. $y - 16 = 88(x - 3)$　　**33.** $0, \pm 2, \pm\frac{5}{4}$　　**35.** $2, 7$　　**37.** $(\frac{1}{2}, \frac{3}{2}), (-\frac{1}{2}, \frac{9}{2})$　　**39.** $2 \times 3 \times 1$
41. 150; $AC(150) = 35 = C'(150)$　　**43.** AR is maximized where $0 = \dfrac{d}{dx}(AR) = \dfrac{x \cdot R'(x) - R(x) \cdot 1}{x^2}$.
This happens when the production level x satisfies $xR'(x) - R(x) = 0$, and hence $R'(x) = R(x)/x = AR$.

45. 38 in.^{2}/s　　**47.** $150{,}853{,}600$ gal/yr　　**49.** $(2, 10)$　　**53.** $\dfrac{1 - 2x \cdot f(x)}{(1 + x^2)^2}$　　**55.** $\frac{1}{8}$
59. $f(x)g(x)h'(x) + f(x)g'(x)h(x) + f'(x)g(x)h(x)$
61. **(a)**　　　　　　　　　　　　　　**(b)** 10.8 mm^2　　**(c)** 2.61 units of light　　**(d)** $-.55$ mm^2/unit of light

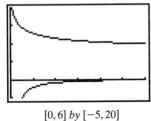

$[0, 6]$ *by* $[-5, 20]$

EXERCISES 3.2, PAGE 227

1. $\dfrac{x^3}{x^3 + 1}$　　**3.** $(x^2 + 4)^5 + 3(x^2 + 4)$　　**5.** $f(x) = x^5, g(x) = x^3 + 8x - 2$　　**7.** $f(x) = \sqrt{x}, g(x) = 4 - x^2$
9. $f(x) = \dfrac{1}{x}, g(x) = x^3 - 5x^2 + 1$　　**11.** $30x(x^2 + 5)^{14}$　　**13.** $6x^2 \cdot 3(x - 1)^2(1) + (x - 1)^3 \cdot 12x$, or

$6x(x - 1)^2(5x - 2)$ **15.** $2(x^3 - 1) \cdot 4(3x^2 + 1)^3(6x) + (3x^2 + 1)^4 \cdot 2(3x^2)$, or $6x(3x^2 + 1)^3(11x^3 + x - 8)$

17. $\dfrac{d}{dx}[4^3(1 - x)^{-3}] = 192(1 - x)^{-4}$ **19.** $3\left(\dfrac{4x - 1}{3x + 1}\right)^2 \cdot \dfrac{(3x + 1) \cdot 4 - (4x - 1) \cdot 3}{(3x + 1)^2}$, or $\dfrac{21(4x - 1)^2}{(3x + 1)^4}$

21. $3\left(\dfrac{4 - x}{x^2}\right)^2 \cdot \dfrac{x^2(-1) - (4 - x) \cdot 2x}{x^4}$, or $\dfrac{3(4 - x)^2(x - 8)}{x^7}$ **23.**

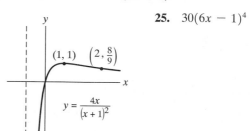

25. $30(6x - 1)^4$

27. $-(1 - x^2)^{-2} \cdot (-2x)$, or $2x(1 - x^2)^{-2}$ **29.** $[4(x^2 - 4)^3 - 2(x^2 - 4)] \cdot (2x)$, or $8x(x^2 - 4)^3 - 4x(x^2 - 4)$

31. $2[(x^2 + 5)^3 + 1]3(x^2 + 5)^2 \cdot (2x)$, or $12x[(x^2 + 5)^3 + 1](x^2 + 5)^2$ **33.** $6(4x + 1)^{1/2}$

35. $[-8(x - 3x^2)^{-3} + \tfrac{1}{2}(x - 3x^2)] \cdot (1 - 6x)$ **37.** $(x^2 + x + 1)^4(6x^2 + 6x + 1)(2x + 1)$

39. $\dfrac{2}{(3 + \sqrt{x})^2} \cdot \dfrac{1}{2\sqrt{x}}$, or $\dfrac{1}{\sqrt{x}(3 + \sqrt{x})^2}$ **41.** $y = 62x - 300$ **43.** $1; 2; 3$ **45.**   **(a)** $\dfrac{dV}{dt} = \dfrac{dV}{dx} \cdot \dfrac{dx}{dt}$ **(b)** 2

47.   **(a)** $\dfrac{dy}{dt}, \dfrac{dP}{dy}, \dfrac{dP}{dt}$ **(b)** $\dfrac{dP}{dt} = \dfrac{dP}{dy} \cdot \dfrac{dy}{dt}$

49.   **(a)** $\dfrac{200(100 - x^2)}{(100 + x^2)^2}$ **(b)** $\dfrac{200[100 - (4 + 2t)^2]}{[100 + (4 + 2t)^2]^2} \cdot (2)$ **(c)**   Falling at the rate of $480 per week

51.   **(a)** $.4 + .0002x$ **(b)**   increasing at the rate of 25 thousand persons per year   **(c)**   rising at the rate of 14 ppm per year   **53.** $x^3 + 1$ **55.** 24 **57.**   The derivative of the composite function $f(g(x))$ is the derivative of the outer function evaluated at the inner function and then multiplied by the derivative of the inner function.

EXERCISES 3.3, PAGE 236

1. $\dfrac{x}{y}$ **3.** $\dfrac{1 + 6x}{5y^4}$ **5.** $\dfrac{2x^3 - x}{2y^3 - y}$ **7.** $\dfrac{1 - 6x^2}{1 - 6y^2}$ **9.** $-\dfrac{y}{x}$ **11.** $-\dfrac{y + 2}{5x}$ **13.** $\dfrac{8 - 3xy^2}{2x^2y}$

15. $\dfrac{x^2(y^3 - 1)}{y^2(1 - x^3)}$ **17.** $-\dfrac{y^2 + 2xy}{x^2 + 2xy}$ **19.** $\tfrac{1}{2}$ **21.** $-\tfrac{8}{3}$ **23.** $-\tfrac{2}{15}$

25. $y - \tfrac{1}{2} = -\tfrac{1}{16}(x - 4), y + \tfrac{1}{2} = \tfrac{1}{16}(x - 4)$ **27.**   **(a)** $\dfrac{2x - x^3 - xy^2}{2y + y^3 + x^2y}$ **(b)** 0 **29.** $-\tfrac{27}{16}$ **31.** $-\dfrac{x^3}{y^3}\dfrac{dx}{dt}$

33. $\dfrac{2x - y}{x}\dfrac{dx}{dt}$ **35.** $\dfrac{2x + 2y}{3y^2 - 2x}\dfrac{dx}{dt}$ **37.** $-\tfrac{15}{8}$ units per second   **39.**   Rising at 3 thousand units per week

41.   Increasing at $20 thousand per month   **43.**   Decreasing at $\tfrac{1}{14}$ L per second   **45.**   **(a)** $x^2 + y^2 = 100$

(b) $\dfrac{dy}{dt} = -4$, so the top of the ladder is falling at the rate of 4 ft/sec.   **47.** $\dfrac{22}{\sqrt{5}}$ ft/sec (or 9.84 ft/sec)

CHAPTER 3 SUPPLEMENTARY EXERCISES, PAGE 239

1. $(4x - 1) \cdot 4(3x + 1)^3(3) + (3x + 1)^4 \cdot 4$, or $4(3x + 1)^3(15x - 2)$
3. $x \cdot 3(x^5 - 1)^2 \cdot 5x^4 + (x^5 - 1)^3 \cdot 1$, or $(x^5 - 1)^2(16x^5 - 1)$
5. $5(x^{1/2} - 1)^4 \cdot 2(x^{1/2} - 2)(\tfrac{1}{2}x^{-1/2}) + (x^{1/2} - 2)^2 \cdot 20(x^{1/2} - 1)^3(\tfrac{1}{2}x^{-1/2})$, or $5x^{-1/2}(x^{1/2} - 1)^3(x^{1/2} - 2)(3x^{1/2} - 5)$
7. $3(x^2 - 1)^3 \cdot 5(x^2 + 1)^4(2x) + (x^2 + 1)^5 \cdot 9(x^2 - 1)^2(2x)$, or $12x(x^2 - 1)^2(x^2 + 1)^4(4x^2 - 1)$
9. $\dfrac{(x - 2) \cdot (2x - 6) - (x^2 - 6x) \cdot 1}{(x - 2)^2}$, or $\dfrac{x^2 - 4x + 12}{(x - 2)^2}$ **11.** $2\left(\dfrac{3 - x^2}{x^3}\right) \cdot \dfrac{x^3 \cdot (-2x) - (3 - x^2) \cdot 3x^2}{x^6}$, or

$\dfrac{2(3 - x^2)(x^2 - 9)}{x^7}$ **13.** $-\tfrac{1}{3}, 3, \tfrac{31}{27}$ **15.** $y + 32 = 176(x + 1)$ **17.** $x = 44$ m, $y = 22$ m

19. $\dfrac{dC}{dt} = \dfrac{dC}{dx}\cdot\dfrac{dx}{dt} = 40\cdot 3 = 120$. Costs are rising $120 per day.   **21.** $0; -\frac{7}{2}$ **23.** $\frac{3}{2}; -\frac{7}{8}$ **25.** $1; -\frac{3}{2}$

27. $\dfrac{3x^2}{x^6 + 1}$ **29.** $\dfrac{2x}{(x^2 + 1)^2 + 1}$ **31.** $\frac{1}{2}\sqrt{1 - x}$ **33.** $\dfrac{3x^2}{2(x^3 + 1)}$ **35.** $-\dfrac{25}{x(25 + x^2)}$

37. $\dfrac{x^{1/2}}{(1 + x^2)^{1/2}}\cdot\frac{1}{2}(x^{-1/2})$, or $\dfrac{1}{2\sqrt{1 + x^2}}$ **39.** **(a)** $\dfrac{dR}{dA}, \dfrac{dA}{dt}, \dfrac{dR}{dx}$, and $\dfrac{dx}{dA}$ **(b)** $\dfrac{dR}{dt} = \dfrac{dR}{dx}\dfrac{dx}{dA}\dfrac{dA}{dt}$

41. **(a)** $-y^{1/3}/x^{1/3}$ **(b)** 1 **43.** -3 **45.** $\frac{3}{5}$ **47.** **(a)** $\dfrac{dy}{dx} = \dfrac{15x^2}{2y}$ **(b)** $\frac{20}{3}$ thousand dollars per thousand

unit increase in production   **(c)** $\dfrac{dy}{dt} = \dfrac{15x^2}{2y}\dfrac{dx}{dt}$ **(d)** 2 thousand dollars per week   **49.** Increasing at the rate of

2.5 units per unit time   **51.** 1.89 m²/year

CHAPTER 4

EXERCISES 4.1, PAGE 247

1. $2^{2x}, 3^{(1/2)x}, 3^{-2x}$ **3.** $2^{2x}, 3^{3x}, 2^{-3x}$ **5.** $2^{-4x}, 2^{9x}, 3^{-2x}$ **7.** $2^x, 3^x, 3^x$ **9.** $3^{2x}, 2^{6x}, 3^{-x}$ **11.** $2^{(1/2)x}, 3^{(4/3)x}$
13. $2^{-2x}, 3^x$ **15.** $\frac{1}{9}$ **17.** 1 **19.** 2 **21.** -1 **23.** $\frac{1}{5}$ **25.** $\frac{5}{2}$ **27.** -1 **29.** 4 **31.** 2^h
33. $2^h - 1$ **35.** $3^x + 1$ **37.**   0.6931   **39.** 2.7

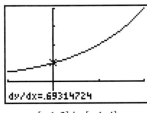

dy/dx=.69314724
$[-1, 2]$ by $[-1, 4]$

EXERCISES 4.2, PAGE 251

1. 1.161, 1.105, 1.10   **3.** 1.005, 1.001, 1.000   **5.** $10e^{10x}$ **7.** **(a)** $4e^{4x}$ **(b)** ke^{kx} **9.** e^{2x}, e^{-x}

11. e^{-6x}, e^{2x} **13.** e^{6x}, e^{2x} **15.** $x = 4$ **17.** $x = 4, -2$ **19.** $xe^x + e^x$ **21.** $\dfrac{e^x}{(1 + e^x)^2}$

23. $20e^x(1 + 5e^x)^3$ **25.** **(a)** 800 g/cm²   **(b)** 14 km   **(c)** -50 g/cm² per km   **(d)** 2 km   **27.** $y = x + 1$

29.   **31.** $\dfrac{d}{dx}(10^x)\Big|_{x=0} \approx 2.3026; \dfrac{d}{dx}(10^x) = m10^x$ where $m = \dfrac{d}{dx}(10^x)\Big|_{x=0}$

$[-1, 3]$ by $[-3, 20]$

EXERCISES 4.3, PAGE 256

1. $8e^{2x}$ **3.** $-e^{-t}$ **5.** $-2e^{-2x} - 2x$ **7.** $3(e^x + e^{-x})^2(e^x - e^{-x})$ **9.** $\frac{2}{3}e^{3+2x}$ **11.** $e^{1/t}(-t^{-2})$
13. $e^{x^2 - 5x + 4}(2x - 5)$ **15.** $60(x^3 + e^{-3x})^3(x^2 - e^{-3x})$ **17.** $2e^{2t} - 4e^{-4t}$ **19.** $-xe^{-x + 2}$

21. $x^3(2e^{2x}) + e^{2x}(3x^2)$ **23.** $\dfrac{(15 + 4x)e^{4x}}{(4 + x)^2}$ **25.** $(-x^{-2} + 2x^{-1} + 6)e^{2x}$ **27.** Max at $x = -2/3$

29. Min at $x = 5/4$ **31.** Max at $x = 9/10$ **33.** $54,366 per year   **35.** **(a)** 45 m/sec   **(b)** 10 m/sec²

(c) 4 sec **(d)** 4 sec **37.** 2 in./week **39.** $.02e^{-2e^{-.01x}}e^{-.01x}$ **41.** $y = Ce^{-4x}$ **43.** $y = e^{-.5x}$
47. 1 **49. (a)** size seems to stabilize near 6 ml **(b)** 3.2 ml **(c)** 7.7 wk **(d)** .97 ml/wk **(e)** 3.7 wk
(f) 1.13 ml/wk

EXERCISES 4.4, PAGE 262

1. $\frac{1}{2}$ **3.** $\ln 4.1$ **5.** $e^{-3.8}$ **7.** -3 **9.** e **11.** 0 **13.** $\frac{1}{2}\ln 5$ **15.** $4 - e^{1/2}$ **17.** $\pm e^{9/2}$
19. $-\dfrac{\ln .5}{.00012}$ **21.** $\frac{5}{3}$ **23.** $\dfrac{e}{3}$ **25.** $3 \ln \frac{9}{2}$ **27.** $\frac{1}{2}e^{8/5}$ **29.** $\frac{1}{2}\ln 4$ **31.** $-\ln\frac{3}{2}$
33. $(-\ln 3, 3 - 3\ln 3)$, minimum **35.** $(\frac{1}{2}\ln\frac{3}{2}, \frac{1}{2})$, minimum **37.** Max at $t = 2\ln 51$ **39.** $\ln 2$
41. The graph of $y = \exp(\ln(x))$ is the same as the graph of $y = x$ for $x > 0$.
43.

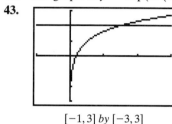

$\frac{1}{5}e^2 \approx 1.4778$

$[-1, 3]$ by $[-3, 3]$

EXERCISES 4.5, PAGE 265

1. $\dfrac{1}{x}$ **3.** $\dfrac{1}{x + 5}$ **5.** $-\dfrac{\ln(x + 1)}{x^2} + \dfrac{1}{x(x + 1)}$ **7.** $\left(\dfrac{1}{x} + 1\right)e^{\ln x + x}$ **9.** $\dfrac{1}{x}$ **11.** $\dfrac{2\ln x}{x} + \dfrac{1}{x}$ **13.** $\dfrac{1}{x}$
15. $\dfrac{\ln x - 2}{(\ln x)^3}$ **17.** $2e^{2x}\ln x + \dfrac{e^{2x}}{x}$ **19.** $\dfrac{5e^{5x}}{e^{5x} + 1}$ **21.** $2(\ln 4)t$ **23.** $\dfrac{6\ln t - 3(\ln t)^2}{t^2}$ **25.** $y = 1$
27. $\left(e^2, \dfrac{2}{e}\right)$ **29.** **31.** $1 + 3\ln 10$ **33.** $R'(x) = \dfrac{45(\ln x - 1)}{(\ln x)^2}$; $R'(20) \approx 10$
35. $\frac{1}{7}$ **37.**

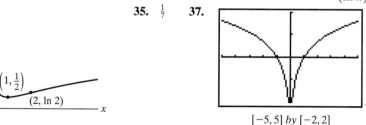

$[-5, 5]$ by $[-2, 2]$

EXERCISES 4.6, PAGE 269

1. $\ln 5x$ **3.** $\ln 3$ **5.** $\ln 2$ **7.** x^2 **9.** $\ln \dfrac{x^5 z^3}{y^{1/2}}$ **11.** $3\ln x$ **13.** $3\ln 3$ **15.** d **17.** d
19. $\dfrac{1}{x + 5} + \dfrac{2}{2x - 1} - \dfrac{1}{4 - x}$ **21.** $\dfrac{2}{1 + x} + \dfrac{3}{2 + x} + \dfrac{4}{3 + x}$ **23.** $5 + \dfrac{1}{x + 4} + \dfrac{3}{3x - 2} - \dfrac{1}{x + 1}$
25. $\dfrac{5}{5x + 1} + \dfrac{4}{4x + 1} + \dfrac{1}{x\ln x} - \dfrac{1}{2x + 1}$ **27.** $\ln(3x + 1)\dfrac{5}{5x + 1} + \ln(5x + 1)\dfrac{3}{3x + 1}$
29. $(x + 1)^4(4x - 1)^2\left(\dfrac{4}{x + 1} + \dfrac{8}{4x - 1}\right)$
31. $\dfrac{(x + 1)(2x + 1)(3x + 1)}{\sqrt{4x + 1}}\left(\dfrac{1}{x + 1} + \dfrac{2}{2x + 1} + \dfrac{3}{3x + 1} - \dfrac{2}{4x + 1}\right)$
33. $2^x\ln 2$ **35.** $x^x[1 + \ln x]$ **37.** $y = cx^k$ **39.** $h = 3, k = \ln 2$

CHAPTER 4: SUPPLEMENTARY EXERCISES, PAGE 271

1. 81   **3.** $\frac{1}{25}$ **5.** 4   **7.** 9   **9.** e^{3x^2} **11.** e^{2x} **13.** $e^{11x} + 7e^x$ **15.** $x = 4$ **17.** $x = -5$

19. $70e^{7x}$ **21.** $e^{x^2} + 2x^2e^{x^2}$ **23.** $e^x \cdot e^{e^x} = e^{x+e^x}$ **25.** $\dfrac{(e^{3x} + 3)(2x - 1) - (x^2 - x + 5)3e^{3x}}{(e^{3x} + 3)^2}$ **27.** $y = Ce^{-x}$

29. $y = 2e^{1.5x}$ **31.**

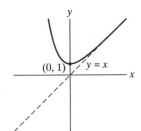

33.

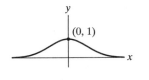

35. $\frac{2}{3}$ **37.** 1   **39.** $\sqrt{5}$

41. $\frac{1}{2} \ln 5$ **43.** $e^{5/2}$ **45.** $e, \dfrac{1}{e}$ **47.** $\dfrac{5}{5x - 7}$ **49.** $\dfrac{2 \ln x}{x}$ **51.** $\dfrac{6x^5 + 12x^3}{x^6 + 3x^4 + 1}$ **53.** $\dfrac{1}{x} + 1 - \dfrac{1}{2(1 + x)}$

55. $\dfrac{1}{x \ln x}$ **57.** $\ln x$ **59.** $\dfrac{e^x}{x} + e^x \ln x$ **61.** $(x^2 + 5)^6(x^3 + 7)^8(x^4 + 9)^{10}\left[\dfrac{12x}{x^2 + 5} + \dfrac{24x^2}{x^3 + 7} + \dfrac{40x^3}{x^4 + 9}\right]$

63. $10^x \ln 10$ **65.**

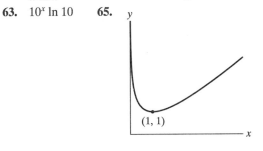

67.

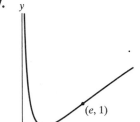

CHAPTER 5

EXERCISES 5.1, PAGE 282

1. **(a)** $P(t) = 3e^{.02t}$ **(b)** 3 million   **(c)** 80,000 people per year   **(d)** 3.52 million   **(e)** 65,000 people per year
(f) 3.5 million   **3.** **(a)** 5000   **(b)** $P'(t) = .2P(t)$ **(c)** 3.5 h   **(d)** 6.9 h   **5.** .017   **7.** 22 yr
9. 27 million cells   **11.** 34.0 million   **13.** **(a)** $P(t) = 8e^{-.021t}$ **(b)** 8 g   **(c)** .021   **(d)** 6.5 g
(e) .021 g/yr   **(f)** 5 g   **(g)** 4 g; 2 g; 1 g   **15.** **(a)** $f'(t) = -.6f(t)$ **(b)** 14.9 mg   **(c)** 1.2 h
17. 30.1 yr   **19.** 176 days   **21.** $f(t) = 8e^{-.014t}$ **23.** **(a)** 8 g   **(b)** 3.5 h   **(c)** .6 g/h   **(d)** 8 h
25. 13,500 days   **27.** 58.3%   **29.** 10,900 years ago   **31.** a–D, b–G, c–E, d–B, e–H, f–F, g–A, h–C

EXERCISES 5.2, PAGE 290

1. **(a)** $5000   **(b)** 4%   **(c)** $7459.12   **(d)** $A'(t) = .04A(t)$ **(e)** $298.36 per year   **(f)** $7000
3. **(a)** $A(t) = 4000e^{.035t}$ **(b)** $A'(t) = .035A(t)$ **(c)** $4290.03   **(d)** 6.4 yr   **(e)** $175 per year
5. $378 per year   **7.** 15.3 yr   **9.** 29.3%   **11.** 17.3 years   **13.** 7.3%   **15.** 2002   **17.** 2006
19. $786.63   **21.** $7985.16   **23.** 15.7 yr   **25.** a–B, b–D, c–G, d–A, e–F, f–E, g–H, h–C   **27.** **(a)** $200
(b) $8 per year   **(c)** 4%   **(d)** 30 yr   **(e)** 30 yr   **(f)** $A'(t)$ is a constant multiple of $A(t)$ since $A'(t) = rA(t)$.

29.

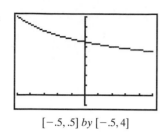

$[-.5, .5]$ *by* $[-.5, 4]$

31. .06

EXERCISES 5.3, PAGE 298

1. 20%, 4%   **3.** 30%, 30%   **5.** 60%, 300%   **7.** $-25\%, -10\%$ **9.** 12.5%   **11.** 5.8 yr
13. $p/(140 - p)$, elastic   **15.** $2p^2/(116 - p^2)$, inelastic   **17.** $p - 2$, elastic   **19.** **(a)** inelastic   **(b)** raised
21. **(a)** elastic   **(b)** increase   **23.** **(a)** 2   **(b)** yes   **29.** **(a)** $p < 2$ **(b)** $p < 2$

EXERCISES 5.4, PAGE 308

1. **(a)** $f'(x) = 10e^{-2x} > 0$, $f(x)$ increasing; $f''(x) = -20e^{-2x} < 0$, $f(x)$ concave down

(b) As x becomes large, $e^{-2x} = \dfrac{1}{e^{2x}}$ approaches 0   **(c)**

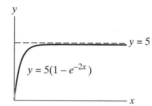

3. $y' = 2e^{-x} = 2 - (2 - 2e^{-x}) = 2 - y$ **5.** $y' = 30e^{-10x} = 30 - (30 - 30e^{-10x}) = 30 - 10y = 10(3 - y)$,
$f(0) = 3(1 - 1) = 0$ **7.** 4.8 h   **11.** **(a)** 2500   **(b)** 500 people/day   **(c)** day 12
(d) day 6 and day 13.5   **(e)** at time 9.78 days   **(f)** $f'(t) = .00004f(t)(10,000 - f(t))$ **(g)** 1000 people per day
13. **(a)**

$[0, 12]$ *by* $[-20, 75]$

(b) 30 units   **(c)** 25 units per hour   **(d)** 9 hr
(e) 65.3 units after 2 hr   **(f)** 4 hr

CHAPTER 5: SUPPLEMENTARY EXERCISES, PAGE 310

1. $29.92e^{-.2x}$ **3.** $5488.12   **5.** .058   **7.** **(a)** $17e^{.018t}$ **(b)** 20.4 million   **(c)** 2011   **9.** **(a)** $36,693
(b) The alternative investment is superior by $3859.   **11.** .02; 60,000 people per year; 5 million people
13. a–F, b–D, c–A, d–G, e–H, f–C, g–B, h–E   **15.** 6%   **17.** 400%   **19.** 3%, decrease   **21.** increase
23. $100(1 - e^{-.083t})$ **25.** **(a)** 400°   **(b)** decreasing at a rate of 100°/sec   **(c)** 17 sec   **(d)** 2 sec

CHAPTER 6

EXERCISES 6.1, PAGE 321

1. $\frac{1}{2}x^2 + C$ **3.** $\frac{1}{3}e^{3x} + C$ **5.** $3x + C$ **7.** $-\frac{1}{4}$ **9.** $\frac{2}{3}$ **11.** -2 **13.** $-\frac{5}{2}$ **15.** $\frac{1}{2}$ **17.** $-\frac{1}{5}$

19. -1 **21.** $\frac{1}{15}$ **23.** $\dfrac{x^3}{3} - \dfrac{x^2}{2} - x + C$ **25.** $4\sqrt{x} - 2x^{3/2} + C$ **27.** $4t + e^{-5t} + \dfrac{e^{2t}}{6} + C$ **29.** $\frac{2}{5}t^{5/2} + C$

31. C **33.** $\dfrac{x^2}{2} + 3$ **35.** $\frac{2}{3}x^{3/2} + x - \frac{28}{3}$ **37.** $2\ln|x| + 2$

39. Testing all three functions reveals that (b) is the only one that works.   **41.**

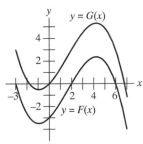

43. $\frac{1}{4}$ **45.** **(a)** $-16t^2 + 96t + 256$ **(b)** 8 s   **(c)** 400 ft   **47.** $P(t) = 60t + t^2 - \frac{1}{12}t^3$
49. $20 - 25e^{-.4t}$ °C   **51.** $-95 + 1.3x + .03x^2 - .0006x^3$ **53.** $5875(e^{.016t} - 1)$
55. $C(x) = 25x^2 + 1000x + 10,000$
57. $F(x) = \frac{1}{2}e^{2x} - e^{-x} + \frac{1}{6}x^3$ **59.** $F(x) = \frac{1}{10}(\frac{1}{4}x^4 - 3x^3 + \frac{5}{2}x^2) + 3x$

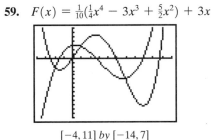

$[-2.4, 1.7]$ *by* $[-10, 10]$ $[-4, 11]$ *by* $[-14, 7]$

EXERCISES 6.2, PAGE 331

1. $.5; .25, .75, 1.25, 1.75$ **3.** $.6; 1.3, 1.9, 2.5, 3.1, 3.7$ **5.** 8.625 **7.** 15.12 **9.** $.077278$ **11.** 40 **13.** 15
15. $5.625; 4.5$ **17.** $1.61321;$ error $= .04241$ **19.** 1.08 L   **21.** 2800 ft   **23.** Increase in the population (in millions) from 1910 to 1950; rate of cigarette consumption t years after 1985; 20 to 50   **25.** Tons of soil eroded during a 5-day period   **29.** 9.5965 **31.** 1.7641 **33.** $.8427$

EXERCISES 6.3, PAGE 342

1. $\displaystyle\int_{1/2}^{2} \frac{1}{x}\,dx$ **3.** $\displaystyle\int_{1}^{3}(1-x)(x-3)\,dx$ **5.**   **7.** 0 **9.** 5 **11.** 30

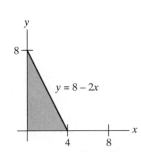

13. $\frac{4}{3}(1 - e^{-3})$ **15.** 14 **17.** $\frac{1}{2}(1 - e^{-10})$ **19.** $\ln 2$ **21.** $1\frac{7}{9}$ **23.** $\frac{1}{5}$ **25.** $3\frac{3}{4}$ **27.** $\frac{115}{6} + \ln\frac{7}{9}$

29. 10 **31.** $2(e^{1/2} - 1)$ **33.** $6\frac{3}{5}$ **35.** Positive **37.** 95.4 trillion
39. (a) 30 ft **(b)**

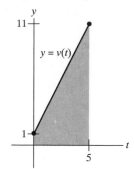

41. (a) $1185.75 **(b)** The area under the marginal cost curve from $x = 2$ to $x = 8$. **43.** The increase in profits resulting from increasing the production level from 44 to 48 units **45. (a)** $368/15 \approx 24.5$ **(b)** The amount the temperature falls during the first 2 h **47.** 2088 million m³ **49.** 10 **51.** $\ln \frac{3}{2}$ **53.** $\frac{80}{3}$ **55.** 2.5452
57. .4636

EXERCISES 6.4, PAGE 352

1. $\displaystyle\int_1^2 f(x)\, dx + \int_3^4 -f(x)\, dx$ **3.** **5.** $\frac{64}{3}$ **7.** $\frac{52}{3}$ **9.** 18 **11.** $\frac{32}{3}$

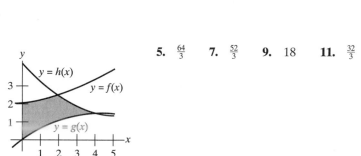

13. $\frac{32}{3}$ **15.** 72 **17.** 108 **19. (a)** $\frac{9}{2}$ **(b)** $\frac{19}{3}$ **(c)** $\frac{79}{6}$ **21.** $\frac{3}{2}$ **23.** $2 + 12 \ln \left(\frac{3}{2}\right)$
25. $\displaystyle\int_0^{20} (76.2e^{.03t} - 50 + 6.03e^{.09t})\, dt$ **27.** No; 20; the additional profit from using the original plan
29. (a) The distance between the two cars after 1 h **(b)** after 2 h **31.** 1.4032 **33.** 1.4293

EXERCISES 6.5 , PAGE 361

1. 3 **3.** 0 **5.** 8 **7.** 55° **9.** $\approx$82 g **11.** $20 **13.** $404.72 **15.** $200 **17.** $25
19. Intersection $(100, 10)$, consumers' surplus = $100, producers' surplus = $250 **21.** $3236.68 **23.** $75,426
25. 13.35 y **27. (b)** $1000e^{-.04t_1}\, \Delta t + 1000e^{-.04t_2}\, \Delta t + \cdots + 1000e^{-.04t_n}\, \Delta t$ **(c)** $f(t) = 1000e^{-.04t}; 0 \le x \le 5$
(d) $\displaystyle\int_0^5 1000e^{-.04t}\, dt$ **(e)** $4531.73 **29.** $\frac{4}{3}\pi r^3$ **31.** $\frac{31\pi}{5}$ **33.** 8π **35.** $\frac{\pi}{2}\left(1 - \frac{1}{e^{2r}}\right)$
37. $n = 4, b = 10, f(x) = x^3$ **39.** $n = 3, b = 7, f(x) = x + e^x$
41. The sum is approximated by $\displaystyle\int_0^3 (3 - x)^2\, dx = 9$. **43. (a)** $\dfrac{1000}{3r}(e^{3r} - 1)$ **(b)** 4.5%
45. (a) $\dfrac{1000}{r}(e^{6r} - 1)$ **(b)** 5%

EXERCISES 6.6, PAGE 371

1. $\frac{1}{6}(x^2 + 4)^6 + C$ **3.** $\frac{1}{4}(x^2 - 5x)^4 + C$ **5.** $e^{5x - 3} + C$ **7.** $\ln |x^3 - 1| + C$ **9.** $\frac{2}{3}(x^3 - 1)^{1/2} + C$
11. $-e^{1/x} + C$ **13.** $\frac{1}{3}\ln |x^3 + 3x + 2| + C$ **15.** $\frac{1}{11}(x^5 - 2x + 1)^{11} + C$ **17.** $\frac{3}{2}\ln |x - 2| + C$
19. $-\frac{1}{2}e^{-x^2} + C$ **21.** $\frac{1}{5}xe^{5x} - \frac{1}{25}e^{5x} + C$ **23.** $\frac{1}{10}x(2x + 1)^5 - \frac{1}{120}(2x + 1)^6 + C$
25. $\frac{2}{3}x(x + 1)^{3/2} - \frac{4}{15}(x + 1)^{5/2} + C$ **27.** $-3xe^{-x} - 3e^{-x} + C$ **29.** $x \ln 3x - x + C$ **31.** $\frac{1}{2}xe^{2x} - \frac{1}{4}e^{2x} + C$

33. $\frac{1}{2}e^{x^2} + C$ **35.** $\frac{2}{3}(2x - 1)(3x - 3)^{1/2} - \frac{8}{27}(3x - 3)^{3/2} + C$ **37.** $-\frac{1}{18}(6x^2 + 9x)^{-6} + C$
39. $4(\sqrt{x}\ln\sqrt{x} - \sqrt{x}) + C$ **41.** $\ln|\ln 5x| + C$ **43.** $x\ln x^4 - 4x + C$ **45.** $\frac{1}{4}(\ln x)^2 + C$
47. $(x^2 + 9)^{1/2} + 3$ **49.** \$4673.47

EXERCISES 6.7, PAGE 378

1. 0   **3.** No limit   **5.** $\frac{1}{4}$ **7.** 2   **9.** 0   **11.** 5   **13.** $\frac{1}{2}$ **15.** 2   **17.** 1
19. Area under the graph from 1 to b is $\frac{5}{14}(14b + 18)^{1/5} - \frac{5}{7}$. This has no limit as $b \to \infty$.   **21.** $\frac{1}{2}$ **23.** $\frac{1}{6}$
25. Divergent   **27.** $\frac{2}{3}$ **29.** 1   **31.** $2e$ **33.** Divergent   **35.** $\frac{1}{2}$ **37.** 2   **39.** $\frac{1}{4}$ **41.** 2   **43.** $\frac{1}{6}$
49. $\dfrac{K}{r}$

EXERCISES 6.8, PAGE 384

1. (a) $\frac{11}{32}$ (b) $\frac{7}{27}$ (c) $\frac{27}{32}$ (d) $\frac{27}{32}$ **3.** $\frac{3}{8}$ **5.** (a) .1813   (b) .4509   (c) .1353   (d) $\frac{1}{2}$
7. (a) $\frac{1}{3}; 0 \le x \le 3$ (b) $\frac{2}{3}$ (c) $\frac{1}{3}$ **9.** (a) .25   (b) 500 acres   **11.** $\frac{22}{25}$ **13.** (a) .3935   (b) .3679
15. .2231   **17.** .8647   **19.** .0211

CHAPTER 6 SUPPLEMENTARY EXERCISES, PAGE 386

1. $-2e^{-x/2} + C$ **3.** $\frac{3}{5}x^5 - x^4 + C$ **5.** $-\frac{2}{3}(4 - x)^{3/2} + C$ **7.** $\frac{3}{4}$ **9.** $\frac{1}{3}\ln 4$ **11.** $\frac{5}{32}$ **13.** 8

15. $\frac{1}{3}(x - 5)^3 - 7$ **17.** (a) $2t^2 + C$ (b) Ce^{4t} (c) $\frac{1}{4}e^{4t} + C$ **19.** $.02x^2 + 150x + 500$ dollars

21. The total quantity of drug (in cubic centimeters) injected during the first 4 min   **23.** 25   **25.** .68571; .69315

27. \$433.33   **29.** 15   **31.** .26; 1.96   **33.** (a) $f(t) = Q - \dfrac{Q}{A}t$ (b) $\dfrac{Q}{2}$

35. (a) The area under the curve $y = \dfrac{1}{1 + t^2}$ from $t = 0$ to $t = 3$ (b) $\dfrac{1}{1 + x^2}$ **39.** $\frac{15}{4}$ **41.** True

43. 65,000 km^3

CHAPTER 7

EXERCISES 7.1, PAGE 395

1. $f(5, 0) = 25, f(5, -2) = 51, f(a, b) = a^2 - 3ab - b^2$ **3.** $g(2, 3, 4) = -2, g(7, 46, 44) = \frac{7}{2}$
7. $C(x, y, z) = 6xy + 10xz + 10yz$ **9.** $f(8, 1) = 40, f(1, 27) = 180, f(8, 27) = 360$
11. $\approx \$50$. \$50 invested at 5% continuously compounded interest will yield \$100 in 13.8 years
13. (a) \$1875   (b) \$2250; yes   **15.**   **17.**

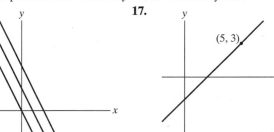

19. $f(x, y) = y - 3x$ **21.** They correspond to the points having the same altitude above sea level.
23. d   **25.** c

EXERCISES 7.2, PAGE 404

1. $5y, 5x$ **3.** $4xe^y, 2x^2e^y$ **5.** $-\dfrac{y^2}{x^2}, \dfrac{2y}{x}$ **7.** $4(2x - y + 5), -2(2x - y + 5)$

9. $(2xe^{3x} + 3x^2e^{3x}) \ln y, x^2e^{3x}/y$ **11.** $\dfrac{2y}{(x + y)^2}, -\dfrac{2x}{(x + y)^2}$ **13.** $\dfrac{3}{2}\sqrt{\dfrac{K}{L}}$ **15.** $\dfrac{2xy}{z}, \dfrac{x^2}{z}, -\dfrac{1 + x^2y}{z^2}$

17. $ze^{yz}, xz^2e^{yz}, x(yz + 1)e^{yz}$ **19.** $1, 3$ **21.** -12

23. $\dfrac{\partial f}{\partial x} = 3x^2y + 2y^2, \dfrac{\partial^2 f}{\partial x^2} = 6xy, \dfrac{\partial f}{\partial y} = x^3 + 4xy, \dfrac{\partial^2 f}{\partial y^2} = 4x, \dfrac{\partial^2 f}{\partial y \partial x} = \dfrac{\partial^2 f}{\partial x \partial y} = 3x^2 + 4y$

25. (a) Marginal productivity of labor $= 480$; of capital $= 40$ **(b)** $480h$ **(c)** production decreases by 240 units.
27. If the price of a bus ride increases and the price of a train ticket remains constant, fewer people will ride the bus. An increase in train-ticket prices coupled with constant bus fare should cause more people to ride the bus.
29. If the average price of video tapes increases and the average price of a VCR remains constant, people will purchase fewer VCRs. An increase in average VCR prices coupled with constant video tape prices should cause a decline in the number of video tapes purchased.

31. $\dfrac{\partial V}{\partial P}(20, 300) = -.06, \dfrac{\partial V}{\partial T}(20, 300) = .004$ **33.** $\dfrac{\partial f}{\partial r} > 0, \dfrac{\partial f}{\partial m} > 0, \dfrac{\partial f}{\partial p} < 0$

35. $\dfrac{\partial^2 f}{\partial x^2} = -\dfrac{45}{4}x^{-5/4}y^{1/4}$. Marginal productivity of labor is decreasing.

EXERCISES 7.3, PAGE 413

1. $(-2, 1)$ **3.** $(26, 11)$ **5.** $(1, -3), (-1, -3)$ **7.** $(\sqrt{5}, 1), (\sqrt{5}, -1), (-\sqrt{5}, 1), (-\sqrt{5}, -1)$ **9.** $(\frac{1}{3}, \frac{4}{3})$
11. Relative minimum; neither relative maximum nor relative minimum
13. Relative maximum; neither relative maximum nor relative minimum; relative maximum   **15.** Neither relative maximum nor relative minimum   **17.** $(0, 0)$ min   **19.** $(-1, -4)$ max   **21.** $(0, -1)$ min   **23.** $(-1, 2)$ max; $(1, 2)$ neither max nor min
25. $(\frac{1}{4}, 2)$ min; $(\frac{1}{4}, -2)$ neither max nor min   **27.** $(\frac{1}{2}, \frac{1}{6}, \frac{1}{2})$ **29.** 14 in. $\times$ 14 in. $\times$ 28 in.   **31.** $x = 120, y = 80$

EXERCISES 7.4, PAGE 422

1. 58 at $x = 6, y = 2, \lambda = 12$ **3.** 13 at $x = 8, y = -3, \lambda = 13$ **5.** $x = \frac{1}{2}, y = 2$ **7.** $5, 5$

9. Base 10 in., height 5 in.   **11.** $F(x, y, \lambda) = 4xy + \lambda(1 - x^2 - y^2); \dfrac{\sqrt{2}}{2} \times \dfrac{\sqrt{2}}{2}$

13. $F(x, y, \lambda) = 3x + 4y + \lambda(18{,}000 - 9x^2 - 4y^2); x = 20, y = 60$
15. (a) $F(x, y, \lambda) = 96x + 162y + \lambda(3456 - 64x^{3/4}y^{1/4}); x = 81, y = 16$ **(b)** $\lambda = 3$ **17.** $x = 12, y = 2, z = 4$
19. $x = 2, y = 3, z = 1$ **21.** $F(x, y, z, \lambda) = 3xy + 2xz + 2yz + \lambda(12 - xyz); x = 2, y = 2, z = 3$
23. $F(x, y, z, \lambda) = xy + 2xz + 2yz + \lambda(32 - xyz); x = y = 4, z = 2$

EXERCISES 7.5, PAGE 430

1. $E = 6.7$ **3.** $E = (2A + B - 6)^2 + (5A + B - 10)^2 + (9A + B - 15)^2$ **5.** $y = 4.5x - 3$
7. $y = -2x + 11.5$ **9.** $y = -1.4x + 8.5$ **11. (a)** $y = .2073x + 2.7$ **(b)** \$4773   **(c)** 2006
13. (a) $y = .497x + 11.2$ **(b)** 22.6 percent   **(c)** 2002   **15. (a)** $y = -4.24x + 22.01$
(b) $y = 8.442$ degrees Celsius

EXERCISES 7.6, PAGE 441

1. ExpReg
$y = a * b^x,$
$a = 2.348685221$
$b = 1.489061785$

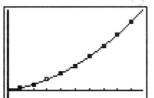
$[-.5, 5.5]$ *by* $[-.167, 20.067]$

3. $z = 10.6 + 3.82x - 1.13y$

5. QuadReg
$y = ax^2 + bx + c$
$a = 1.971212121$
$b = 4.181212121$
$c = 1.216666667$

$[.2, 9.8]$ *by* $[-24.736, 230.936]$

7. QuadReg
$y = ax^2 + bx + c$
$a = -.5224941725$
$b = -2.011818182$
$c = 3.706759907$

$[-6, 6]$ *by* $[-23.55, 9.95]$

9. $y = (92.065)(.9755)^x$ **11.** $y = -4.37x^2 + 17,391.90x - 17,298,092.83$

EXERCISES 7.7, PAGE 449

1. $e^2 - 2e + 1$ **3.** $2 - e^{-2} - e^2$ **5.** $309\frac{3}{8}$ **7.** $\frac{5}{3}$ **9.** $\frac{38}{3}$ **11.** $e^{-5} + e^{-2} - e^{-3} - e^{-4}$ **13.** $9\frac{1}{3}$

CHAPTER 7: SUPPLEMENTARY EXERCISES, PAGE 450

1. $2, \frac{5}{6}, 0$ **3.** $\approx 19.94.$ Ten dollars increases to 20 dollars in 11.5 y.   **5.** $6x + y, x + 10y$ **7.** $\frac{1}{y}e^{x/y}, -\frac{x}{y^2}e^{x/y}$

9. $3x^2, -z^2, -2yz$ **11.** $6, 1$ **13.** $20x^3 - 12xy, 6y^2, -6x^2, -6x^2$ **15.** $-201, 5.5.$ At the level $p = 25, t = 10,000,$
an increase in price of \$1 will result in a loss in sales of approximately
201 calculators, and an increase in advertising of \$1 will result in the sale of approximately 5.5 additional calculators.
17. $(3, 2)$ **19.** $(0, 1), (-2, 1)$ **21.** Min at $(2, 3)$ **23.** Min at $(1, 4)$; neither max nor min at $(-1, 4)$
25. $20; x = 3, y = -1$ **27.** $x = \frac{1}{2}, y = \frac{3}{2}, z = 2$
29. $F(x, y, \lambda) = xy + \lambda(40 - 2x - y); x = 10, y = 20$ **31.** $y = \frac{5}{2}x - \frac{5}{3}$ **33.** $y = -2x + 1$ **35.** 5160
37. 40

CHAPTER 8

EXERCISES 8.1, PAGE 457

1. $\frac{\pi}{6}, \frac{2\pi}{3}, \frac{7\pi}{4}$ **3.** $\frac{5\pi}{2}, -\frac{7\pi}{6}, -\frac{\pi}{2}$ **5.** 4π **7.** $\frac{7\pi}{2}$ **9.** -3π **11.** $\frac{2\pi}{3}$

13.

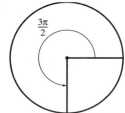

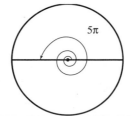

15.

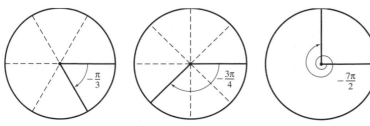

17.

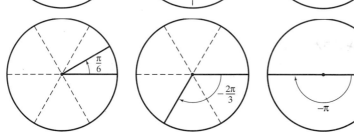

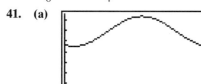

EXERCISES 8.2, PAGE 463

1. $\sin t = \dfrac{1}{2}, \cos t = \dfrac{\sqrt 3}{2}$ **3.** $\sin t = \dfrac{2}{\sqrt{13}}, \cos t = \dfrac{3}{\sqrt{13}}$ **5.** $\sin t = \tfrac{12}{13}, \cos t = \tfrac{5}{13}$

7. $\sin t = \dfrac{1}{\sqrt 5}, \cos t = -\dfrac{2}{\sqrt 5}$ **9.** $\sin t = \dfrac{\sqrt 2}{2}, \cos t = -\dfrac{\sqrt 2}{2}$ **11.** $\sin t = -.8, \cos t = -.6$ **13.** .4

15. 3.59   **17.** 10.86   **19.** $b = 1.31, c = 2.73$ **21.** $\dfrac{\pi}{6}$ **23.** $\dfrac{3\pi}{4}$ **25.** $\dfrac{5\pi}{8}$ **27.** $\dfrac{\pi}{4}$ **29.** $\dfrac{\pi}{3}$

31. $-\dfrac{\pi}{6}$ **33.** $\dfrac{\pi}{4}$ **35.** Here $\cos t$ decreases from 1 to -1.   **37.** $0, 0, 1, -1$ **39.** $.2, .98, .98, -.2$

41. **(a)**   **(b)** 46°   **(c)** 45° coldest, 73° warmest   **(d)** January 26   **(e)** July 27
                                          (f) October 27 and April 27

$[0, 365]$ by $[-10, 75]$

$[0, 365]$ by $[-10, 75]$

EXERCISES 8.3, PAGE 472

1. $4\cos 4t$ **3.** $4\cos t$ **5.** $-6\sin 3t$ **7.** $1 - \pi\sin\pi t$ **9.** $-\cos(\pi - t)$ **11.** $-3\cos^2 t \sin t$

13. $\dfrac{\cos\sqrt{x-1}}{2\sqrt{x-1}}$ **15.** $\dfrac{\cos(x-1)}{2\sqrt{\sin(x-1)}}$ **17.** $8(1+\cos t)^7\cdot(-\sin t)$ **19.** $-6x^2\cos x^3\sin x^3$

21. $e^x(\sin x + \cos x)$ **23.** $2\cos(2x)\cos(3x) - 3\sin(2x)\sin(3x)$ **25.** $\cos^{-2} t$ **27.** $-\dfrac{\sin t}{\cos t}$ **29.** $\dfrac{\cos(\ln t)}{t}$

31. -3 **33.** $y = 2$ **35.** $\tfrac12\sin 2x + C$ **37.** $-\dfrac{\cos(4x+1)}{4} + C$

39. **(a)** max $= 120$ at $0, \dfrac{\pi}{3}$; min $= 80$ at $\dfrac{\pi}{6}, \dfrac{\pi}{2}$ **(b)** 57   **41.** 0   **43.** **(a)** 69°   **(b)** increasing 1.6°/wk
(c) weeks 6 and 44   **(d)** weeks 28 and 48   **(e)** week 25, week 51   **(f)** week 12, week 38

EXERCISES 8.4, PAGE 478

1. $\sec t = \dfrac{\text{hypotenuse}}{\text{adjacent}}$ **3.** $\tan t = \frac{5}{12},\ \sec t = \frac{13}{12}$ **5.** $\tan t = -\dfrac{1}{2},\ \sec t = -\dfrac{\sqrt{5}}{2}$ **7.** $\tan t = -1,\ \sec t = -\sqrt{2}$

9. $\tan t = \frac{4}{3},\ \sec t = -\frac{5}{3}$ **11.** $75 \tan (.7) \approx 63$ ft **13.** $\tan t \sec t$ **15.** $-\csc^2 t$ **17.** $4 \sec^2(4t)$

19. $-3 \sec^2(\pi - x)$ **21.** $4(2x + 1) \sec^2(x^2 + x + 3)$ **23.** $\dfrac{\sec^2 \sqrt{x}}{2\sqrt{x}}$ **25.** $\tan x + x \sec^2 x$

27. $2 \tan x \sec^2 x$ **29.** $6[1 + \tan (2t)]^2 \sec^2 (2t)$ **31.** $\sec t$

CHAPTER 8: SUPPLEMENTARY EXERCISES, PAGE 480

1. $\dfrac{3\pi}{2}$ **3.** $-\dfrac{3\pi}{4}$ **5.** **7.** $\frac{4}{5}, \frac{3}{5}, \frac{4}{3}$ **9.** $-.8, -.6, \frac{4}{3}$ **11.** $\pm\dfrac{2\sqrt{6}}{5}$

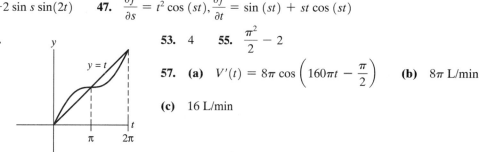

13. $\dfrac{\pi}{4}, \dfrac{5\pi}{4}, -\dfrac{3\pi}{4}, -\dfrac{7\pi}{4}$ **15.** Negative **17.** 16.3 ft **19.** $3 \cos t$ **21.** $(\cos \sqrt{t}) \cdot \frac{1}{2}t^{-1/2}$

23. $x^3 \cos x + 3x^2 \sin x$ **25.** $-\dfrac{2 \sin(3x)\sin(2x) + 3 \cos(2x)\cos(3x)}{\sin^2 (3x)}$ **27.** $-12 \cos^2(4x)\sin(4x)$

29. $[\sec^2 (x^4 + x^2)](4x^3 + 2x)$ **31.** $\cos(\tan x) \sec^2 x$ **33.** $\sin x \sec^2 x + \sin x$ **35.** $\cot x$

37. $4e^{3x} \sin^3 x \cos x + 3e^{3x} \sin^4 x$ **39.** $\dfrac{\tan(3t) \cos t - 3 \sin t \sec^2(3t)}{\tan^2 (3t)}$ **41.** $e^{\tan t}\sec^2 t$ **43.** $2(\cos^2 t - \sin^2 t)$

45. $\dfrac{\partial f}{\partial s} = \cos s \cos (2t),\ \dfrac{\partial f}{\partial t} = -2 \sin s \sin(2t)$ **47.** $\dfrac{\partial f}{\partial s} = t^2 \cos (st),\ \dfrac{\partial f}{\partial t} = \sin (st) + st \cos (st)$

49. $y - 1 = 2\left(t - \dfrac{\pi}{4}\right)$ **51.** **53.** 4 **55.** $\dfrac{\pi^2}{2} - 2$

57. **(a)** $V'(t) = 8\pi \cos \left(160\pi t - \dfrac{\pi}{2}\right)$ **(b)** 8π L/min

(c) 16 L/min

Index